Natural Science in Archaeology

Series Editors: B. Herrmann · G. A. Wagner

Springer-Verlag Berlin Heidelberg GmbH

Günther A. Wagner

Age Determination of Young Rocks and Artifacts

Physical and Chemical Clocks in Quaternary Geology and Archaeology

With 177 Figures

Springer

Author and Series Editor
Professor Dr. Günther A. Wagner
Institute of Archaeometry
Max Planck Institute of Nuclear Physics
Saupfercheckweg 1
D-69117 Heidelberg, Germany
E-mail: gwagner@goanna.mpi-hd.mpg.de

Series Editor
Professor Dr. Bernd Herrmann
Institute of Anthropology
University of Göttingen
Bürgerstrasse 50
D-37073 Göttingen, Germany
E-mail: bherrma@gwdg.de

Translator
Dr. Solveig Schiegl
Institute of Archaeometry
Max Planck Institute of Nuclear Physics
Saupfercheckweg 1
D-69117 Heidelberg, Germany

Title of the original German edition:
Günther A. Wagner: Altersbestimmung von jungen Gesteinen und Artefakten
Ferdinand Enke Verlag, Stuttgart 1995

ISBN 978-3-642-08331-0 ISBN 978-3-662-03676-1 (eBook)
DOI 10.1007/978-3-662-03676-1

Library of Congress Cataloging-in-Publication Data
Wagner, Günther A. [Altersbestimmung von jungen Gesteinen und Artefakten. English] Age determination of young rocks and artifacts : physical and chemical clocks in Quaternary geology and archaeology / Günther A. Wagner. p. cm. (Natural science in archaeology) Includes bibliographical references and index.
1. Archaeological dating. 2. Geology, Stratigraphic – Quaternary. I. Title. II. Series.
CC78.W34 1998 930.1'028'5 – dc21 98-014454

Originally published by Springer-Verlag Berlin Heidelberg New York in 1998.
Softcover reprint of the hardcover 1st edition 1998
Coverdesign: design & production, Heidelberg
Typesetting: Fotosatz-Service Köhler OHG, Würzburg

SPIN: 10556176 32/3020 – 5 4 3 2 1 0 – Printed of acid-free paper

Preface

Dating the Quaternary, which covers approximately the last 2 million years, has experienced considerable progress over the past few decades. On the one hand, this resulted from the necessity to obtain a valid age concept for this period which had seen tremendous environmental changes and the advent of the genus *Homo*. On the other hand, instrumental improvements, such as the introduction of highly sensitive analytical techniques, gave rise to physical and chemical innovations in the field of dating. This rapid methodological development is still in full progress.

The broad spectrum of chronometric methods applicable to young rocks and artifacts also becomes increasingly intricate to the specialist. Hence, it is my goal to present a comprehensive, state-of-the-art summary of these methods. This book is essentially designed as an aid for scientists who feel a demand for dating tasks falling into this period, i.e., Quaternary geologists and archaeologists in the broadest sense. Since it has been developed from a course of lectures for students of geological and archaeological sciences, held at the University of Heidelberg, it certainly shall serve as an introduction for students of these disciplines. Furthermore it is intended for all those interested in questions dealing with the temporal record of prehistoric epochs and paleoecological upheavals during the "ice age". Chronometric methods are predominantly of a physical or chemical nature, and are applied to geological and archaeological questions. A multidisciplinary constellation like this can only be successful when the analyst possesses a thorough understanding of the problem and the ability to transpose it methodologically. Vice versa, besides knowledge of the appropriate dating technique for a specific material type, the user must be able to critically interpret the data. Owing to its transdisciplinary nature, I have tried to present this subject matter intelligibly.

The breadth of the subject requires a selection which certainly may not in each case be completely satisfactory. The individual dating techniques are treated in coherent groups. As a rule, these groups begin with a general introduction to the physical and chemical background. For the sake of a balanced presentation, the length of discussion of the individual methods is roughly related to their relevance. For a deeper understanding, the reader is referred to the detailed list of references. The description of the individual chronometric methods is followed by a fixed scheme comprising an introductory summary, methodological basis, practical aspects and representative examples of application. Also, some climatically controlled geological and biological phenomena, not precisely of a physical or chemical nature, but nevertheless of great importance to the chronology of the Pleistocene and Holocene, supplement this treatise. This book can be approached in two ways. Owing to its organized design into methodologically coherent groups, information can be obtained on a certain dating technique, its potential and limitations. Alternatively, the reader who is not sufficiently acquainted with the individual dating techniques, but has a particular interest in the dating of a certain sample material, can approach this subject via the chapter on *materials*. Each section provides, for all relevant rock and artifact materials, the appropriate dating methods and their age ranges, which are described in detail in the special chapters.

I would like to take this opportunity to express my gratitude to the late Professors Wolfgang Gentner and Josef Zähringer, who aroused my interest in the field of chronometry during the 1960s. Under Gentner's far-reaching initiative, the archaeometry group was established at the Max-Plank-Institut für Kernphysik. An offspring of this institution is represented by the Forschungsstelle Archäometrie (Research Laboratory of Archaeometry) of the Heidelberger Akademie der Wissenschaften founded in 1989.

The first edition of this book was published in German by the Ferdinand Enke Verlag, Stuttgart. This revised edition in English was improved through the translation by Dr. Solveig Schiegl. During its preparation, I received essential information and helpful comments from many sides and would like to express my deep gratitude to Prof. Karl Dietrich Adam, Dr. Roland Gläser, Kürsad Gögen, Dr. Ulrich Hambach, Dr. Raymond Jonckheere, Dr. Bernd Kromer, Dr. Andreas Lang, Dr. Marco Spurk, Dr. Mario Trieloff, Clemens Woda, Dr. Irmtrud Wagner and Prof. Ludwig Zöller. I would also like to thank Dr. Paul van den Bogaard, Kürsad Gögen, Dr. Yeter Göksu, Dr. Bernd Kromer, Ralf Kuhn, Dr. Andreas Lang, Prof. Augusto Mangini, Prof. Joseph W. Michels, Uwe Rieser, Dr. Irmtrud Wagner and Prof. John A. Westgate for supplying or drafting of the figures. Many colleagues and publishing companies gave their kind permission to reproduce figures from their work. I feel deeply indebted to all of them. I would also like to thank

Veronika Träumer for the photographical work and Dieter Braun for the literature search. Finally, I want to thank the Springer-Verlag, in particular Dr. Wolfgang Engel, for the concessions in the design of this book.

Heidelberg, March 1998 Günther A. Wagner

Contents

Abbreviations

Units and Symbols

mass m

g	gram (1 g = 10^{-3} kg)
t	ton (10^3 kg)

time t

s	second
min	minute
d	day
h	hour
a	year (1 a ≈ 365.25 days ≈ 31.56 Ms)

length s

m	meter

energy E

J	Joule (1 J = 1 kg m^2 s^{-2})
eV	electron volt (1 eV = 1.602 × 10^{-19} J)

temperature T

°C	temperature in degrees Celsius
K	absolute temperature in Kelvin

energy dose D

Gy	Gray (1 Gy = 1 J kg^{-1})
rad	Rad (1 rad = 10^{-2} Gy)

radioactivity [N]

Bq	Becquerel (1 Bq = 1 disintegration s^{-1})

magnetic induction (magnetic field within matter) B

T	Tesla (10^{-4} T = 1 Gauss)

symbols (before units) for order of magnitude

n (nano)	10^{-9}
µ (micro)	10^{-6}
m (milli)	10^{-3}
c (centi)	10^{-2}
k (kilo)	10^{3}
M (mega)	10^{6}

Symbols

a	reaction rate coefficient ($= \tau^{-1}$) [s^{-1}] or [a^{-1}]
a_h	hydration rate [μm^2 ka^{-1}]
a_o	frequency factor [s^{-1}]
[A]	educt concentration
[B]	product concentration
d	mass thickness [g cm^{-2}]
D	diffusion coefficient [cm^2 s^{-1}]
E	activation energy [eV]
k	Boltzmann constant (1.381×10^{-23} J K^{-1} = 8.617×10^{-5} eV K^{-1})
l	mean path length [g cm^{-2}]
λ	decay constant [a^{-1}]
K	reaction equilibrium constant
M	thickness of hydration rim [µm]
N	quantity of atoms
[N]	activity (= dN/dt) [s^{-1}] or [min^{-1}]
ND	natural dose [Gy]
NDR	natural dose rate [mGy a^{-1}]
P	production rate [atoms g^{-1} a^{-1}]
ρ	mass density [g cm^{-3}] or areal density [cm^{-2}]
σ	standard deviation
$\overline{\sigma}$	standard error
t	age [Ma], [ka] or [a]
$t_{1/2}$	half-life (= $0.693/\lambda$ or 0.693/a) [a]
τ	mean lifetime (=$1/\lambda$ or 1/a) [a]
v	denudation rate [g cm^{-2} a^{-1}] or reaction rate [a^{-1}]

1 Introduction

The Quaternary, which covers approximately the last 2 million years, represents a particular period in the Earth's history. It is characterized by dramatic environmental changes. Cold and warm periods repeatedly alternated. Enormous ice masses which covered vast regions of the northern continents melted off and recurred. Which factors caused these tremendous climatic changes? Early Man entered this scenario. Was his evolution somehow triggered by the severe environmental stresses or not? In the course of the warm period lasting 11,500 years, in which we are living, highly advanced civilizations eventually developed. Does the reconstruction of the paleoenvironment allow the prediction of future climatic and ecological developments? Sound answers to these questions depend not least on the degree to which chronometry succeeds in resolving the processes of the past. Considerable progress has been achieved therein, especially during recent years.

As in any other historical discipline, Quaternary geology and prehistory have to arrange the transmitted facts in their correct chronological order. Only then do the facts make sense. Processes and recurrent patterns become obvious. The classic stratigraphic methods allow the establishment of chronological sequences of past events on the basis of the lithologic features and the fossil records of rocks. A relative chronology can be accomplished herewith, although there is an increasing uncertainty with growing spatial distance between the events to be correlated (Fig. 1). The relatively dated events on the basis of the stratigraphic method have to be linked to the chronometric time scale in order to grasp the pace of evolutionary processes. *Chronometry* is the measurement of time by splitting it up into equal time units, i.e., seconds. One second (s) is defined by 9,192,631,770 oscillations of the ^{133}Cs atom. In Quaternary geology and archaeological sciences it is common to use the chronometric units year "a" as well as its multiples "ka" (1000 a) and – less frequently – "Ma" (10^6 a). Astronomically, a year is defined by one revolution of the Earth around the sun.

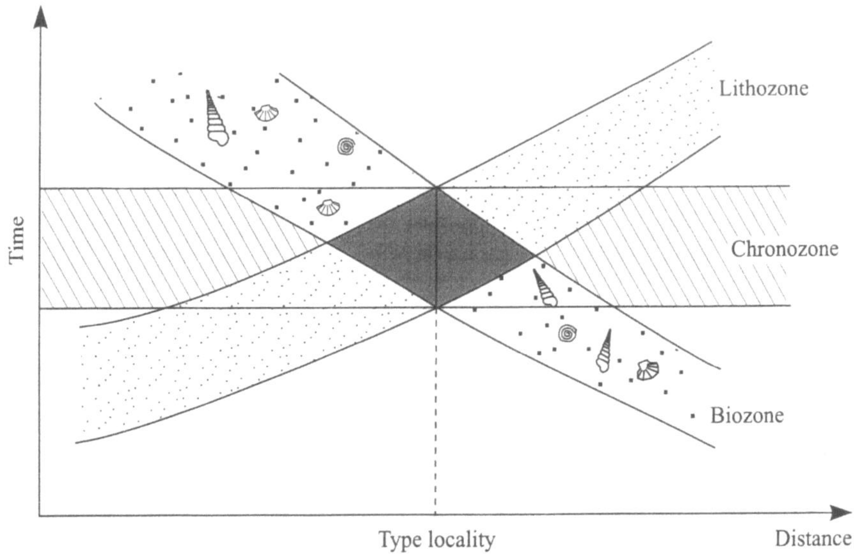

Fig. 1. Three distinct categories have to be distinguished in stratigraphic classification of strata: *lithostratigraphy* is based on the lithologic character, *biostratigraphy* is based on the fossil content and *chronostratigraphy* is based on the age. Litho- and biostratigraphic units, here schematically manifested as zones, are defined by a stratotype at a type locality. Both boundaries of a *chronozone* are synchronous, whereas those of *litho-* and *biozones* may be diachronous since the lithology and the fossil content may behave time-transgressively. Consequently, the litho- and biozones as defined for a distinct type locality may show with respect to the chronozone a more or less strong divergence at other localities. The reciprocal slope of the lines reflects the spreading rate of the ecological changes. The time interval represented by a litho- or a biozone can vary from place to place. Principally, this also applies to the culturally defined archaeological zones

For the last 5 millennia, chronometric dating has been accomplished by the written tradition of astronomical observations. The earliest astronomic calendar had been employed in ancient Egypt. The Egyptian solar calendar year began with the first early rise of the star Sirius (Egyptian *Sothis*) preceding the sunrise. The civilian Egyptian year covered 365 days. An intercalary year, which would compensate for the difference from the $365^1/_4$ day solar year, did not exist. Hence, after four years the civilian calendar preceded the seasonal occurrences (e.g. Nile inundation) that are governed by the solar year by 1 day. After 1460 years (Sothic cycle), the civilian calendar again agreed with the solar calendar. Through the observation of the early rise of Sirius at the civilian New Year in 139 AD the Egyptian calendar is linked to the Roman and thus to the Christian calendar. The written tradition of the early rise of Sirius at a certain calendar day in the seventh year of the reign of Sesostris III sets this event in the year 1870 (± 6) BC. On the basis of the lists of the Egyptian rulers it is then possible to date the First Dynasty to 3000 years BC within an uncertainty of 100 years. Mesopotamia, too, possessed an astronomic calendar, although dating not so far back. Prehistoric chronologies of the Mediterranean region and Europe can be stratigraphically linked to these ancient calendars.

For a long time, the prehistoric and glacial periods which fall beyond these historic astronomical calendars have been considered to be hardly quantifiable. With the development of the radiocarbon method, starting in the early 1950s, a comprehensive set of data of the last 30 ka has been collected. On the basis of ^{14}C dates a number of archaeological conceptions were revolutionized (Renfrew 1976). For instance, nowadays we know that ceramic production in central Europe, which came along with Neolithic sedentariness, did not begin in the early 3rd millennium BC – as had been deduced from comparative stratigraphy – but already during the second half of the 6th millennium. Of course, the ^{14}C ages demanded a calibration by means of tree-ring data. The next dating technique coming into play was the K–Ar method in the 1950s. This method allowed the dating of tephra horizons older than several 100 ka. A dating gap between the two methods existed for a long time, and included the Middle Paleolithic as well as a considerable period of the Early Paleolithic – i.e. the Middle and Young Pleistocene. The last few decades brought the invention of various new dating methods which tried to fill this gap. These promising methodological developments are still in full operation.

Time-dependent physical and chemical processes leading to quantifiable changes are suitable for time measurement. The time dependence of the individual processes has to be known. A further prerequisite to the age determination of a geological or archaeological event is the coincidence between the event and the start of a time dependent process – i.e. the clock has to be reset. In particular radioactivity, which is a strictly time-dependent process, is used for dating. Distinguishable as radioactive nuclides are: primordial nuclides that have existed at least since the formation of the Earth; nuclides generated within the radioactive decay series; and cosmogenic nuclides. Of further interest to chronometry are the kinetics of chemical reactions, the variation of the Earth's magnetic field, and climatically triggered processes. This division of the time-dependent processes is followed in this book.

The term *archaeochronometry* refers to the whole complex of physical and chemical dating methods covering the period of the last two million years and applies to both archaeological and geological problems. It is worth emphasizing that in this context the term "chronometry" is preferable to the frequently used term "chronology", for chronology refers to the linkage of stratigraphic records *and* chronometrically established ages (Bowen 1978; Harland et al. 1990). This relation is expressed by the formula:

stratigraphy + chronometry = chronology

In addition to the division based on physical and chemical phenomena, sometimes the distinction between *relative* and *absolute* dating methods can be found. The stratigraphic approach makes relative dating feasible and permits statements about whether a stratum is older, of the same age or younger than another one. On the other hand, the term *absolute dating methods* is not really justified because of the uncertainties associated with any of these methods. Therefore it will be avoided in the following. Instead, such terms as *chronometric, numeric* or *quantitative* dating will be employed. Furthermore, dating methods which do not rely on other dating methods have to be distinguished from those which do. For example, thermoluminescence dating represents an independent method, whereas archaeomagnetism belongs to the dependent methods because it requires the calibration of the temporal variations of the Earth's magnetic field. On the other hand, there are methods which can be conducted both in an independent and dependent way. For example, in the case of the hydration dating method of obsidian, the hydration rate can be either experimentally determined or delineated from known ages.

1.1 Terminology: Age and Date

A chronometric age is defined as the time interval in years which has passed since a certain geological or archaeological event and a defined date in the present. Usually the age refers to the date of the measurement. Because of the time interval between the date of the measurement and the present this interval has sensu stricto to be added to the measured age, if the age refers to the actual date. This means that, since its measurement and publication, the age of the dated event increases every year by 1 year. However this way of looking at the age of an event is of no importance to geochronology, because the time interval between measurement and present is negligible compared to the error of the age. For example, it is insignificant whether the statement that the Nördlinger Ries impact crater was formed 15.1 ± 0.5 Ma ago becomes known in the year of publication (Gentner et al. 1961) or only in the year 1997 with a delay of 0.000036 Ma. In contrast, in more recent age periods covering the Holocene with the development of the Neolithic and metal-processing cultures the age errors may be comparable or smaller than the time which has passed since the age measurement. So for instance, the statement that a tree-ring chronology dating back 9928 years and accurate to 1 year – established on the basis of South German oak trees (Becker et al. 1991) – is dependent on the year of the latest measured ring, which in this case is 1990. At the end of 1997 the chronology would comprise seven additional years, that is, it begins already 9935 (7938 rings from the BC period plus 1997 rings from the AD period) tree-ring years ago. In such cases it is confusing to operate with the information on the explicit age, which is only valid for the year of reference.

One can avoid this difficulty inherent to low ages by reporting the fixed calendar-based *date* of an event instead its age. Herewith the data become numerically independent of the current viewpoint. For this reason an identity card, for example, does not state the explicit biological age but the date of birth of a person. Dates are given in the form of years before Christ ("BC") and after Christ ("AD"). In the case of the dendrochronological example, mentioned above, the sequence of tree rings reaches back to 7938 BC – note that the year 0 does not exist in the calendar – and the individual growth ring corresponds to a fixed calendar year, regardless of whether asserted in 1990 or in 1998.

So far it has been implicitly assumed that the years obtained by the chronometric methods and the solar years are equal in respect to their length. This does not apply for all methods because there are properties inherent in them that may invalidate this assumption. For instance, in the case of the ^{14}C method, radiocarbon and calendar years have to be distinguished. On the basis of changes of the atmospheric ^{14}C content radiocarbon years can apparently be shorter or longer than calendar years. Therefore ^{14}C ages need a calibration. Conventionally, radiocarbon years are given as "BP" (i. e. before present). The calendar year obtained from the radiocarbon year by calibration is marked by the symbol "cal" and is presented either as a date "cal AD" or "cal BC", respectively, or as an age "cal BP". The notation "BP" defines the "present" as "1950 AD" and is – apart from ^{14}C ages – also frequently used for dendrological, varve and ice-layer ages. The indication of an age in years "before present" causes misunderstanding if uncritically applied to other dating methods. It has to be emphasized that the term "before present" does not mean "present" in the sense of the actual present, but by definition the year 1950, which approximately marks the advent of modern chronometric methods. This indication was introduced because of the special difficulties inherent in the radiocarbon method. Hence it is justified only in the case of ^{14}C ages and age comparisons with ^{14}C ages. An age indication referring to the actual present – i. e. "today" – would not be convenient because it would grow continuously with time. The situation becomes thoroughly confusing, if "BP" unnecessarily refers to another year than 1950, as for instance to the year 1990 in Johnsen et al. (1992). In order to avoid such problems it is advisable for young-age data (<12 ka) to present them as calendar date "BC/AD".

1.2
Natural Radioactivity: The Physical Basis for Dating

The physical phenomenon of radioactivity is the basis for chronometry in geology and archaeology. Its rate is purely governed by nuclear properties and, therewith, insensitive to environmental parameters, such as temperature, pressure and chemical conditions. This is a unique advantage over other time-dependent processes used as "clocks". Since its discovery in 1896 by Henry Becquerel, radioactivity has gradually revolutionized the geological and prehistoric time-scales, first in geology (Holmes 1937) and then in archaeology (Renfrew 1976). Owing to its importance it seems appropriate to introduce some fundamentals about radioactivity.

Natural matter is made up of chemical elements. An *element* is a chemical substance consisting of atoms with a characteristic number of protons – the atomic number – in the nucleus. For example, all atoms of the element carbon (symbol C) have 6 protons. If electrically neutral, the nucleus is surrounded by an equal number of orbiting electrons in the shell, because the positively charged protons and the negatively charged electrons compensate each other. If there is a deficiency or surplus of electrons in the shell the atom becomes a positively or negatively charged *ion*, respectively.

In addition to protons the nucleus also contains neutrons which have nearly the same mass as protons but carry no charge. The sum of protons and neutrons in the nucleus is called the mass number of the atom. A *nuclide* is an atomic species with a characteristic number of protons and neutrons. The mass number is indicated as a superscript to the upper left, whereas the atomic number is sometimes given as a subscript to the lower left of the elemental symbol. For example, the nuclide $^{14}_{6}C$ has 6 protons and 8 neutrons and, thus, the mass number is 14. It is sufficient to write simply ^{14}C, since the element carbon is already specified by the atomic number 6.

Isotopes are two or more nuclides which belong to the same element, i.e., they have the same number of protons, but differ by their number of neutrons. For example, the three carbon isotopes ^{12}C, ^{13}C and ^{14}C all have 6 protons, but 6, 7 or 8 neutrons, respectively. Since isotopes belong to the same element they show essentially identical chemical – but not physical – behavior.

One distinguishes stable nuclides and radioactive nuclides. The latter ones disintegrate spontaneously and transmute into a nuclide of a different element – a process called *radioactivity*. In other words, a radioactive parent nuclide decays into a radiogenic daughter nuclide.

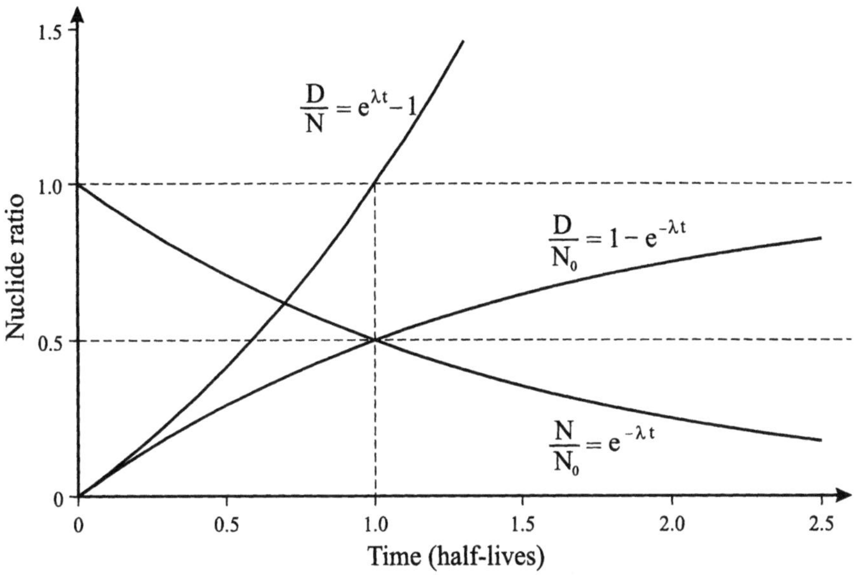

Fig. 2. Decay of the radioactive parent nuclide (amount N) to the radiogenic daughter nuclide (amount D). N – normalized to its initial amount N_0 – decreases exponentially with time t (normalized to half-life $t_{1/2}$), whereas D increases to the same extent; the ratio D/N increases exponentially

Radioactivity is a statistical process. Its quantitative treatment requires a sufficient number of radioactive nuclides N because it is impossible to predict the event when a single nucleus disintegrates. The activity dN/dt (measured in Becquerel Bq [s^{-1}]) is defined as the fraction dN of the unstable nuclide N (number of atoms with initial amount N_0) that disintegrates within the time interval dt into the stable daughter atoms D ($= N_0 - N$)

$$\frac{dN}{dt} = -\lambda N = -\frac{\ln 2}{t_{1/2}} \times N \tag{1}$$

whereby λ [a^{-1}] is the decay constant and $t_{1/2}$ [a] is the half-life ($= 0.693/\lambda$).

By integration one obtains (Fig. 2)

$$\frac{N}{N_0} = e^{-\lambda t} \tag{2}$$

and substituting ($N_0 - D$) for N

$$\frac{D}{N_0} = 1 - e^{-\lambda t} \tag{3}$$

and ($D + N$) for N_0

$$\frac{D}{N} = e^{\lambda t} - 1 \qquad (4)$$

The age t [a] can be derived, depending on whether N_0, N or D is known, according to the following equations:

$$t = \frac{1}{\lambda} \times \ln\left(\frac{D}{N} + 1\right) \qquad (5)$$

$$t = \frac{1}{\lambda} \times \ln\left(\frac{N_0}{N}\right) \qquad (6)$$

The use of a radioactive system for age determination presupposes that neither the parent nor the daughter nuclide is lost or gained except through the decay process itself, in other words, the condition of a *closed system*.

Several types of radioactive decay can be distinguished (Fig. 3) and for some nuclides – as for instance ^{40}K – two types of decay coexist (dual decay mechanism):

- *Alpha-decay:* decay under the emission of an α-particle which is a ^{4}He nucleus

  $$^{238}_{92}U \longrightarrow {}^{234}_{90}Th + {}^{4}_{2}He$$

- *Beta-decay:* decay under the emission of a β^--particle which is an electron e^- whereby a neutron is converted into a proton

  $$^{14}_{6}C \longrightarrow {}^{14}_{7}N + e^-$$

- *Electron capture:* decay under the capture of an extranuclear, orbiting electron from the innermost atomic shell (K-shell) whereby a nuclear proton is converted into a neutron

  $$^{40}_{19}K + e^- \longrightarrow {}^{40}_{18}Ar$$

  similar in its effect is the *β^+-decay* with the emission of one β^+-particle, i.e. one positron e^+

  $$^{26}_{13}Al \longrightarrow {}^{26}_{12}Mg + e^-$$

- *Spontaneous fission:* decay under splitting into two heavy nuclides and 2–3 neutrons n (example: $^{238}_{92}U \longrightarrow {}^{142}_{56}Ba + {}^{93}_{36}Kr + 3n$)
  Most of the types of radioactive decay are accompanied by the emission of energetically discrete γ-rays – a type of electromagnetic radiation coming from excited nuclei.

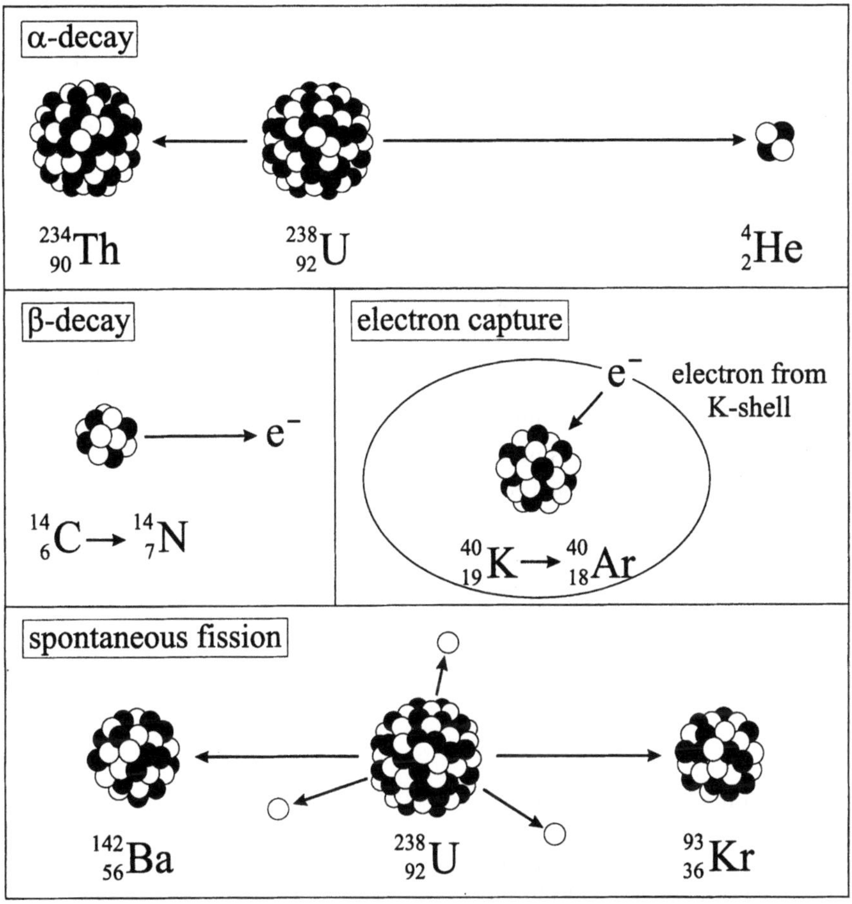

Fig. 3. For radiometric dating four types of natural radioactivity are used: α-decay, β-decay, electron capture and spontaneous fission (white: neutrons, black: protons)

Instead of being stable the radiogenic daughter nuclide may be radioactive and decay itself. If several such radioactive daughter nuclides follow each other, a decay chain is formed until finally a stable end member is reached. This is, for instance, the case in the decay chain starting from ^{238}U and ending at ^{206}Pb involving several steps of α- or β-decay. In such decay chains a balance between production and decay of the interim members is established after a certain time (radioactive or *secular equilibrium*) in which for all radioactive members N_1, N_2, N_3 etc. the activity dN/dt becomes the same:

$$\lambda_1 N_1 = \lambda_2 N_2 = \lambda_3 N_3 = \dots \text{ etc.} \tag{7}$$

Table 1. Nuclides used in Quaternary chronometry

Element	Nuclide	Relevant Origin	Decay	Energy (keV)	Half-life
Hydrogen	^{3}H	Cosmogenic Anthropogenic	β^-	18.6	12.43 a
Helium	^{3}He	Cosmogenic	Stable		
Beryllium	^{10}Be	Cosmogenic	β^-	555	1.51 Ma
Carbon	^{14}C	Cosmogenic	β^-	158	5730 a
Neon	^{21}Ne	Cosmogenic	Stable		
Aluminum	^{26}Al	Cosmogenic	β^+	1170	705 ka
Silicon	^{32}Si	Cosmogenic	β^-	≈100	140 a
Chlorine	^{36}Cl	Cosmogenic	β^-	714	301 ka
Argon	^{39}Ar	Cosmogenic	β^-	565	269 a
Potassium	^{40}K	Primordial	β^- EC	1330 1460	1397 Ma 11930 Ma
Calcium	^{41}Ca	Cosmogenic	EC	427	103 ka
Krypton	^{81}Kr	Cosmogenic	EC	276	210 ka
Lead	^{210}Pb	Radiogenic	β^-	17	22.3 a
Radium	^{226}Ra ^{228}Ra	Radiogenic Radiogenic	α β^-	4784 39	1.60 ka 5.75 a
Thorium	^{230}Th ^{232}Th	Radiogenic Primordial	α α	4688 4012	75.4 ka 14050 Ma
Protactinium	^{231}Pa	Radiogenic	α	5012	32.8 ka
Uranium	^{234}U ^{235}U ^{238}U	Radiogenic Primordial Primordial	α α α SF	4775 4400 4197 2×10^5	245 ka 704 Ma 4468 Ma 8.2×10^{15} a

α, alpha; β^-, electron beta; β^+, positron beta; EC, electron capture; SF, spontaneous fission.

In nature various kinds of radioactive nuclides occur and many of them can be used for chronometry (Table 1). Because of their origin these nuclides can be divided into various categories whereby a nuclide may be formed by several of these processes:

- *Primordial origin:* the nuclide is still present from the time of nucleosynthesis and, thus, is older than the formation of the Earth (example: ^{40}K). Consequently, such nuclides have correspondingly long half-lives.
- *Radiogenic origin:* the nuclide is formed radiogenically within the decay chains of uranium and thorium (example: ^{231}Pa). These nuclides have half-lives less than 250 ka.

- *Cosmogenic origin:* the nuclide is formed by the interaction of cosmic rays with the atmosphere and the Earth's surface (example: ^{14}C). Cosmogenic nuclides have various half-lives.
- *Anthropogenic (technogenic) origin:* the nuclide is produced in nuclear plants and explosions (example: ^{3}H).
- *Nucleogenic origin:* the nuclide is formed by nuclear reactions induced by neutrons deriving from natural (α, n) reactions (example: ^{26}Al).

1.3 Error: Precision and Accuracy

Instruments and persons involved in physical measurements are imperfect. Therefore, each measurement results in a certain degree of uncertainty. The measured value x is only an approximation of the sought after, true value which is and remains unknown. The possible deviation of the measured from the true value is estimated in terms of an error. The error $\pm \overline{\sigma}$ describes an interval from $(x - \overline{\sigma})$ to $(x + \overline{\sigma})$ surrounding the measured value x so that the true value is expected with a certain probability within that interval (*confidence interval*). Consequently, the error is an integral and indispensable component when quoting measured data, such as $x \pm \overline{\sigma}$. Strictly speaking, an age without an error quotation has zero probability. Obviously, it makes no sense to report data without their error.

The sources of two types of error need to be distinguished, the *random* error $\overline{\sigma}_r$ and the *systematic* error $\overline{\sigma}_s$ (Fig. 4). A random error is caused by random inaccuracies when reading an instrument and by statistical processes such as, for instance, the radioactive decay. Random errors are recognized from repeated observations of the same phenomenon that cluster statistically around a central value. Random errors are characterized by the term *precision* (or reproducibility), i.e., a high precision implies a small random error. The random error is assessed from the frequency distribution of the single observations. Systematic errors, on the other hand, may be caused by instrumental deficiencies, environmental influences and personal inadequacies. Any systematic error source influences the single measurements in the same direction and degree and, consequently, cannot be recognized by repeated observations. Systematic errors are generally difficult to identify and quantify. They are described by the term *correctness* (or reliability). High correctness implies low systematic errors. The result of a measurement may be equally unprecise *and* correct or, vice versa, precise *and* incorrect. In both cases the result would be inaccurate. The term *accuracy* comprises precision plus correctness (Fig. 4), and the total error $\overline{\sigma}$ is calculated according to

$$\overline{\sigma} = \sqrt{\overline{\sigma}_r^2 + \overline{\sigma}_s^2} \tag{8}$$

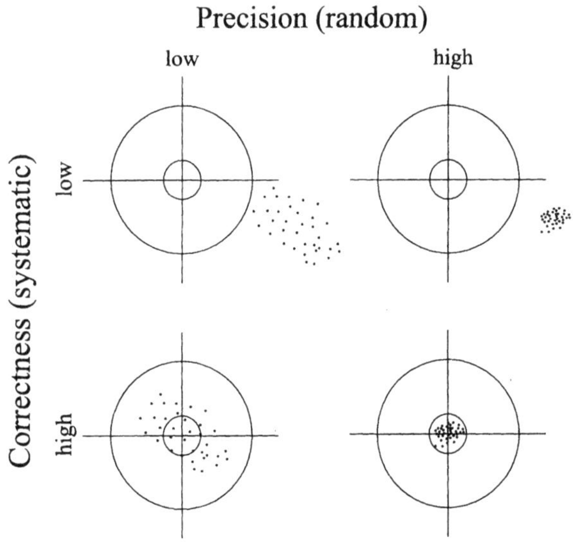

Fig. 4. Random and systematic measuring errors need to be distinguished from one another. Correctness (or reliability) means minor systematic deviation (*below*). Precision means high reproducibility (on the *right*). Accuracy comprises precision combined with correctness. A result may be correct but unprecise or incorrect but precise; however, in both cases it still would be inaccurate

The inappropriate use of the terms precision, correctness and accuracy in chronological literature has frequently caused confusion. For instance, the high precision of uncalibrated, conventional ^{14}C ages is often mistaken for accuracy. But, although precise, these ages are not accurate owing to past ^{14}C variation in the atmosphere, and only after dendrochronological correction (*calibration*) do they become accurate. Contrarily, thermoluminescence ages are usually correct, but for various physical reasons they are of limited precision.

Errors quoted for chronometric data represent mostly precision, in other words, an error $\overline{\sigma}$ quoted in a publication may represent $\overline{\sigma}_r$, only. In some cases systematic error sources are also discussed, but as mentioned before, often they can hardly be estimated. Generally, the analytical quality of published data can be judged from the care with which authors comment on their various error sources. In any case, sufficient information should be given to make clear which of the random as well as the systematic errors are taken into account and which are not. Standardized formats for the calculation and presentation of errors exist only for few dating methods. They facilitate the data assessment and comparison.

The calculation of random errors requires that the variable x follows certain laws of frequency distribution. In dating practice the most important is the *Gauss* (or normal) distribution. In this unimodal distribution the random observations, let us say many replicate determinations of an age t, center symmetrically and bell-shaped around a maximum value (Fig. 5). To test the hypothesis that a series of real data is consistent with a

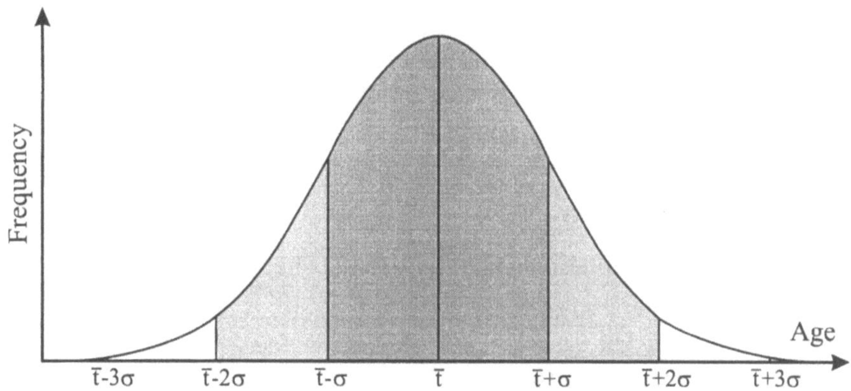

Fig. 5. Gaussian or normal frequency distribution (bell-shaped curve) of random events, such as repeatedly measured age results t_i, around the arithmetic average $\bar{t}$. The standard deviation σ is a measure for the scatter of the single result t_i. The true value *t* falls within the interval $\bar{t} - \sigma$ and $\bar{t} + \sigma$ with a probability of 68.3 %, and within the interval $\bar{t} - 2\sigma$ and $\bar{t} + 2\sigma$ with a probability of 95.4 %

Gaussian distribution the χ^2-test (Chi-square test) is used (e.g., Ward and Wilson 1978). To give an example, the fission-track ages of single, cogenetic zircon grains from the same tuff horizon should exhibit a Gaussian distribution scattering around the age of the eruption. If the tuff contains additionally a component of detrital zircon grains with significantly higher ages, the discrete ages form a bimodal frequency distribution. The application of the χ^2-test would reveal the incompatibility with an unimodal Gaussian distribution.

For the case of Gaussian distribution, the confidence interval around the *arithmetic mean value* $\bar{x}$ is defined by the *standard error* $\bar{\sigma}$ – to be exact $\bar{\sigma}_r$ – which is derived from the *standard deviation* σ_r. This last quantity is a measure for the scatter of the single x_i values. If the measurement is repeated *n* times, $\bar{x}$ is defined as

$$\bar{x} = \frac{1}{n} \sum_{i=1}^{n} x_i \qquad (9)$$

The standard deviation σ_r is calculated according to

$$\sigma_r = \sqrt{\frac{1}{(n-1)} \sum_{i=1}^{n} (\bar{x} - x_i)^2} \qquad (10)$$

and the standard error $\bar{\sigma}_r$ according to

$$\bar{\sigma}_r = \frac{\sigma_r}{\sqrt{n}} \qquad (11)$$

The ratio $\bar{\sigma}/\bar{x}$ is called the relative error.

The counting of random, discrete events, such as radioactive decay, follows the *Poisson* distribution. This frequency distribution applies when the probability of the event is low, but the number n of trials is large. The mean value $\bar{x}$ is calculated as

$$\bar{x} = \frac{N}{n} \tag{12}$$

where N is the total number of counted events. The standard deviation σ_r is

$$\sigma_r = \sqrt{\bar{x}} \tag{13}$$

the standard error $\bar{\sigma}_r$ becomes

$$\bar{\sigma}_r = \frac{\sigma_r}{\sqrt{n}} \tag{14}$$

and the relative error $\bar{\sigma}_r / \bar{x}$ is

$$\frac{\bar{\sigma}_r}{\bar{x}} = \frac{\sqrt{N}}{N} \left(= \frac{1}{\sqrt{N}} \right) \tag{15}$$

As already mentioned, the systematic error $\bar{\sigma}_s$ can neither be evaluated nor improved by measurement repetition. In dating practice commonly one standard error, i.e., $1\bar{\sigma}$ (in most cases only $\bar{\sigma}_r$ instead of $\bar{\sigma}$), of a mean age $\bar{t}$ is used for the confidence interval. Then, in the age interval from $(\bar{t} - 1\bar{\sigma})$ to $(\bar{t} + 1\bar{\sigma})$ the true age is expected with 68.3% probability, i.e., the probability that the true value falls outside this interval is still roughly 1/3. If $\bar{\sigma}$ is doubled, the true age is expected with 95.4% probability within the interval from $(\bar{t} - 2\bar{\sigma})$ to $(\bar{t} + 2\bar{\sigma})$, and if tripled, with 99.7% probability within the interval from $(\bar{t} - 3\bar{\sigma})$ to $(\bar{t} + 3\bar{\sigma})$.

Instead of the arithmetic mean $\bar{x}$ (Eq. 9) of several values x_i, the weighted (or pooled) mean $\bar{x}_w$ is commonly preferred if the individual values x_i have different standard deviations σ_i. This is done in order to give greater weight to values x_i with higher accuracy. The weighted mean $\bar{x}_w$ is calculated according to

$$\bar{x}_w = \sum(x_i/\sigma_i^2)/\sum(1/\sigma_i^2), \tag{16}$$

its random error $\bar{\sigma}_{rw}$ is calculated from the individual random σ_{ri}

$$\bar{\sigma}_{rw}^2 = 1/\sum(1/\sigma_{ri}^2) \tag{17}$$

and – if the individual systematic σ_{rs} can be evaluated – its systematic error $\bar{\sigma}_{sw}$ becomes

$$\bar{\sigma}_{sw} = \sum(\sigma_{si}/\sigma_i^2)/\sum(1/\sigma_i^2)\,. \tag{18}$$

Table 2. Thermoluminescence dates of a medieval kiln from Lübeck, Germany

Sample	Fraction	Age t_i (a)	Standard deviation (a) σ_{ri} σ_{si} σ_i	Mean age (a) with error (a) $\bar{t}_w(\bar{\sigma}_{rw}$ $\bar{\sigma}_{sw}$ $\bar{\sigma}_w)$
K179 T	fine-grain	836	± 35 ± 63 ± 72	
	quartz	721	± 43 ± 80 ± 91	
				779 (± 21 ± 70 ± 73)
K180 T	fine-grain	733	± 36 ± 68 ± 77	
	quartz	830	± 85 ± 76 ± 114	

When comparing ages with each other it is said that they are *significantly different* if they do not overlap within their 2 σ-confidence intervals.

Example 1. Two samples of burned clay lining, taken from a medieval kiln at Lübeck, were dated with thermoluminescence, whereby the ages of fine-grain as well as quartz separates were determined, resulting all together in four TL ages (Wagner 1980c). The data in Table 2 indicate, that there is no significant difference between the single ages t_i since they agree within their 2 σ_r bars. Treating all single ages as one age context, Eqs. (16) to (18) yield a weighted mean age $\bar{t}_w = 779$ a with the random error ± 21 a (1 $\bar{\sigma}_{rw}$) and the systematic error ± 70 a (1 $\bar{\sigma}_{sw}$). According to Eq. (8), the accuracy is ± 73 a (1 $\bar{\sigma}_w$) or, expressed as relative error (Eq. 11), ± 9.4%. The TL age was determined in 1979, so that the TL date becomes 1200 AD ± 73 a. The true age of the kiln is expected to be between 1127 and 1273 AD with 68.3% probability and between 1054 and 1346 AD with 95.4% probability.

Example 2. A heated obsidian artifact (sample 075E) from the site Sierra El Inga, Ecuador, has been dated with fission tracks (Miller and Wagner 1981). Altogether 70 spontaneous (total number *N*) and 2500 (total number *N*) induced fission tracks were counted. Since the frequency of fission track counting obeys the Poisson distribution, the relative standard errors of these counts are, according to Eq. (15), 12 and 2%, respectively. The fission track age is derived from the ratio of fossil to induced tracks and, thus, its random error is calculated from these values following the rule of error propagation, which is the same as Eq. (8). The result is 2060 a ± 12.2% ($\bar{\sigma}_r/\bar{x}$) or 2060 a ± 250 a ($\bar{\sigma}_r$), i.e., the true age of heating this artifact is expected with 68.3% probability within the time-span 1810 to 2310 a ago or, expressed as a date, between 330 BC and 170 AD.

As in other physical measurements, errors cannot not be avoided in numerical age determination and are, actually, part of the analytical result.

Of course, the analyst will try to keep his errors under control. On the other hand, the required age resolution in a geological or archaeological application may be beyond the reach of the method. It is always advisable for anyone interested in dating applications to enter a dialog with the dating specialist before sampling and before submitting the samples to the laboratory. This avoids misunderstanding and unnecessary irritation.

1.4 The Classification of the Quaternary

The Quaternary covers the period of nearly the last 2 million years of the Earth's history. By international convention the Pliocene–Pleistocene-boundary has been delimited at the stratotype Vrica near Crotone in Calabria, Italy, at the base of a homogeneous argillaceous sediment, which concordantly overlies the sapropel layer E (Aquirre and Pasini 1985). This facies change is indicative of a change to cooler climate. Magnetostratigraphically the boundary coincides approximately with the end of the Olduvai subchron 1.8 Ma ago (Hilgen 1991; McDougall et al. 1992). It should be noted that the Pliocene–Pleistocene-boundary is alternatively assigned to the first climatic depression that occurred around the Gauss–Matuyama boundary 2.6 Ma ago (Azzaroli et al. 1988; Hilgen 1991; Partridge 1997). Subdivisions of the Quaternary are the *Pleistocene*, also termed the Ice Age, and the *Holocene*, the so-called postglacial period, comprising 11.5 ka (Johnsen et al. 1992; Goslar et al. 1995; Björk et al. 1996; Spurk et al. 1998), and lasting until the present. The Pleistocene again is subdivided into Lower, Middle and Upper Pleistocene with the boundaries delimited by the Matuyama–Brunhes boundary 778 ka ago (Singer and Pringle 1996; Tauxe et al. 1996) and by the beginning of the marine oxygen isotope stage 5e (equivalent to the beginning of the Eemian-interglacial) 128 ka ago (Edwards et al. 1986/87). The Holocene can be subdivided by means of pollen analysis.

In addition to this geological division of the Quaternary based on climatically induced litho-, pedo-, morpho- and biostratigraphic characteristics, there is the possibility of an archaeological division based on the materials and production techniques of the artifacts. The Early Paleolithic covers the period until the end of penultimate glacial period ca. 128 ka ago. The succeeding Middle Paleolithic lasted in Europe up to ~40 ka ago. The subsequent Late Paleolithic lasted until the end of the last glacial period 11.5 ka ago. Both Mesolithic and Neolithic belong to the postglacial period, during which the Neolithic with the "Bandkeramik" culture started approximately 5500 BC in central Europe. In this region the earliest copper items

Table 3. Geological and archaeological division of the Quaternary in central Europe

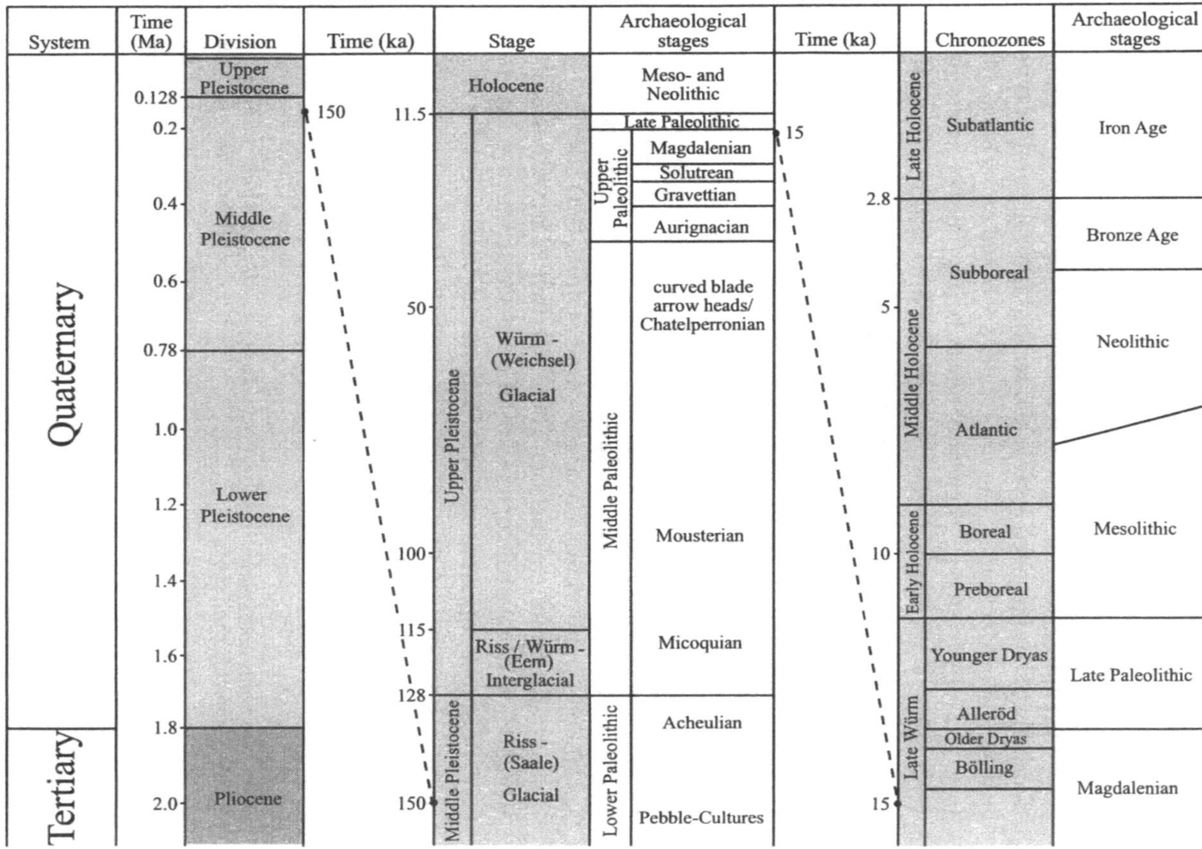

appear since approximately 4000 BC, still in the Neolithic. In this region the Bronze Age began ~2300 BC and was succeeded by the Iron Age around 850 BC.

The geological and archaeological division of the Quaternary outlined above is summarized in Table 3. It is noteworthy that the sequence of individual epochs has been established stratigraphically and/or typologically based on artifact types. The time scales are based on physical and for the last 12 ka mostly on dendrochronological dating. Since stratigraphically or typologically recognized epochs may be time-transgressive from one region to another (e.g., the postglacial period begins earlier in central Europe than in Scandinavia, or the Bronze Age starts in the Aegean region earlier than in central Europe), chronologies are valid only within given regions, such as central Europe in this case.

The division of the Quaternary on the basis of continental sediments is problematic since Quaternary stratigraphic sequences are temporally and spatially incomplete (Fig. 6). Fortunately, there exists a much more complete Quaternary chronology based on deep-sea sediments. In contrast to the continental deposits, sediments on the deep ocean floor form time-continuously and uniformly across large regions, thus allowing for large-scale stratigraphic correlation. There are two extremely useful features which allow the global correlation of deep-sea sediments; these are their *oxygen isotope signature* and *paleomagnetic direction*. The varying oxygen isotopic composition of foraminifera shells in these sediments is derived from that of the ocean water which reflects changes of the global ice volume and thus is closely related to the global climatic fluctuation during

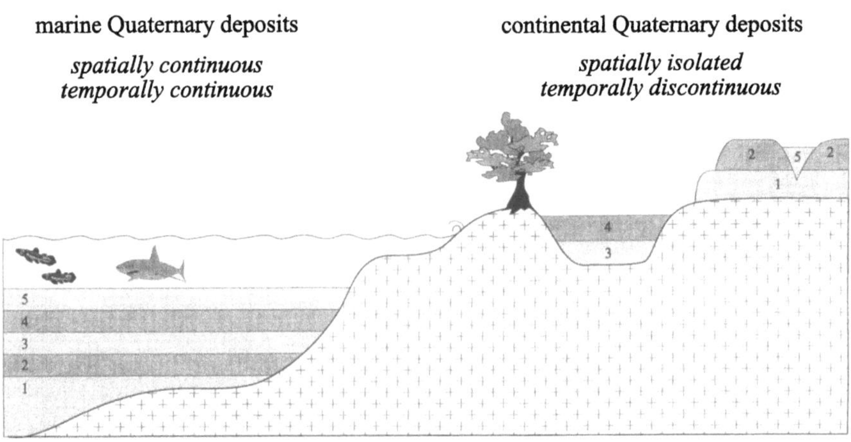

Fig. 6. Schematic comparison of marine and continental Quaternary deposits. The successive sedimentary layers (*1–5*) in the deep-sea are continuous, whereas their chronological equivalents on the continent show temporal and spatial gaps

the Quaternary period (Sect. 10.2). The paleomagnetic field reversals caused by the changing polarity of the Earth's magnetic dipole field are recorded by deep-sea sediments. Since the polarity reversals affect the magnetic field worldwide and simultaneously, paleomagnetism is an outstanding stratigraphic tool (Chap. 9). Since continental sediments also record the polarity changes, magnetostratigraphy allows the linkage of continental to deep-sea chronology.

2 Materials

In the present chapter geological, archaeological and historical materials will be listed that can be analyzed by means of the dating methods which are the content of this book. This introduction to the various materials and their characteristic properties will allow material-oriented access to the book. The chronological significance of the individual materials will be briefly discussed with reference to the appropriate dating methods. Their potentials and limitations, especially the age ranges to which these techniques are applicable, will be outlined. In the subsequent chapters a detailed description of the dating methods including the appropriate sampling strategies will be given.

2.1 Volcanites

Volcanic rocks are of great importance to Quaternary chronology. They are formed in volcanic eruptions of short duration under high temperatures and subsequent fast cooling. Such conditions of formation make a good basis for chronometric dating. Apart from the possibility of reconstructing the temporal sequence of volcanic events, their dating is significant to sedimentary sequences if volcanic rocks are intercalated. Tuff horizons of considerable regional extension represent especially important time markers for the tephrochronological age bracketing of horizons containing Paleolithic records. Recent volcanism is geographically linked to certain belts which coincide with geologic plate boundaries.

2.1.1 Basalt

This fine-grained dark-colored volcanic rock forms by solidification of 800–1100 °C hot magmas which chemically consist approximately of 50 % of SiO_2. Mineralogically, basalt contains predominantly plagioclase and pyroxene, more rarely olivine, iron ores, sanidine, biotite and glass, both in the matrix and as phenocrysts. According to their composition, alkali basalt and tholeiitic basalt are distinguished. Transitions exist to the SiO_2-richer volcanic rocks phonolite (containing orthoclase, nepheline and others) and andesite (containing plagioclase, pyroxene, biotite, hornblende, quartz). Basalts form slag cones, lava fields, flows and tubes. Most chronometric methods applied to basalts and related rocks determine the time of solidification or the immediately following cooling.

Basalt (> 1 ka) can be dated by means of the *K–Ar* technique (Sect. 3.1). Suitable are whole-rock samples and potassium feldspar phenocrysts. An important contribution of the *K–Ar* dating of basalts is the calibration of the paleomagnetic polarity time scale. Because of the high precision attainable, the *K–Ar* ages form the basis for the chronometric frame of the Early and Middle Pleistocene. Experiments to date basalts of young age by means of the *U–He* method (Sect. 3.2) resulted in age underestimates because of loss of helium. The occasional occurrence of titanite (> 100 ka) allows the application of the *fission track* method (Sect. 6.1). Furthermore, this technique can be applied in the dating of basaltic deep-sea glass (> some ka) that forms crusts on pillow lava in the active rift zones of the mid-oceanic ridges. The glass was generated through rapid cooling, when the hot lava came into contact with the cold ocean water. If basalt contains feldspar phenocrysts, glass or quartz xenocryts, the *thermoluminescence (TL)* method (Sect. 7.1) can be applied (1 ka – several hundred ka). Young volcanic rocks, especially those of the mid-oceanic ridges, frequently display radioactive disequilibrium (excess of daughter). Hence the uranium-series dating techniques $^{230}Th/^{238}U$ (1 ka – 300 ka), $^{231}Pa/^{235}U$ (> 1 ka – 130 ka) and $^{226}Ra/^{230}Th$ (<10 ka) are suitable (Sect. 4). On the surfaces of basalt flows (ka–Ma) the eruption age can be determined by means of exposure dating with the ^{3}He (Sect. 5.2), ^{10}Be (Sect. 5.3), ^{21}Ne (Sect. 5.5) and ^{36}Cl methods (Sect. 5.8) on condition that no erosion has taken place since solidification. Since the cosmogenic production rates become reduced by erosion or by soil cover on the surface, these ages are mostly minimum ages. Basalts and other volcanic rocks contain relatively high admixtures of ferromagnetic minerals which are thermoremanently magnetized during cooling of the hot rock. Thus they can be exploited for *paleomagnetic dating* (Chap. 9). The episodic nature of the volcanism excludes a continuous recording of the magnetic field parameters. The age of basaltic glass can also be deduced from *hydration rinds* (Sect. 8.2) on the rock surface.

2.1.2 Obsidian Flows

The dark-colored volcanic glass obsidian forms during the rapid cooling of melts of rhyolitic composition and consists of ca. 3/4 of SiO_2 with admixtures of Al, Na, K, Ca, Fe and Mg oxides. In addition obsidian contains 0.1 – 0.3 % of water as well. Pitchstone is a water-rich (5 – 9 %) volcanic glass with pitch-like luster. Pumice is a light-colored vesicular glass. Obsidians of different geologic provenance vary in chemistry; on the other hand the ones originating from the same lava flow are commonly of chemically homogeneous composition. Obsidian consists of an optically isotropic, weakly translucent glass matrix and frequently contains tiny inclusions of ore minerals (microlites) and vesicles. Obsidian occurs as independent flows, as inclusions in tuffs, and at the surface and the contact of lava flows, i.e., at spots with rapid cooling. Geographically, the occurrences of obsidian are widely spread. It is found in all major volcanic belts on the Earth. Representing a valuable raw material for prehistoric stone tools, it has been mined at many sites (Sect. 2.7.3).

For the dating of obsidian flows (> 1 ka) *fission tracks* (Sect. 6.1) are suitable. High contents of microlites or vesicles are disturbing. Thermal healing of fission tracks results in too young ages. Obsidian (< 1 Ma) with fresh surfaces can be dated by means of *hydration* (Sect. 8.2). This application is affected by strong temperature variations of the surfaces, especially in the case of insolation. The suitability of obsidian for the *K – Ar* method (Sect. 3.1) is limited by argon loss. Rhyolitic rocks (> 10 ka) may be datable by the *U – Th – Pb* method (Sect. 4.1.9).

2.1.3 Tephra

Loose volcanic ejecta are called tephra, whereas consolidated ones are called volcanic tuff. Volcanic ash is the term for dust-like to fine-grained ejecta. Explosive volcanic eruptions produce ash clouds, which are blown over large distances and deposited within short time spans. Depending on the distance from the eruption site, proximal and distal tephra are distinguished. Ash horizons intercalated in sediments that can be characterized on a petrologic, geochemical and geochronological basis represent excellent time markers and form the tephrochronological frame for sedimentary sequences. Tephra contain – in addition to juvenile volcanic material such as glass – xenolithic rocks torn away from the site of explosion. During the deposition or redeposition of tephra frequently older detrital portions gets admixed as well. Therefore, the dating of tephra demands careful sampling for primary volcanic material. If grain-discrete dating techniques are applied, such as the *Ar–Ar laser* and *fission track* methods, distinctive age components can be split up.

Of paramount importance to Quaternary geology is *K–Ar* dating (Sect. 3.1) of tephra (>1 ka). Especially interesting are sequences, which contain in addition to paleoclimatic indicators fossils of early hominids and artifacts as well. Our knowledge on the age of the early hominids in East Africa is founded nearly exclusively on *K–Ar* data. By means of the *Ar–Ar laser* technique the ages of discrete feldspar crystals (sanidine) can be precisely determined. Provided the ash layers (>10 ka) contain glass fragments or titanite they also can be dated by means of *fission tracks* (Sect. 6.1). For checking whether discrete titanite grains are of juvenile volcanic or of detrital origin, analysis of discrete grains (>10^5 a) serves well here, in addition to mineralogical criteria. Mica flakes from tuff (>1 ka) can be dated by means of the *alpha-recoil-track* technique (Sect. 6.2). The *TL* method (Sect. 7.1) bears a considerable potential for the dating of fine-grained glass in volcanic ashes (1–400 ka), in particular also distal ash deposits, which is a prerequisite for correlation over long distances. The *electron-spin-resonance* (*ESR*) method (Sect. 7.3) is suitable for the dating of the feldspar, quartz and glass fractions of tephra (10 ka–1 Ma). Of the volcanic tuffs the hot-deposited ignimbrites are suitable for *paleomagnetic* dating (Chap. 9). Cold, i.e., below the Curie point, deposited tuffs are paleomagnetically datable, if they gain their characteristic remanence chemically after the deposition. In volcanites from active subduction zones the *^{10}Be* content (Sect. 5.3) bears information on the mass balance and rate of magmatic processes.

2.1.4 Xenoliths and Baked Contacts

In volcanic eruptions rock fragments of the crystalline basement are enclosed by chance in the melt and heated up to the magma temperature. Material from the wall rocks of the vent or the covering lid is more or less thermally affected and torn away. These rocks, which had been carried within the magma to the surface can be dated under certain circumstances. If the temperatures are sufficiently high for a complete degassing of argon, the *K–Ar* method (>10 ka) can be employed (Sect. 3.1). It is advisable to check whether the resetting had been complete by means of single-grain-dating. Since in zircon, titanite and apatite *fission tracks* (Sect. 6.1) anneal to a certain extent, again analysis on discrete grains should be carried out for dating (>100 ka). *^{10}Be–^{26}Al* exposure ages (Sects. 5.3 and 5.6) can be determined on quartz grains of surface rocks, which have been suddenly carried from the depths to the surface during volcanic eruption (>10 ka–Ma). The heating (>400 °C) of already existent rocks associated with volcanic activities (1–some 100 ka) enable their *TL* and *ESR* dating (Sects. 7.1 and 7.3, respectively). Fractions of quartz and feldspar are suitable. Fritted contacts, e.g. in soils beneath a lava flow, can be exploited for *TL* dating, in which the fine-grain fraction (4–11 µm) can be used.

2.1.5 Polymetallic Sulfides

Associated with active mid-oceanic ridges are submarine exhalations of hydrothermal solutions. The hot solutions gain their abundance of metals (Cu, Pb, Zn, Fe, Au, Ag) through the leaching of the oceanic crust. At the exhalation sites polymetallic sulfides are deposited building meter-high chimneys on the ocean floor. The ores are of economic interest. During formation of the sulfide ores ^{210}Pb is separated from its radioactive precursor ^{226}Ra in the uranium series. The time-dependent decrease of the *^{210}Pb excess* (Sect. 4.1.6) in the sulfide ore can be exploited for dating (several a–100 a) the ore formation. Barite-containing hydrothermal vent chimneys (1–15 a) in active mid-ocean ridges can be dated by *$^{228}Th/^{228}Ra$* (Sect. 4.1.8).

2.2 Impactites

If big meteorites collide with the Earth they hit the surface with non-reduced cosmic velocity. The kinetic energy liberated during the collision causes the explosive formation of craters. The affected rocks suffer high shock wave pressures and temperatures. They become brecciated, metamorphosed, molten and evaporated. Rocks altered in this way by an impact are called impactites, especially if they contain glass. Dating of impactites is of general interest because of the frequency with which such disasters struck the Earth. Only a few impact events are known from the Quaternary period.

2.2.1 Tektites

Tektites are attributed to huge meteorite impacts, even if the details of their formation process are not yet fully understood. Tektites are natural glasses of extremely low water content, derived from a fully molten stage at high temperatures. They may reach the size of pigeon eggs and display an immense variability in shape. Their color ranges from bottle-green to dark brown. Chemically they consist of ca. 3/4 SiO_2, and the remaining portion is made up of Al, Na, K, Ca, Fe and Mg oxides. They occur within four strewn fields (North America, Bohemia–Moravia–Lusatia, Ivory Coast and Southeast Asia with Australia). The relationship to a source crater – as the one between the Bohemian – Moravian moldavites and the Nördlinger Ries crater – is not clear in the case of all the occurrences. Only the two youngest strewn fields, the one extending from Southeast Asia to Australia and the one in Ivory Coast, fall into the age range of the Quaternary. Tektites can be dated by means of both the *K–Ar* (Sect. 3.1) and the *fission track* methods (Sect. 6.1). Both methods can also be used to determine the ages of the mm-sized microtektites from deep-sea sediments, which form a significant stratigraphic marker horizon in the drill cores taken within the strewn fields.

2.2.2 Impact Glass

Impact or crater glasses are situated close to the impact site. Their chemical composition resembles that of the initial material of the target rock. Compared to volcanic glasses, the duration of the melted stage is very short in the case of impact glasses. Therefore they are not always fully degassed and frequently contain unmelted relicts, a fact which causes difficulties in the application of (>100 ka) the *K–Ar* method (Sect. 3.1). By means of *fission tracks* (Sect. 6.1) impact glasses of some young (>1 ka) impact craters were dated. As in volcanic glasses, annealing effects of the fission tracks are observable in impact glasses also. The *TL* method too bears a potential in the dating of impact glass (Sect. 7.1).

2.2.3 Ejecta

Associated with giant meteorite impacts are high pressures and temperatures, which extinguish the *TL* signals in the affected rocks. Hence, the impact can be dated by means of the intensity of the newly generated *TL* (Sect. 7.1). In addition, rock masses deriving from the depth of the crater, where they had been shielded from cosmic rays, become ejected to the Earth's surface. The exposure ages of the ejecta and the crater wall determine the time of the impact. For carbonates (1 ka–1 Ma) the ^{36}Cl method (Sect. 5.8) and for quartz (10–several Ma) the ^{10}Be–^{26}Al method (Sects. 5.3 and 5.6) are applied.

2.3 Fault Breccia and Pseudotachylite

In the brittle Earth's crust tensions are reduced by means of faulting. The rock material within faults is strongly affected by the pressure and is crushed and ground (mylonite, gouge). For the tectonic development of rocks and, in particular, the evaluation of earthquake recurrence intervals in seismically active regions, age determination of such fractures is essential. The *ESR* dating (Sect. 7.3) of mylonite (1 ka – 1 Ma) exploits the pressure sensitivity of the *ESR* signals in quartz. If the mylonitization process is associated with a strong temperature increase, a glass-like rock mass forms, the so-called pseudotachylite, which is known from the friction plains of big mountain slides, for example in the Alps and the Himalayas. Pseudotachylites can be dated by means of *fission tracks* on the glass phase (>10 ka; Sect. 6.1), and with *TL* (Sect. 7.1) on relicts of quartz and feldspar as well as the glass phase (1 – 100 ka). During fault movement the uranium-rich mineral carnotite may be formed on the fault surface (10 – 350 ka) enabling ^{230}Th and ^{231}Pa dating (Sect. 4.3).

2.4 Fulgurite

If lightning strikes surfaces consisting of quartz sand, tube-shaped and root-shaped rock melts, the so-called fulgurites are formed. Fossil fulgurites are of interest for geomorphologic and paleoclimatic reasons. Principally they are datable (1 ka – 100 ka) by means of *radiation dosimetry* methods (Chap. 7).

2.5 Sediments

Sediments are classified on the basis of their formation into clastic (through mechanical deposition), chemical (through precipitation) and organic (of biogenic origin) sediments. Depending on the transporting agent, one distinguishes between aeolian (wind), aquatic (water) and glacial (ice) sediments. A special group represent the archaeosediments formed under anthropogenic influence.

The classic stratigraphic divisions of the Quaternary are based on continental sedimentary sequences in periglacial areas. In central Europe, there are two such divisions, one with four alpine glaciations (Günz, Mindel, Riss and Würm) and one with three Nordic glaciations (Elster, Saale and Weichsel). Their mutual correlation is still problematic. The main obstacles to such correlation are frequent time gaps as well as multiple facies changes over short distances in the sedimentary record of Quaternary deposits (Fig. 6). The continuously deposited deep-sea sediments reveal considerably more climatic cycles compared to the continental deposits, and allow worldwide correlation across all oceans. It is therefore desirable to establish a standard division on the basis of deep-sea sediments and to transfer it to continental deposits. Such marine–continental correlation involves, however, new difficulties.

An obvious strategy to transfer the marine division to the continent is chronostratigraphy (Fig. 1). Through as much as possible complete and accurate chronometric dating of both marine and continental sedimentary sequences, it should be possible to achieve a consistent, global chronology for the Quaternary. Thereby the problem arises that sediments belong to the group of rocks that are most difficult to date numerically, but lately enormous methodical progress has been and still is being achieved in Quaternary chronometry. Even aeolian or aquatic clastic sediments, which for long time were considered as undatable, can now be dated.

A further cogent reason for the necessity to date Quaternary sediments is the fact that archaeological finds are unearthed from them, so that the age of the sediment indirectly permits the assessment of the archaeological age.

2.5.1 Loess

Loess is a yellow-gray, aeolian dust sediment. The grain sizes mainly fall into the range of coarse silt (20–63 μm). Mineralogically predominating are small grains of quartz and feldspar. Loess is formed in the periglacial region and represents an indicator of a dry–cold steppe climate. Loess is of widespread occurrence and forms almost unstratified blankets up to several meters in thickness. The loess units belonging to individual glacial periods are separated by interglacial paleosols. The correlation of the loess and paleosol sequences in time and space is the concern of pedostratigraphy. Because of erosion processes such sequences frequently display incomplete preservation, thus they require a chronometric screening. Of special interest for dating are loess horizons with archaeological finds.

The extensive bleaching of *TL* (Sect. 7.1) during aeolian transport is an excellent prerequisite for dating. The polymineral fine-grain fraction is mostly used. For the last glacial–interglacial cycle, i.e., the last ~100 ka, *TL* ages are considered to be reliable, but in the case of older loess there is a tendency toward age underestimation depending on the mineralogical loess composition and the method applied. Because of its high light sensitivity the signal of the *optically stimulated luminescence* (*OSL*) method also (Sect. 7.2) is suitable for dating. The *ESR* method provides another possibility of dating loess (<1 Ma; Sect. 7.3). The ^{10}Be content (Sect. 5.3) in loess–paleosol sequences represents a sensitive climatic indicator and thus serves the indirect dating of loess profiles. A natural remanent magnetization is frequently observable in loess, which allows its *paleomagnetic* dating (Chap. 9) by means of geomagnetic inversion and secular variation.

2.5.2 Sand (Aeolian)

Dunes are sand hills piled up by the wind. Mineralogically they consist predominantly of quartz and more rarely of feldspars of grain sizes in the range of 63 µm – 2 mm. They occur in areas of weak vegetation at seashores, riverbanks as well as in the dry inner continental regions, and thus represent important environmental indicators. With transition to a more humid climate the dunes become consolidated by vegetation resulting in soil formation on their surface. Factors weakening the vegetation, such as drier climate, deforestation or overgrazing, reactivate the formation of dunes. Of special interest within this context are the late glacial inland dunes of the Upper Rhine Graben. They formed on the blank gravel plains of the Rhine at the end of the last Glacial. With beginning forestation their formation stopped. Prehistoric and historic deforestation anew caused dune movements and are recorded through archaeological finds. Consequently, the dating of dunes is not only of paleoclimatic but also geoarchaeological interest.

The intensive bleaching of aeolian quartz and feldspar grains is a good prerequisite for *luminescence* dating (Sects. 7.1 and 7.2). In use are monomineral fractions (mostly in the range of 100 – 200 µm) of quartz or potassium feldspar. The last aeolian redeposition of the sand grains is dated (100 a – 100 ka). Thinkable, although not yet tested, is *ESR* dating (Sect. 7.3) of dunes.

2.5.3 Sand (Aquatic)

Sand deposits in water are formed under different facies conditions: fluvial in rivers, glaciofluvial through meltwater in the foreland of glaciers, limnic in continental lakes, littoral in the coastal area and marine in the open sea. The main mineral components of aquatic sands are quartz and feldspar. Among Quaternary sediments, sands are widely spread in stratigraphic profiles, thus the clarification of their temporal placement is extraordinarily desirable. In addition, sands may bear finds of the Stone Age.

Until a few years ago, sands were considered chronometrically undatable. The situation has improved because of the development of *luminescence* techniques (Sects. 7.1 and 7.2), and quite a lot of useful age data (100 a – 100 ka) already exist. The bleaching of aquatic sands is less intensive compared to aeolian ones, and it is smaller for the *TL* signal than for the *OSL* signal. In individual aquatic depositional milieus different light conditions prevail, e.g., coastal sands of the shallow water are more strongly bleached than glaciofluvial sands, and appropriate fundamental investigations on their suitability for luminescence dating are in progress.

2.5.4 Alluvium

Alluvial sediments are young, unconsolidated sediments that are transported and deposited by streams or other bodies of running water. Mostly they are of polymineral composition and of grain size predominantly in the silt fraction (2 – 63 µm). Alluvial sediments occur in stream beds, flood plains and fans. They constitute a considerable portion of the Pleistocene and Holocene deposits. Often they are associated with archaeological layers. Alluvial deposits are important for the understanding of the geomorphic development. Of special concern within this context is the question of their age.

Beside the ^{14}C method (Sect. 5.4), which is applicable to synsedimentary organic components (< 40 ka), *luminescence* dating (Sects. 7.1 and 7.2) is suitable for alluvial sediments (100 a – 100 ka), if the minerals have been exposed sufficiently intensively or long to daylight during their transportation or deposition. It has to be assumed that the light conditions are strongly facies-dependent and alluvial sediments are variably well suited for luminescence dating. Generally, the more light-sensitive *OSL* signal is preferable to the *TL* signal for dating.

2.5.5 Colluvium and Talus

At the base of gentle slopes one finds masses of heterogeneous material that was deposited downward by rainwash, sheetwash or continuous creep. The term colluvium is applied to such deposits. Depending on the material and processes involved, colluvia may vary in grain size, but silt-sized sediments prevail. Man's activities, such as deforestation and agriculture, trigger soil erosion and thus contribute greatly to the formation of colluvia further downslope. Colluvial deposits give evidence of past ecosystems and allow the reconstruction of ancient, intensely worked landforms, but such studies require reliable dating of these sediments. Archaeological finds are common in colluvia. The term "talus" is used for more steeply sloping accumulations of coarse, unweathered rock fragments transported by rolling and falling.

The ^{14}C method (Sect. 5.4) is applicable to organic components and finds in colluvia (<40 ka). *OSL* dating (Sect. 7.2) turns out to be extremely useful for fine-gained colluvial sediments (100 a–100 ka) in various geoarchaeological settings. Feldspar-bearing rock fragments, found in talus deposits at the slope base, may have been exposed to light during their transportation, allowing *OSL* dating of their bleached – and afterward light-protected – surfaces. Alluvial fans with stabilized geomorphic surfaces (ka–Ma) are datable with the *cosmogenic nuclides* ^{10}Be and ^{26}Al (Sects. 5.3 and 5.6, respectively).

2.5.6 Limnic Sediments

Lake sediments belong to the most important climatic indicators of the continental Quaternary, especially in respect to the Upper Pleistocene and the Holocene. Similarly to deep-sea sediments, the deposition occurs continuously. Admittedly, the period of time is limited by the sedimentary filling up, however, a limnic sequence may cover some 100 ka. Climatic changes are expressed in the height of the lake level, in the lithology and the *pollen record* (Sect. 10.2.4), through which the lake sediments can be correlated with other deposits. In periglacial regions, seasonal depositional changes may give rise to annual layers, the *varves*. Varved sediments are invaluable archives for climatic proxy data with high time resolution. The magnetic *susceptibility* (Chap. 9) can be used for the correlation as well, since it depends on the proportion of ferromagnetic mineral compounds which again is controlled by the environment. In addition, calm conditions of deposition favor the acquisition of detritic remanence. As a result of their fast deposition, limnic sediments – in contrast to deep-sea sediments – record the *secular variation*, through which they become paleomagnetically datable. There have been attempts to use ^{10}Be exposure dating (ka–Ma) on quartz from beach conglomerates (Sect. 5.3). The age of calcium carbonate precipitated in continental lakes can be determined by the $^{230}Th/^{234}U$ (<350 ka; Sect. 4.1.1), $^{231}Pa/^{235}U$ (<150 ka; Sect. 4.1.3) and ^{14}C (<40 ka; Sect. 5.4) methods. The $^{230}Th/^{234}U$ and ^{36}Cl methods (Sect. 5.8) are used for the dating of limnic evaporites also. In limnic sediments (>10 ka) the ^{10}Be analysis (Sect. 5.3) permits predictions on the transportation balance and the age of deposition. The potential of *luminescence* techniques (Sects. 7.1 and 7.2) for the dating of clastic limnic deposits has not yet been systematically tested. If limnic sediments bear authigenic organic matter they can be dated by ^{14}C (< 40 ka; Sect. 5.4).

2.5.7 Glacial Sediments

Moraines, till, glacial debris and others belong to the group of sediments transported by glaciers. Their dating is of basic importance in Quaternary geology. Quartz of moraine blocks and other glacially worked surfaces, which were shielded from cosmic rays as long as they were situated at depth and which came to the surface by glacial activity, can be dated with the *cosmogenic nuclides* ^{3}He, ^{10}Be, ^{21}Ne and ^{26}Al (Sects. 5.2, 5.3, 5.5 and 5.6, respectively) in the age range of ka–Ma. If glaciers cross obsidian flows, thereby transporting away obsidian fragments, the *hydration* (Sect. 8.2) in tension cracks of the affected obsidian (10–100 ka) can be exploited for dating the movements of the glaciers. In semiarid high mountainous areas, coatings resembling desert varnish may form on exposed moraine boulders (Sect. 2.6.3), and this phenomenon allows *cation* dating (Sect. 8.6) of glacial episodes (10–100 ka). Through the thickness of *weathering crusts* (Sects. 2.6.4 and 8.1) on glacial debris (10 ka–Ma) a relative dating of glacial activities is possible. In this case, basalts and other fine-grained rocks are suitable.

2.5.8 Archaeological Sediments

By archaeological sediments one understands the deposits which have been formed under anthropogenic influence. In this sense archaeosediments are regarded as a subgroup of geological sediments in whose formation cultural influences also contributed, in addition to natural processes, and enclosed artifacts are considered part of the sediment. Archaeological sediments have generally received little attention from archaeologists. Only recently are they becoming increasingly included as an important source of information. They are indispensable in the reconstruction of the physical and biological environment of a site. They can be helpful in solving chronological questions and play a major role in site as well as landscape reconstruction. Deposits (< 40 ka) rich in organic components can be dated by means of ^{14}C (Sect. 5.4). Inorganic deposits which have been exposed to daylight, directly at the surface or at lower depths because of anthropoturbation, e.g. through trampling, are suitable for *luminescence* dating (Sects. 7.1 and 7.2).

Because of its high light sensitivity the *OSL* is especially promising for the dating of archaeological sediments (100 a – 100 ka).

2.5.9
Calcareous Cave Deposits

At the outflow of groundwater in caves through degassing of surplus CO_2, calcareous sinter crusts are precipitated. In cave systems the ceilings, floors as well as walls are coated by stalactites, stalagmites and sinter tapestries. The calcareous precipitates in caves mostly consist of calcite, more rarely of aragonite. The surplus CO_2 of the groundwater derives from the soil where it is formed by microbiological processes. Therefore calcareous sinters are preferentially formed under the warm and humid climatic conditions of the interglacials. They are thus paleoclimate indicators. Calcareous sinters often contain, in addition to chemical calcareous precipitates, clastic components, which were carried in by means of water and wind. Caves and springs are frequently visited by animals and men. Hence, such places belong to the most important paleontological and archaeological sites.

To the chronologist the dating of cave sediments including their archaeological finds represents an important, but frequently very challenging task. Among cave sinters, the stalagmites are most suitable for dating, since they grow up vertically in layers and not radially as the stalactites do (Fig. 7). The calcite component of the calcareous sinters

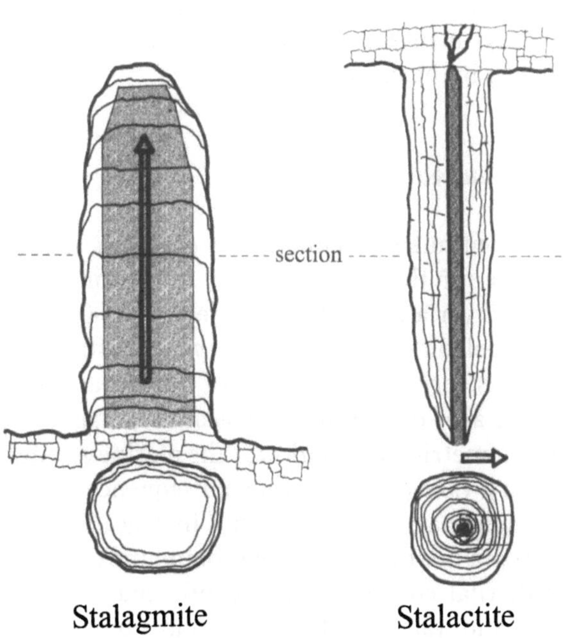

Fig. 7. Schematic depiction of stalagmitic and stalactitic growth pattern (indicated by *arrows*). The growth zones in stalagmites are thicker and more clearly expressed than in stalactites. The areas suitable for allowing optimal sampling are *shaded*. (After Geyh 1983)

incorporates ~0.1–1 μg g^{-1} uranium, but no thorium, in the course of its crystallization, so that the $^{230}Th/^{234}U$ ratio (<350 ka) and more rarely (as a result of the low uranium content) the $^{231}Pa/^{235}U$ ratio(<150 ka) can be exploited for dating (Chap. 4). The *ESR* method (Sect. 7.3) permits the dating of the sinter formation (<0.5 Ma) and is frequently conducted in association with uranium-series dating. The *TL* method (Sect. 7.1) also bears an interesting potential for the dating of calcareous sinters (<1 Ma). In the case of calcareous sinters, like the groundwater, the carbon partly derives from the atmospheric and biospheric reservoirs, which permits ^{14}C dating (Sect. 5.4; <40 ka). The remanent magnetization inherent in cave sediments and calcareous sinters offers a potential for *paleomagnetic* dating (Chap. 9).

2.5.10 Travertine

At the mouths of calcareous springs changes of pressure and temperature result in calcareous precipitates. These sinter terraces, called travertine or calcareous tufa, preferably form under humid climatic conditions. The travertine deposits of central Europe, which partly reach conspicuous thickness, are therefore assigned to the warmer periods of the Quaternary. The important archaeological finds of the travertine terraces, such as those of Bilzingsleben, Weimar-Ehringsdorf and Stuttgart-Cannstatt, Germany, demonstrate that such springs must have been attractive to early Man. The remarks mentioned previously regarding datability of cave sinters, apply equally in the present context.

2.5.11 Deep-Sea Sediments

The continually accumulating deep-sea sediments are almost an ideal archive for the Earth's history far beyond the Quaternary. The *oxygen isotopic* variation in deep-sea sediments (Sect. 10.2.2), combined with the micropaleontology and the *magnetostratigraphy* based on polarity changes, forms the backbone of Quaternary stratigraphy. Linked to radiometric ages of continental volcanites, magnetostratigraphy furnishes the chronometric time marks for the marine stratigraphic system of the Quaternary. Conversely, continental sequences can be coupled to this system through magnetostratigraphic linkage. The pelagic depositions on the deep-sea floor consist of two compounds of different origin: a terrigenous or detrital compound, i.e., continental weathering products carried by wind and water into the ocean, and a biogenic compound formed by biological–chemical precipitation from the ocean water. The biogenic fraction consists predominantly of carbonaceous and siliceous skeletons of marine unicellular organisms. Depending on the microfaunal and -floral habitat, a distinction is made between planktonic (upper meters below the sea level) and benthic (at the ocean bottom) origin of the skeletons. Abundant types of pelagic sediments are the calcareous globigerina ooze and the non-calcareous terrigenous deep-sea clay. Sampling is conducted by coring, and it must be added that cores up to 10 m in length and ca. 10 cm in thickness can be obtained.

The non-calcareous sediment proportion is dated by the methods of excess ^{230}Th (<350 ka) and ^{231}Pa (<150 ka) (Chap. 4). Using sediment cores, the rate and age of the deposition can be determined. The data do not merely serve the chronometric calibration of the $\delta^{18}O$-curve, but also reveal processes of erosion and turbation that reflect the climatically triggered changes in deep-sea water circulation. From the cosmogenic ^{10}Be content of deep-sea sediments (>10 ka) the sedimentation rate can be read as well (Sect. 5.3). This technique is also applied in the determination of the growth rates of manganese nodules. These concretions of hydrated iron–manganese oxides with clay grow on the ocean floor and are of economic interest because of their polymetallic richness. Their growth rate can also be measured by ^{26}Al (>1 ka; Sect. 5.6). For the determination of the deposition rate of subrecent deep-sea sediments ^{32}Si ($<10^3$ a) and ^{210}Pb ($<10^2$ a) analyses (Sects. 5.7 and 4.1.6, respectively) are also employed. In deep-sea sediments the polarity changes of the magnetic field are continuously recorded. Their remanent magnetization is predominantly caused by small admixtures of aeolian-derived grains that slowly sink to the ocean floor where they become magnetically aligned in the sediments. The polarity pattern and the intensity changes are employed for *paleomagnetic* dating (Chap. 9).

2.5.12 Marine Phosphorite

Marine phosphorites reveal paleo-oceanographic and paleoclimatic processes, and thus are worthwhile dating. In the course of phosphorite formation apatite becomes enriched in uranium, so that the $^{234}U/^{238}U$, $^{230}Th/^{234}U$ and $^{231}Pa/^{235}U$ ratios (Chap. 4) can be used for dating Middle and Lower Pleistocene phosphorites. However, the majority of phosphorites are older than the time span under consideration.

2.6 Weathering Products

On exposure to atmospheric and biotic agents the rock surface undergoes physical and chemical changes. Chemical weathering processes play a dominant role in the presence of humidity and increased temperatures. They are complex and include hydrolysis, diffusion, oxidation, carbonation, dissolution and ion exchange, frequently assisted by biogenic agencies. Since humid and warm environmental conditions enhance chemical reaction rates, zones with intensive chemical weathering represent important paleoclimatic indicators. During the Quaternary, the interglacials and to a lesser degree the interstadials represent such periods of increased chemical weathering. Early Man's activities are often associated with periods of improved climate. The age of fossil weathering phenomena is of interest to paleoclimatology, stratigraphy, and archaeology.

2.6.1 Soils

The mechanical and chemical rock degradation leads under assistance of organisms to soil formation. In addition to rock and mineral debris, soils contain organic components, above all humic acids. Soil formation is strongly dependent on climatic conditions and needs several centuries or even millennia, during which the rock surface is exposed to weathering. Fossil soils (paleosols) are of importance for the identification of interglacial and interstadial climatic conditions. The correlation of fossil soils is the basis of the pedostratigraphy. Because of their climatic and stratigraphic significance, soils greatly demand chronometric dating which, however, is difficult to achieve. The main obstacle is the complexity of soil formation.

Paleosols contain carbon in variable fractions. The organic proportion consisting of humic acids and micro-remnants is suitable, although with limitations, for ^{14}C dating (Sect. 5.4) of soils (< 40 ka). ^{14}C ages gained from humic acids can be interpreted, at best, as average ages of soil formation because of the complexity of pedochemical genesis. By means of the ^{10}Be content (Sect. 5.3), statements concerning the age of soil formation (> 10 ka) can be made. Through bioturbation of soils quartz and feldspar grains are brought to the surface and become bleached, so that the dating of soil formation should be possible by means of *luminescence* techniques (Sects. 7.1 and 7.2). Also, *uranium-series* dating is applicable to soils (Sect. 4.1.2), in particular to pedogenic carbonates (Sect. 4.3).

2.6.2
Caliche and Calcrete

Under semiarid climatic conditions calcitic precipitates from vadose waters form calcareous horizons or banks near the surface, such structures are called caliche or calcrete. As a secondary deposition, they cement pre-existing materials, such as sand, gravel or soils. They occur more or less parallel to the surface. During the formation of the calcium carbonate, uranium is preferably incorporated and fractionated from thorium. This allows the application of *$^{230}Th/^{234}U$* dating (Sect. 4.1.1). Since the carbon in the calcium carbonates (<40 ka) that precipitated from the groundwater derives partly from the atmospheric and biospheric reservoirs, *^{14}C* dating (Sect. 5.4) is applicable. The fact that the deposition of the secondary carbonates changes the radiometric dose rate to which sand grains are exposed allows the employment of *TL* dating (Sect. 7.1.3).

2.6.3
Desert Varnish

On natural and artificial rock surfaces, which have been exposed to the atmosphere under arid or semiarid climatic conditions for a long time, a stable dark coating gradually forms by precipitation, assisted by the microflora. It is known as desert varnish and consists of iron–manganese compounds. In addition to their occurrence in deserts, desert varnish-like weathering crusts are found in semiarid mountainous areas (Sect. 2.5.7). The age of the desert varnish represents the exposure time of a rock surface, and in arid and semiarid areas it permits statements on the stabilization of the land surfaces and especially on the minimum age of pediment plains, alluvial fans, debris fans, gravel terraces, aeolian deposits and moraines in high mountains. Attempts have been made to date such coatings (1–100 ka) on the basis of the *cation ratio* (Sect. 8.6). Since the desert varnish, at least in the Californian Mojave Desert, contains small amounts (<2%) of organic substances, *^{14}C* dating can be applied (<40 ka; Sect. 5.4).

2.6.4
Weathering Rinds and Patina

The surfaces of fine-grained basalt fragments, but also of silices, such as chert and flint, frequently show *weathering rinds*, also known as patina. Through the thickness of *weathering rinds* (Sect. 8.1) on glacial debris (1 ka–<1 Ma) a relative dating of the glacial activities is possible. To what extent this method is suitable in order to determine the time since when the surfaces of silices were exposed to weathering (Sect. 2.7.2) remains questionable.

2.6.5 Diffusion Fronts

During weathering, chemical components move through natural and artificial rock surfaces by diffusion, resulting in microscopically thin diffusion fronts. The front separates the fresh unchanged material from the chemically altered rind. The thickness of the diffusion rind grows with the exposure time of the surface, which allows dating. Surface ages of volcanic and artificial glasses (100 a – <1 Ma) can be determined by the *hydration* method (Sect. 8.2). Dating by means of *fluorine diffusion* (Sect. 8.4) has been tried on surfaces of siliceous rocks (1 ka – <10 ka). Bricks with mortar attached (70 a – 4 ka) are suitable for dating by means of *calcium diffusion* (Sect. 8.5). The last two techniques, however, still need practical testing.

2.7 Inorganic Artifacts

Inorganic artifacts are among the most abundant objects found by archaeologists. The majority of chronological frameworks rests on their typological development, which is especially true for ceramics. Thus, there is a great demand for an independent chronometry of the artifacts. Their suitability for physical or chemical dating is rather variable. Metals are hardly datable. In the case of stone tools the situation has considerably improved during recent years. More promising are all kind of objects made of burned clay, although the attainable accuracy of measurement does not always satisfy chronological demands.

The term *age* needs to be specified. Concerning inorganic artifacts one has to clearly distinguish between the age of formation of the material, the age of manufacture, the age of burial, and if necessary the age of other events, such as heating or bleaching. Depending on the type of material and the method applied different dates are determined, even for the same object.

2.7.1 Stone Artifacts (General)

Stone implements belong to the most important prehistoric findings. Their direct dating is especially desirable for the Paleolithic period beyond the time range of the ^{14}C-method. Unfortunately, their chronometric fixation causes considerable difficulties. Exceptions are silex and obsidian artifacts, which will be treated separately. In some cases however, an age determination can be achieved.

Stone tools and flakes within ashes of prehistoric fireplaces and in burned layers (>100 ka) might be so intensively heated that the *fission tracks* (Sect. 6.1) in apatite, zircon or titanite are completely erased and thus their heating age can be determined. *Luminescence* (Sects. 7.1 and 7.2) can also be applied on heated silicates and carbonates (100 a – 100 ka). It also has a great potential for dating the last light exposure of rock surfaces, which may be relevant for architectural structures and stone tools made of marble and silicates. After the burial of stone tools and flakes in the soil, *fluorine diffusion* (Sect. 8.4) into the fresh rock surface starts. The thickness of the *fluorine diffusion* rind reflects the relative burial age. This technique may become applicable to surfaces of siliceous rocks (1 – 10 ka), such as volcanites, plutonites and arkoses. Stone artifacts bearing desert varnish on the worked faces (1 – 100 ka) may be dated by the *cation* method (Sect. 8.6). One can also think of dating stone tools by means of *cosmogenic nuclides* (Chap. 5), if they have been abruptly either exposed to or shielded from cosmic radiation.

2.7.2 Flint and Chert (Silex)

The term *silex* (plural: *silices*) is used for a series of compact, hard rocks consisting predominantly of silica. Representatives are agate, chalcedony, flint, hornstein, jasper, siliceous shale, radiolarite, siliceous sinter and others. The classification, which unfortunately is not uniformly applied, is based on macroscopic and microtextural criteria, but also on the geological provenance. From the mineralogical point of view silices are complex materials which consist of cryptocrystalline quartz (<30 μm in grain size) and amorphous silica (>90% SiO_2) admixed with carbonaceous, organic and hydrous compounds. Flint, a sedimentary rock, occurs mainly in the form of concretions in limestone and dolomite. Its considerable hardness (Mohs 6 – 7) made flint an important raw material for the production of sharp-edged implements during the long period from the Lower Paleolithic to the Neolithic. The systematics of Paleolithic cultures is founded essentially on the typology of flint artifacts. Flint belongs to the most durable archaeological materials.

A *weathering rind*, the patina, that macroscopically differs from the fresh interior, forms on the flint surface during burial. Attempts have been made to exploit the thickness of *weathering rinds* for dating artifact surfaces. This approach proved to be very problematic because of the

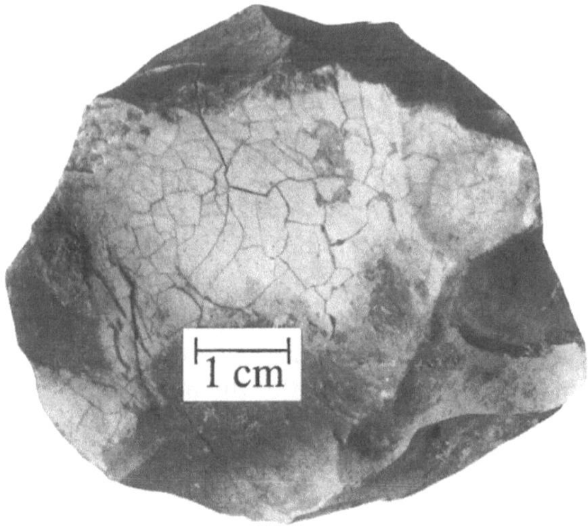

Fig. 8. Paleolithic flint artifact from Berigoule in Vaucluse, France, showing craquelation. This crack pattern is indicative of heating

manifold soil-chemical influences on the weathering process (Rottländer 1989). *TL* and *ESR* (Sects. 7.1 and 7.3) are suitable for the dating in the range of 1–100 ka, provided the artifacts were subjected to prehistoric heating, which indeed applies for a considerable portion of flint artifacts. Indicators of heating may be delicate crack patterns, the so-called craquelation (Fig. 8), color changes, which may be whitish, reddish or blackish depending on the original material and the redox conditions, as well as fracturing (Hahn 1991). The height of the *TL* and *ESR* signals also serve as indicators of heating. The reason for the abundance of burned flint artifacts possibly lies in an intended amelioration of the splitting characteristics of the rock (Griffiths et al. 1986).

2.7.3 Obsidian

Prehistoric cultures with access to obsidian (Sect. 2.1.2), either directly by mining or indirectly through trading, abundantly used this raw material because of its conspicuous hardness (Mohs 7; it exceeds the hardness of chromium-nickel-bearing steel). It has been widely traded from its geological sources over thousands of kilometers. Since the Lower Paleolithic it was used in the manufacture of blades, arrowheads, scrapers and other sharp cutting and pointed implements. Because of the marked brittleness, obsidian tools easily break and become hardly usable. Hence they are frequently found in the form of broken fragments in archaeological places. Flakes are found abundantly on working sites, whereas cores are less abundant (Fig. 9).

Fig. 9. Obsidian core from Sehitemin/Kars, eastern Anatolia

In the course of prehistoric tool manufacture by knapping, fresh obsidian surfaces (>100 a–100 ka) were created that can be dated by means of the *hydration* method (Sect. 8.2). This method is advantageous because extensive sample series can be investigated without too much cost and operating time. Obsidian tools and flakes have been frequently subjected to a heat treatment during the production process for unknown reasons, but accidental or intended heating also took place during usage of the tools. Provided that the heating was high enough to anneal the pre-existing *fission tracks* (Sect. 6.1), this event (>1 ka) can be dated on the base of the fission tracks produced afterward.

2.7.4 Tektite Glass

Occasionally, tektites (Sect. 2.2.1) were used by prehistoric Man as a raw material for pointed and cutting stone tools. A good example of this are the tools produced from moldavites found in Upper Paleolithic layers in Willendorf in the Wachau, Austria. In Indochina tektites are known from the Bronze Age (Bezborodov 1975). Freshly produced surfaces of such glass artifacts (1–100 ka) should be datable by means of *hydration* (Sect. 8.2). On the other hand, heated tektite tools (>1 ka) can be dated by the *fission track* method (Sect. 6.1).

2.7.5 Petroglyphs

The thin rinds of desert varnish (Sect. 2.6.3) on rock surfaces are excellently suited for the production of durable drawings. The dark layer is slightly perforated by carving or hammering, as a result of which the brighter, fresh interior of the host rock becomes visible. Petroglyphs, as this kind of rock drawings are called, are handed down from many cultures, as e.g., the famous petroglyphs along the silk road on the Karakorum Highway. During the engraving the fresh rock-surface gets exposed, on which desert varnish anew forms. The *cation method* (Sect. 8.6) enables the dating of this secondarily formed desert varnish (1–100 ka) and herewith also of the rock drawing.

2.7.6 Mortar

Mortar is produced from limestone ($CaCO_3$). At first the CO_2 is removed by burning and quicklime, CaO, remains. Then the latter is slaked by water to form $Ca(OH)_2$. During drying the slaked lime paste reacts with the atmospheric CO_2, thus forming the $CaCO_3$ of the mortar. This reaction constitutes the essential step for *^{14}C* dating (Sect. 5.4), since recent carbon is incorporated in this way. In a mortar produced in this way, all of the inorganic carbon should derive from atmospheric CO_2. Provided that no further carbon exchange with the atmosphere takes place after hardening, the construction date of buildings can be determined from the decaying ^{14}C content of the mortar (>100 a). *Calcium diffusion* (Sect. 8.5) from the mortar into the brick is observable as a front and thus yields an appropriate means to determine the age of brickwork (70 a–4 ka). Suitable, although not yet applied, may be the *TL* (Sect. 7.1) and *ESR* (Sect. 7.3) for dating the renewed formation of the $CaCO_3$ in the mortar.

2.7.7 Ceramics and Bricks

Ceramic technology looks back on a history which has lasted nearly 10,000 years. Ceramic is burned from clay with the admixture of temper at temperatures >600 °C. As a result of the raw material available, the chemical and mineralogical composition of the ceramics is highly variable. The high fragility ensures that ceramics are short-lived, so that soon after their production ceramic sherds become embedded in the ground. Stylistically, pottery was subjected to rapid changes. Both features, short-livedness and typology, together with the abundant occurrence make potsherds into extremely valuable objects for archaeological chronology. Consequently, chronometric age determination on ceramic objects is desirable.

The *TL* method (Sect. 7.1) assesses the date of burning, i.e., the date of the ceramic production (>100 a). The fine-grain and the quartz-inclusion fractions are used. Clearly, the age resolution of 6–10 % is not sufficient for the majority of chronological problems, however there are many instances where the dating of ceramics and burned bricks can offer meaningful and advantageous solutions despite the low precision. *TL* is well suited for authenticity tests of allegedly ancient ceramic objects (Fig. 10). *OSL* (Sect. 7.2) was successfully applied to the quartz temper of ceramic sherds (>100 a) in the course of which a higher precision of measurement was achieved compared to *TL*. Provided that chaff or cereal grains as temper have become baked in the ceramics, the ^{14}C method (Sect. 5.4) can be conducted for ceramics dating (>100 a). As a result of the small amounts of organic carbon in ceramics, in practice this application has been possible only

Fig. 10. Pre-Columbian bowl after a sample was collected (small *drill hole* within the *dark field* on the bottom) for the purpose of a *TL* authenticity testing. The result turned out to be in agreement with the stylistic age of ca. 1000 a

through *acceleration mass spectrometry (AMS)*. Ceramics dating was attempted by means of *alpha recoil tracks* in muscovite inclusions (Sect. 6.2). Commonly, ceramic clays are fired at 700–800 °C, and during cooling below the Curie temperature the prevailing magnetic field is thermoremanently frozen in. *Archaeomagnetic dating* (Chap. 9) of ceramics uses the intensity and – in case of known position of the object during firing – the direction of the magnetic field. The determined date directly corresponds to the manufacture of the object. Air-dried and burned bricks, the surfaces of which have been covered with a calcium-bearing binding agent (mortar, plaster or stucco), are suitable materials for *calcium diffusion* dating (Sect. 8.5). It is not the material itself which is dated by this means, but the time of the erection of the brickwork.

2.7.8
Kilns, Burned Soil and Stones

Heated clay objects, such as kiln walls, mud linings of kilns (Fig. 11), burned clay remnants, and burned soil horizons, can be treated analogously to ceramics for dating. Of particular interest are heated stones and archaeo-

Fig. 11. Kiln with mud lining in a Roman villa at Bad Kreuznach, Germany. The last firing event was established by *TL* dating. (Wagner 1980b)

metallurgical furnace remains, which are otherwise hardly datable. Generally, brick-red baked appearances of earthy materials indicate high firing temperatures. The application of *TL* (Sect. 7.1) provides the date of the last heating (>400 °C). In the case of burned rocks, quartz and feldspar grains are used. The fact that kilns and hearths, unlike ceramic sherds, frequently occur in situ, i.e., they still display their initial position as during firing, permits the application of *archaeomagnetic dating* (Chap. 9). This allows the employment not only of the intensity of the thermoremanent magnetization, but also its inclination and declination. However, it has to be ascertained that the samples have been exposed to sufficiently high firing temperatures (>500 °C).

2.7.9 Artificial Glass

Ancient glass displays a widely varying chemical composition because it can be produced from various sorts of raw materials. Most types of glass belong to the group of alkali–lime glasses. The classical raw materials used in glass production are plant ashes (potash) and quartz sand. Besides the potassium-rich and sodium-rich plant ashes, soda had already been used during antiquity. The high alkali content of glass results in a lowering of the melting point. Hues were achieved through various admixtures of special color. Apart from a few glass beads of the Sixth Dynasty, vessels made of artificial glass are known since the middle of the second millennium BC from Mesopotamia and slightly later from Egypt. The art of glass technology presumably developed from the glazing techniques. From the Near East the art of glass making spread into the Mediterranean region. In northern Italy glass has been known since the 5th century BC. The glass industry underwent enormous development during Roman times and once more in the Middle Ages.

Under the influence of moisture, ancient glass suffers more or less from weathering. *Glass layers* (Sect. 8.3) are formed in the weathering crusts on the glass surface (>100 a), the number of which possibly corresponds to the duration of burial. Medieval and modern glasses are apparently best suited for this dating technique. *Hydration* (Sect. 8.2) also has been tried for determining the age of artificial glass (>10 a). Although in principle feasible, *fission track* dating (Sect. 6.1) of artificial glass (>1 ka) is laborious because of the relatively low ages and the usually low uranium contents. However, uranium glass (Fig. 12), which was produced in Bohemia since the middle of the 19th century and bears up to ~1% uranium, is better suited for fission track dating.

Fig. 12. Nineteenth century Bohemian spa-glass, which owes its green-yellow fluorescence to an admixture of ca. 1% uranium-oxide

2.7.10
Vitrified Forts

In western and southern Europe there are numerous defense structures known as vitrified forts. Usually they are ascribed to the Celtic tribes. The rock boulders of the walls are partially molten and fused by a glassy slag. Presumably they were created by destructive burning of structures which had been erected of rock and wood. The glass matrix is very heterogeneous, contains frequently crystalline relicts and its formation requires burning temperatures >1000 °C. The dating of the glass phase and the heated rock relics may be achieved by *TL* (Sect. 7.1).

2.7.11
Metallurgical Slags

Slags are often the only remains of archaeometallurgical activities. They represent an important source of information on the actual developments of pyrotechnology. Their temporal placement, however, commonly presents difficulties, since the stratigraphic situation of metallurgical sites is often disturbed.

Charcoal finds which frequently accompany slags can be dated by *¹⁴C* (Sect. 5.4). The data must be treated with caution if their temporal relation to the slag remains obscure, since smelting of ores has often been carried out at the same site over long periods. However, tiny remains of charcoal to which the ^{14}C *AMS* technique can be applied are often enclosed in slags, enabling the dating of the wood used as fuel. On the other hand, the analysis of the slag material itself offers the advantage of dating the archaeometallurgical activities directly. There have been attempts to apply the *TL* (Sect. 7.1) to slags. Principally, the glass as well as the fayalite and pyroxene phases of the slag are suitable for this purpose. Unfortunately, the dating experiments on slag components did not get beyond their incipient stage. More promising is *TL* dating on heated quartz inclusions or on relicts of burned clay adhering to the slag.

2.7.12
Lead Pigments and Alloys

White lead ($PbCO_3$) was the most important white pigment until its replacement by titanium white (TiO_2) during the last century. Therefore modern art forgers deliberately applied lead white again. Whether lead-bearing pigments and alloys are older or younger than ca. 100 a can be determined by means of the radioactive lead isotope ^{210}Pb (Sect. 4.1.6) – an intermediate member of the ^{238}U-decay chain – through which recent forgeries can be revealed.

2.8
Plant Remains

Geological strata frequently contain more or less degraded plant remains, which allow conclusions in respect of paleoclimatic and paleo-ecological conditions. In some sediments, as e.g. peat, the plant component is predominant. In cultural occupation layers the prominent plant-derived material is charcoal. Plant residues are also common in waste dumps and as food relics in vessels, hearths and ashes. Wood fragments are preserved in timber constructions and post holes. Plant remains are of manifold interest to archaeology and history of architecture. Commonly ^{14}C (Sect. 5.4) dating is applied. In such cases, attention must be paid to the fact that the age data relate to the sample material, but do not necessarily date the geological or archaeological event in question.

2.8.1 Wood

Wood consists of cellulose, carbohydrates and lignin. Especially in respect to trunks it has to be taken into account that trees are long-lived plants which can reach ages of many hundred years. Each year as they are growing a new ring is added. The annual addition of one ring allows *dendrochronological* dating (Sect. 10.1.2). In central Europe usually oaks and pines (<12 ka) are dated. This technique requires at least 100 consecutive rings. Because of its unequaled accuracy (sharp to the year) this method forms the basis of the Holocene calendar. Wood (<40 ka) can also be dated by ^{14}C (Sect. 5.4). At the end of each annual growth period the latest annual ring is cut off from further exchange with the atmosphere, hence the ^{14}C age of each ring determines the growth year. This phenomenon which is referred to as "old-wood effect" can cause considerable uncertainties, for instance, if a building is to be dated by means of timber. If at least 50 consecutive tree rings (<11 ka) do exist, the more accurate ^{14}C dating by means of the wiggle-matching technique can be applied. The organic matrix of the wood contains minor amounts of amino acids, hence wood dating (10–100 ka) has been tried by *racemization* (Sect. 8.8).

2.8.2 Charcoal

Charcoal represents an abundant find material in archaeological layers. It is well suited for ^{14}C dating (Sect. 5.4). However, the old-wood effect can lead to considerable uncertainties if the archaeological context is to be dated purely on the basis of charcoal fragments. After burning the older wood core of the trunk preferentially remains as charcoal, which can simulate context ages up to a few 100 years too old. It is advisable to determine the wood species in order to assess this risk. Charred, short-lived twigs are preferable to massive pieces.

2.8.3
Seeds and Grains

Seeds and grains occurring in aceramic Neolithic strata testify to the agricultural development in the Near East. The exact temporal placement of these beginnings and its spread outward are questions long pursued by ^{14}C dating (Sect. 5.4). These plant materials offer the advantage of being annual. The *AMS* technique is suitable for the analysis of single grains and therefore enables an independent test of the stratigraphic fixation of a grain that may be disturbed by bioturbation.

2.8.4
Pollen and Spores

Flowering plants produce pollen. On the other hand, ferns and mosses generate spores. These plant materials can be species-specific, having sizes of usually ca. 30 μm, and show in general good preservation because of their resistant waxy coats, especially under reducing soil conditions. The qualitative and quantitative composition of pollen and spores (*pollen analysis*, Sect. 10.2.4) assembled in a layer reflects the plant association as well as the climatic condition of the paleo-environment. The time-dependent record is presented in pollen diagrams. Defined pollen zones play an important role in the biostratigraphy of the Quaternary and especially of the Holocene. The application of the ^{14}C *AMS* technique (Sect. 5.4) on pollen grains (<40 ka) themselves permits the direct age determination of these "index fossils".

2.8.5
Phytoliths

These microscopic mineral bodies are secreted by living plants in order to support the organic tissue, especially in grasses. They often consist of opaline silica. They are found as resistant microfossils in natural and cultural sediments. Phytoliths exhibit a wide variety of shapes that are often characteristic of the species. Analogously to pollen, their identification allows conclusions about the former plant assemblages, in particular the cultivation of cereals. Since silica phytoliths contain remnants of the original plant tissue (<40 ka) encapsulated within them, they can be dated by ^{14}C *AMS* (Sect. 5.4). The attempt has also been made to date the opal matrix of phytoliths (1–100 ka) with *TL* (Sect. 7.1).

2.8.6 Paper and Textiles

The plant fibers of textiles are suitable for the *^{14}C* method (Sect. 5.4) because the raw materials derived from plants have a short life span and are processed within a few years after being harvested. Hence, the *^{14}C* date corresponds to the moment of the textile production. With respect to paper, the possibility must be considered that waste materials (e.g. old rags) have frequently been recycled for the paper production. Depending on the production method, the *^{14}C* age may be several 100 years too old.

2.8.7 Peat and Sapropels

These unconsolidated deposits of plant remains are indicative of humid and warmer periods, i.e., for interglacials and interstadials. In addition, their richness in pollen permits important paleobotanic conclusions. The *^{14}C* method (Sect. 5.4) is directly applicable to them (< 40 ka). The analysis of organic sedimentary compounds requires special care to ascertain their contemporaneous nature. Suitable materials are peat and sapropel deposits but also sediments bearing lower proportions of organic components. Pleistocene peat (< 350 ka) falling beyond the age range covered by *^{14}C* can be dated by the *$^{230}Th/^{234}U$* technique (Sect. 4.1.1).

2.8.8 Organic Remains in Vessels, on Stone Tools and Rock Paintings

Ceramic vessels and stone implements which were brought into contact with organic materials during their production or use commonly retain small traces of them on the surface. Observable are residues both of faunal origin, such as blood, muscle fibers, hair, feathers, fats, and of floral origin, such as resins, starch and plant tissues. The organic materials are rather resistant to secondary alteration. Since the yields of samples are small, the dating of organic remains (< 40 ka) requires the *^{14}C AMS* technique (Sect. 5.4).

2.8.9 Wine

Wine stored in barrels or bottles is cut off from further exchange with the biosphere and the hydrosphere, so that it is possible in principle to deduce the age of the wine from the decay of *tritium* (Sect. 5.1). But as in the case of groundwater, there is an interference because of the strong, anthropogenic increase in the tritium level. However, for wine samples, the general decrease in the tritium from thermonuclear explosions – analogous to radiocarbon – could be well monitored since the 1960s, and thus for this period an indirect age determination of the vintage seems to be possible.

2.8.10 Diatoms

These tiny, single-celled plants grow in both marine and freshwater environments. They secrete opaline walls that may accumulate in large amounts as sediments. The siliceous diatom ooze and diatomite is found on lake and ocean bottoms. Its dating is of chronological and paleoclimatic significance. Chronometric age determination of diatoms has been attempted by *TL* (1 ka – 100 ka; Sect. 7.1), ^{226}Ra dating (<10 ka; Sect. 4.1.7) and ^{32}Si (50 – 500 a; Sect. 5.7).

2.9 Animal Remains

Fossil animal remains occur almost exclusively in the form of calcareous or apatitic skeleton parts associated with greater or lesser amounts of organic substances. They are found in marine as well as continental Quaternary sediments. Archaeological layers frequently contain bones and teeth. There is a paramount interest in the chronometric dating of fossil animal remains, in most cases focusing on chronological questions, but nevertheless also related to the age of the find itself, e.g., a fossil hominid bone. Apart from fossil material, teeth can also be dated in vivo.

2.9.1 Bones and Antlers

The bones of living vertebrates contain approximately 25 – 30 % organic material, which consists predominately of the protein collagen. The inorganic component consists of carbonate hydroxyapatite, a carbonaceous calcium phosphate. After death and burial in sediments the hydrolytic degradation of the organic component begins. Facilitated by the high porosity of bone, the inorganic components are in chemical exchange with the sedimentary environment. These processes are strongly influenced by the ambient chemical and hydrological milieu as well as the temperature. Bones of Paleolithic find horizons, frequently representing remainders of prey, are of direct interest for dating. Furthermore, the age of human remains is directly relevant to the reconstruction of hominid evolution. Thus bones represent a common and important dating material.

For long time, bones (< 40 ka) have been considered of little use for the application of *^{14}C* dating (Sect. 5.4). They contain predominantly inorganically bound carbon, which is liable to exchange reactions with the groundwater. The organically fixed carbon in bones is, despite diagenetic alteration, almost unaffected by exchange processes. The *AMS* has the advantage over the β-counting technique of permitting the investigation of individual organic components instead of non-differentiated collagen.

The diagenetic uranium uptake by fossil bones – namely up to 100-fold compared to the bone in vivo – takes place post-mortem during their burial in the sediments (Millard and Hedges 1995). The question about the time function of this secondary uranium uptake is crucial to all dating methods being based on this element. A relatively fast increase in uranium, reaching saturation (model of early uptake), is mostly favored. For bones (< 350 ka) from geochemically and biologically inactive, dry sediments, reliable *$^{230}Th/^{234}U$* ages (Sect. 4.1.1) are to be expected. It is advantageous to cross-check *$^{230}Th/^{234}U$* ages by means of *$^{231}Pa/^{235}U$* dating (<150 ka; Sect. 4.1.3). *U–He* dating (Sect. 3.2) has been tried on some Quaternary and Tertiary bones (>10 ka). *ESR* dating (Sect. 7.3) on bones (1–100 ka) is still in the testing stage, and there is no doubt about the lesser suitability of bone compared to tooth enamel, presumably as a result of its relatively low proportion of mineral material. Through the development of the *AMS*, the determination of *^{41}Ca* (Sect. 5.10) on bones became technically possible, which might lead to a useful dating method. After death, collagen degrades to free amino acids, which are subjected to *racemization* (Sect. 8.8). The age of the bone (recent–1 Ma) can be deduced from the degree of racemization. Nowadays, the *fluorine–uranium–nitrogen test* (Sect. 8.7) on bones is only of minor importance. Nevertheless, F, U and N analyses are available from more than 1000 fossil and recent skeleton samples, and results on the relative age sequence have been achieved.

2.9.2 Teeth

The dating of teeth offers interesting applications to the Middle and Lower Paleolithic as well as to Quaternary geology. Tooth enamel consists almost purely of hydroxyapatite $[Ca_5(PO_4)_3(OH)]$. Dental cementum and dentine contain higher amounts of organic matter. Like bones, fossil teeth take up uranium almost exclusively during their burial. Therefore the problems connected to the time-function of the uranium uptake are entirely analogous, although not so marked because of the lower degree of porosity. Teeth – mostly of vertebrates – can be dated by means of their *$^{230}Th/^{234}U$* (Sect. 4.1.1; < 350 ka) and *$^{231}Pa/^{235}U$* (Sect. 4.1.3; < 150 ka) ratios. It must be added that concordance of both ages eliminates the uncertainty associated with the uranium uptake. The mass spectrometric determination of *$^{230}Th/^{234}U$* allows the dating of both the dentine and the enamel fraction of the same teeth. This procedure increases the reliability of the age measurement. The mineral component hydroxyapatite of the dental enamel (1 – some 100 ka) is suitable for *ESR* dating (Sect. 7.3), whereas the cementum and dentine are not suitable for this method because of their higher content of organic matter. The *racemization* (Sect. 8.8) of amino acids is applied to fossil teeth up to 3 Ma old. The D-aspartic acid of recent teeth is suited for the determination of the individual's age. The *fluorine-uranium-nitrogen test* (Sect. 8.7) on teeth is of minor interest.

2.9.3 Corals

These sessile marine invertebrates live in tropical oceans less than 50 m below the surface. Additional requirements are purity and salinity of the water. They produce external skeletons of calcium carbonate, predominantly aragonitic, and form colonies. Reef-building corals bear testimony to sea level changes and tectonic movements. Corals can be dated by counting their *annual growth layers*, a method reaching back to 10 ka. Corals incorporate uranium, but neither thorium nor protactinium, into their aragonitic skeleton, hence, ages up to 350 ka are measurable by the uranium-series methods. The recrystallization of the metastable aragonite into calcite may disturb the radiometric system. The *$^{230}Th/^{234}U$* (Sect. 4.1.1) method and, more rarely, the *$^{231}Pa/^{235}U$* (Sect. 4.1.3) method are applied. Through mass spectrometric *$^{230}Th/^{234}U$* analysis, age data of high precision are being gained, the reliability of which has been proved by their correspondence with growth layer counting in the age range below 10 ka. In combination with *^{14}C* measurements (Sect. 5.4), results fundamental to the history of the climate and the ocean circulation have been obtained from corals. The *ESR* method (Sect. 7.3) is also successful for dating corals (1 ka – < 1 Ma), frequently in combination with uranium-series dating. The fact that corals (> 10 ka) possess a good helium retention and show insensitivity toward secondary uranium mobilization, offers the best presupposition for *U–He* dating (Sect. 3.2). Dating attempts on corals using *racemization* (Sect. 8.8) have been only partially successful, and a better understanding of the kinetic and geochemical characteristics of amino acids in this material is needed.

2.9.4
Foraminifera

Foraminifera are unicellular protozoans with a perforated skeleton mainly consisting of calcite. Most foraminifera are marine, either planktonic or benthic. The fossil foraminifera, predominantly planktonic *Globigerina*, embedded in deep-sea sediments are excellent paleoclimatic indicators. Above all, they preserve in their shells the oxygen isotopic composition of the ocean water and thus reflect the development of the ice volume during the Quaternary.

The climatically controlled, continuous $^{18}O/^{16}O$ stratigraphy (Sect. 10.2.2) of *Globigerina* permits the worldwide correlation of deep-sea sediments and forms the basis of marine Quaternary chronology. During the formation of their calcitic shells, the foraminifera take up uranium, but no thorium, from the ocean water. The gradually increasing $^{230}Th/^{234}U$ (<350 ka; Sect. 4.1.1) and $^{231}Pa/^{235}U$ (<150 ka; Sect. 4.1.3) ratios can be exploited for the age determination. To gain information on the deep water circulation in the ocean and its relationship to the paleoclimate during the last 20 ka, planktonic and benthic shells of foraminifera (<50 ka) have been dated by ^{14}C *AMS* (Sect. 5.4). Foraminifera (<100 ka) can be dated using *ESR* (Sect. 7.3). The constant temperature of around 2–4 °C on the deep-sea floor favors the application (<5 Ma) of *racemization* (Sect. 8.8).

2.9.5
Mollusk Shells

During the Quaternary abundant representatives of the mollusks are pelecypods (mussels) and gastropods (snails). Their biostratigraphic importance makes mollusk shells especially interesting dating materials. They are found both in marine and continental deposits. Their shells are secreted from the mantle of the mollusks. They consist of the organic matrix (conchiolin), in which calcium carbonate is intercalated as a major constituent. ^{14}C dating (Sect. 5.4) of marine mollusks (<40 ka) has been conducted almost exclusively on the inorganic component. Quite frequently, the shells of non-marine mollusks, essentially continental freshwater snails, are not suitable for ^{14}C dating, because they nutritionally take up 'dead' carbon of old limestone. During their growth, but mainly postmortem, the mollusks fix uranium in their calcitic or aragonitic shells. This fact makes both marine and limnic mollusk shells feasible for the $^{230}Th/^{234}U$ (<350 ka; Sect. 4.1.1) and $^{231}Pa/^{235}U$ (<150 ka; Sect. 4.1.3) methods. *ESR* dating (Sect. 7.3) is applicable, but not equally well to all genera (1–100 ka). Only few *TL* investigations of mollusk shells exist (Sect. 7.1). Mollusks

(>10 ka) have been dated by *U–He* (Sect. 3.2). The shells of mollusks and snails (<1 Ma) contain ca. 0.02% proteins forming thin layers between the calcareous matrix, and thus are suitable for *racemization* dating (Sect. 8.8).

2.9.6 Eggshell

The shells of eggs laid by flightless birds, such as ostrich and emu, are commonly found at Paleolithic sites in Africa, Asia and Australia. The eggs served as food sources. From their shells water containers and beads were produced. Apart from the archaeological use eggshells allow paleoecological reconstruction. They contain in their carbonate matrix about 3% organic component. Both the inorganic as well as the organic carbon components may be utilized for ^{14}C dating (Sect. 5.4). The organic matter, primarily protein, can be used for *racemization* dating (Sect. 8.8).

2.10 Water and Ice

2.10.1 Ocean Water

The circulation pattern of the oceans exerts an essential influence upon the climate. Through the application of the ^{39}Ar method (Sect. 5.9) the mixing (<1.2 ka) of the oceans can be investigated. An improved understanding of ocean circulation can be achieved by the combination of ^{3}H, ^{14}C, ^{32}Si and ^{85}Kr measurements (Chap. 5).

2.10.2 Groundwater

The age determination of groundwater belongs to the most problematic ^{14}C applications (Sect. 5.4), because its carbon content is biogenic only to a certain fraction, i.e., ultimately of cosmogenic origin. Furthermore, there is a permanent chemical exchange between the groundwater, the aquifer and other bodies of water. Nevertheless, ^{14}C analysis may furnish information on the age, genesis and hydrodynamic aspects of groundwater. The ^{3}H determination (Sect. 5.1) of the bomb-tritium produced at the beginning of the 1960s enables – frequently in combination with ^{14}C measurements – investigations of the rate and direction of flow as well as the catchment area, but also the age of recent groundwater (<30 a). The surface water infiltrating the soil introduces atmospheric ^{39}Ar (Sect. 5.9) into the groundwater, thus permitting its dating (<1.2 ka). More seldom are groundwater studies employing the nuclides ^{32}Si and ^{81}Kr (Sects. 5.7 and 5.11, respectively).

2.10.3 Glacier Ice

Ice deposits – especially the ice caps of Greenland and Antarctica originating at least from the Middle Pleistocene and continuously growing since then – represent an important archive for atmospheric climatic changes during the last 250 ka. Comparable to the deep-sea sediments, they bear long-time records of the Quaternary environmental changes, in particular for the continental domain. In addition to the snowfall, dust particles and air are incorporated into the ice year after year, so that not only local phenomena but also global features are reflected in the ice. Regarded as sediment, ice does not stay in situ because of its plastic flow. The reconstruction of the depositional history of glaciers thus requires glaciological modelling.

Samples of glacial material are taken in form of drill cores down to 3 km depth. The annual *ice layers* of the last 14 ka can be directly counted on the basis of annual changes in the $\delta^{18}O$ (Sect. 10.1.3), dust and acid contents. The dating of older ice can be achieved by means of ^{14}C analysis (Sect. 5.4) of the atmospheric carbon dioxide in the gas bubbles within the ice. The *tritium* content (Sect. 5.1) in the snow deposits (<50 a) around the South Pole displays a strong seasonal variation. Counting the maxima furnishes ages of snow layers within an accuracy of 1 year. When the glacial ice was formed, atmospheric ^{39}Ar was enclosed. The ^{39}Ar age (Sect. 5.9) defines the compaction stage of the snow (<1 ka). Glacier ice (<1 ka and 50 ka – 1 Ma) is also dated by ^{32}Si and ^{81}Kr (Sects. 5.7 and 5.11, respectively). The Quaternary climatic development can be read from $^{18}O/^{16}O$ and $^{2}H/^{1}H$ (Sects. 10.2.2 and 10.2.3) manifested in the ice. These isotopic variations are also helpful in the stratigraphic correlation of ice cores as well as in the synchronization with the deep-sea sequences. In this way the ice cores can be indirectly dated. In ice cores the ^{10}Be input (Sect. 5.3) is strongly climate-controlled and correlates negatively with the $^{18}O/^{16}O$ ratio.

3 Radiogenic Noble Gases

Two Quaternary chronometric methods, the K–Ar and the U–He clock, are based on radiogenic noble gas nuclides. The K–Ar method uses the decay of the potassium isotope ^{40}K to the argon isotope ^{40}Ar and the U–He method rests on the accumulation of the radiogenic helium isotope ^{4}He which is produced during the α-decays within the uranium and thorium decay series. Noble gases are chemically inert and are not incorporated into the crystal lattice. Consequently, the daughter nuclides (^{4}He and ^{40}Ar) should be absent during mineral formation. Thus, their in the sample produced and accumulated amount (*radiogenic* component) is directly related to the time elapsed since the formation of the crystal. When sufficiently heated, any noble gas already contained in the crystal should be completely driven out, thus resetting the noble gas clock. In this case one dates the time of last degassing. It presupposes that the noble gas was fully retained in the mineral after the event to be dated and that no loss occurred.

In practice such ideal conditions are not always fulfilled. Minerals may bear in addition to the *radiogenic* component an *excess* as well as an *atmospheric* component of the daughter nuclide which both become incorporated by other processes (e.g., noble gas diffusion) than that of the in situ decay of the respective parent nuclide. Furthermore, a defined thermal event may not result in complete degassing so that some fraction of the previously formed daughter nuclide remains as an *inherited* component in the subsequent system. If undetected these contaminant components would cause the obtained noble gas age to be an overestimate of the event of interest. This is a problem particularly for Quaternary samples in view of the minute fractions of daughter produced in such a comparatively short time span. Another difficulty is the possible *leakage* of the radiogenically produced noble gas from the mineral. Any such loss results in an underestimate of the noble gas age. The loss increases with ambient temperature. The smaller helium atom is more susceptible to leakage than argon. The ability to retain noble gases differs among minerals. Also during weathering and secondary alteration of the rocks noble gases may be lost from the system and, therefore, such samples should be omitted from dating studies.

The practicability of sensitive and precise mass spectrometric analysis of noble gases is advantageous for the chronometric application of both of these radioactive clocks. The *mass spectrometer* is an instrument to analyze the isotopic abundance of elements and, thus, is a basic tool for isotopic chronometry. In the case of noble gas analysis, the gases are first extracted from the sample, which is achieved by heating and melting at temperatures up to 2000 °C under high vacuum. Then, the noble gas of interest needs to be separated by various purification steps from all the other extracted gases before it is introduced into the mass analyzing system of the spectrometer. At the ion source the atoms are ionized. The ions are focused and accelerated by electrostatic lenses. During the passage of the ion beam through a magnetic field, the ions are more or less deflected depending on their mass – more precisely on their charge/mass ratio. Finally, the mass-separated ions are recorded at the ion detector.

The quantitative calibration of the spectrometer can be performed with the isotope dilution technique in which an exactly known volume of a mono-isotopic spike or tracer gas is added and analyzed together with the purified sample gas. The amount of noble gas isotopes is given in cm^3 (STP) g^{-1}, whereby the acronym *STP* stands for *s*tandardized *t*emperature (0° C) and *p*ressure (1 atm) conditions of the gas volume (1 mol of gas has a volume of 22.4×10^3 cm^3 at STP and contains 6.02×10^{23} molecules). The measuring apparatus introduces some atmospheric contamination which can be corrected for. Because of the very small fraction of the radiogenic component in the total signal of ^4He or ^{40}Ar, the mass spectrometric analysis of the noble gas isotopes requires very high precision (<1%).

3.1 Potassium–Argon

In its general conception potassium–argon comprises several dating techniques. In addition to the conventional K–Ar technique ($^{40}Ar/^{40}K$), these are the argon–argon ($^{40}Ar/^{39}Ar$) and the argon–argon laser techniques. They all rest upon the radioactive decay of the potassium isotope ^{40}K to the argon isotope ^{40}Ar, so that the amount of accumulated radiogenic argon – in relation to the potassium content – is a measure of the age of the sample. The various techniques differ by their extraction treatment, analytical procedures, applicability, accuracy and in their geologic information.

The ^{40}K–^{40}Ar decay is a rare type of natural radioactivity, namely an electron capture, whereby the nucleus captures an electron from the innermost orbit (Fig. 3). This type had been theoretically predicted for the first time in 1937 by von Weizsäcker for ^{40}K in order to explain the high atmospheric argon content of ca. 1%, of which more than 99% consists of the isotope ^{40}Ar. A much lower content would be expected on the basis of theoretical reasoning about nuclear synthesis. Von Weizsäcker further concluded that, owing to the postulated $^{40}K-^{40}Ar$ decay, old potassium-rich minerals should have an increased content of ^{40}Ar. Aldrich and Nier (1948) succeeded in proving this prediction by mass spectrometric analyses. The first K–Ar ages were determined a few years later by Smits and Gentner (1950) on Tertiary potassium salts from Buggingen, Germany. Today, the K–Ar method, which covers the whole age range from the beginning of the solar system to the Holocene, is undoubtedly one of the most important chronometric dating tools.

Potassium is with 2.1% geochemical abundance the eighth most common element in the Earth's crust and occurs widely in rock-forming minerals, such as feldspar, hornblende and mica. This is a favorable condition for potassium–argon dating. However, for the ages less than 2×10^6 a the long half-life ($\sim 1.2 \times 10^{10}$ a) of $^{40}K-^{40}Ar$ decay within a mineral sample results in only very low amounts of radiogenic argon $^{40}Ar_{rad}$ in relation to the omnipresent atmospheric argon $^{40}Ar_{atm}$ and the commonly occurring excess argon. Therefore, potassium-rich minerals, such as sanidine, are preferred in Quaternary applications. The ultrasensitive analytical technology developed lately allows the determination of ages down to few ka on samples on a milligram scale.

The K–Ar clock dates events of complete degassing. In the Quaternary context such events are the solidification of volcanic minerals and rocks as well as the volcanic contact heating of preexisting minerals. In both cases this means the time of volcanic eruption. An important material in this connection are the widespread tephra layers for which K–Ar dating yields excellent tephrochronological time marks, which in particular are important for intercalated early hominid-bearing sedimentary layers. Furthermore, K–Ar dating delivers the chronometric calibration of the oxygen-isotope curve and the geomagnetic polarity time scale. An age precision of better than 1% can be obtained with the argon–argon technique.

3.1.1 Methodological Basis

Natural potassium consists of the three isotopes ^{39}K (93.2581%), ^{40}K (0.01167%) and ^{41}K (6.7302%) with a known abundance ratio. The isotope ^{40}K is radioactive and decays through a dual mechanism (Sect. 1.2). One branch leads under *electron capture* to the argon isotope ^{40}Ar and the other one under β^-*-emission* to the calcium isotope ^{40}Ca, whereby the branching ratio R ($=\lambda_e/\lambda_\beta$) amounts to 0.1171. The total decay constant λ for ^{40}K is 5.543×10^{-10} a^{-1}, the decay constants λ_e for electron capture and λ_β for β^--emission are 0.581×10^{-10} a^{-1}, and 4.962×10^{-10} a^{-1}, respectively, whereby $\lambda=\lambda_e+\lambda_\beta$.

The noble gas argon is the third most abundant gas (0.934%) in the atmosphere and consists of the three isotopes ^{36}Ar (0.337%), ^{38}Ar (0.063%) and ^{40}Ar (99.600%). The relatively great atmospheric abundance of argon (more precisely of ^{40}Ar) is the result of the radiogenic ^{40}Ar production from potassium in the Earth's crust as well as mantle and the escape into the atmosphere.

Under the fundamental assumptions that, first, no $^{40}Ar_{rad}$ was present in the sample at the moment of the last complete degassing and, secondly, the subsequently produced $^{40}Ar_{rad}$ remained quantitatively in the sample, the K–Ar age t [a] is calculated according to

$$t = 1/\lambda \times \ln\left[1 + (1 + 1/R) \times (^{40}Ar_{rad}/^{40}K)\right] \quad (19)$$

When the numeric values for λ and R are inserted, and the amount of $^{40}Ar_{rad}$ is expressed as [cm³(STP)] and that of K as grams [g], it follows that

$$t = 1.8041 \times 10^9 \times \ln\left[1 + 142.63 \times {}^{40}Ar_{rad}/K\right] \quad (20)$$

For ages < 2 Ma, Eq. (20) simplifies within negligible error to the linear relation

$$t = 2.573 \times 10^{12} \times (^{40}Ar_{rad}/K) \quad (21)$$

The age determination according to Eqs. (19–21), which involves mass spectrometric argon and atomic absorption potassium analysis, is termed *conventional K–Ar dating.*

Samples as well as the equipment used for argon analysis contain as contaminant *atmospheric argon* $^{40}Ar_{atm}$, which needs to be subtracted from the total ^{40}Ar in order to obtain the radiogenic fraction $^{40}Ar_{rad}$. This correction is possible owing to the known isotopic abundance of atmospheric argon $(^{40}Ar/^{36}Ar)_{atm} = 295.5$. Thus, the radiogenic argon $^{40}Ar_{rad}$ is found by measuring the isotope ^{36}Ar in addition to ^{40}Ar according to the relation $^{40}Ar_{rad} = {}^{40}Ar - 295.5 \times {}^{36}Ar$. Atmospheric argon contamination may become a problem especially for geologically young samples, since its relative fraction becomes very high in view of the smallness of the radiogenic argon component.

Argon as a noble gas is chemically inert. Therefore, it should not enter the crystal lattice during mineral formation, and any argon present in a mineral should be driven out during sufficient heating. This basic absence of initial ^{40}Ar – apart from the atmospheric contaminant – is a favorable condition for the use of the K–Ar clock. However, there are cases where a nonatmospheric *extraneous* ^{40}Ar component is present, i.e., the resulting K–Ar age would be an overestimate. Such nonatmospheric argon contamination can be introduced into the mineral essentially as *excess argon*, which was incorporated into minerals by processes (e.g., diffusion) other than the in situ ^{40}K decay, and *inherited argon*, which has been produced by the ^{40}K decay within mineral grains (e.g., xenocrysts, incomplete degassing) before the event being dated (Dalrymple and Lanphere 1969). Both components, excess and inherited argon, may cause serious problems when dating Quaternary volcanites. They need to be identified and overcome, which is achieved by the single grain and isochron dating techniques.

After the event being dated the K–Ar system must stay closed. Partial loss of argon results in an apparent lowering of the K–Ar age. The leaking of argon from minerals may happen continuously (e.g., weathering) or episodically (e.g., thermal overprint). The ability of argon retention differs among minerals. It decreases with increasing ambient temperature. In unweathered Quaternary volcanites that cooled quickly to surface temperature after eruption and stayed cool afterwards, argon loss is rarely observed. If present, the problem of argon loss can be recognized and corrected for by the step-heating plateau technique.

The younger the rock the less abundant is the radiogenic ^{40}Ar and the larger is the fraction of the contaminant argon sources. Therefore, for age determination of Pleistocene and Holocene samples, special K–Ar techniques were developed which allow the analysis of minute proportions of radiogenic argon in the total argon.

Cassignol Technique. This technique is a variant of the conventional dating technique. The argon isotopes are measured without reference to an added ^{38}Ar spike gas (Cassignol and Gillot 1982; Guillou et al. 1997). The mass spectrometer is calibrated before each measurement with an exactly known volume of atmospheric argon, so that the partial pressures of the three argon isotopes can be directly inferred from the intensity of the measuring signals. The main advantages of the technique are improved accuracy of the measurement and the opportunity to include the ^{38}Ar from the sample for the atmospheric argon correction. The drawback is the increased technical expenditure. The technique is especially suited for very young samples down to a few thousand years, and an accuracy better than ± 1 ka for Holocene volcanic rocks and minerals has been reached; however, several grams of sample material is required.

Argon–Argon Technique. The ^{39}Ar–^{40}Ar technique differs from the conventional K–Ar technique by the detection procedure for potassium

(Merrihue and Turner 1976). Instead of determining potassium chemically the sample is irradiated with fast neutrons (neutron energy >1 MeV). Through the nuclear reaction $^{39}K(n,p)^{39}Ar$ the argon isotope ^{39}Ar is produced from the main potassium isotope ^{39}K. The abundance of the artificial isotope ^{39}Ar is determined together with ^{40}Ar and the other natural argon isotopes with the mass spectrometer. Because of its short half-life of 269 a, natural samples contain no ^{39}Ar, so that the neutron induced ^{39}Ar is proportional to the potassium content and the ratio $^{40}Ar_{rad}/^{39}Ar$ is proportional to $^{40}Ar_{rad}/^{40}K$. The age equation is analogous to Eq. (19)

$$t = 1/\lambda \times \ln[1 + (^{40}Ar_{rad}/^{39}Ar) \times J] \qquad (22)$$

whereby the factor $J = e^{\lambda \times t'} - 1)/(^{40}Ar_{rad}/^{39}Ar)'$ is derived from an age monitor (standard of known age t') which is irradiated together with the sample to be dated. The determination of potassium in terms of an argon isotope brings essential advantages over the conventional dating technique:

1. Instead of explicit quantities, only argon isotope *ratios* need to be measured, which improves the accuracy.
2. Only *one* aliquot is required for the measurement and not two separate aliquots for potassium and argon analysis, a benefit that is important in the case of potassium inhomogeneity.
3. Samples that experienced argon loss are generally useless for the conventional technique. This can be overcome with the argon–argon technique by *stepwise heating* and differential degassing the sample whereby for each step the $^{40}Ar/^{39}Ar$ ratio of the argon already released is measured. The corresponding ages display an age spectrum (Fig. 13). The basis behind this procedure is the discrimination between the argon fractions of leaky or retentive behavior. As the temperature rises argon is released from increasingly stable phases. The shape of the age spectrum reflects this behavior. For samples which had previously experienced argon loss the spectrum will increase and eventually reach a plateau value which is interpreted as the true, undisturbed age. A decreasing spectrum may indicate excess argon implying that this component is bound at relatively weak crystallochemical positions within the sample.

The argon–argon technique requires considerable analytical effort. Another shortcoming is the irradiation induced recoil effect of ca. 0.1 μm range that occurs when the ^{39}Ar nuclei produced emit their protons and that may cause disturbing redistribution or even loss of the ^{39}Ar. Problems may also be caused by flux-gradients and the energy distribution of the neutrons in the nuclear reactor.

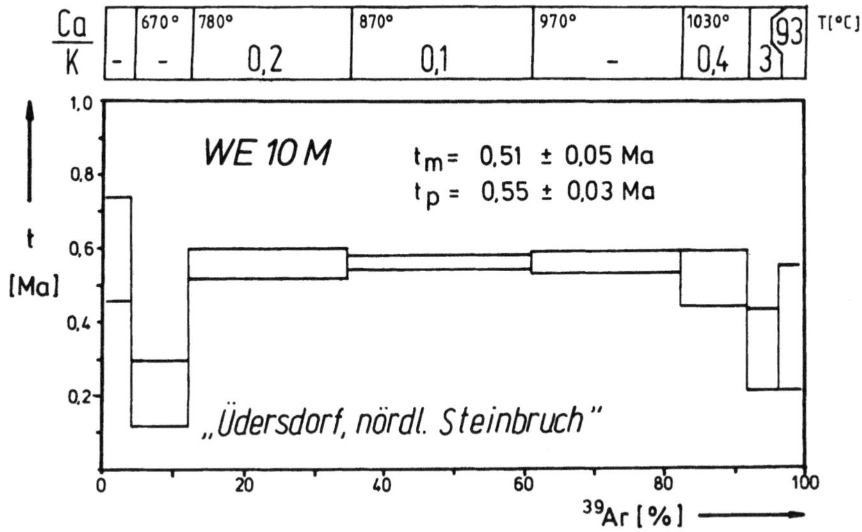

Fig. 13. $^{40}Ar/^{39}Ar$ age spectrum of the groundmass of basalt lava from Üdersdorf, Eifel region, Germany, as a function of argon degassing during stepwise heating (from Fuhrmann and Lippolt 1987). The attainment of the plateau indicates undisturbed K–Ar systematics with the correct age value t_p being in accordance with the conventional K–Ar age t_m of this sample. The boxes above indicate the temperature and the Ca/K ratio of the degassed mineral fractions

Laser Technique. This dating technique (more precisely ^{39}Ar–^{40}Ar laser technique) is a variant of the argon–argon technique. The heating is achieved under a focused laser beam. The sample is either incrementally heated and finally fused – analogous to the plateau technique – or without gradual degassing directly fused and analyzed (York et al. 1981). The main advantage of this technique over the other K–Ar dating procedures is its ability of analyzing single grains in the submilligram range. The discrete grain probing, in connection with the isochron technique (see below), allows the detection of contaminating argon components and thus has a great potential for the dating of young volcanics (Fig. 14).

Argon–Argon Isochron Technique. In order to test whether the K–Ar system is disturbed by excess argon, isochron diagrams are frequently used. They presuppose various cogenetic fractions from the same rock to be dated, but which have different potassium contents. In the isochron diagram the ratio $^{40}Ar/^{36}Ar$ is plotted against the ratio $^{40}K/^{36}Ar$ (conventional technique) or against $^{39}Ar/^{36}Ar$ (argon–argon technique; Fig. 15). All data points of the subsamples lie on a straight line whose slope defines the age and whose intercept with the $^{40}Ar/^{36}Ar$ axis reveals the initial $^{40}Ar/^{36}Ar$ ratio. If no

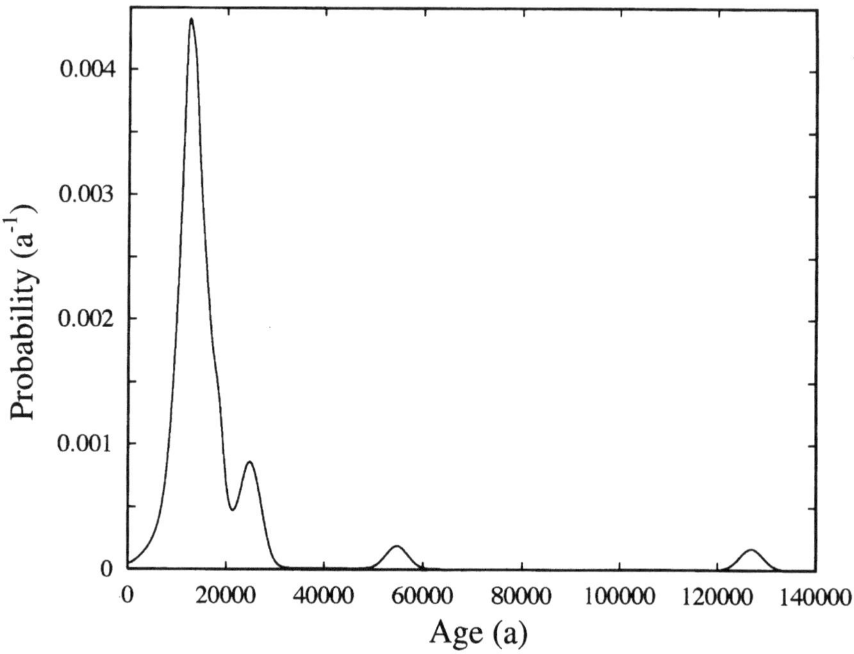

Fig. 14. Probability density diagram of single grain $^{40}Ar/^{39}Ar$ ages determined on sanidine phenocrysts from Laacher See Tephra, Eifel area, Germany (from van den Bogaard 1995). The discrete maximum at 12.9 ka dates the eruption. The maxima with higher ages at 25, 55 and 127 ka probably represent three earlier magma emplacement and crystallization events

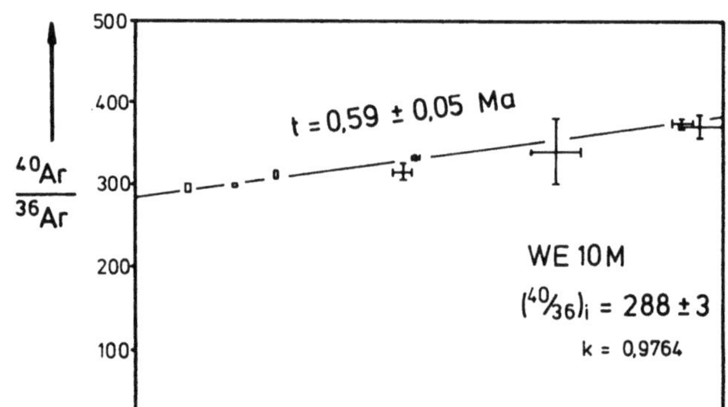

Fig. 15. Isochron diagram of $^{40}Ar/^{36}Ar$ versus $^{39}Ar/^{36}Ar$ of the groundmass of basalt lava from Üdersdorf (cf. Fig. 13; from Fuhrmann and Lippolt 1987). The intercept with the $^{40}Ar/^{36}Ar$ axis indicates the isotopic composition $(^{40}Ar/^{36}Ar)_i$ of the non-radiogenic argon component. The slope of the straight line yields the isochron age t

extraneous argon is present, the isochron intercept is at the atmospheric ratio of 295.5 and if present at a higher value. $^{36}Ar/^{40}Ar$ versus $^{39}Ar/^{40}Ar$ diagrams are also used. In this case, straight lines of negative slope define systems of fixed age. Usually isochron plots are applied in connection with argon- argon laser single grain analysis. In such a case it is important to probe a sufficiently large number of crystals in order to discriminate between xenocrysts (older crystals incorporated into the volcanic rock), phenocrysts (crystals that solidified from the volcanic melt before eruption) and the crystals formed during the volcanic eruption (Chen et al. 1996).

3.1.2
Practical Aspects

Owing to the danger of argon loss the rock samples must be fresh and unweathered. Also rocks which exhibit signs of hydrothermal changes must be avoided. Suitable for dating are total rocks, such as basalts and acidic volcanic glasses, as well as single crystals, in particular potassium-bearing feldspars from tephra. In combination with Rb–Sr dating, also leucite, taken from the middle Pleistocene Villa Seni Tuff in Italy, was analyzed by ^{40}Ar–^{39}Ar (Radicati et al. 1981). Due to the possibility of excess argon, it is advisable in geologically young rocks to separate various minerals from the same rock for K–Ar dating; it is not enough to simply collect several total rock samples from the same lava flow (McDougall et al. 1969).

The sample size depends on the mineral to be dated, the expected age and the dating technique. Quaternary basalts are usually collected as hand-sized pieces. On the other hand, in case of potassium-rich minerals, such as sanidine from tephra layers, the dating can be carried out on mm-sized single grains. For sample preparation the usual mineral separation techniques are used. Too small sieving fractions may be susceptible to argon loss (Flisch 1986). Often hand picking of the mineral grains under the binocular is the optimal procedure in order to get a concentrate of high purity. Sanidine and plagioclase may bear clay covers that need to be removed by washing in hydrofluoric acid. During processing, the sample material must never be exposed to hydrochloric acid, because HCl^+ with the mass numbers 36 and 38 interferes with the argon isotopes of the same mass during mass spectrometric analysis.

In the conventional K–Ar technique potassium is mostly determined with flame photometry or atomic absorption, occasionally also with wet chemistry, X-ray fluorescence or neutron activation and rarely with mass spectrometry. For argon analysis the sample is fused under high vacuum at temperatures of 1300–2000 °C, whereby the argon is released together with the other gases. After extraction and purification the argon enters the mass spectrometer. The isotopic composition and content of argon is analyzed in the spectrometer with the isotope dilution technique, whereby a calibrated amount of ^{38}Ar spike gas is added to the mass spectrometer and measured together with the argon released from the sample. Because natural argon contains only negligibly little of this isotope, the ^{38}Ar signal is directly proportional to the amount of the argon spike, so that the amount $^{40}Ar_{rad}$ can be derived from the ratio $^{40}Ar_{rad}/^{38}Ar$. The atmospheric argon contamination is assessed from its known ratio $^{40}Ar/^{36}Ar$ and subtracted.

For the argon–argon techniques the sample must be irradiated – together with an age monitor – in a nuclear reactor with a high neutron fluence (ca. 10^{17} n cm^{-2}). During extraction argon is progressively released while the temperature is increased stepwise until final fusion. After each degassing step the gas is purified and isotopically analyzed, and the corresponding argon–argon-age is computed. In the laser variant the heating is achieved with a laser beam. As age standard the 450-ka-old BB-6 basalt may be used (Jäger et al. 1985).

With conventional K–Ar dating an accuracy of 2–4% can be obtained, however, for geologically very young samples the errors are often larger. According to Flisch (1986), a 7500-a-old sanidine with 10% potassium can be accurately dated to ±10% (1 σ). The quoted precision of $^{40}Ar/^{39}Ar$ plateau ages and isochron ages is often better than 1%. However, one has to be aware of the difference between precision and accuracy (Sect. 1.3). The uncertainty in the age monitors is unlikely to be better than 1% and such uncertainty needs to be taken into account when assessing accuracy, rendering such precisely quoted age values less accurate.

3.1.3 Application

In the Quaternary, K–Ar age determination is restricted to volcanic rocks and minerals. It dates the last, complete resetting of the K–Ar system during eruption, such as rock solidification, mineral formation and thermal degassing. The direct K–Ar dating of Quaternary sediments (in contrast to older, glauconite-bearing sediments) is not possible. However, if sediments are intercalated with tephra layers or volcanic flows, K–Ar dating of the volcanics can establish a chronometric frame for the whole stratigraphic suite.

Basalt. Young basalts are determined with the K–Ar method mostly as whole rock samples. Only fresh, unweathered material is appropriate. Due to low argon retentivity, glassy fractions are less suited than crystalline components. The basalt specimen should be compact and fine-grained, so that sufficiently homogeneous aliquots are available for the potassium as well as the argon analysis. For total rock samples with very heterogeneous potassium and grain-size distribution Horn and Müller-Sohnius (1988) have proposed an etching technique in which the fine-grained, potassium-rich matrix is removed with hydrofluoric acid. One of the first attempts to date basaltic lava with K–Ar is that by von Koenigswald et al. (1961) who studied the massive flow underlying Bed I at Olduvai, Tanzania, and found an age of 1.3 ± 0.1 Ma, which is an underestimate probably because of argon loss.

K–Ar dating of basalt is important for the calibration of the geomagnetic polarity time scale (Sect. 9.1). Astronomical tuning (Sect. 10.2.1) for the $\delta^{18}O$-chronology of deep-sea sediments indicates that the prevailing date for the Matuyama–Brunhes boundary at 730 ka ago needs to be shifted backwards to ca. 780 ka ago. In order to date the Matuyama–Brunhes boundary with K–Ar as precisely as possible, four specimens of basaltic lava were collected from the transition zone of reverse to normal magnetization in the caldera of the volcano Haleakala on the island of Maui (Baksi et al. 1992). Incremental heating $^{40}Ar/^{39}Ar$ analysis was performed on whole rock samples. Three samples exhibited consistent age plateaus, however, with minor argon loss. The isochron diagram showed an initial $^{40}Ar/^{36}Ar$ slightly higher than the atmospheric ratio, indicating an excess argon contamination. Therefore, the isochron age was chosen as the better approximation of the crystallization event than the plateau ages. The mean age of three lava flows was calculated as 783 ± 11 ka, which agrees with the astronomical dating of the Matuyama–Brunhes boundary. This result has been confirmed by Singer and Pringle (1996) who investigated eight basaltic to andesitic lavas from the Matuyama-Brunhes transition taken at different localities in Chile, Tahiti, La Palma and again in Maui. The analyses were performed predominantly on phenocryst-free groundmass separates or phenocryst-poor whole rock samples, using incremental $^{40}Ar/^{39}Ar$ analysis combined with isochron plotting. A weighted mean of 778.7 ± 1.9 ka was obtained, whereby the quoted error reflects the 1 $\bar{\sigma}$ precision.

When reversed paleomagnetic polarity was observed in 1967 in the young basaltic flows from Laschamp and Olby, Chaine de Puys, France, it was believed that a reversal event had been detected whose K–Ar age was estimated as at most 20 ka (Bonhommet and Zähringer 1969). As later studies on basalts of the same age but from different localities could not confirm such an event, renewed dating of both lava flows became desirable. Cassignol and Gillot (1982) found with their spike-less K–Ar technique ages of 44 ± 5 ka and 49 ± 5 ka for total rock, and 37 ± 5 ka and 44 ± 5 ka for separated ground-matrix samples from Laschamp and Olby, respectively (Fig. 16). The higher total rock ages were interpreted as too old owing to inclusions of early crystallized minerals and xenoliths, so that the K–Ar

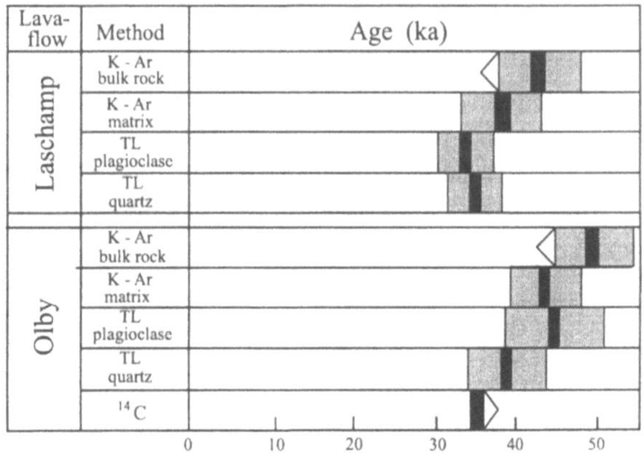

Fig. 16. Age determination of the reverse-magnetized lava flows of Laschamp and Olby in the Auvergne, France. The K–Ar ages are in agreement with the TL and ^{14}C dates. (After Cassignol and Gillot 1982)

ages of the matrix were considered as representative of the eruption. Independent thermoluminescence dating on quartz crystals collected from lava-baked contacts yielded ages between 32 and 40 ka (Sect. 7.1.3; Valladas and Gillot 1978; Guerin and Valladas 1980). Accordingly, the paleomagnetic Laschamp excursion seems to have taken place sometime between 50 and 30 ka ago (Sect. 9.1).

Tephra and Tuff. Of particular significance in the Quaternary chronometry is the K–Ar dating of tuffaceous horizons which are intercalated with sedimentary sequences. The K–Ar ages yield invaluable tephrochronological and stratigraphic time-marks. Our knowledge about the early hominids, their evolution in and spreading out of Africa rests essentially on K–Ar data of such sequences. However, tephra layers in sediments are commonly reworked and contaminated with older mineral detritus of various provenances. In such a case, bulk K–Ar or Ar–Ar dating of mineral concentrates yields due to this admixture an integrated age that is too old with respect to the event of volcanic eruption. For this reason, but also for the recognition as well as correction of excess argon, single crystal probing is required in order to identify the various components of different ages. This is optimally achieved with the grain-discrete Ar–Ar laser technique. Suitable mineral phases are mainly potassium-bearing feldspars, such as sanidine and plagioclase, but also biotite, hornblende and acidic glass shards. The high sensitivity of the technique allows the analysis of even single sanidine grains from Holocene tephra (Chen et al. 1996).

One of the most renowned sites with early hominid fossils is the Olduvai Gorge in Tanzania. The ca. 100-m-thick Plio–Pleistocene sequence of limnic, fluvial and alluvial sediments contains numerous tephra horizons and lava flows. In bed I remains of *Australopithecus boisei* (= "*Zinjan-*

thropus") were uncovered together with stone tools. The tephra layers of bed I exhibit a normal paleomagnetic direction although bed I belongs magnetostratigraphically to the inverse Matuyama chron. They form the *locus typicus* for the Olduvai subchron. Continuous efforts over three decades to date this key section aroused, apart from methodological discussions on the reliability of the K-Ar as well as the fission track clocks, controversial views on the age of bed I. Finally, single grain Ar-Ar laser-fusion dating succeeded in establishing a detailed and reliable chronology (Walter et al. 1991). Various old populations of feldspar grains were observed, and the juvenile tuff component could be identified and distinguished from older, reworked detritus contamination. Ages were calculated as weighted means over several juvenile grains each, the errors represent 1 $\overline{\sigma}$ precision. They range between 1.798 ± 0.004 (Tuff IB) and 1.749 ± 0.007 Ma (Tuff IF) for the middle to upper layers of bed I, including the fossil hominids and signs of significant climatic change (Fig. 17). The results for Tuff IA also imply that the base of the Olduvai subchron is 1.98 - 2.01 Ma old, about 0.1 Ma earlier than previously assumed.

The earliest fossil hominids (*Australopithecus afarensis*) hitherto discovered in East Africa are apparently at least 3.9 Ma old. The overlying Pliocene Cindery Tuff, Middle Awash in Ethiopia, was determined with Ar-Ar on plagioclase and glass as well as with fission tracks on zircon and yielded ages of 3.8 - 4.0 Ma (Hall et al. 1984). At Aramis, in the same region, the Gaala Vitric Tuff which occurs stratigraphically below the hominid fossils, was dated by single-grain Ar-Ar laser fusion. Despite dominant contamination with older grains an age of 4.387 ± 0.031 Ma was calculated from the juvenile feldspar fraction, placing an upper age limit on these earliest hominids (WoldeGabriel et al. 1994). In southern Ethiopia, in the Konso-Gardula region, Acheulian stone tools were uncovered from Plio-Pleistocene sediments containing several tuff horizons (Asfaw et al. 1992). Ar-Ar single feldspar grain dating of these horizons places this earliest occurrence of the Acheulian in Africa around 1.4 Ma ago. The oldest stone tools so far discovered are those from the Oldowan Industrial Complex at Gona, Ethiopia. This occurrence is dated between 2.6 - 2.5 Ma based on single-feldspar laser fusion Ar-Ar analyses and magnetostratigraphy on underlying and overlying tephra beds (Semaw et al. 1997).

As to the question when and how *Homo erectus* migrated out of Africa, K-Ar data are of special interest. The earliest fossil trace of this species in Africa occurs in the Koobi Fora region, Kenya, at 1.8 Ma ago (McDougall 1985). According to K-Ar evidence *Homo erectus* seems to have appeared practically at the same time in Southeast Asia. Swisher et al. (1994) presented Ar-Ar laser-incremental ages on hornblende separated from pumice of two *Pithecanthropus* sites in Java. The analyses yielded well-defined

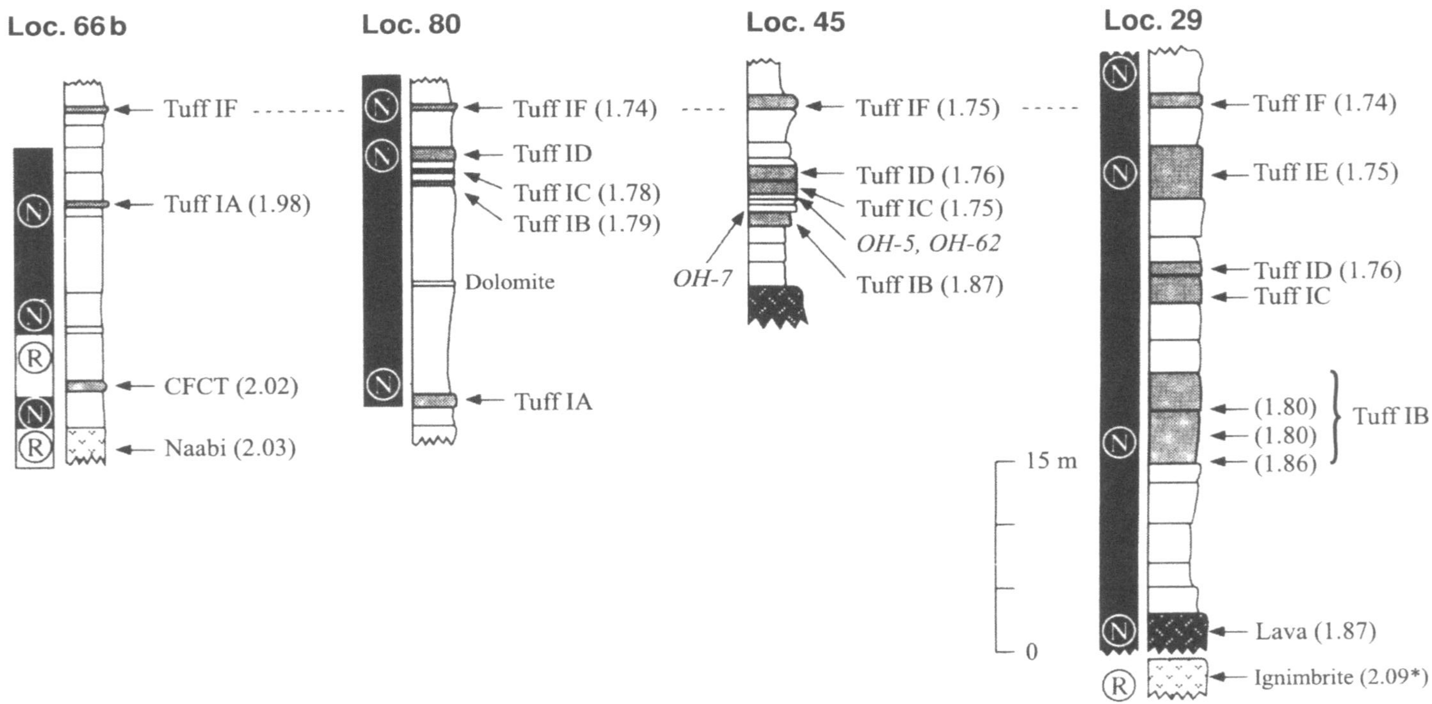

Fig. 17. Four key bed I sections in the Plio–Pleistocene sequence of Olduvai Gorge, Tanzania. Numerous tuffs and lavas are intercalated in lacustrine, fluvial and alluvial sediments. In bed I early hominids and artifacts were unearthed (find layers are marked by the symbol *OH*). The single-crystal laser-fusion ages (in Ma) of the tuff horizons are given in *parentheses*. In bed I normal paleomagnetic polarity *N* (Olduvai subchron) occurs within the reversed Matuyama chron (*R*). (Reprinted with permission from Nature; Walter et al. 1991; copyright 1991, Macmillan Magazines Limited)

plateaus with weighted means of 1.81 ± 0.04 and 1.66 ± 0.04 Ma. However, the stratigraphic position of the former hominid finds in relation to the pumice is not certain.

In central Europe, the Middle Rhine area with the Eifel volcanic field plays a key role in Quaternary and Paleolithic chronology (Fig. 18). The numerous tephra occurrences supply excellent tephrochronological marker horizons which are frequently also connected with archaeological sections. Following the first attempts by Frechen and Lippolt (1965), many conventional as well as argon–argon ages were determined on tephra and on basalt flows in this area (Fig. 19). The main difficulties were argon loss, excess argon and phenocryst as well as xenocryst contamination. According to the K–Ar data, the main volcanic activity falls into the Middle Pleistocene. The last major maar-type explosion was that of the Laacher See with a weighted mean age of 12.9 ± 0.56 ka (Fig. 14) derived by isochron evaluation of single sanidine grain analysis (van den Bogaard 1995). This result is of particular importance since the Laacher See Tephra was widely spread across Europe.

Obsidian. Attempts to date obsidian with K–Ar (Kaneoka 1969) have been impeded by argon loss which correlates with the degree of hydration for rhyolitic glasses. Only fresh, unweathered obsidian material with water contents less than 0.5% seems to be appropriate, as was demonstrated for an obsidian from the Wada Toge Pass, Japan, with a K–Ar age of 0.95 ± 0.10 Ma. Apparently, the K–Ar dating of rhyolitic glass from a Quaternary tuff at Bori, India, presented no difficulties; the glass was accompanied by early Acheulian artifacts. The 50 – 100 µm fraction yielded an age of 538 ± 47 ka (Horn et al. 1993).

Tektites and Impact Glass. The high-temperature formation of tektites is combined with complete degassing, so that the K–Ar clock dates such events of extraterrestrial impact. Conventional K–Ar ages have been determined on numerous specimens of tektites, among them the Quaternary tektites from the Ivory Coast and Australasian strewnfields (Schaeffer 1966). The $^{40}Ar/^{39}Ar$ technique has also been applied to Australasian tektites (Storzer et al. 1984). Generally, there is a good agreement between K–Ar and fission track ages of tektites (Sect. 6.1.3). The high contents of atmospheric argon – included during the glass formation – may disturb the K–Ar dating of impact glass, as, for instance, encountered for the Darwin Crater, Tasmania, that gave a K–Ar age of 700 ± 80 ka (Gentner et al. 1973). The K–Ar age of 1.3 ± 0.3 Ma for impact glass from the Bosumtwi Crater, Ghana, agrees with that of Ivory Coast tektites and supports their origin from this crater (Gentner et al. 1964, 1969).

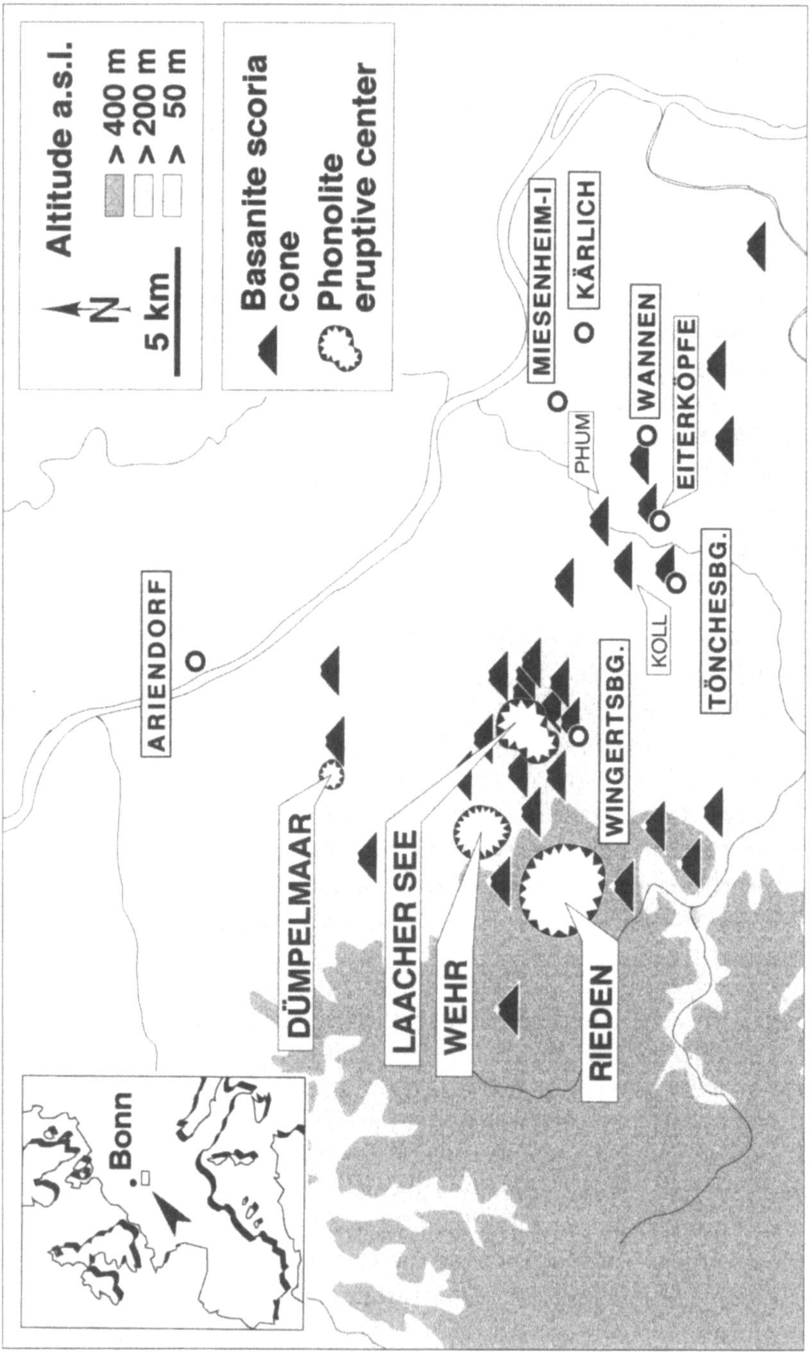

Fig. 18. A schematic map of the Quaternary volcanic field of the East Eifel, Germany. (van den Bogaard and Schmincke 1990, courtesy of P. van den Bogaard)

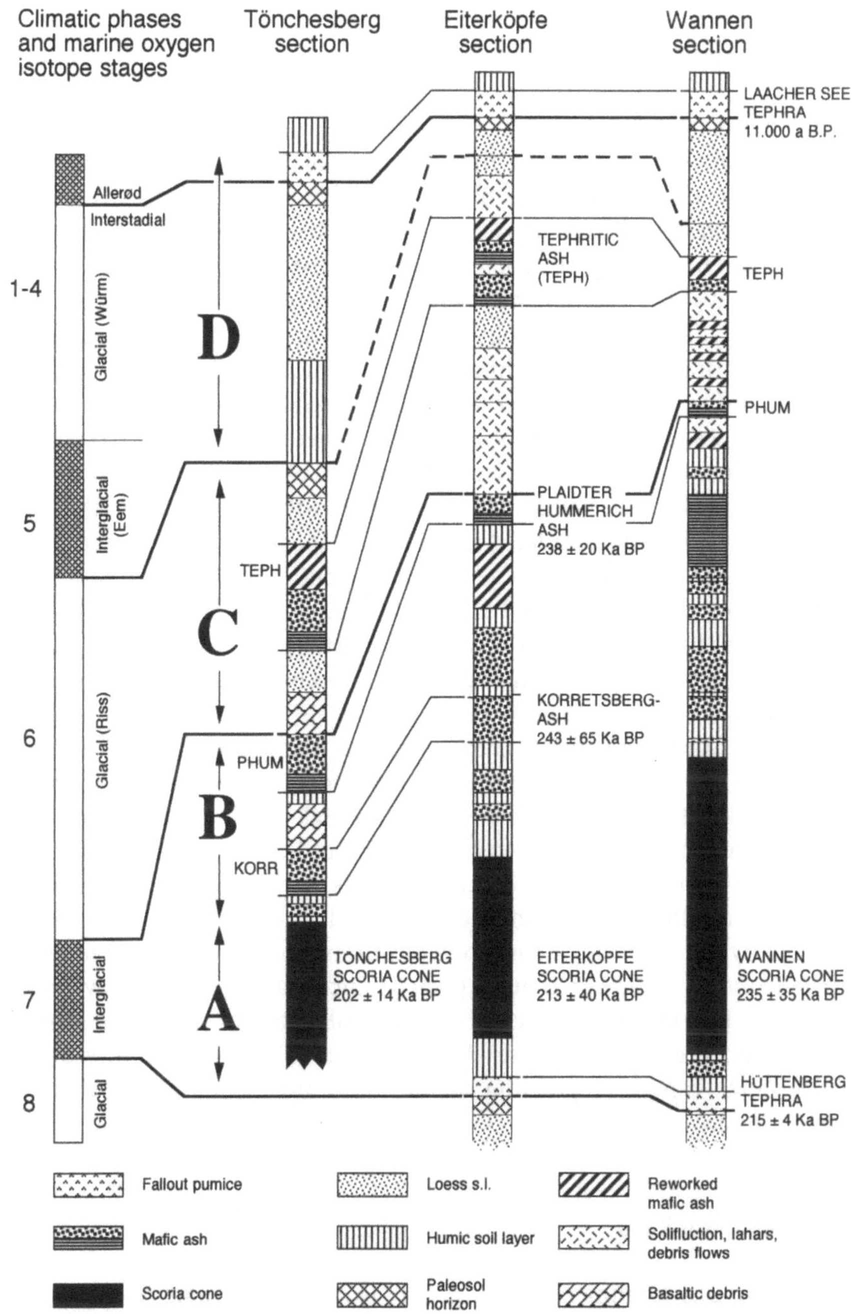

Fig. 19. Tephrostratigraphic correlation of several important Quaternary profiles of the East Eifel (cf. Fig. 18). These profiles contain important Paleolithic find horizons. The correlation with the marine $\delta^{18}O$ stages is based on the $^{40}Ar-^{39}Ar$ ages. (van den Bogaard and Schmincke 1990, courtesy of P. van den Bogaard)

3.2 Uranium–Helium

When it was found that α-particles, released during uranium and thorium decay, are helium nuclei, the first attempts of radiometric dating were undertaken. Already in 1905, only 9 years after Becquerel discovered the phenomenon of radioactivity, Strutt tried to determine a radiometric age of a mineral for the first time, using its accumulated helium content (Strutt 1905). This was the beginning of radiometric chronometry. The U–He method was leading geochronology during the first half of the 20th century. Isotopes of uranium and thorium are parent nuclides of radioactive decay series (cf. Fig. 22). They decay through various radioactive interim members to stable lead isotopes, whereby several α- and β–particles are emitted. Since α-particles are ^{4}He nuclei, atoms of the noble gas helium are produced. If all these helium atoms are kept within the system, the total amount of helium accumulated is a measure for the system's age. Certain parallels exist to the uranium series dating methods (Chap. 4), which are also based on decay chains and exploit the disturbance of radioactive equilibrium. Because the emission of the α-particle causes recoiling of the emitting nucleus, the uranium–helium method is related with the α-recoil track dating technique (Sect. 6.2).

The facts that the α-decay is frequent and that the analytical detection of helium is feasible, are preconditions for uranium–helium dating which make it attractive. However, additional conditions must be fulfilled, such as the complete retention of the radiogenic helium as well as the closed system behavior of all members within the decay chain. In geologic situations both conditions often do not hold, as was already realized by the pioneers of chronometry. Uranium–helium ages frequently turned out to be lower than expected. Encouraging results have been obtained for Quaternary corals and mollusks. In addition, bones, basalts and ore minerals have also been dated. The applicability of the U–He clock ranges, depending on the uranium content, from 10 ka to some 100 Ma.

3.2.1 Methodological Basis

The noble gas helium consists of two isotopes, ^{3}He and ^{4}He, with an atmospheric abundance ratio of 1.384×10^{-6} (Sect. 5.2.1). The high ^{4}He fraction is caused by the radioactive decay of uranium and thorium, both of which are enriched in the Earth's crust. In the Earth's mantle, in contrast, helium still has its primordial ^{3}He/^{4}He ratio of around 10^{-4}. To a large extent the radiogenic helium diffuses from the Earth's crust into the atmosphere, from where in turn it may escape into extraterrestrial space. If retained in minerals the growing content of radiogenic helium becomes a radiometric clock.

Natural uranium is composed of the two isotopes ^{238}U with 99.3 % and ^{235}U with 0.7 % abundance, whereas thorium consists only of ^{232}Th. Each α-decay within the three chains starting from uranium and thorium produces one ^{4}He atom. Actually, the numbers of helium atoms, formed in each complete cascade of decays, are 8 in the ^{238}U, 7 in the ^{235}U and 6 in the ^{232}Th series (cf. Fig. 22). Because thorium and its radioactive descendants also contribute to helium production, it is more correct to use the term (U + Th)–He instead of simply U–He method. For age determination, the radiogenic helium as well as the uranium and thorium must be quantitatively analyzed. Obviously, in order to apply the method, one has to know the helium production rate for given uranium and thorium contents. However, this is not as easy as it might seem due to the phenomenon of *radioactive equilibrium* (Sect. 1.2). After being disturbed the equilibrium is gradually established again within about five half-lives of the daughter nuclide and full production rate is reached only under the condition of radioactive equilibrium. Then, the ^{4}He production rate for both uranium isotopes and their descendants is 12×10^{-8} cm^3 (STP) per µg U and Ma, and for thorium 2.9×10^{-8} cm^3 (STP) per µg Th and Ma. Because of the low natural $^{235}U/^{238}U$ ratio (~0.007) the helium contribution from the ^{235}U chain is small. With Th/U ratios of 3 to 4, which are geochemically frequent, a significant helium contribution comes from the ^{232}Th series. The adjustment of equilibrium within the ^{238}U series is controlled by the long-lived members – essentially ^{234}U and ^{230}Th. To build up the equilibrium takes for the uranium descendants about 400 ka or even 1.3 Ma, if the initial $[^{234}U]/[^{238}U]$ activity ratio (Sect. 4.1.4) is different from one. Therefore, in disturbed systems the helium production follows mathematically complex functions (Fig. 20) which require iterative solutions. Obviously, the problem applies in particular to Quaternary samples whose ages are similar to the half-lives. If the thorium content is negligible, as for instance in corals, the mathematical difficulty is reduced.

The major problem for the U–He clock is, however, its tendency to open system behavior. It manifests itself in helium diffusion and in the mobility of uranium as well as the radioactive descendants. Helium loss is the main cause for the commonly observed age underestimates. The helium retentivity depends on the mineral species (Lippolt and Weigel 1988). The relatively small helium atom leaks out easily from the crystal lattice. With growing radiation damage the helium diffusion is enhanced. If grain diameters are comparable with the α-range of ~15 µm, for geometric reasons

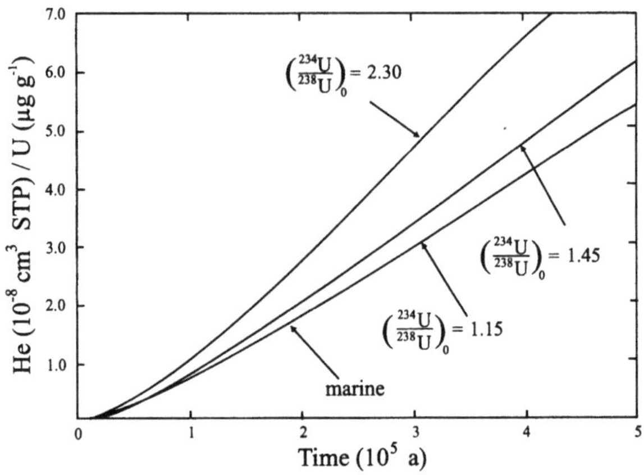

Fig. 20. He/U ratio as a function of the age for different initial activity ratios $(^{234}U/^{238}U)_0$. During the first 400 ka, ^{230}Th and the ^{234}U descendants build up until equilibrium is reached. Excess ^{234}U decays fully within 1.3 Ma. Subsequently, helium is produced at a constant rate, i.e., the production curves then follow linear and parallel straight lines. (Reprinted with permission from Science; Fanale and Schaeffer 1965; copyright 1965, American Association for the Advancement of Science)

the helium atom may be catapulted out of the grain into the pore liquid, from where it escapes to the atmosphere.

The mobility of uranium and its descendants is governed by their physicochemical position within the mineral. Crystallochemically bound uranium is relatively resistant to exchange, but secondarily incorporated uranium – fossil bones, for instance, take up uranium from the groundwater – is more susceptible. Diagenetic recrystallization may cause chemical exchange, as takes place in corals, when the metastable aragonite recrystallizes to stable calcite. Furthermore, during each event of α-decay the remaining nucleus suffers a recoil causing strong radiation damage of 10^{-2} µm; consequently, with the progress of the decay cascade the daughter nuclei are situated in increasingly damaged positions of weakened chemical resistance and easier leachability. This is, actually, the reason why groundwater has $[^{234}U]/[^{238}U]$ activity ratios larger than one. Especially susceptible to diffusion is the noble gas ^{222}Rn ($t_{1/2}=3.8$ d) in a porous matrix.

3.2.2 Practical Aspects

Aragonitic fossils, such as corals and mollusks, are suitable for U–He dating. A few grams of material are sufficient. The presence of calcite indicates partial recrystallization and thus the danger of helium loss. The aragonite is separated by hand-picking under the binocular and is cleaned from detrital contaminants in an ultrasonic bath. Helium is analyzed with mass spectrometry using the isotope dilution technique. The ^{4}He content is given in 10^{-8} (STP) cm^3 g^{-1} sample material. Due to its much lower atmospheric abundance the atmospheric ^{4}He contamination in samples is much less serious than that by ^{40}Ar. The ^{4}He contamination may either be identified and corrected for by determining ^{3}He through the known atmospheric $^{3}He/^{4}He$ ratio or by ^{20}Ne through the atmospheric $^{4}He/^{20}Ne$ ratio (Bender 1973). The contents of uranium and thorium can be determined in different ways, mostly by isotope dilution mass spectrometry or by neutron activation. The $^{234}U/^{238}U$ ratio is measured by α-spectrometry or by mass spectrometry. In order to assess possible helium loss the texture, grain-size and porosity of the sample material need to be investigated, which is best achieved with a scanning electron microscope. Although U–He ages can be determined with high precision, their accuracy may be rather low due to uncertainties as to open system behavior, which frequently results in age underestimates.

3.2.3 Application

Corals. Corals incorporate a few µg g^{-1} uranium in their aragonitic skeleton. Because of the danger of system opening the aragonite must not be recrystallized to calcite. Otherwise, aragonite has good helium retentivity and is insensitive to secondary uranium mobilization, hence corals are suitable for U–He dating, as was demonstrated for Pleistocene corals by Fanale and Schaeffer (1965). In that study a nonradiogenic excess helium component was observed that must have already been present initially and was probably caused by detritus contamination. The marine $^{234}U/^{238}U$ ratio with which uranium was initially built into the aragonite structure needs to be determined and considered in the age evaluation (Fig. 20). Bender (1973) was able to show that U–He dating is reliable for Pleistocene corals if possible loss of helium and radon is taken into account.

Mollusks. After death and burial, mollusks take up uranium from the groundwater and place it into their aragonitic shells. Provided the uptake occurs rather fast in comparison to the geologic age of the fossils, U–He dating should be applicable. Of course, this also presupposes that the uranium, once adsorbed, is geochemically fixed and that the radiogenic helium is quantitatively stored in the mollusk shells. For Quaternary mollusk shells, collected from tectonically raised beach terraces in California,

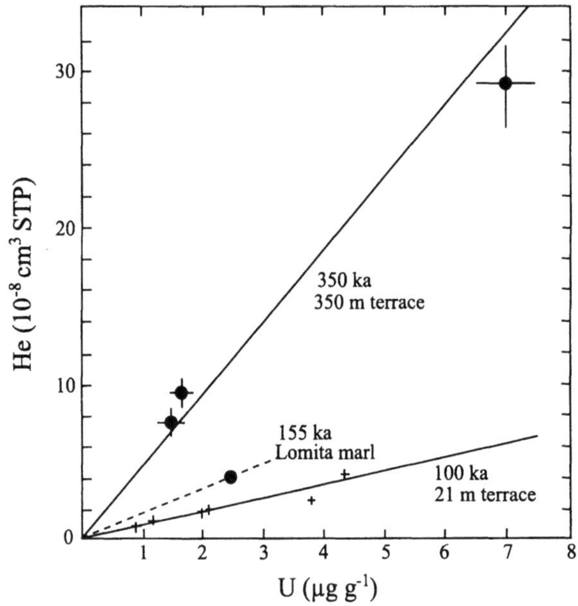

Fig. 21. U–He dating of Quaternary mollusk shells from tectonically raised beach terraces in California. The U–He data of mollusks containing different amounts of uranium, but originating from the same terrace lie on isochrons. (Reprinted with permission from Science; Fanale and Schaeffer 1965; copyright 1965, American Association for the Advancement of Science)

Fanale and Schaeffer (1965) were able to show that these conditions seem to hold true. The U–He ages for mollusks from the same terrace agree with each other within their error limits, in spite of varying uranium contents (Fig. 21). Ages of 100 ± 20 ka for the 21 m terrace and 350 ± 50 ka for the 375 m terrace were found. Furthermore, these ages are consistent with the $^{230}Th/^{234}U$ data. As for corals, the $^{234}U/^{238}U$ ratio needs to be taken into account. Before generalizing these encouraging results, further experience on the reliability of U–He for mollusks needs to be gathered.

Bones. The reliability of U–He dating was tested also on Quaternary and Tertiary bones (Turekian et al. 1970). The ages determined turned out to be generally too young. Fossil bones take up uranium from the groundwater and retain it in their porous phosphatic matrix. The uranium precipitation is enhanced under the reducing conditions caused by the decomposition of the organic bone matter. During this period (up to 10 ka), the uranium content in the bone increases by several orders of magnitude. Since this duration might be significantly long in relation to the total age the U–He age calculation has to rely on model assumptions about the timing of uranium uptake. Moreover, the adsorbed uranium and its radioactive descendants are physiochemically only weakly bound to the bone material and, thus, are susceptible to steady exchange reactions with the groundwater. In addition to all these uncertainties, helium as well as radon loss further contribute to the open system behavior. Therefore, bone has to be considered as unsuitable for U–He dating.

Basalt. During extrusion, when basaltic lava degasses and, therewith, also loses its helium, the U–He clock is reset. After solidification, the radioactive decay of uranium and thorium, contained in the basalt, again produces helium whose amount should be a measure of the time elapsed since the eruption. When attempting to date whole rock samples of young basalts from the Madeira and Porto Santa Islands, Ferreira et al. (1975) ascertained systematic helium loss. The U–He ages range from 0.4 to 4.6 Ma and turned out to be, without exception, significantly lower than the K–Ar ages of the same samples. More promising for U–He dating may be uranium-bearing mineral inclusions.

4 Uranium Series

This general term comprises several closely related dating methods, based on the radiometric disequilibrium within the radioactive decay series arising from the two uranium isotopes ^{238}U and ^{235}U. Synonymous are the terms *decay series*, *disequilibrium* or, to some extent, *uranium–thorium* methods of dating. The individual techniques are named after the daughter/parent nuclide pairs, as e.g. $^{230}Th/^{234}U$, $^{231}Pa/^{235}U$ and $^{234}U/^{238}U$. A comprehensive description of the various techniques and their applications is found in the collected edition by Ivanovich and Harmon (1992).

Both uranium isotopes are radioactive and decay over a chain of intermediate daughters to stable isotopes of lead. In closed systems an equilibrium state between all radioactive nuclides of the decay chain will be established. In this state of secular equilibrium all radioactive nuclides possess equal activity. If such a system is disturbed it will take time for the secular equilibrium state to be reestablished. The moment of the disturbance can be determined from the degree to which the equilibrium is reestablished. In natural systems the disturbance occurs through geochemical fractionation. The fractionation is predominantly caused through differences in the solubility of the decay products during weathering, transport, mineral formation and sedimentation processes. In this way exogenetic geochemical processes become datable by the uranium series techniques.

The first approaches to estimating ages by means of radioactive disequilibrium trace back to the observations of radium and ionium (^{230}Th) excesses in marine sediments (Joly 1908; Pettersson 1937; Piggot and Urry 1942). With improving detection techniques in the middle of the 1950s, the actual development of the uranium series methods of dating began, and it must be added that the major interest was dedicated to the deep-sea sediments, corals and terrestrial carbonates. As soon as it was realized that corals and deep-sea sediments bear the record of climatic history, uranium series ages became an essential support of marine Quaternary chronology. The change from the α-spectrometric to the mass spectrometric analysis, which requires smaller sample size (by a factor of ca. 10^{-2}) and provides higher age precision (Edwards et al. 1986/87), brought the uranium series dating methods a further impetus. Nowadays, its rapid rate of development opens new applications.

The uranium series methods can be used for dating over time periods ranging from a few years to ca. 1 Ma, thus covering essentially the Middle and Upper Pleistocene. The age ranges of the individual methods of dating, however, are quite distinct, since the time period until establishment of the secular equilibrium is dependent on the half-life of the particular daughter nuclide. The upper limit of the datable time period is reached when the degree of re-establishment does not differ significantly anymore from the secular equilibrium state. In other words, datable are merely systems which have not yet regained complete re-establishment of their secular equilibrium state. As a rule of thumb, the datable time period roughly equals five half-lives. In the case of the $^{230}Th/^{234}U$ method (^{230}Th half-life 75.4 ka), for example, the upper limit of the datable time period is ~350 ka; for the more sensitive mass spectrometric analysis – instead of α-counting – it is possibly ~500 ka, corresponding roughly to seven half-lives (Edwards et al. 1986/87).

The uranium series methods are versatile and applicable to many kinds of material. The uranium content should not be less than 0.1 $\mu g\ g^{-1}$. In respect to marine environments, worth mentioning are deep-sea sediments, manganese nodules and coral reefs. As to terrestrial sediments, attention is drawn especially to secondary carbonates of caves and springs. Furthermore, pedogenic calcareous deposits are suitable for dating, as well as marls and calcareous lake-sediments. Important additional applications are fossil bones, teeth and mollusk shells. After all, peat occurrences are datable too. By reason of the diversity of materials suitable for uranium series dating, the data obtained from these techniques contributed much to marine as well as continental Quaternary chronologies, and in particular to the chronometry of Paleolithic deposits.

The accuracy of an uranium series date is not merely dependent on the analytical precision of the measurement, since it is significantly affected by difficulties arising from sample contamination and the opening of the radiometric system. In order to reduce and assess such influences, it is advisable to conduct the sampling procedure in the presence of a specialist or at least under consultation with him. Under favorable circumstances, e.g., in the case of mass spectrometric $^{230}Th/^{234}U$ dates, age precision better than 1% can be obtained. In most cases, however, the dating errors are definitely larger.

Radioactive Equilibrium. The physical fundamental of the uranium series dating methods is the phenomenon of radioactive equilibrium. It is observed in the decay chains starting from uranium and thorium. Naturally occurring uranium consists of the two radioactive isotopes ^{238}U (99.3%) and ^{235}U (0.7%). These parent nuclides decay to daughter nuclides, which in their turn are radioactive, until after prolonged decay cascades the stable isotopes of lead ^{206}Pb and ^{207}Pb, respectively, are ultimately formed (Fig. 22). The activity *[N]* of an unstable nuclide (with number N of atoms and decay-constant λ) is defined as the number of decaying atoms *dN* within the time interval *dt*, thus $[N] = dN/dt = \lambda \times N$ (Sect. 1.2). The term radioactive equilibrium means the state in which for each radioactive member within the chain the activities [N] become equal to each other. In

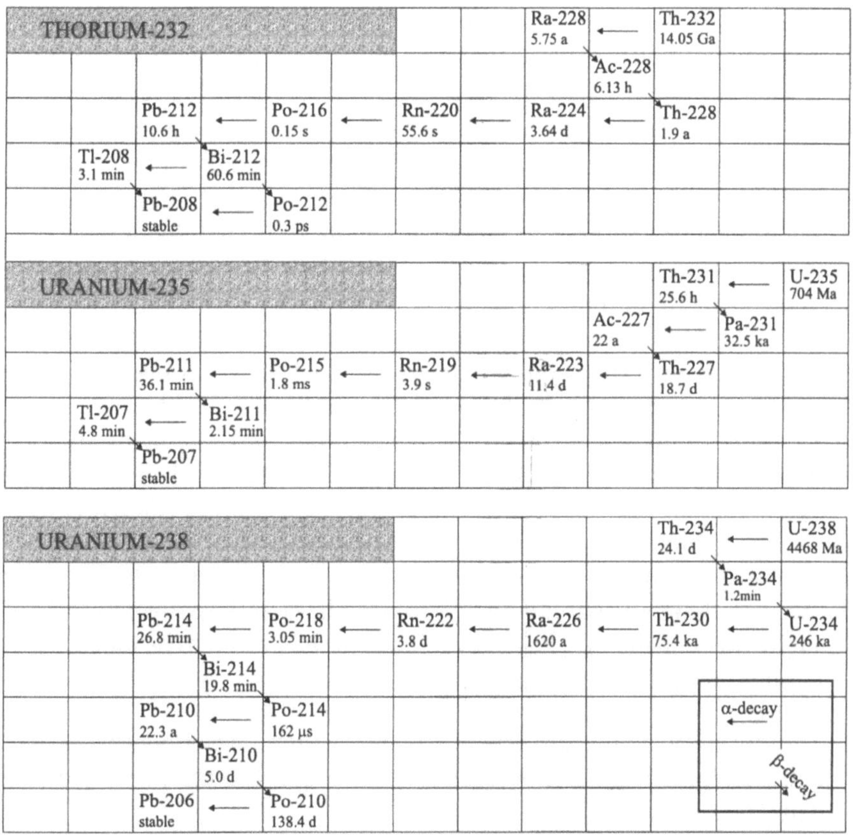

Fig. 22. ^{232}Th, ^{235}U and ^{238}U are parent nuclides of radioactive decay chains. They disintegrate ultimately, via several radioactive daughters of different half-lives, to the stable lead isotopes ^{208}Pb, ^{207}Pb and ^{206}Pb, respectively. An equilibrium is gradually established in the decay chains, at which the rates of radiogenic production and radioactive disintegration balance each other for each interim member

this state, for each intermediate member the number of disintegrating atoms is compensated by the newly formed radiogenic atoms through the decay of its parent. Then, the activity ratio of daughter D and parent N equals one, i.e., [D]/[N] = 1 and the number N of atoms of each nuclide is proportional to its half-life, i.e., $D/t_{1/2,D} = N/t_{1/2,N}$.

If uranium is firmly tied into a mineral, the decay products gradually and successively build up until the establishment of equilibrium. The time span necessary until the equilibrium state is reached is determined by the intermediate daughters with the longest half-lives; in the ^{238}U series these are the long-lived daughter products ^{234}U with 245 ka and ^{230}Th with 75.4 ka and in the ^{235}U series ^{231}Pa with its half-life of 32.8 ka. The establishment of the equilibrium state furthermore requires a closed system which means that there is neither an influx nor a loss of parents and daughters. In nature most unweathered minerals and rocks represent closed systems, in which radioactive equilibrium persists as a result of their geologically great ages.

Disequilibrium as Clock. The state of equilibrium can be disturbed by chemical (weathering, precipitation, magmatic differentiation), physical (adsorption) and biological processes (calcareous secretion) as parent or daughter nuclides are added or leached. Differential solubility fractionates uranium and thorium especially effectively, since as opposed to thorium, uranium is liable to dissolution, particularly under oxidizing conditions. In the case of such disturbance of the system, a more or less complete fractionation between the daughter and parent nuclides takes place (Fig. 23). After the disturbance has ended, as for example after sedimentation and new closure of the system, a radioactive equilibrium is gradually attained again. The rate with which a nuclide reestablishes its equilibrium state depends on its half-life. Thus it takes at least $5 \times t_{1/2,D}$, until the daughter nuclide is practically in equilibrium with the parent nuclide. This time-dependent process allows to date the moment of disturbance. Well suited for dating are daughter/parent nuclide pairs characterized by $t_{1/2,N} \gg t_{1/2,D}$, whereby the half-life $t_{1/2,D}$ has to be similar to the period to be dated.

Requirements for dating are:

1. The duration of the disturbance of the system (fractionation of the daughter/parent nuclide pair) is short in respect to the half-life of the daughter.
2. The initial abundance of the daughter at closure is, unless known, negligibly small.
3. After its disturbance the radioactive system remains closed, thus all changes in respect to the abundance of parent and daughter nuclides are exclusively the result of radioactive decay.

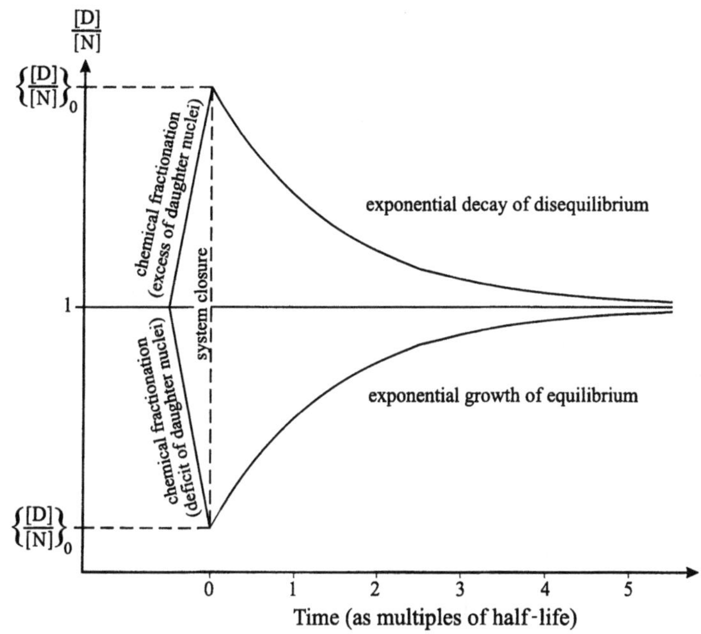

Fig. 23. A radiometric system in radioactive equilibrium possesses an activity ratio [D]/[N] = 1 between the radioactive mother nuclide [N] and the radiogenic daughter nuclide [D]. The equilibrium can be disturbed by chemical, physical and biological fractionation processes, resulting in addition or loss of N or D. In such opened systems, [D]/[N] increases or decreases, i.e., at the time ($t = 0$), when the system is closed again, there exists either a daughter excess or deficit relative to the mother, and hence $([D]/[N])_0 \neq 1$. After termination of the disturbance, i.e., closure of the system, radioactive equilibrium is gradually re-established. The closure time of the system can be dated from the time-dependent re-equilibration, provided $([D]/[N])_0$ is known

Provided these presuppositions are fulfilled, the total activity of the daughter [D] of the system may consist of three components

$$[D] = [D_e] + [D_d] + [D_c] \qquad (23)$$

where $[D_e]$ is the activity of the excess, decaying fraction, already incorporated in the system at the beginning and not supported by the parent, $[D_d]$ is the activity of the deficient, growing daughter fraction supported by the parent, and $[D_c]$ is the activity of the inherited contaminating fraction which is already and stays in equilibrium (Fig. 24). It is essential to consider these individual components separately. Because the components are associated with different mineral phases they can be split up by means of mineralogical and chemical separation.

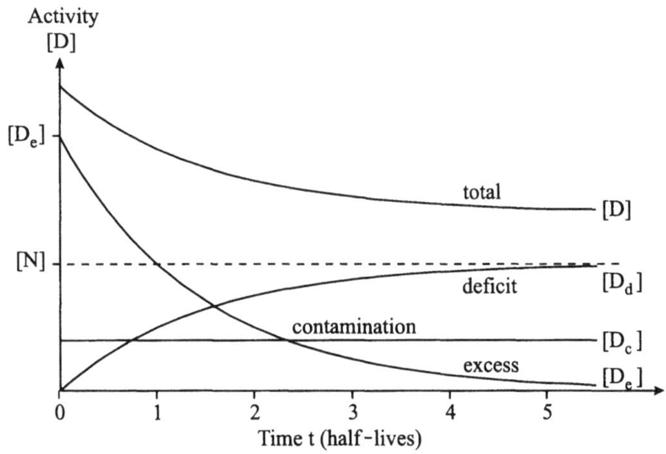

Fig. 24. The activity of the daughter nuclide [D] in a radiometric system prior to its equilibrium state consists of three components: *excess* fraction $[D_e]$, *deficit* fraction $[D_d]$ and inherited *contamination* fraction $[D_c]$. $[D_e]$ and $[D_d]$ decrease or increase, respectively, until radioactive equilibrium is attained at the activity level of the parent nuclide [N]; $[D_c]$ is in equilibrium from the very beginning and remains constant

The daughter activity [D] changes as a function of the age t

$$[D] = [D_e(0)] \times e^{-\lambda t} + [N](1-e^{-\lambda t}) + [D_c] \tag{24}$$

where λ is the decay constant of the daughter nuclide. The first term in Eq. (24) expresses the exponentially declining activity of the initial excess component $D_e(0)$ of the daughter. The second term describes the exponential activity increase of the deficient daughter until equality with the parent activity is re-established. Only the first two terms are time-dependent. The third term is a fixed value that has to be subtracted. The activity ratio freed of contamination

$$([D] - [D_c])/[N] = ([D_e(0)]/[N]) \times e^{-\lambda t} + 1-e^{-\lambda t} \tag{25}$$

approaches after some half-lives the equilibrium value (= 1), so that the age t of the disturbance can be determined.

The uranium series dating methods can be divided into two groups, depending on whether the disturbance resulted in an activity quotient $[D]/[N] < 1$ (daughter deficiency) or > 1 (daughter excess). A daughter deficiency arises from a disturbance where either the parent nuclide is incorporated without its daughter, e.g., uranium (^{234}U) without thorium (^{230}Th) in carbonates, or a loss of daughters occurs. After closure an age-dependent build up of the daughter nuclide takes place until the reestablishment of equilibrium with the parent nuclide (Fig. 23, lower curve). Daughter excess results from incorporation of daughter substance in surplus to the equilibrium level during the disturbance, e.g., thorium (^{230}Th) without uranium (^{234}U) in deep-sea sediments. In this case the daughter concentration declines until the equilibrium level is reached (Fig. 23, upper curve).

Uranium series dating utilizes the measurement of the activity ratio [D]/[N] or the abundance ratio D/N of daughter/parent nuclide pairs. The analysis of the actual present activities or concentrations of the nuclides generally does not cause difficulties. However, apart from the measurement of the actual values, the age determination also requires the knowledge of the activities $[D_e(0)]$ and $[D_c]$ or concentrations $D_e(0)$ and D_c, which are rarely measurable directly. Hence, estimations or models have to be assumed. In this concept lies one of the major difficulties of the uranium series methods of dating, which will be discussed in more detail in the following descriptions of the individual dating methods.

Another weak point of uranium series dating is the requirement of a closed radiometric system over the full time span after the disturbance of the equilibrium. The geochemical processes, which cause the disturbance and thus provide the basis for the dating in the first place, often continue to affect the system. In buried bones, for example, streams of groundwater, migrating through the pore volume, can cause a secondary mobilization of the parent and daughter nuclides. This ambiguity will be discussed in more detail below.

Detection Techniques. The daughter/parent nuclide pairs are usually analyzed by means of α-spectrometry or mass spectrometry as activity ratio [D]/[N] or abundance ratio D/N, respectively. Rarely is γ-spectrometry employed. Summary descriptions are found in the collected editions by Ivanovich and Murray (1992) and Chen et al. (1992).

The dating requires the precise measurement of the nuclides ^{238}U, ^{235}U, ^{234}U, ^{232}Th, ^{230}Th, ^{228}Th, ^{231}Pa, ^{228}Ra, ^{226}Ra and ^{210}Pb. Of primary interest are not the abundances but the activities of these nuclides, hence, activity measurements appear to be favorable. Since all these nuclides are α-emitters, *α-spectrometry* is employed as the standard method. The α-spectrometric analysis requires laborious preparation steps. A $^{232}U-^{228}Th$ spike needs to be added to monitor the different yields of uranium and thorium. The measurement of the α-activity is carried out in the surface-barrier detector. The processed sample and the detector are situated in a vacuum chamber to prevent loss of energy caused by air. In the detector the α-particles induce an electronic signal proportional to their energy loss, which after amplification is recorded in a multichannel analyzer. The precision of the measurement depends on the counting rate, which in turn depends on the uranium content, the age and the available amount of sample material, and the detector noise. A precision of 1% requires 10,000 counting events, which may take hours, days or weeks.

Thermo-Ionization Mass Spectrometry (TIMS). TIMS provides a considerably more sensitive and precise detection of the nuclides than α-counting (Edwards et al. 1986/87). Instead of recording their decay events, the content of the nuclides is measured directly. The abundance N of a nuclide is related to its activity (dN/dt) by Eq. (1). The purified sample material is applied to the heating filament of the ionization chamber and thermally ionized. In the solid-state mass spectrometer with high resolution the ions are separated according to their masses (Fig. 25) and counted by the sensitive detector. High precision of <1% is obtainable. Small amounts of ca. 10 ng per nuclide are sufficient. Due to these obvious advantages it can be expected that mass spectrometry will more and more replace α-spectrometric detection in the field of uranium series dating.

Gamma Spectrometry. Gamma spectrometry, too, is occasionally used for activity measurement of the nuclides or one of their short-lived equilibrated daughter products (Reyss et al. 1978). It has the advantage of being non-

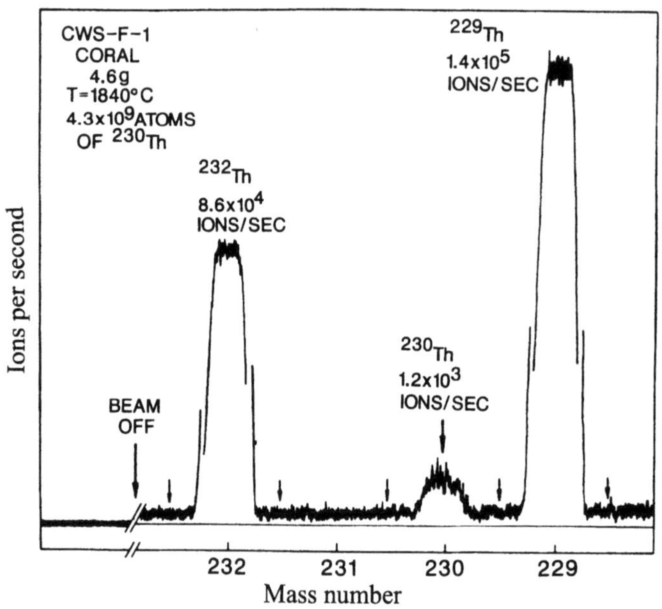

Fig. 25. Thorium mass spectrum of an 845-a-old coral. ^{230}Th and ^{232}Th are the natural isotopes, ^{229}Th was added as spike. The *scale* of ^{229}Th and ^{232}Th is shortened (Reprinted from Edwards et al. 1986/87; copyright 1986, with kind permission from Elsevier Science Ltd., The Boulevard, Langford Lane, Kidlington OX5 IGB, UK)

destructive – an important prerequisite for precious samples. The sample is – without processing – directly exposed to a Ge-detector. However, the counting geometry, attenuation and energy-variable background of the γ-rays cause major difficulties. Comparatively high uranium contents and large amounts of primary material are required. The error estimates can be fairly large (Latham 1997). Neutron activation and particle tracks (Crawford et al. 1985; Kasuya and Ikeda 1991) have also been utilized for the analysis of certain radionuclides and activities.

4.1 Methods

4.1.1 Thorium-230/Uranium-234

The thorium-230/uranium-234 technique is the most often and successfully applied one among the uranium series dating techniques (Barnes et al. 1956). $^{230}Th/^{234}U$ is applicable to a wide spectrum of samples, reaching from marine carbonates over freshwater calcium carbonates, bones, teeth and peat deposits to volcanites. The method is based on the fractionated incorporation of the parent nuclide ^{234}U without its radioactive daughter ^{230}Th during the formation of the material. Because calcium carbonate is precipitated in the ocean (e.g., corals) and groundwater (e.g., cave sinters and travertine), it adopts uranium but not thorium in significant amounts, as a result of the comparatively high solubility of uranium in water. In contrast, fossil bones, teeth and mollusk shells take up uranium from the groundwater essentially after their deposition. Hence, beyond doubt, the requirement of the closed system is not strictly fulfilled in these materials. The model for the dating of peat proceeds from the assumption that the uranium is adsorbed by organic matter from the groundwater during the peat formation, whereas after deposition the overlying layer inhibits any further exchange between peat and groundwater.

Complications inherent in $^{230}Th/^{234}U$ dating have various sources:

1. During its incorporation the parent nuclide ^{234}U is not necessarily in equilibrium with its radioactive predecessor ^{238}U. The $^{234}U/^{238}U$ ratio – above all in terrestrial waters (cf. the $^{234}U/^{238}U$ method) – may considerably deviate from equilibrium. Hence, the $^{234}U/^{238}U$ ratio must be known and taken into consideration for age calculation.
2. The requirement of the model, that the sediment is free of ^{230}Th at the time of deposition, is frequently not fulfilled. Thorium, and consequently its isotope ^{230}Th also, may enter the sediment with the detritus. By means of

the quasi-stable ^{232}Th, this unwanted ^{230}Th contamination, being unsupported by uranium decay, is assessed and subtracted. Minor thorium contamination, indicated by $[^{230}Th]/[^{232}Th] > 20$, does not require correction. A suitable, although time-consuming technique for the recognition and avoidance of initial ^{230}Th contamination is the analysis of chemically or mineralogically separated fractions, in combination with an isochron plot in form of a $^{230}Th/^{232}Th$ vs. $^{234}U/^{232}Th$ diagram (Fig. 26).

3. The system may experience a secondary opening as a result of the geochemical mobility of the uranium. The age determination of bones, teeth, mollusk shells and organic sediments can be strongly affected by uranium exchange with the groundwater. If the mathematical function describing the uranium uptake with time is known, ages can be determined for open systems as well. For bones, for example, models of early, linear and late uranium uptake are assumed.

Well suited are marine calcium carbonates, above all corals and oolites. However, in the case of corals, the absence of aragonite/calcite recrystallization, which could have opened the radiometric system, has to be ascertained. Mollusk shells are less suitable since they are liable to uranium exchange. It must be added that large and dense shells are preferable. As far

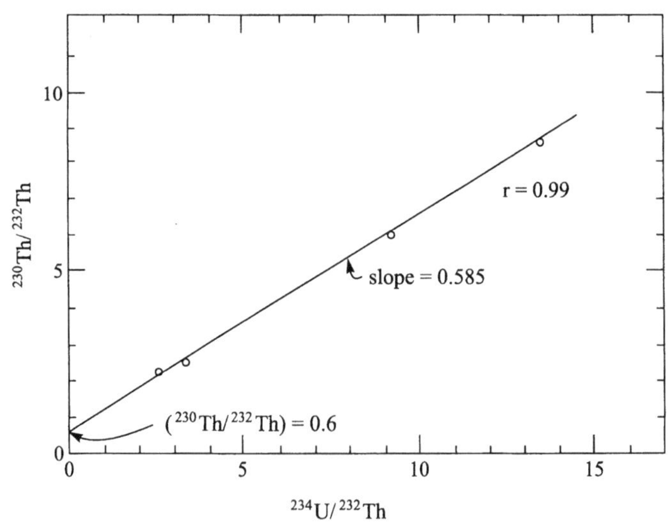

Fig. 26. Isochron diagram of $^{230}Th/^{232}Th$ and $^{234}U/^{232}Th$ analyses of contaminated travertine samples from the Mousterian site Tata, Hungary. The *intercept* with the *$^{230}Th/^{232}Th$ axis* represents the initial ratio $(^{230}Th/^{232}Th)_0$ of the detrital component. The travertine age of 99.4 ± 0.1 ka is calculated from the *slope* of the *straight line*. (Reprinted with permission from Nature; Schwarcz and Skoflek 1982; copyright 1982, Macmillan Magazines Ltd.)

as terrestrial carbonates are concerned, cave sinter and travertine have been proved to be well datable, however, they have frequently inherited detrital thorium. Also calcareous tufa, inorganic marl and limnic calcite carbonate are suitable. Due to the uranium uptake from groundwater, bones, in particular the porous portions, and teeth often cause problems for age determination. Calcitic samples (30 mg each) from the Devils Hole vein in Nevada were mass spectrometrically analyzed for their $^{230}Th/^{234}U$ ages and yielded results ranging between 60 and 560 ka (cf. Fig. 175) with 2σ-precision better than 1 ka for the youngest samples and 50 ka for the oldest ones (Ludwig et al. 1992). The age concordance with $^{234}U/^{238}U$ dates, which has been found in this case, is suggestive of a closed U–Th system in these calcites, which were precipitated while they were continuously below the water table. Evaporites too can be subjected to the $^{230}Th/^{234}U$ method, using samples of different U/Th ratios from the same deposit in combination with isochron plotting (Phillips et al. 1993). Peat often causes difficulties because of detrital thorium and of uranium mobility.

The time range covered by the α-spectrometric $^{230}Th/^{234}U$ technique ranges from several ka to 350 ka, by TIMS detection from ca. 10 a – 550 ka. The precision of the age data declines with decreasing uranium content and increasing age. The precision achieved by mass spectrometry dating is five to ten times better and the required sample size is two to three orders of magnitude lower compared to α-spectrometric analysis.

4.1.2 Uranium Trend

In Quaternary sediments and soils, the radiometric system $^{230}Th/^{234}U$ often has not remained geochemically closed. Attempts have been made to convert this disadvantage into an advantage and to utilize the time-dependent isotopic fractionation within an open system as the basis for the age determination. Such an open-system method, based on the disequilibrium of the $^{238}U–^{234}U–^{230}Th$ system, is the uranium-trend method (Rosholt et al. 1966). In this way, sediments and soils can be dated of which the mineralogical, physical and chemical characteristics are sufficiently variable and which are exposed to a chemical exchange with a steady stream of water (e.g., groundwater). The water usually contains some amount of dissolved uranium whose daughter products ^{234}U and ^{230}Th are fractionally adsorbed in the sediment, that is, to variable degrees in different spots within the sediment depending on the physical properties and chemical composition of the solid matrix material. Adsorption and radioactive decay give rise to a characteristic temporal development (trend) of the isotopic pattern. The data pairs $([^{234}U]-[^{238}U])/[^{238}U]$ and $([^{238}U]-[^{230}Th])/[^{238}U]$ for several samples (of different characteristics) belonging to a sedimentary unit are represented in a U-trend diagram, from which the age is empirically calculated. The complex correlation requires calibration with sediments of known age (Rosholt et al. 1985; Muhs et al. 1989). This method of dating has been applied to alluvial, colluvial as well as marine sediments, to carbonate-bearing coastal terraces, loess and glacial sediments being up to 700 ka old.

4.1.3 Protactinium-231/Uranium-235

Protactinium is geochemically related to thorium, and thus the dating principle of this method (Sackett 1960) is analogous to the one of $^{230}Th/^{234}U$. As calcium carbonates are precipitated from the ocean water or groundwater, uranium without protactinium is incorporated, and therefore ^{231}Pa is deficient with respect to its radioactive predecessor ^{235}U. The age can be obtained from the growth of the $^{231}Pa/^{235}U$ ratio. The difficulties mentioned in connection with the $^{230}Th/^{234}U$ technique do apply here as well, with the exception that the $^{234}U/^{238}U$ activity ratio of the sample is of no consequence for the age determination. The applicable time span reaches from several ka to ~150 ka. This technique is less often utilized than $^{230}Th/^{234}U$, since due to the smaller abundance of the ^{235}U ($^{235}U/^{238}U = 0.0073$) the nuclides of the ^{235}U-decay chain, being in equilibrium, possess a smaller activity compared to ^{238}U ($[^{235}U]/[^{238}U] = 0.046$). Hence, this method of dating requires higher uranium contents ($> 3\ \mu g\ g^{-1}$). This method may be essentially applied to the same kinds of material as the $^{230}Th/^{234}U$ technique. The achievement of coincidental ages by both methods is indicative of correct model assumptions, because protactinium reacts very sensitively to diagenetic opening of the system (Hille 1979). Therefore the technique represents an important completion to $^{230}Th/^{234}U$ dating. A combination of both techniques is $^{231}Pa/^{230}Th$ dating (Rosholt 1957). Thermo-ionization mass spectrometry has also been successfully applied to $^{231}Pa/^{235}U$ dating of Barbados corals and Devils Hole calcites (Edwards et al. 1997). Apart from improved precision by more than ten times compared to α-counting, the TIMS technique extends the datable age range to ~250 ka.

4.1.4 Uranium-234/Uranium-238

The uranium isotope ^{234}U, which is formed by α-decay via ^{234}Th from ^{238}U occupies a location in the crystal lattice which has suffered radiation damage as a result of α-recoil. Hence, it is less strongly bound compared to its predecessor isotope, thus being more liable to leaching during weathering.

Consequently, the $^{234}U/^{238}U$ activity ratio in surface waters is increased, i.e., >1. In surface waters it ranges widely up to 12 (Kronfeld et al. 1994). In recent ocean water it amounts to 1.144 ± 0.004 (Chen et al. 1986) and is somewhat larger in more than 150-ka-old seawater (Bard et al. 1991). As a consequence, inorganic and biogenic precipitation products from such water incorporate uranium enriched in ^{234}U. Provided the initial $^{234}U/^{238}U$ ratio is sufficiently large and known, the decrease in excess ^{234}U enables age determinations up to 1 Ma (Thurber 1962). Marine sample materials, such as foraminifera, corals and manganese nodules, are datable. Attention has to be paid to the prerequisite of the closed system, i.e., that it has not been reopened by aragonite/calcite recrystallization. In contrast to ocean water, the initial $^{234}U/^{238}U$ activity ratios of groundwater are unknown, and this complicates the application to freshwater secondary calcareous deposits. Samples of calcite taken from the Devils Hole vein, Nevada, and having been subjected to mass spectrometric analysis provided $^{234}U/^{238}U$ ages of 390 to 570 ka with a 2σ-precision of ~20 ka. The initial $^{234}U/^{238}U$ ratios (~2.7) underwent only minor change during the extended period of precipitation (Ludwig et al. 1992).

4.1.5
Excess Thorium-230 (Ionium) and Protactinium-231

These two techniques are based on the precipitation of ^{230}Th and ^{231}Pa from the ocean water and their fixation in the deep-sea sediments (Pettersson 1937; Sackett 1960). Due to the long residence (400 ka) of uranium in the ocean water, the oceanic production rates of ^{230}Th and ^{231}Pa are constant. Because both thorium and protactinium are insoluble, the radiogenic ^{230}Th and ^{231}Pa nuclei are quickly adsorbed onto sinking sedimentary particles and thus get incorporated into deep-sea sediments. There, they give rise to an excess component of ^{230}Th and ^{231}Pa, being unsupported by parent nuclides, and decay gradually (Fig. 24). From this age-dependent decline of the ^{230}Th and ^{231}Pa contents the contamination component D_c inherited in the detrital grains needs to be subtracted. The initial ^{230}Th and ^{231}Pa concentrations, required for age calculation, are assumed to remain constant over time, although this is not self-evident, since they do not depend only on the production rate in the water column. Obviously, they vary with the sedimentation rate and the detrital fraction of the sediment. The incorporation of thorium in the sediment can be tested by the ^{232}Th concentration ($^{230}Th/^{232}Th$ method). However, this presumes that both thorium isotopes show geochemically and sedimentologically identical behavior. Since the ^{230}Th and ^{231}Pa surplus decreases with growing age, it decreases downward in the sedimentary column. This enables the determination of the deposition rate as well as the age of deep-sea sediments and manganese nodules. Due to the lower activity, ^{231}Pa is more rarely used than ^{230}Th. The datable periods reach back to 300 ka for ^{230}Th and 150 ka for ^{231}Pa. The data precision is around ±10 to 20%. The accuracy of the age, however, depends essentially on the extent to which the geochemical and sedimentological model assumptions are realized. The combined excess $^{231}Pa/^{230}Th$ technique brings certain advantages (Sackett 1960; Rosholt et al. 1961).

In young basalts from active zones of the mid-oceanic-ridges excess ^{230}Th (5–40%) and ^{231}Pa (100–200%) with respect to the parent nuclides ^{238}U and ^{235}U, respectively, have been observed (Goldstein et al. 1993). The radioactive disequilibrium is generated during partial melting of the magma. Provided that the provenance and the formation process of magmas remain essentially unchanged over several 100 ka and the residence times of magmas until eruption are short (< 5 ka), the $^{230}Th/^{238}U$ and $^{231}Pa/^{235}U$ ratios can be utilized for the dating of the basalt eruption.

4.1.6 Lead-210

The short-lived lead isotope ^{210}Pb ($t_{1/2}$ = 22.3 a) is formed via several very short-lived intermediate members from ^{222}Rn – a noble gas nuclide. The latter is a decay product of radium and escapes from the rock surface into the atmosphere, where it decays to ^{210}Pb. The residence time of ^{210}Pb in the troposphere is days to months, before it is washed out by precipitation and incorporated into sediments on the Earth's surface. If the sedimentary input at a certain place remains constant over several half-lives, the deposition ages and rates can be deduced from the decay of the excess ^{210}Pb – after subtraction of the radium-supported detritus component (Goldberg 1963). This technique is mostly applied to glacier ice, corals as well as estuary, lake and peat deposits formed within the last 150 a (Appleby and Oldfield 1992). The accuracy of the age determination depends critically on the sedimentological and geochemical constraints in respect to the initial activity and the closed system behavior.

Lead-bearing objects, too, are subjected to the excess ^{210}Pb dating method (Keisch et al. 1967). Lead ores most often contain traces of uranium, being in equilibrium with its decay products ^{222}Rn and ^{210}Pb. During the smelting of lead ores, ^{210}Pb together with the lead is extracted from the ore – without the parent nuclides, hence, the extracted lead contains unsupported ^{210}Pb, which decays within several half-lives. The ^{210}Pb activity can be used for the distinction of old lead (> 100 a), which is radioactively dead and younger, active lead. Excess ^{210}Pb in hydrothermal sulfides associated with active mid-oceanic ridges enables the dating of ore formation.

4.1.7 Radium-226

Radium, a geochemically mobile element, is brought with the surface waters into the ocean. There it becomes enriched in the phytoplankton, prominently in diatoms, thus the radium isotope ^{226}Ra ($t_{1/2}$ = 1620 a) is not in equilibrium with its radioactive predecessors (Piggot and Urry 1942). The diatoms sink to the sea-floor where the age-dependent decline of the excess ^{226}Ra permits the determination of the deposition rates of < 10-ka-old sediments. Major uncertainties are caused by the initial content and the geochemical fractionation of the radium. The total activity of ^{226}Ra has to be reduced by the ^{230}Th-supported ^{226}Ra component of the sediment. In most marine sediments ^{230}Th decreases with depth while ^{226}Ra increases toward equilibrium with ^{230}Th resulting in a subsurface maximum of ^{226}Ra. This phenomenon enables the dating of barite which grows in the sediment and incorporates unsupported ^{226}Ra together with the barium from the pore-water. With this approach sedimentation rates (2–3 cm ka^{-1}) of Holocene marine sediments were obtained for equatorial Pacific deposits (Paytan et al. 1996). Magmatic processes also cause fractionation, which is evident from excessive ^{226}Ra/^{230}Th ratios (100–400 %) in recent basalts of the mid-ocean rift zones and enables the dating of recent volcanites (Volpe and Goldstein 1993; Voltaggio et al. 1995).

4.1.8 Thorium-228/Radium-228

This technique is based on two radionuclides that belong to the ^{232}Th decay chain. ^{228}Th is the daughter of ^{228}Ra and has a half-life of 1.9 a, i.e., radioactive equilibrium is already reached after 10–15 a, limiting the dating range to this short time scale. Therefore, only very recent events may be dated with this technique. Samples with barite-rich phases are suited due to the chemical similarity between radium and barium. Actually, the only objects to which the technique seems applicable are barite-containing hydrothermal vent chimneys in active mid-ocean ridges (Reyes et al. 1995). Such barites bear only little ^{232}Th so that essentially all ^{228}Th comes from the excess ^{228}Ra.

4.1.9 Lead-206, -207, -208/Uranium, Thorium

Apart from utilizing the disequilibria within the uranium and thorium decay chains, the accumulation of the stable end-products, the lead isotopes ^{206}Pb, ^{207}Pb and ^{208}Pb, has been exploited for dating Quaternary rocks; this method is also known as the *U–Th–Pb method* (Getty and DePaolo 1995). The techniques are based on ^{238}U, ^{235}U and ^{232}Th, which finally disintegrate through their respective decay chains to the above-mentioned lead isotopes, respectively. They belong to the classic geochronologic methods for dating rocks ranging in age back to the formation of the Earth 4560 Ma ago. The major difficulty, when extending these techniques to ages less than 2 Ma, is the detection of very small amounts of radiogenic lead in presence of the disturbing mass fractionation of the lead isotopes caused by the thermal ionization in the mass spectrometer. This problem has been solved by a special correction procedure. Another uncertainty is introduced by possible disequilibria in the decay chains, but seems to be of minor significance in view of the total accuracy available. The U–Th–Pb method was successfully tested on rhyolite from Alder Creek at Cobb Mountain, California. An isochron age, using albite, sanidine, ilmenite and matrix separates, yielded 1.03 ± 0.10 Ma in agreement with K–Ar data (Getty and DePaolo 1995).

4.2 Practical Aspects

The radioactive systems that are fundamental to the uranium series methods are sensitive towards sedimentological, diagenetic and geochemical processes. Therefore the understanding of these processes and appropriate sampling procedures are prerequisites for reliable and geologically meaningful ages. In general, the aspects subsequently listed should be taken into account, when making a choice of samples:

1. In sediments the nuclide indicative of the age usually consists of two components: an authigenic one of disturbed equilibrium and an unwanted allogenic (detrital) one. The sample should be taken as much as possible in favor of the authigenic component (e.g., no calcareous sinter contaminated by clay).
2. The sample material should be as compact and impermeable to groundwater as possible, otherwise there is the risk of an open system to uranium and thorium exchange (e.g., porous bones must be excluded). Merely in the case of uranium trend dating, open system behavior is favorable.
3. Samples displaying diagenetic alteration or traces of weathering are to be avoided, because they indicate opening of the system (e.g., because of the danger of uranium mobilization, primary aragonitic material must not be recrystallized to calcite).

Information relevant to the uranium abundance and sample size is provided for the individual materials in Sect. 4.3.

The α-spectrometric analysis requires the separation of uranium and thorium from the sample matrix as well as from each other beforehand, which necessitates consumptive steps to achieve complete dissolution of the sample, purification and chromatographic separation on ion-exchange columns (Lally 1992). The preparation procedure varies in detail according to the sample material. For the determination of the yield after separation and measurement so-called spikes are added to the sample solution; these are accurately known amounts of certain artificial radio nuclides, e.g., ^{232}U and ^{228}Th. Insoluble residues, which in the case of carbonates may derive from detrital silicates, are separately treated in order to evaluate their allogenic uranium and thorium contamination. By evaporation or electrolytic precipitation, U and Th have to be plated onto the surface of disks of high-grade steel or platinum, where the layers must be ultra-thin to avoid energy loss of the α-particles. The sample processing for TIMS analysis is similar, but much less amounts (~10 ng of the extracted, purified nuclide) are sufficient. The TIMS technique is superior to α-counting because of its high detection sensitivity and measurement precision of less than 1%.

4.3 Application

Deep-Sea Sediments. The sedimentary deposits on the ocean bottom consist of terrigenous detrital and oceanic biogenic components. As far as their nuclides are concerned, the components display different behavior. The detritus, being geologically old, is in radioactive equilibrium. Its inherited contents of ^{238}U, ^{235}U, ^{230}Th, ^{232}Th and ^{235}Pa are of *allogenic* origin, i.e., they do not derive from the geochemical cycle of the ocean. During their residence in the oceanic water column, the fine-grained suspended particles adsorb additionally an *authigenic*, i.e., derived from ocean water, ^{230}Th and ^{231}Pa component. In contrast to the soluble uranium, with its long residence time of 400 ka in the ocean water, thorium and protactinium are strongly particle-active, thus the uranium daughters ^{230}Th and ^{231}Pa precipitate comparatively fast (~10 a) and sink adsorbed to clay particles to the sea floor. In deep-sea sediments they cause a ^{230}Th and ^{231}Pa *excess* being unsupported by parent nuclides and decay with time to the sediment's equilibrium level. On the other hand, the biogenic carbonates of the deep-sea sediments are characterized by a ^{230}Th and ^{231}Pa *deficit*. When planktonic foraminifera secrete their carbonaceous shells, uranium, but neither thorium nor protactinium is incorporated. After their death the shells settle slowly to the ocean floor and are embedded in the pelagic sediment. In these fossil foraminifera the ^{230}Th and ^{231}Pa contents continuously increase until they reach equilibrium with their uranium parents. Both types of equilibrium disturbances – excess as well as deficit of daughter – may occur side by side in a sediment, giving rise to complex time functions of the daughter/parent ratios (Fig. 24). If identified, both can be used for dating deep-sea deposits.

The non-carbonaceous component of the sediment is dated by the excess ^{230}Th and ^{231}Pa, for which 0.5 g of deep-sea sediment is required (Broecker and van Donk 1970; Ku 1976). The initial activities are estimated on the assumption that the rates of nuclide production and sedimentation have been constant. In sediment cores the rate and age of deposition can be deduced from the decreasing activity with increasing depth (Fig. 27), employing both nuclides and their mutual $^{230}Th/^{231}Pa$ ratio. ^{230}Th and ^{231}Pa ages of up to 300 and 150 ka, respectively, have been obtained. Such data played an important role in the temporal calibration of the oxygen isotope curve derived from foraminifera of the same sediment core. A constant sedimentation rate, as it appears for example in Fig. 27, is disturbed by erosion and redeposition of the deep-sea sediments. If such disturbance ap-

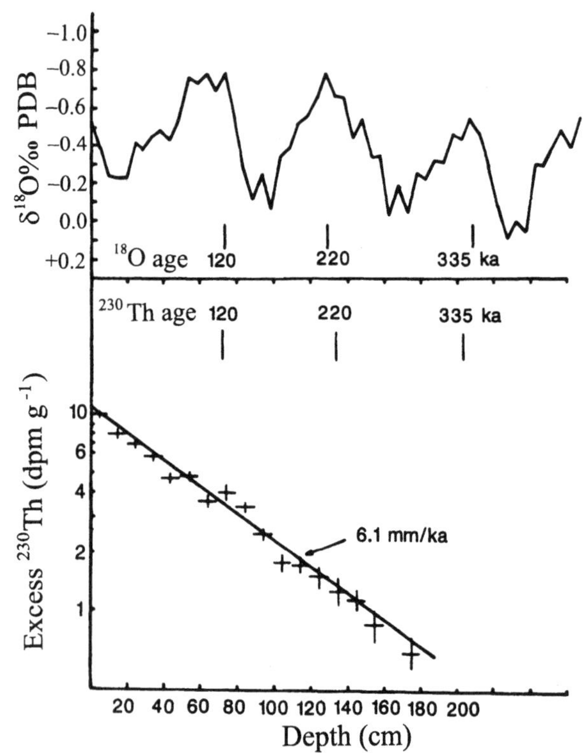

Fig. 27. Comparison of $\delta^{18}O$ variation and ^{230}Th activity (disintegrations min^{-1} g^{-1}) along a deep-sea sediment core from the Indian Ocean. The *^{230}Th depth profile* shows a constant sedimentation rate of 6.1 mm ka^{-1} for the last 300 ka. (After Mangini 1986)

pears simultaneously in several deep-sea cores, it may furnish information on climatically controlled changes of the deep oceanic circulation. Both excess techniques are also applied to the determination of the growth rates of manganese nodules (Ku and Broecker 1967). Their excess ^{230}Th and ^{231}Pa contents provide a means to deduce the growth rates ranging from 0.001 to 1 mm ka^{-1} (cf. Sect. 5.3.3). Eisenhauer et al. (1992) demonstrated by high resolution ^{230}Th analysis of Pacific manganese crusts that the growth rates are climatically controlled: during warm periods the crusts grow faster than in cold periods (Fig. 28).

The biogenic, carbonaceous component of deep-sea sediments can be dated by deficit ^{230}Th and ^{231}Pa. The dominantly planktonic foraminifers *Globigerina* retain the oxygen-isotopic composition of the ocean water in their carbonaceous shells and thus reflect the development of the ice volume during the Quaternary (Sect. 10.2.2). Hence, dating of foraminiferal deposits is highly desirable. However, dating of the foraminifers on the

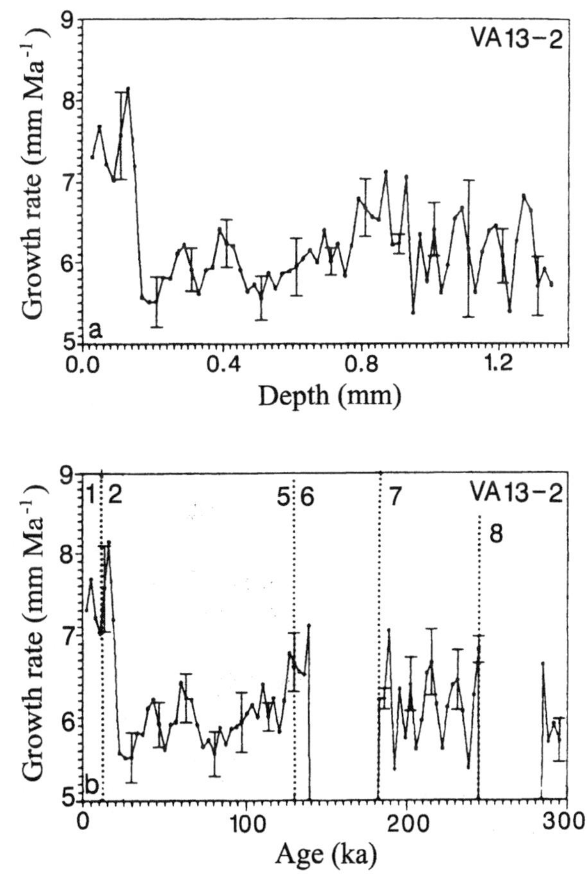

Fig. 28. Growth rate of a manganese crust from the Pacific Ocean as a function of crust thickness (*upper diagram*) and age (*lower diagram*). Highly resolving (0.02 mm) analysis of ^{230}Th shows that the growth is associated with interglacials (in particular $^{18}O/^{16}O$ stages 5 and 1). During the glacial periods ($^{18}O/^{16}O$ stages 8 and 6), the growth comes to a standstill. (Reprinted from Eisenhauer et al. 1992; copyright 1992, with kind permission from Elsevier Science Ltd., The Boulevard, Langford Lane, Kidlington OX5 IGB, UK)

basis of the increasing $^{230}Th/^{234}U$ and $^{231}Pa/^{235}U$ ratios is complicated by low uranium contents ~0.1 μg g^{-1} (Henderson and O'Nions 1995). Algae on the shallow edges and tops of coral reefs produce fine-grained aragonite with uranium contents far in excess of those found in foraminifers, but similar to or even somewhat higher than those of corals. These aragonite grains are transported off the banks and settle, together with other sedimentary components, such as foraminifers, to the sea floor. The relatively high U content and the low detrital Th content render this aragonite material well suited for TIMS $^{230}Th/^{234}U$ dating (Slowey et al. 1996). For cores recovered from crests of ridges, which extend down the slope of the Little Bahama Bank, these investigators reported such aragonite ages, combined with oxygen isotope data of foraminifers. They obtained 120–127 ka for the last interglacial and 189–190 ka for the late stage of the penultimate interglacial.

Corals. Corals secrete a carbonaceous (aragonitic) skeleton from the ocean water that contains uranium, typically 2–3 μg g^{-1}, but no thorium and protactinium. Their initial $^{230}Th/^{238}U$ ratios of 10^{-6} are equivalent to an apparent age of ~8 a (Eisenhauer et al. 1993). After its formation calcite carbonate of corals remains largely excluded from chemical exchange. However, under humid conditions, the aragonite is liable to recrystallization to calcite. On this occasion, the system may become open, limiting the age determination. For dating, $^{230}Th/^{234}U$ is most often employed, whereas $^{234}U/^{238}U$ and $^{231}Pa/^{235}U$ are employed more rarely. Sample sizes of 0.5–3 g are sufficient for TIMS measurements, but α-spectrometric analysis requires at least 2 g. Reliable $^{230}Th/^{234}U$ and $^{231}Pa/^{235}U$ TIMS ages with a 2σ-precision of a few percent have been obtained from corals reaching back to 200 ka. The uranium and thorium mobilization as well as the $^{234}U/^{238}U$ variation in the ocean water represent limiting factors to the applicability to older corals (Bard et al. 1991; Hamelin et al. 1991; Henderson et al. 1993). The high precision of coral dating permits both the correlation of the sea-level variation with the Milankovitch cycles of solar insolation (Sect. 10.2) and the backward extension of the ^{14}C calibration curve (Bard et al. 1990b, 1992; cf. Sect. 5.4.1).

Since the first uranium series dating on corals from Barbados (Mesollela et al. 1969) this West Indian island has become a classical test area with respect to eustatic sea-level changes during the past. Since the recent sea level is representative of a high, interglacial level, one expects to find corals of past sea levels down to ~100 m depth. But the island underwent tectonic uplift synchronously. Hence, fossil seashores are also located above the present sea level. TIMS $^{230}Th/^{234}U$ and $^{231}Pa/^{235}U$ studies on Barbados demonstrate that during the last 200 ka the sea level reached maxima at ~195,

~125, ~105 as well as ~85 ka ago (Bard et al. 1990a; Edwards et al. 1997). The minimum sea level (118 m below present) dates back to 19 ka during the high glacial period. In addition, precise $^{230}Th/^{234}U$ TIMS dating of corals from Sumba Island, Indonesia, revealed the major sea level high stands over the last 350 ka (Bard et al. 1996a). During the last retreat of the ice sheets, 19 – 18 ka ago, the sea level started to rise again, with two sudden pulses of meltwater input 13 and 11 ka ago (Fig. 29), based on TIMS uranium series and AMS ^{14}C ages obtained from coral reefs at tectonically more stable sites in New Guinea (Edwards et al. 1993) and Tahiti (Bard et al. 1996b). The development of the sea level during the Holocene has been investigated on the Abrolhos coral reefs, West Australia (Eisenhauer et al. 1993). Zhu et al. (1993) were able to show by means of highly precise TIMS data of Abrolhos corals that the sea level maximum of the last interglacial lasted from 134 to 116 ka ago.

When comparing $^{230}Th/^{234}U$ and ^{14}C ages on corals of New Guinea, Edwards et al. (1993) observed systematic differences in age as a function of the drill core depths (Fig. 29). In respect to the samples of the greatest

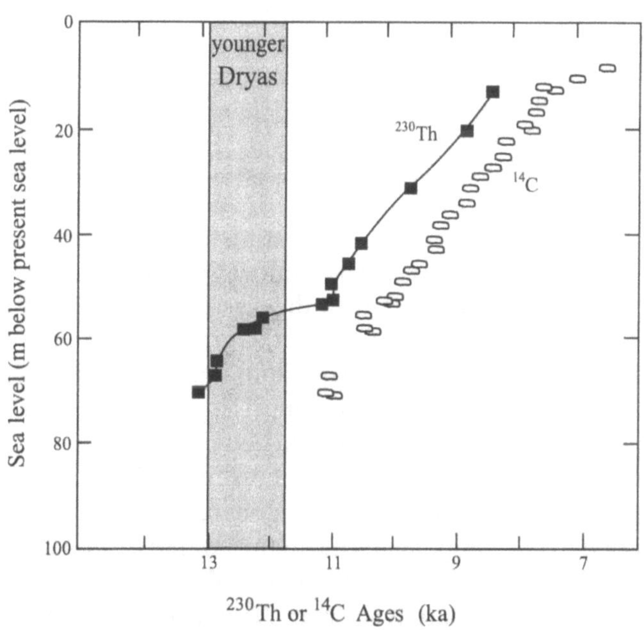

Fig. 29. $^{230}Th/^{234}U$ and conventional ^{14}C ages of corals from New Guinea in relation to drill-core depth plotted as fossil sea level. The width of the symbols represents the 2σ-error. The curve fitting the *^{230}Th* ages traces the post-glacial rise of sea level. (Reprinted with permission from Science; Edwards et al. 1993; copyright 1993, American Association for the Advancement of Science)

depths, the ^{14}C ages are ~2 ka too young. Between 12 and 11 ka (^{230}Th ages) this difference in age reduces to ~1 ka. The decreasing difference in age reflects a rapid and strong decline in the atmospheric ^{14}C content (by 15%) and falls into a period of slowly rising sea level. Edwards et al. (1993) link these phenomena causally: due to the climatic deterioration in the Younger Dryas the melting of the ice caps slowed down; consequently, the rise of sea level progressed more slowly and the salinity of the ocean water increased; hence, the deep oceanic circulation became stronger, thereby bringing water from the depth to the surface and setting free ^{14}C-aged CO_2 into the atmosphere.

Marine Phosphorite. The phosphorites become enriched in uranium during the formation of apatite. Hence the uranium daughters ^{230}Th and ^{231}Pa are deficient and gradually increase until the radioactive equilibrium is established (Burnett and Veeh 1992). Also the $^{234}U/^{238}U$ ratio can be utilized for age determination. The risk of a geochemical opening of the apatite system can be excluded, provided the $^{234}U/^{238}U$, $^{230}Th/^{234}U$ and $^{231}Pa/^{235}U$ ratios yield concordant ages, as O'Brien et al. (1986) were able to demonstrate for Middle and Upper Pleistocene phosphorites of the East Australian continental shelf.

Secondary Carbonates. Secondary carbonates incorporate ~0.1–1 μg g^{-1} uranium, but almost no thorium during crystallization. After crystallization the system remains geochemically closed and the radioactive equilibrium starts to build up, thus the $^{230}Th/^{234}U$ and more rarely (because of the low uranium content) the $^{231}Pa/^{235}U$ dating techniques can be applied. Methodological problems are caused by initial ^{230}Th being incorporated as thorium contamination together with the clastic component of the sediment, and the variable $^{234}U/^{238}U$ ratio of the groundwater. Porous secondary carbonates bear the risk of later uranium mobilization. The age precision of the α-spectrometric data comprises mostly 5–10% and that of the mass spectrometric data is 1% or even better.

During sampling, white and transparent carbonates should be preferred to varieties showing brownish and dirty-gray coloration owing to contamination by a detrital component. Due to the layered carbonate precipitation – which finds expression in a radial growth in stalactites and stalagmites (cf. Fig. 7) – the formation age may show strong gradients even over short distances within the sinter samples. This phenomenon has to be considered during the sampling procedure. Small amounts of samples (<1 g), being sufficient for mass spectrometry, permit sampling in 1-mm intervals, which equal 10–40 a (Dorale et al. 1992), and thus result in an improvement of the temporal resolution. The larger amounts of 10–20 g required for α-

spectrometry, mostly do not permit resolution of this quality. Baker et al. (1993b) succeeded by means of high-precision TIMS analysis in reconstructing the annual rhythm of the growth of Holocene stalactites. The annual growth layers possess a width of 0.05 mm and can be recognized under the fluorescence microscope. From their thickness and fluorescence intensity important paleoclimatic information is probably deducible, in analogy to the annual rings of trees.

When interpreting age data, the spatial and temporal relation between the sinter sample and the interest-arousing find has to be taken into account. For example, a sinter crust attached to a bone surface does not provide an age of the bone itself, but solely furnishes a lower age limit (terminus ante quem) for the deposition of the bone, whereby the innermost crust layer comes closest to the age of the bone. On the other hand, a stalactite fragment, which broke off the cave ridge and got into the sediment, determines only a maximum age (terminus post quem) of the embedding sediment. Complications concerning the correlation between the dated sample and the find whose date is actually wanted arose in the case of the famous hominid skull of the Petralona cave, located on the northern Greek peninsula Chalkidiki. The skull presumably represents a transition form from *Homo erectus* to Neanderthal Man. A controversial debate about the age with various positions between 700 and 160 ka came up. The skull was found cemented to the cave wall by a brown calcareous crust. The skull encrustation was dated 198 ± 40 ka by ESR (Hennig et al. 1981; Sect. 7.3.3), however, the available material was not enough for α-spectrometric $^{230}Th/^{234}U$ dating. On the basis of trace elemental analysis the contemporaneity of the skull encrustation and the upper layer of a multi-layered calcite crust on the cave floor was proved. $^{230}Th/^{234}U$ dating was performed on

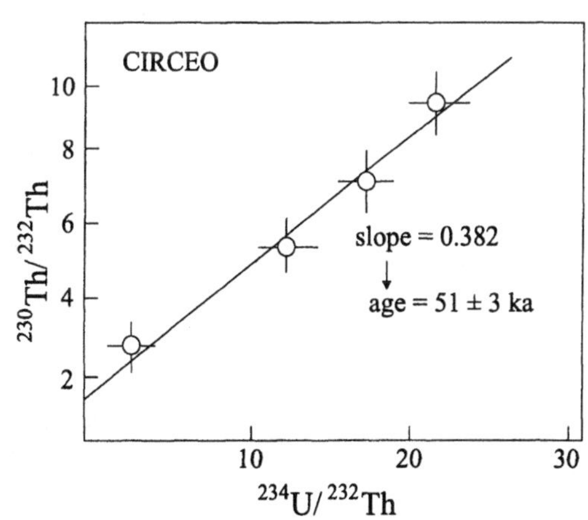

Fig. 30. $^{230}Th/^{232}Th–^{234}U/^{232}Th$ isochron (age: 51 ± 3 a) of calcareous sinter crust attached to the cranium of Neanderthal Man from Grotta Guattari, central Italy. The different fractions were gained through leaching. (Reprinted from Schwarcz 1989; copyright 1989, with kind permission from Elsevier Science Ltd., The Boulevard, Langford Lane, Kidlington OX5 IGB, UK)

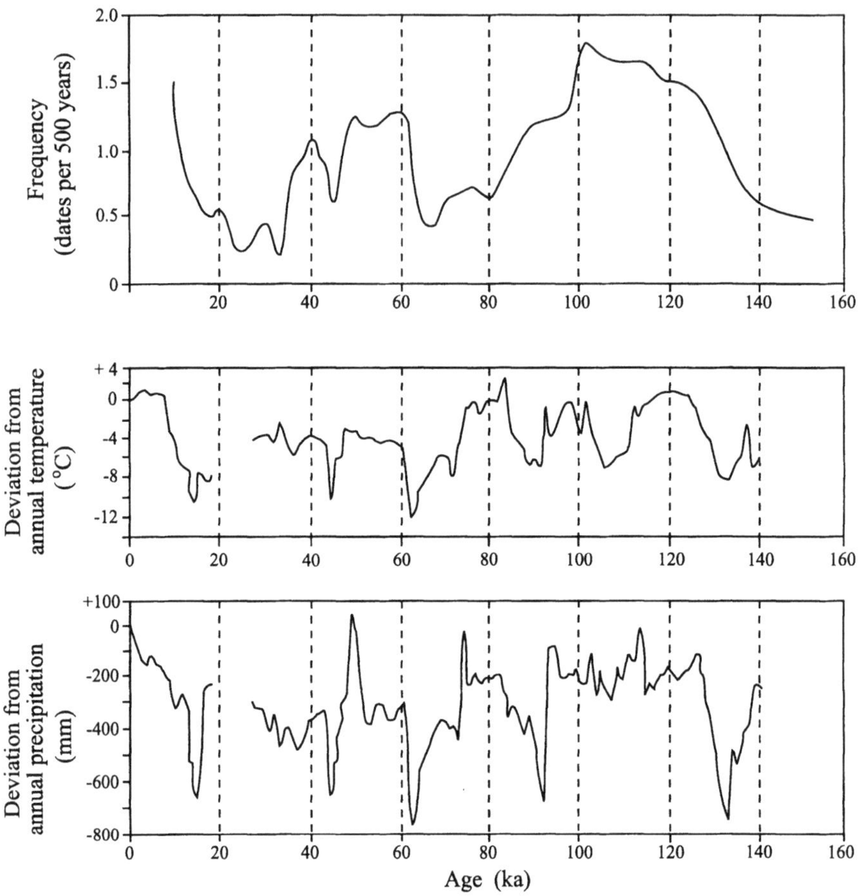

Fig. 31. Frequency of calcareous sinters according to $^{230}Th/^{234}U$ data in west and north Europe during the age range 20–150 ka (after Baker et al. 1993a). The frequency distribution shows clear parallels with the climatic development (temperature, precipitation) during the last glacial cycle

the three layers of this crust (Latham and Schwarcz 1992). The upper layer yielded 160 ± 27 ka, the middle and lower layers > 350 ka each. The $^{230}Th/^{234}U$ age of the upper layer is in good agreement with the ESR age. A minimum age of 160–200 ka for the Petralona skull can be solely deduced from these age determinations, for the time being. In the case of the Petralona skull the $^{230}Th/^{234}U$ age was corrected for detrital ^{230}Th using assumed values. Application of the $^{230}Th/^{232}Th$–$^{234}U/^{232}Th$ isochron technique permits the determination of the initial $^{230}Th/^{232}Th$ ratio, resulting in higher age-precision. Schwarcz (1989) used this technique for the dating of the sinter crust on the cranium of the classic Neanderthal Man at the

Grotta Guattari at Monte Circeo, central Italy. The encrustation on the cranium consisted of a brighter inner and a dark brown outer layer. The outer layer provided 16 ka. Several sub-samples of the inner layer, containing detrital contamination of different U/Th ratios, were obtained through fractionated leaching and analyzed. From the slope of the straight line (isochron) through the data points in the $^{230}Th/^{232}Th$–$^{234}U/^{232}Th$ diagram the age is calculated as 51 ± 3 ka (Fig. 30). This age is in agreement with the ESR age of mammal teeth (Grün and Stringer 1991). To gain the paleoclimatic information inherent in sinter formations Baker et al. (1993a) listed more than 500 $^{230}Th/^{234}U$ data on samples from western and northern Europe covering the age range between 20 and 150 ka. The frequency distribution of the ages (Fig. 31) reflects periods with increased temperatures and humidity.

As examples for $^{230}Th/^{234}U$ dating of travertine, the Thuringian profiles Weimar-Ehringsdorf and Bilzingsleben, with important finds of hominids, are worth mentioning (Brunnacker et al. 1983). These sites are also referred to in connection with ESR dating (Sect. 7.3.3). The $^{230}Th/^{234}U$ data show that the travertine complex of Weimar-Ehringsdorf (Fig. 32) was deposited in two main episodes, namely, the Lower Travertine containing pre-Neanderthal remnants during the penultimate interglacial 200 ka ago and the Upper Travertine during the last interglacial 120 ka ago (Blackwell and Schwarcz 1986). TIMS $^{230}Th/^{234}U$ analysis of the Lower Travertine yielded an isochron age of of 246–279 ka (Frank 1996). The Middle Pleistocene travertine deposits of Bilzingsleben contain early remains of hominids along with Lower Paleolithic tools. There is no general agreement on the classification of the hominid remains as *Homo erectus* or as early *Homo sapiens*. Therefore much importance is attached to the numeric dating of this site. Recent $^{230}Th/^{234}U$ ages of the travertine of Bilzingsleben (Fig. 33) are close or beyond the methodological upper age limit of 350 ka (Schwarcz et al. 1988). This value is considered a minimal age for the travertine formation, and from the $^{234}U/^{238}U$ ratio (1.29 ± 0.10) the maximum age is estimated to 440 ± 130 ka. The ESR ages of ~400 ka also fall within this age span, so that the temperate period manifested in the travertine of Bilzingsleben correlates at least with the $\delta^{18}O$ stage 9, but more likely with stages 11 or even 13.

Limnic Carbonates. Precipitation of calcareous carbonate in lakes can be dated by means of $^{230}Th/^{234}U$ and $^{231}Pa/^{235}U$, provided that the carbonate inherited uranium, but no thorium and protactinium, and that after deposition these elements were not subjected to chemical exchange. The main complication is caused by the common presence of clastic components in limnic sediments (Lin et al. 1996). In order to reduce this detrital ^{230}Th contamination and, if necessary, to correct for it, mechanical and chemical

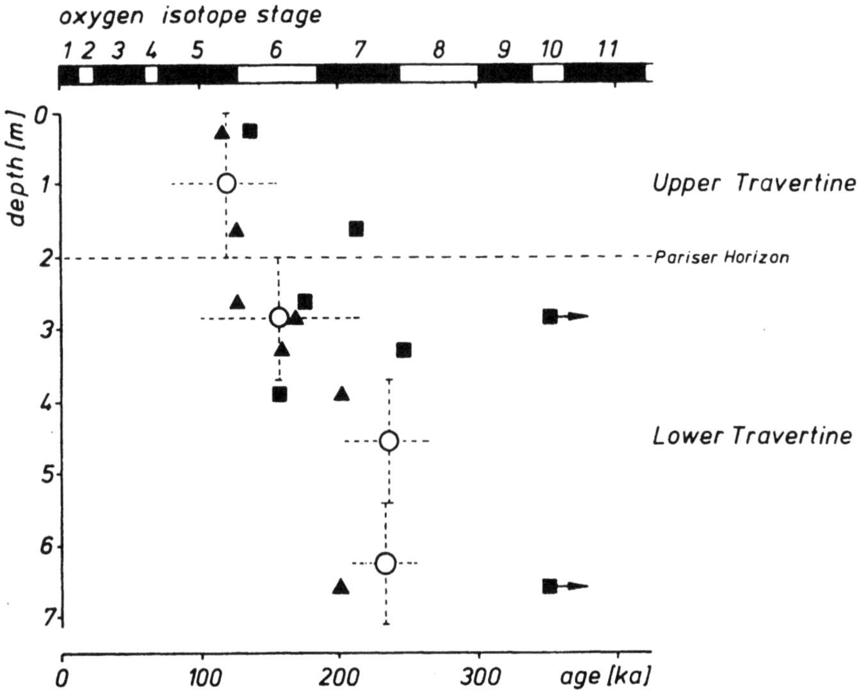

Fig. 32. Uranium series and ESR ages of travertine from Weimar-Ehringsdorf, central Germany (from Grün and Stringer 1991); *triangles* ESR ages; *rectangles* uranium series ages after Brunnacker et al. (1983); *circles* uranium series ages after Blackwell and Schwarcz (1986)

separation procedures as well as serial analyses associated with isochron plotting have been used (Kaufman 1993). An exemplification of this application is the Paleolithic site associated with Acheulian implements in the Pan deposits of Rooidam, South Africa (Szabo and Butzer 1979). Although concordant $^{230}Th/^{234}U$ and $^{231}Pa/^{235}U$ ages have been found, the results are stratigraphically inconsistent, being related to recrystallization of aragonite occurring along with secondary opening of the system. On the other hand, consistent $^{230}Th/^{234}U$ ages were determined on evaporites from playa sediments of the central Asian Qaidam basin (Phillips et al. 1993). There were also successful approaches to dating diagenetic minerals in limnic sediments with $^{230}Th/^{234}U$, as investigations of carbonates, zeolites, and phosphates of the Lake Magadi in the East African Rift have shown (Goetz and Hillaire-Marcel 1992).

Caliche and Calcrete. During formation of these secondary carbonates uranium is incorporated preferably to thorium, allowing for the application of the $^{230}Th/^{234}U$ dating method (Ku et al. 1979). Dating of such pedogenic car-

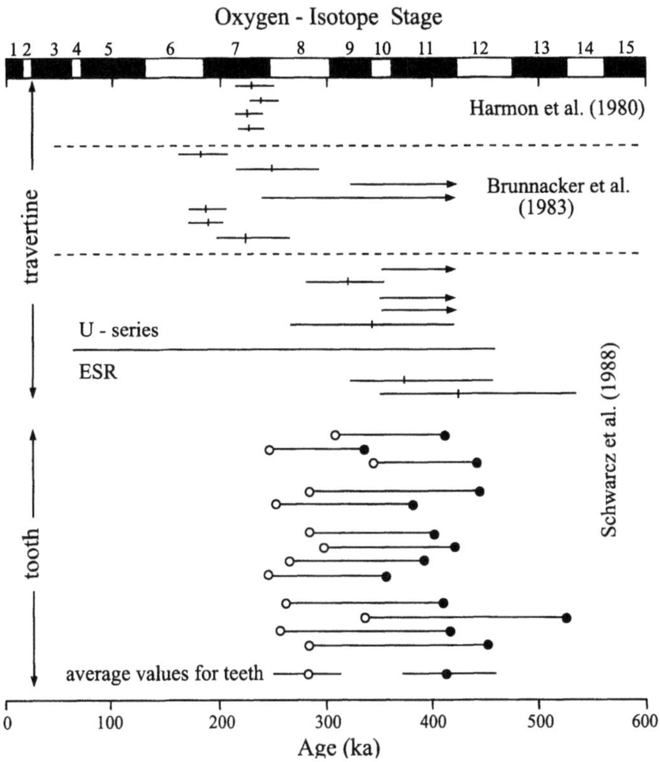

Fig. 33. Uranium series ages of travertine and ESR ages of travertine and tooth enamel from Bilzingsleben, central Germany (after Schwarcz et al. 1988); *open* and *filled circles* ESR ages with early and linear uranium uptake into teeth, respectively

bonates was conducted, for instance, by Knauss (1981) on Californian soil. The high detrital component in the soil necessitates corrections for the initial ^{230}Th. Under humid soil conditions geochemical exchange and continual growth of the carbonates add to the possible complications.

Mollusk Shells. Despite minor uptake of uranium (0.1 µg g^{-1}) by mollusks during the growth of their calcitic or aragonitic shells, the major uranium enrichment takes place post-mortem. This results in a deficit in radioactive daughter products ^{230}Th and ^{231}Pa and hence makes this kind of material suitable for $^{230}Th/^{234}U$ and $^{231}Pa/^{235}U$ dating. However, the requirement for a closed system is markedly disturbed owing to this secondary uranium uptake after burial. Nevertheless, in principle, the dating can be carried out, provided the temporal development of the uranium enrichment can be taken as known. Depending on the chosen model of early, linear or late uranium uptake, systematically different low, middle or high ages are obtainable for the same daughter/parent ratio. If recrystallization of aragonite is observed, uranium and thorium mobilization has to be expected. This demands reservation in respect to the use of uranium series ages for mollusk shells. Uncertainties caused by opening of the system can be largely eliminated, if concordant $^{230}Th/^{234}U$ and $^{231}Pa/^{235}U$ ages are found. It is suggested that independent dating methods, such as ^{14}C, ESR and racemization should be included.

Both marine and limnic mollusk shells were studied using this techniques. As an example, results on fossil shells from the Santa Cruz Terrace, California Pacific coast, are presented. Application of X-ray diffraction showed that the shells contained 97% aragonite, ruling out recrystallization. The uranium content of fossil shells exceeded that of recent shells by a factor 10–20, which is indicative of uranium uptake. Individual shells were dated by both methods, $^{230}Th/^{234}U$ and $^{231}Pa/^{235}U$ (Szabo 1980). Despite the concordance of both ages the data of locality Ano Nuevo Point (22 ± 3 ka and 19 ± 2 ka) cannot be accepted – in contrast to locality Santa Cruz (114 ± 20 ka) – from the geological point of view, whereby the reasons for the age underestimation remain open. For big, non-contaminated mollusk shells of the Holstein interglacial, Sarnthein et al. (1986) determined $^{230}Th/^{234}U$ ages between 350 ka and < 370 ka, which were confirmed by ESR dates.

Bones and Teeth. Similar or even more pronounced, compared to the dating of mollusk shells, are the complications linked to the application of uranium series techniques to fossil bones and teeth. Buried bones take up uranium by complex diagenetic processes (Millard and Hedges 1995). The understanding of the time-function of this uranium uptake is crucial for age evaluation. Especially porous bone components are liable to uranium adsorption from the groundwater. Since the exact temporal development of the uranium incorporation remains unknown, models have to be utilized at the expense of the accuracy of the age determination. In most cases a relatively fast uranium increase towards the state of saturation is preferred (early uptake model). Furthermore, radiogenic daughter products may be released through chemical exchange with the groundwater, causing additional complications for dating. The required sample sizes comprise 10–20 g.

If bones are recovered from dry sediments, providing a geochemically and biologically inactive milieu, reliable $^{230}Th/^{234}U$ ages are more likely to be obtained, as was demonstrated by Rae et al. (1987) for vertebrate bones from the Upper Paleolithic site Little Holy Cave, Wales. The mean $^{230}Th/^{234}U$ age of 16.5 ± 1.4 ka is approximately comparable to the ^{14}C and racemization data obtained from the same bones. As in the case of mollusk shells, the

$^{230}Th/^{234}U$ ages should preferably be verified by $^{231}Pa/^{235}U$ data, as was demonstrated, for example, for the Middle Pleistocene fauna of the Lower Paleolithic site Cornelia, South Africa ($^{230}Th/^{234}U$ age 290 (+100, −40) ka and $^{231}Pa/^{235}U$ age >150 ka) by Szabo (1979). In the eastern Alps, Leitner-Wild and Steffan (1993) found concordant late glacial $^{230}Th/^{234}U$ and $^{231}Pa/^{235}U$ ages on fossil bones of cave bears, also supported by ^{14}C data.

By mass spectrometric $^{230}Th/^{234}U$ dating on mammal teeth from the Middle Pleistocene sites Tabun, Qafzeh and Skhul, Israel, McDermott et al. (1993) were able to confirm the existence of anatomically modern humans (*Homo sapiens sapiens*) for the Levant already before 100 ± 5 ka, contemporary with the Neanderthal Man. This result supports earlier, although less precise TL ages on burned flint as well as ESR data of teeth (Sects. 7.1.3 and 7.3.3, respectively). The analyses were carried out on mechanically separated dentine and enamel subsamples of the same tooth, each comprising 50–100 mg. Important to note is the fact that the ages are in agreement although both subsamples differ strongly in the uranium content, being suggestive of an early uranium uptake soon after the deposition and a closed system. $^{230}Th/^{234}U$ dating on teeth – in combination with ESR, TL, OSL and amino acid racemization – contributed also to the question when the earliest anatomically modern humans appeared in Africa: applied to Middle Stone Age sites at Katanga, Zaire, these methods yielded dates older than 89 (+22, −15) ka (Brooks et al. 1995). For an early form of *Homo sapiens*, unearthed from Pleistocene cave deposits at Jinniushan, Liaoning Province, China, Chen et al. (1994) reported $^{230}Th/^{234}U$ and ESR ages of about 200 ka determined on fossil animal teeth, making it as old as some of the latest Chinese *Homo erectus*. This raises the possibility of the coexistence of both human forms.

Peat. If Pleistocene peat is to be dated beyond the age range covered by ^{14}C, the $^{230}Th/^{234}U$ technique can be employed. Peat adsorbs uranium (up to 20 $\mu g\ g^{-1}$) from groundwater. On the other hand, thorium is insoluble in groundwater, consequently, at the time of the peat formation no ^{230}Th should be present. The actual content of ^{230}Th is then traceable to the decay of ^{234}U and thus a measure of the age of the peat. Detrital components in peat cause thorium contamination and thus an initial ^{230}Th content. The dating of peat requires a closed radiometric system after its formation, i.e., no further uranium uptake from the environment. Overlying sediments cause compaction and thus largely prevent chemical exchange with the groundwater. To minimize the influences of secondary uranium uptake the peat layers should not be thinner than 10–20 cm and the samples should be taken from the central part of the layer. $^{230}Th/^{234}U$ dates, for instance, were obtained for peat deposits of Tenagi Philippon, northern Greece, and Fenit, Ireland, ascribing both of them with 122 (+15, −14) ka and 118 (+9, −8) ka, respectively, to the last interglacial (Heijnis and van der Plicht 1992).

Lake and Estuary Deposits. Recent sediments of estuaries and lakes are important archives of environmental changes, in particular those induced by industrial pollution. This is achieved by various geochemical analyses combined with accurate age determination of the sedimentary layers. However, because of postdepositional mixing by physical and biological agents, it is desirable to have reliable information on the age and sedimentation rates of such layers. Excess ^{210}Pb is able to provide this information over the time range of the last 100–150 years. This has, for instance, been attempted for the sediments in the Sabine-Neches estuary, Texas (Ravichandran et al. 1995), where sedimentation rates of 4–5 mm a^{-1} were estimated. In addition to the ^{210}Pb analysis used in that study, the radioisotopes ^{239}Pu and ^{240}Pu, released from atmospheric testing of nuclear weapons during the 1950s and early 1960s, were employed as time-markers.

Volcanites. Recent volcanites frequently show radioactive disequilibrium (Gill et al. 1992). This is induced by chemical fractionation during partial melting and crystallization, since uranium, thorium and radium have different distribution coefficients between melt and mineral phases. Even for a magma, which is in radioactive equilibrium and possesses a homogeneous $^{230}Th/^{232}Th$ ratio, the solidifying phases are not necessarily in equilibrium. Depending on their U/Th ratio, the equilibrium between ^{238}U and ^{230}Th as well as between ^{235}U and ^{231}Pa is more or less disturbed (Picket and Murell 1997). The re-establishment of the equilibrium as a function of time permits age determination of Quaternary volcanites up to 350 ka. Using this method, which requires the analysis on mineral separates, Condomines and Allegre (1980) succeeded in dating individual lava flows of Stromboli, Italy. For Lower Pleistocene volcanic rocks from the Alban Hills, Italy, Voltaggio et al. (1994) found $^{230}Th/^{234}U$ ages between 33 and 11 ka. Reid et al. (1997) used the $^{230}Th/^{238}U$ ion microprobe technique to date individual zircon grains from rhyolites associated with Long Valley Caldera, California. A $^{226}Ra/^{230}Th$ technique for dating potassic volcanic rocks from Vulcano Island, Italy, which are less than 5 ka old, was described by Voltaggio et al (1995). The $^{226}Ra/^{230}Th$ crystallization age of anorthoclase of recent phonolites from the Mt. Erebus, Antarctica, was determined as 2380 a (Reagan et al. 1992). The U–Th–Pb method has a certain, not yet fully scrutinized potential for dating Pleistocene volcanites.

The excess $^{230}Th/^{238}U$, $^{231}Pa/^{235}U$ and $^{226}Ra/^{230}Th$ ratios in basalts from active rift zones of the mid-oceanic ridges (MORB) were exploited for the dating of eruption events. The radioactive disequilibria are ascribed to partial melting processes in the upper mantle. Goldstein et al. (1993) conducted mass spectrometry on basalts of the East Pacific Ridge and the Juan de Fuca Ridge and obtained concordant $^{230}Th/^{238}U$ and $^{231}Pa/^{235}U$ ages for the last 130 ka (Fig. 34). The required sample sizes comprised several grams. MORB-glasses were studied by Volpe and Goldstein (1993) and yielded concordant $^{226}Ra/^{230}Th$ and $^{230}Th/^{238}U$ ages. This result led the investigators to the conclusion that the $^{226}Ra/^{230}Th$ technique permits a resolution of ± 100 a for episodic volcanic events within an age range of < 10 ka.

Faults. Uranium series age determination has also been applied to faults. During tectonic movement along faults minerals are formed, coating the fault surfaces. One of them, the uranium vanadate carnotite, has been tested for $^{230}Th/^{234}U$ and $^{231}Pa/^{235}U$ dating. Such dates, ranging from 40 to 260 ka, have been reported for various carnotite samples from young faults in the Ashelim and Maale Adumim areas, Israel (Kaufman et al. 1995). The concordance between the two dating techniques was used to check if the carnotites acted as a closed chemical system, which was the case for the majority of them.

Polymetallic Sulfides. In the course of hydrothermal sulfide formation associated with active mid-oceanic ridges, ^{210}Pb is separated from its radioactive predecessor ^{226}Ra. The time-dependent decrease of the excess ^{210}Pb in the sulfide can be used for the dating of the ore formation. In this way it was shown for the Juan de Fuca Ridge that the associated massive ore vents grow concentrically within a few decades, with rates up to several cm a^{-1} (Kim and McMutry 1991). Barite-rich chimneys, which are however not common in mid-ocean ridge sites, can be dated with the $^{228}Th/^{228}Ra$ tech-

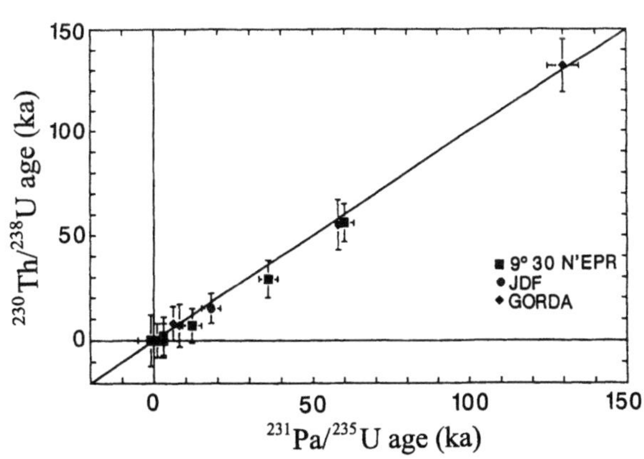

Fig. 34. Concordant $^{230}Th/^{238}U$ and $^{231}Pa/^{235}U$ ages of basalt from the East Pacific Ridge (*EPR*), the Juan de Fuca Ridge (*JDF*) and the *Gorda* Ridge. (Reprinted from Goldstein et al. 1993; copyright 1993, with kind permission from Elsevier Science Ltd., The Boulevard, Langford Lane, Kidlington OX5 IGB, UK)

nique. For hydrothermal vents from the Juan de Fuca Ridge vertical and radial growth rates of chimneys in the age range 1–15 a have been obtained with this technique (Reyes et al. 1995).

Lead Pigments and Alloys. On basis of the radioactive lead isotope ^{210}Pb it is discernible whether lead-bearing pigments and metal alloys and hence, the objects produced with these materials, are younger or older than ca. 100 a (Keisch et al. 1967), a fact which enables the detection of recent forgeries. After metallurgic extraction of lead from its ore, ^{210}Pb is no longer supported by its long-lived predecessor ^{226}Ra. The ^{226}Ra constitutes a contaminant which is not extracted except in minor traces, hence after several half-lives the excess ^{210}Pb content of the lead decays to a new, low equilibrium level. White lead (lead carbonate) represented the most important white pigment until its replacement by titanium white (titanic oxide) during the last century. Therefore modern art forgers intentionally applied white lead. However, recent lead is recognizable by its radioactivity. Keisch (1968) carried out $^{210}Pb/^{226}Ra$ measurements on several questionable paintings – among them "Jesus and his disciples in Emmaus", which had been discovered by Van Meegeren in 1937 and was claimed to be a genuine Vermeer, a famous artist of the 17th century. Keisch (1968) could clarify beyond doubt that the supposed "Vermeer paintings" were forgeries from the 20th century. For these investigations ~30 mg white lead were sampled from insignificant areas of the painting. Similar authenticity tests can also be conducted on lead-bearing pieces of art, including lead bronzes.

5 Cosmogenic Nuclides

The Earth is exposed to a steady flux of cosmic rays which interact with the atmosphere and reach the Earth's surface in a strongly attenuated and modified form. The primary cosmic radiation consists of extraterrestrial, highly energetic particles, mainly hydrogen and helium nuclei. The primary nuclei collide and react with the atoms – essentially nitrogen and oxygen – of the upper atmosphere and are already used up at about 20 km height. These nuclear reactions yield the lower energetic, secondary cosmic radiation that consists of subatomic particles. The charged particles – predominantly protons, neutrons, and muons – of the secondary cosmic rays collide for their part with the atmospheric atoms, whereby they lose their energy and slow down. Some of them reach the rock surface where they are stopped short. As to the kind of nuclear reactions, these particles induce spallation of the target atoms and their chance of being captured by them increases with decreasing velocity (Fig. 35).

Both radioactive and stable nuclides are formed. For dating purposes, the radioactive cosmogenic nuclides ^{3}H, ^{10}Be, ^{14}C, ^{26}Al, ^{32}Si, ^{36}Cl, ^{39}Ar, ^{41}Ca and ^{81}Kr, and the stable cosmogenic nuclides ^{3}He and ^{21}Ne are important. According to the domain of formation, one distinguishes cosmogenic nuclides produced in the air (the *atmospheric* or *meteoric* nuclides) from those produced in situ in the uppermost meters of the rock surface (Lal and Peters 1967). Because of nuclear reactions the cosmic rays become absorbed with increasing path length. The *mean path length l* of particle radiation is defined as the range within which the number of particles is reduced to $1/e$ (= 36.8%), and is expressed as mass thickness [g cm^{-2}]. For the secondary cosmic protons and neutrons, l amounts to about 150 g cm^{-2}. If the density [g cm^{-3}] of the penetrated matter is known, the mean path length can be directly calculated as range [cm], i.e., l is about 55 cm in rocks with 2.7 g cm^{-3} density. Because of the continuous decrease in cosmic rays along their path, the production rate of cosmogenic nuclides in the rock surface is several 10^2 times less frequent than in the atmosphere. The exact knowledge of the production rates of the various cosmogenic nuclides is a pre-

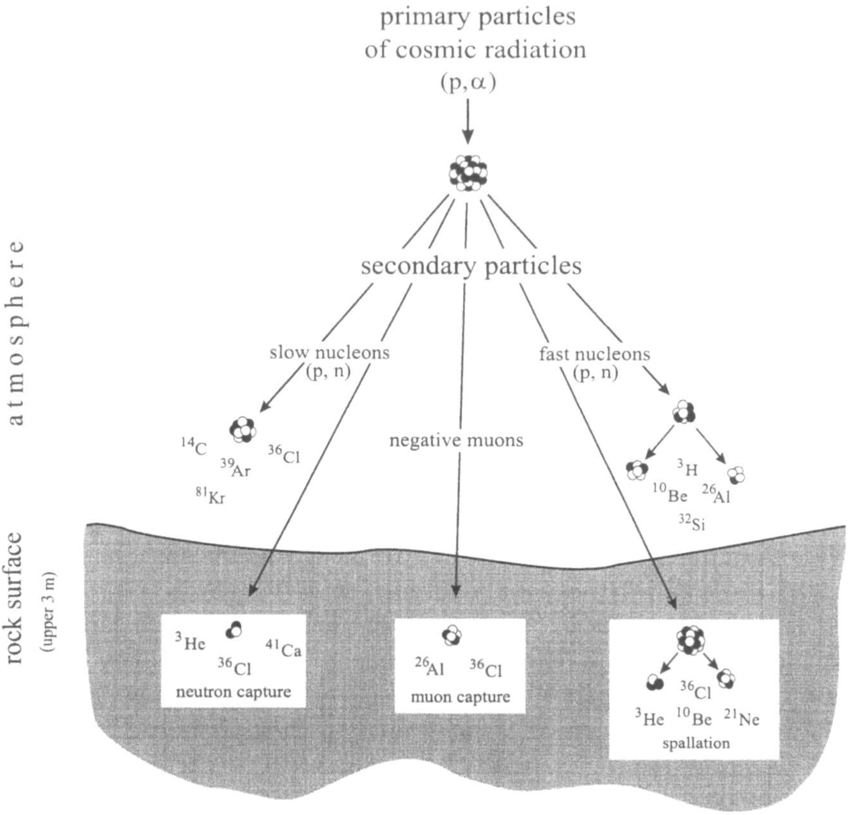

Fig. 35. Production of cosmogenic nuclides by nuclear reactions of the cosmic ray nucleonic component with the atoms of the atmosphere and the rock surface

requisite for their dating application. These rates are derived by experimental measurements as well as theoretical calculations. The results of both approaches agree well with each other (Masarik and Reedy 1995).

The charged particles of the primary cosmic rays are deflected in the Earth's magnetic field (Sect. 9.1). The terrestrial dipole field acts as a magnetic shield for cosmic rays. Only particles directed parallel to the terrestrial magnetic field can enter straight into the atmosphere. This is the case near the geomagnetic poles. At lower geomagnetic latitudes, that is near the equator, the magnetic deflection of the primary cosmic radiation is most effective. The decrease in the cosmic radiation intensity from the poles to the equator is known as *latitude effect*. The latitude effect is reduced by the atmospheric shielding: at 4 km height it amounts to about 25%, but at sea level only to 7% difference in the production rates of cosmogenic nuclides between polar and equatorial positions. The fast mixing within the air diminishes the latitude dependence of atmospheric cosmogenic nuclides further.

In addition to the geomagnetic latitude effect, an *altitude effect* also needs to be taken into account for the cosmogenic in situ production which increases strongly with the altitude. This effect is caused by the fact already mentioned that the intensity of cosmic rays decreases with increasing penetration depth into the atmosphere. For instance, the in situ production rates of ^{10}Be and ^{26}Al increase, related to the sea level rate, by factors of 2.3, 10 and 31, at 1000, 3000 and 5000 m above sea level, respectively. This effect might have potential applications for the reconstruction of the uplift history of mountain ranges (Lal 1986).

The possible production of cosmogenic nuclides in the terrestrial environment was already conceived in the 1930s (Grosse 1934), but experimental confirmation had to wait for another 10 years before naturally occurring radiocarbon was detected (Libby 1946; Anderson et al. 1947).

Atmospheric Production. The atmospheric cosmogenic nuclides ^{3}H, ^{10}Be, ^{14}C, ^{26}Al, ^{32}Si, ^{36}Cl, ^{39}Ar and ^{81}Kr are produced in nuclear reactions of the cosmic radiation with atoms of the atmosphere, mainly with nitrogen, oxygen and argon. Through exchange cycles, these nuclides may enter into other reservoirs, such as the biosphere, the hydrosphere and the lithosphere. Within the atmosphere, the concentration of a radioactive cosmogenic nuclide attains an equilibrium level N_g between production and loss due to decay and exchange. In the other reservoirs – apart from the lithosphere surface – the situation is quite different: there, cosmogenic nuclides are fed in, but not formed, and of course decay. A stationary equilibrium level N_g appears when the exchange net supply and decay of the cosmogenic nuclide compensate each other. This requires fast and thorough mixing of the reservoir in relation to the half-life. When, without further supply, a subsystem is removed from such a reservoir (for example, by death of an organism or by deposition of a sediment), the nuclide content starts to decrease because of radioactive decay. This strictly time-dependent process makes it possible to date the moment of removal, i.e., the age t [a] of the removed system, if the remaining amount N of the nuclide is known (Fig. 36), according to

$$t = \ln (N_g/N)/\lambda \qquad (26)$$

The given model assumptions are often disturbed in nature. Temporal and spatial variations of the initial concentration N_0 within the reservoir frequently occur and may be caused by changing production and supply rates as well as by incomplete and slow mixing. Although such conditions may impede the dating, the behavior of the cosmogenic nuclides as tracers give,

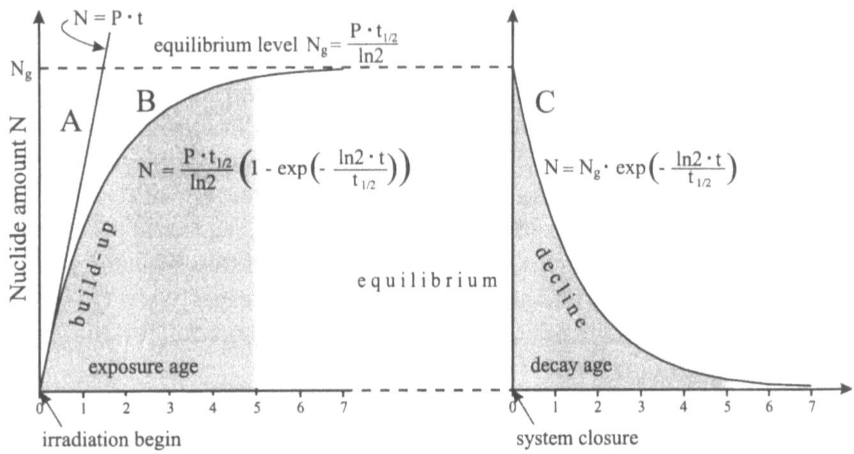

Fig. 36. In the upper meters of the rock surface the amount N of in situ produced cosmogenic nuclides builds up with increasing irradiation time t. The linear growth of the stable nuclides (function A)has to be distinguished from the exponential saturation of the radioactive nuclides (function B). The radioactive nuclides tend to attain an equilibrium level N_g. If the rock is shielded from radiation, N decreases exponentially (function C). Surface exposure ages can be evaluated in the build-up phase and decay ages in the decline phase after system closure

on the other hand, interesting insights into the underlying geophysical phenomena, such as the temporal variation of the cosmic radiation (Geiss et al. 1962) and the Earth's magnetic field, the circulation of the atmosphere and the oceans. When dating groundwater systems with cosmogenic nuclides, one has to be aware of a possible *nucleogenic* contribution to the nuclide under consideration that is induced by neutrons deriving from natural (α, n) reactions (Andrews et al. 1989). Within the last 50 years some of the nuclides (e.g. ^3H, ^{14}C) were injected in large amounts into the atmosphere by nuclear bomb tests and the nuclear industry (*anthropogenic* contribution).

In Situ Production. Cosmic rays produce the cosmogenic nuclides ^3He, ^{10}Be, ^{21}Ne, ^{26}Al, ^{36}Cl and ^{41}Ca in the rock surface, mainly by nuclear reactions with silicon atoms. Because of the mean path length of 150 g cm^{-2} for secondary protons and neutrons, only the upper few meters are involved. Beyond that depth, down to about 1 km, the more penetrating muons can induce cosmogenic nuclides. The production depth-profile $P(d)$ is characterized by the production rate P_0 [atoms g^{-1} a^{-1}] at the surface (i.e., P for depth d = 0 g cm^{-2}) and the mean path length l according to

$$P(d) = P_0 \times e^{-d/l} \tag{27}$$

The nuclide concentration N [atoms g^{-1}] grows gradually with the duration t of exposure (Fig. 36), for stable nuclides (^3He) according to

$$N = P \times t \tag{28}$$

and for radioactive nuclides according to

$$N = (1 - e^{-\lambda \times t}) \times P/\lambda \tag{29}$$

The exposure age of a sample, which had been at first completely shielded from and later was suddenly exposed to cosmic radiation, derives from Eq. (28) for stable cosmogenic nuclides according to

$$t = N/P \tag{30}$$

and from Eq. (29) for radioactive cosmogenic nuclides according to

$$t = -\ln(1 - \lambda \times N/P)/\lambda \tag{31}$$

The amount of radioactive nuclides gradually approaches the equilibrium level N_g

$$N_g = P/\lambda \tag{32}$$

that is practically after about five half-lives reached. If the surface is subjected simultaneously to denudation, the level N_g is diminished, depending on the denudation rate v, according to

$$N_g = P/(\lambda + v/l) \tag{33}$$

In this equation, the denudation apparently increases the decay rate of the nuclide by the amount of v/l. If not taken into account, the denudation lowers the exposure age, so that it yields a lower limit for the exposure time. The denudation rate can be derived from the in situ cosmogenic nuclides according to

$$v = (P/N_g - \lambda) \times l. \tag{34}$$

Therefore, the in situ produced cosmogenic nuclides potentially provide quantitative information not only about landform ages, but also about rates of landscape evolution (Bierman 1994; Cerling and Craig 1994). The rates of terrestrial deposition may also be assessed from their radioactive cosmogenic nuclide record (Lal 1991). Temporal covering of a rock surface by snow, ice, volcanic ash or other materials reduces the production rate and eventually requires correction, corresponding to the thickness of the covering layer.

If a sample, which has already attained its level N_g, is suddenly shielded from cosmic radiation, so that the production of the nuclide is suspended (Fig. 36), the nuclide content N declines according to

$$N = N_g \times e^{-\lambda \times t} \quad (35)$$

and the beginning of shielding can be dated analogously to Eq. (26).

The chronometric application of an in situ cosmogenic nuclide may be impeded by contamination with its own kind, but from different sources:

1. *Atmospheric cosmogenic* production, which is more abundant; the contaminating contribution comes to the ground with the precipitation (e.g., ^{10}Be);
2. *Radiogenic* production (e.g., ^{3}He);
3. *Nucleogenic* production by neutron-induced processes whereby the neutrons derive from natural (α, n) reactions (e.g., ^{3}He);
4. *Anthropogenic* production (e.g., ^{3}H).
5. A *primordial* component (only for stable nuclides), present since the formation of the Earth (e.g., ^{3}He).

Accelerator Mass Spectrometry (AMS). Most cosmogenic nuclides – in particular those produced in situ – occur in extremely low concentrations that require very sensitive detection techniques. Until the 1970s, the only available technique was β-counting. With radioactivity $[N]$ one determines the number of decays dN in the time interval dt, i.e., $[N] = dN/dt$, rather than the amount N of the radioactive nuclide directly. From the relation $[N] = \lambda \times N$ it becomes clear that the activity determination becomes inefficient for small λ, that is, for long-lived nuclides. For example, if a ^{14}C sample is counted for half a week, only one millionth of the ^{14}C atoms actually present is allowed for! Therefore, it was great progress when accelerator mass spectrometry was introduced (Benett et al. 1977; Muller 1977) by which N is directly measured. Quantities as low as 10^5 atoms can be detected, even if the main isotope is 10^{15} times more abundant in the sample. The AMS technique allows the analysis of minute samples, for instance, the ^{14}C measurement of 0.1 mg of extracted carbon, that is, 10^{-4} of the amount required for β-counting (Tuniz 1996). In addition, with the ultrasensitive AMS technique the dating range of the cosmogenic nuclide clocks should be greater. In principle, ^{14}C dating back to 90-ka-old samples – instead of 50 ka with β-counting – should become possible, but in practice this is unrealistic due to omnipresent levels of contamination with modern carbon. The analytical precision, attainable with AMS, is around 0.3% for ^{14}C and 1 to 3% for ^{10}Be, ^{26}Al and ^{36}Cl (Wölfli 1987). The instrumental revolution by AMS nevertheless bears a great, not yet fully grasped potential for geology and archaeology.

For the AMS technique, particle accelerators are utilized as large mass spectrometers (Fig. 37). For atoms to be accelerated they need to be electrically charged. The ionization is achieved in the ion source. The negatively charged ions are initially accelerated, and focused with magnetic lenses.

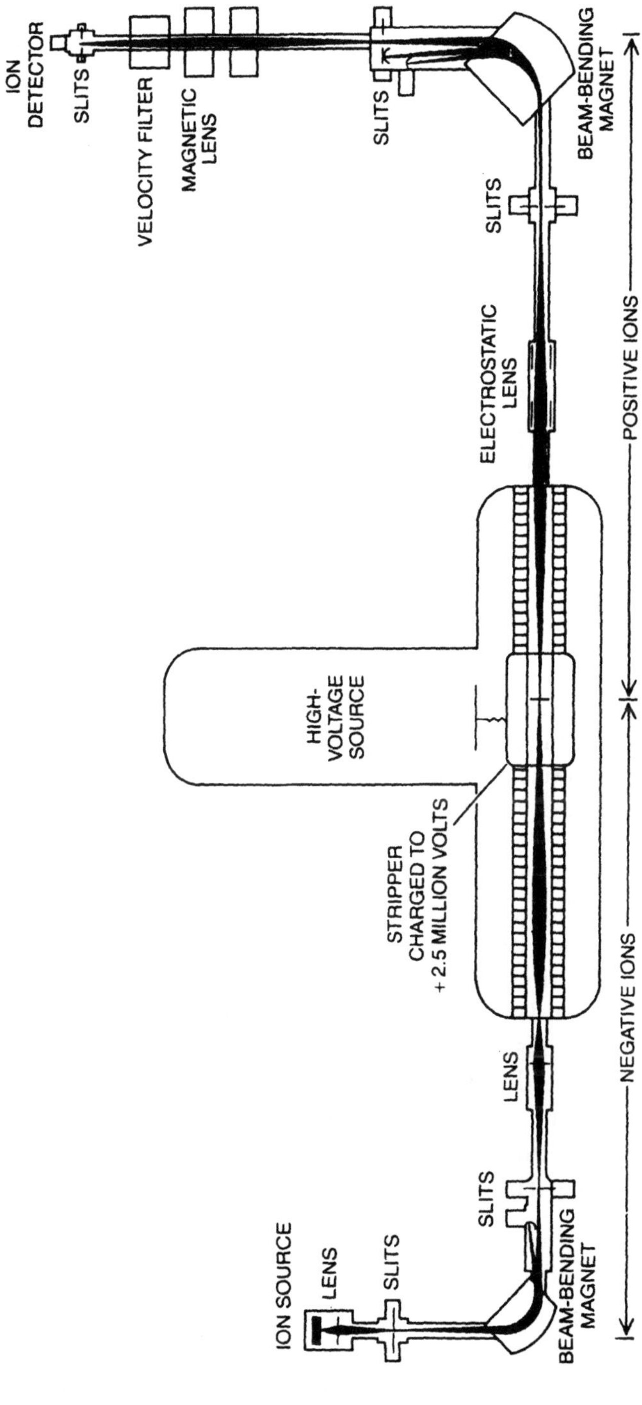

Fig. 37. Principal components of an accelerator mass spectrometer. (After Hedges and Gowlett 1986)

In an electrostatic field they are pulled by the positive pole and thus accelerated to energies of ca. 10 MeV. When arriving at the pole, the ions pass a stripper, which removes electrons from them converting them into positive ions which now become pushed away and further accelerated by the positive pole. The stripper also has the advantage of breaking up contaminating molecules, contained in the particle beam, into their atomic constituents. The highly energetic ion beam passes a magnetic field in which the ions are separated according to their masses. Lighter ions are more deflected than the heavier ones. By placing slits in the widened beam, ions of defined mass can be singled out and quantitatively recorded in the detector. The ionization chamber or surface barrier detectors are able to discriminate between ions of same mass and velocity, but of different atomic number, for example, between ^{14}C and ^{14}N. The AMS analytical detection limit of a nuclide is often higher than its level of contamination, even after extensive extraction and purification procedures, so that the analytical potential of the AMS technique is not fully employed.

5.1 Tritium (Hydrogen-3)

Natural tritium – radioactive hydrogen-3 with a half-life of 12.43 a – is produced in the atmosphere by cosmic rays. A strong anthropogenic contribution to tritium in nature originates from the nuclear bomb explosions between 1954 and 1962. Additionally, technogenic tritium is released from the nuclear industry. The atmospheric tritium combines with oxygen to give water and comes into the hydrologic cycle with the precipitation. When cut off from further supply the radioactive decay of ^{3}H enables the dating of water and ice. However, the tritium system is largely open, so that model assumptions are needed. The short half-life allows age determination up to about 60 a. The first attempts of ^{3}H dating go back to the 1950s (Kaufman and Libby 1954). Because of the anthropogenic contamination, cosmogenic ^{3}H has hardly any more significance for dating. On the other hand, the bomb tritium is used as a tracer signal in the hydrologic cycle. A variant of tritium dating of water is the $^{3}H-^{3}He$ method, in which the accumulation of the tritium daughter helium-3 is exploited as a clock (Jenkins et al. 1972).

5.1.1 Methodological Basis

Hydrogen is one of the two major constituents of the hydrosphere. Natural hydrogen consists – apart from cosmogenic ^{3}H – of the stable isotopes ^{1}H

(99.985%) and ^2H (0.015%). Tritium is radioactive and disintegrates under emission of an electron to ^3He. In nature, ^3H is mainly produced in the stratosphere by cosmic ray-induced spallation of nitrogen and oxygen atoms. The average production rate is 0.25 ^3H atoms cm^{-2} s^{-1}. After oxidation tritium enters the water cycle as ^3H_2O. Because of the latitude effect of cosmic radiation, the $^3H/H$ ratio of the precipitation varies from 25×10^{-18} in high geographic latitudes to 4×10^{-18} near the equator. The tritium content of the precipitation increases from the coast landward (*continental effect*), since the fraction of the tritium-poor ocean water within the atmospheric water decreases over continents. Seasonal variations with greater tritium content in the summer rain have also been observed (Brown 1970). These strong spatial and temporal changes make the chronometric utilization of tritium more difficult.

Model tritium dating assumes a constant initial content of cosmogenic ^3H in the precipitation, allowing the determination of the formation age of water and ice (Eq. 26). Dating according to this simple model is, however, not only disturbed by the varying initial ^3H values, but also by the open behavior of hydrologic systems and by the anthropogenic tritium contamination. At best, it succeeded for samples from the time before the nuclear explosions. The introduction of the bomb tritium into the hydrologic systems during the late 1950s and early 1960s exceeded the cosmogenic ^3H level by several orders of magnitude at times. Since the stopping of the aboveground nuclear tests in 1962, this tritium peak has decayed. It is utilized as a tracer for the path and time scale of hydrologic systems (Münnich 1968). In the mean time the tritium released by the nuclear industry has surpassed the remaining bomb tritium (Hebert 1990).

^3H–^3He Method. The gas component in near-surface waters is steadily exchanged with the atmosphere, so that its $^3He/^4He$ ratio is identical with the atmospheric ratio of 1.384×10^{-6}. If a body of water sinks from the ocean or lake surface to greater depth, it begins to accumulate ^3He formed by the ^3H-disintegration. The age t [a] of the sunk water body can be calculated from the radiogenic ^3He, the ^4He and the ^3H contents according to

$$t = 17.9 \times \ln\left[1 + (^3He - {}^4He \times 1.384 \times 10^{-6})/^3H\right] \quad (36)$$

Such data, recorded as depth profiles, are of interest for the dynamic behavior of oceans, groundwater and lakes during the last 50 years or so. The age determination assumes the existence of a closed system which holds true – if at all – only for short periods (several years). Water mixing and degassing disturb the simple model for dating and complicate the interpretation.

5.1.2
Practical Aspects

The water is sampled in bottles, typically of 1–2 l volume. In order to prevent isotopic exchange with the air, the bottles must be tightly sealed. Usually, the tritium activity is determined with the liquid scintillation technique, which requires electrolytic enrichment of the tritium. A scintillation phosphor is then added to the sample. The scintillation events, induced by the β-rays, are counted with a photomultiplier. The counting time lasts hours or days, depending on the precision required. In particular for smaller samples, techniques with higher detection sensitivity, such as proportional counting, mass spectrometry and AMS, are used.

5.1.3
Application

Water. Cosmogenic ^{3}H plays no role in dating water at present because of heavy contamination with tritium from the nuclear tests (in the late 1950 s and early 1960 s) and from the nuclear industry. The tritium content in water from the pre-contamination time, that is older than 50 a, is already too low for dating through tritium decay. On the other hand, the bomb tritium - often in combination with bomb ^{14}C - enables hydrodynamic studies, such as the flow rate and direction as well as the provenance of young groundwater. After residing in the atmosphere for about 1 a, bomb tritium enters the soil as precipitation (Sonntag 1980). Depending on soil type and depth of the water table, it takes up to several years before the tritium reaches the aquifer. For example, Siegenthaler et al. (1970) were able to estimate a model age of 13 a for water of the Funtenen spring near Meiringen, Switzerland, from the ^{3}H difference between the precipitation input and the spring output (Fig. 38). Actually, ^{3}H analysis is more important for hydrodynamic and hydrogenetic studies of aquifers than for explicit age determination.

For the investigation of vertical movement and mixing of water in oceans and lakes the ^{3}H and the $^{3}H-^{3}He$ methods are used, often combined, and preferably on depth profiles. In Lake Erie, for instance, $^{3}H-^{3}He$ ages of a few days near the surface and a 100 days at depth were obtained (Torgersen et al. 1977). Within such short periods one may safely assume that the $^{3}H-^{3}He$ system stayed closed, so that the last rejuvenation of the water at the surface can be directly dated. Ocean water circulates slowly, resulting in mixing and helium degassing. The oceanographic interpretation of depth profiles requires demanding model calculations (Jouzel 1989).

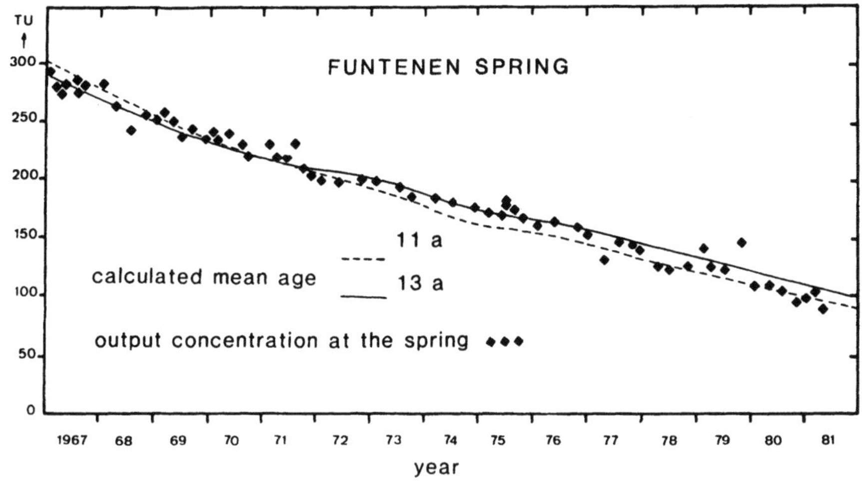

Fig. 38. Tritium content (TU = 10^{-18} $^3H/H$) in water from Funtenen Spring, Switzerland (after Oeschger and Schotterer 1986). The data are in good agreement with the model ages based on the ^3H content of the precipitation in the catchment area

Snow. The tritium content of the snow precipitation near the south pole varies with the seasons. The annual tritium peak falls into the winter, when the tropopause over the polar regions disappears and the stratospheric tritium is introduced into the troposphere (Jouzel 1989). By counting the tritium maxima, the annual snow layers can be accurately dated back to about 1950.

Wine. Once stored and sealed in barrels or bottles, wine is excluded from further exchange with the biosphere and hydrosphere. This event – implicitly that of wine-making – can be derived from the tritium decay. However, the great increase in the tritium level due to nuclear tests disturbs the application of the ^3H clock similarly to groundwater. Thus, the general decline of the bomb tritium – analogous to the bomb radiocarbon – since the mid-1960s is well reflected by the wine of consecutive vintages, as demonstrated by Schönhofer (1989) for Austrian wines. Consequently, tritium may serve as an independent age estimate for the pretended vintage given on the bottle's label.

5.2 Helium-3

Cosmogenic ^3He is produced in situ in rocks at the Earth's surface. In contrast to other cosmogenic nuclides that are used for dating, ^3He is not radioactive. It builds up in the rocks proportionally to the cosmic ray exposure. For known ^3He production, the time (*exposure age*) which a rock sample spent at the surface can be obtained from the accumulated cosmogenic ^3He. Since the ^3He nuclei do not disintegrate and thus do not approach an equilibrium level, the dating range of the ^3He clock stretches widely from 250 a to about 1 Ma. With the ^3He method, surface exposure ages and denudation rates of olivine- or quartz-bearing rocks can be determined.

The first detection of cosmogenic ^3He in terrestrial rocks was accomplished by Kurz (1986a) with Pleistocene basalts from Haleakala volcano on Hawaii. Since then several promising attempts have been undertaken to develop the method for dating young volcanic rocks and moraines. The insufficient knowledge of the ^3He production rate, which requires independently dated samples, and the temporal variation in the ^3He production cause problems.

5.2.1 Methodological Basis

The noble gas helium occurs naturally as ^4He and ^3He. Both isotopes are stable. The much more abundant (almost 100 %) isotope ^4He is radiogenically released as an α-particle, which is a ^4He nucleus, in the uranium and thorium decay series (Sect. 3.2; Chap. 4), whereas ^3He is both cosmogenically and, as a tritium daughter, radiogenically produced. In addition to the cosmogenic and the radiogenic origin, there are two more sources: nucleogenic and primordial ^3He. Neutron-induced nuclear reactions, such as $^6Li(n, \alpha)^3H$ with decay to ^3He, produce nucleogenic ^3He. Apart from these formation processes, both isotopes were always present on Earth as primordial helium. These various helium sources differ in isotopic composition. Natural helium is a mixture of various compounds. Depending on the mixing ratio, its isotopic composition varies. In the Earth's mantle the $^3He/^4He$ ratio ranges from several 10^{-4} to 1.3×10^{-5} (Craig and Lupton 1976). This isotopic composition is still close to that of the primordial helium (3×10^{-4}, Geiss et al. 1972), because the Earth's mantle is low in uranium and thorium and thus low in radiogenic helium dilution. In contrast, the Earth's crust is enriched in uranium and thorium, so that the large radiogenic ^4He component lowers the $^3He/^4He$ ratio of the crust to about 2×10^{-8} (Mamyrin and Tolstikhin 1984). As a noble gas, helium is not chemically bound. It diffuses easily from the crust into the atmosphere. The atmospheric $^3He/^4He$ ratio amounts to 1.384×10^{-6}.

The most important process of the cosmogenic in situ production of ^3He at shallow depths (< 2 m) beneath the rock surface is spallation of heavy nuclei, such as O, Si, Mg and Fe. Thermal neutron capture of ^6Li according to the nuclear reaction $^6Li(n, \alpha)^3H$ with consequent decay of ^3H to ^3He may also contribute, particularly in the presence of lithium. The $^3He/^4He$ ratios caused by these cosmogenic processes are close to 0.1, i.e., higher by many orders of magnitude than those of mantle, atmosphere and crust.

The ^3He production rate was derived on a radiocarbon-dated lava flow from Hualalai and Mauna Loa volcanoes in Hawaii (Kurz 1986b; Kurz et al. 1990). The well preserved flow structure on the rock surface indicated the complete absence of erosion since the solidification of the lava. The cosmogenic production rate $^3He_c(0)$ at sea level and 37° latitude turned out to be 125 ± 30 ^3He atoms g^{-1} a^{-1} and, therewith, belongs to the highest among the cosmogenic nuclides. With increasing depth below surface the production rate declines – as expected – exponentially. Following Eq. (27), the effects of altitude and depth of rock are taken into account, so that the corrected production rate $^3He_c(d)$ becomes

$$^3He_c(d) = {^3He_c}(0) \times e^{-(d-1030)/l} \qquad (37)$$

whereby d [g cm^{-2}] is the path length traversed by cosmic rays (1030 g cm^{-2} at sea level, < 1030 g cm^{-2} above sea level, > 1030 g cm^{-2} for soil cover at sea level). An unsolved problem is the temporal variation of the ^3He production that stays probably within ± 30 %.

Basalts often bear an inherited helium component with a certain isotopic ratio $(^3He/^4He)_i$ in addition to the cosmogenic component. Inherited helium is of magmatic origin and, in olivine, is held primarily in fluid inclusions. During mechanical crushing under vacuum only inherited helium is set free, but during melting, a mixture of inherited and cosmogenic helium is released. The helium freed in both steps is analyzed. Because the amount of cosmogenic ^4He is insignificant with respect to inherited ^4He_i, the inherited ^3He_i can be calculated according to Kurz (1986b) from the ^4He_m amount released by the melting

$$^3He_i = {^4He_m} \times (^3He/^4He)_i \qquad (38)$$

For the determination of the cosmogenic ^3He_c, the inherited ^3He_i needs to be subtracted from the total ^3He_m

$$^3He_c = {^3He_m} - {^4He_m} \times (^3He/^4He)_i \qquad (39)$$

As for the other noble gases, the isotopic composition of helium can be measured with high sensitivity and precision by mass spectrometry.

By accompanying neon isotope analyses it has been shown that cosmogenic ^3He is sufficiently retained in olivine (Anthony and Poths 1992). Also quartz grains of appropriate size (≥ 0.5 mm) seem to retain their ^3He, at least for exposure times less than 100 ka at Antarctic temperatures (Brook et al. 1995), whereas microcrystalline quartz is poor in retention, in particular under hot climatic conditions (Trull et al. 1995). The reliability of ^3He ages depends crucially on the independent assessment of the ^3He production rate. Furthermore, the production rate may vary in the course of time.

5.2.2
Practical Aspects

Both, olivine- and quartz-bearing rocks are suited for ^3He dating. The samples are collected at a very shallow (few cm) and defined depth below the rock surface. For dating it is important that the surface is not eroded, as may be evident from certain criteria, such as well preserved flow structures or chilling crusts on basaltic flows. Olivine and clinopyroxene (0.2–0.4 mm in size) or quartz grains (>0.5 mm) are separated from the ground rocks. The helium is analyzed using gas mass spectrometry. First, the mineral grains are crushed under vacuum in order to release the helium inherited from the magma and to measure – after purification – its isotopic composition $(^3He/^4He)_i$. Then the sample is heated stepwise until melting, with corresponding helium isotopic analysis. The sensitivity of the mass spectrometric ^3He detection is about 10^5 atoms. Possible contamination with nucleogenic ^3He may require correction, in particular for quartz (Trull et al. 1995).

5.2.3
Application

Basalt. On various basaltic lava flows of the Hualalai and Mauna Loa volcanoes in Hawaii, dated independently between 600 a and 14 ka with ^{14}C, Kurz et al. (1990) measured the cosmogenic ^3He component. The samples were collected in the upper 1–4 cm of the lava flows. To make sure that no erosion had taken place since solidification only spots with completely preserved flow structures were selected, on the arid, leeward flanks of the volcanoes. Olivine was separated for the ^3He analysis. The ^3He_c content, corrected to sea level, correlates with the ^{14}C age (Fig. 39). The variation of the data is partly caused by erosion and by temporal covering with soil or volcanic ash. Diffusive helium loss is excluded for olivine. All these processes would lower the ^3He_c content. Despite these uncertainties, the data reveal a temporal variation of the ^3He_c production rate over the last 10 ka, being lower from 7 until 2 ka ago than before and thereafter. This minimum is connected with the increased geomagnetic dipole field at that period.

The reliability of ^3He dating was demonstrated for young basalt flows of the Potrillo volcanic field in southern New Mexico (Anthony and Poths

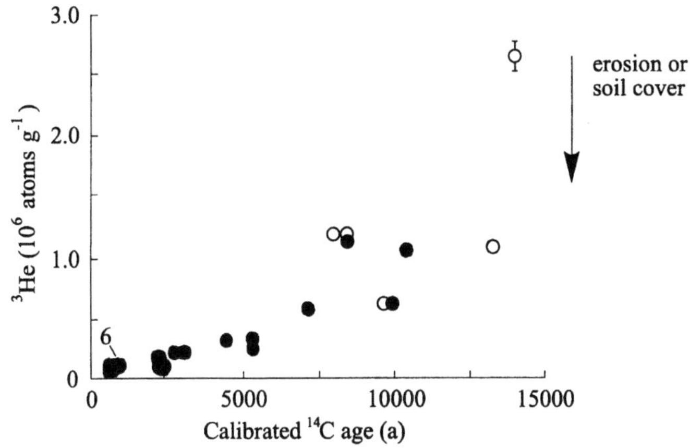

Fig. 39. The cosmogenic ^3He_c content of Hawaiian basalt flows – corrected to sea level – correlates with the ^{14}C ages of the eruptions. Erosion or soil covering reduces the ^3He_c content. Samples for which erosion or soil covering has to be assumed as well as those of questionable age are marked by *open symbols*. (Reprinted from Kurz et al. 1990; copyright 1990, with kind permission from Elsevier Science Ltd., The Boulevard, Langford Lane, Kidlington OX5 IGB, UK)

1992). The specimens were taken from the upper 3 cm below the surface, which still had its primary flow structure, chilling ring flow features and cooling rinds, indicating the absence of erosion. Mineral separates were a mixture of olivine and little clinopyroxene. The inherited helium in the basalts had $(^3He/^4He)_i$ ratios eight times larger than the atmospheric ratio. The ^3He_c production rate of 434 atoms g^{-1} a^{-1} at 39° northern latitude and 1445 m elevation was reduced to the actual sample position (32.2° N, 1284–1387 m elevation and 3–6 cm sampling depth). The various lava flows yielded ^3He ages between 15.5 ± 2 and 85 ± 7 ka. The data showed good reproducibility and agreed with K–Ar as well as geologic age determinations. Another example of surface exposure dating is that on mafic volcanics from Piton de la Fournaise volcano, Réunion, by Staudacher and Allegre (1993). In addition to the ^3He-method, they employed the ^{21}Ne-method also (Sect. 5.5). With both methods concordant ages were obtained for two eruption phases 62 ± 4 and 3.4 ± 1 ka ago, with an intermediate caldera collapse 23.8 ± 2 ka ago.

Moraines. If glacial till contains quartz- or olivine-bearing rocks, ^3He dating may be applied. This has been shown for Quaternary drift sheets in McMurdo Sound, Antarctica, where ^3He was employed in combination with ^{10}Be and ^{26}Al dating (Brook et al. 1995). Samples were taken from the largest clasts of quartz-bearing granites, metamorphites and sandstones, besides olivine-bearing basalts, from the moraine crest. Clasts in this position presumably did not suffer postdepositional mobilization. The specimens were cut at depths of 2 to 6 cm. The ^{3}He content of the quartz was contaminated with a nucleogenic component which required appropriate correction. The mean ^{3}He exposure ages turned out as 252 ± 62 ka for the older drift sheet and 39 ± 8 ka for the younger one, clearly distinguishing these two events. The results are supported by the ^{10}Be and ^{26}Al data.

5.3 Beryllium-10

Radioactive beryllium-10 is a cosmogenic nuclide that is produced in the atmosphere as well as in situ at the rock surface. The atmospheric ^{10}Be is adsorbed on aerosol particles and rains out with them onto the Earth's surface where it is incorporated into sediments. The ^{10}Be decrease due to the radioactive decay is a measure for the sedimentation age in the time range of 1 ka to 10 Ma. The in situ ^{10}Be is used for exposure dating of quartz- or olivine-bearing rocks.

The beginning of the ^{10}Be method goes back several decades (Arnold 1956; Amin et al. 1966). The method made great progress by the introduction of the ultrasensitive ^{10}Be detection with the AMS technique. As a dating nuclide, ^{10}Be was originally applied mainly to deep-sea sediments and manganese nodules, but was limited, however, by temporal variation in its initial content. So far, the unsatisfactory knowledge of the exogenic beryllium system has prevented its widespread application to continental sediments. However, semiquantitative information on the age of soils and on sedimentary processes is possible. The climate-controlled ^{10}Be signal in sediments and glaciers allows inferences regarding the paleoclimate. Inversely, the ^{10}Be record of loess sections and ice cores can serve to date such sequences stratigraphically (*^{10}Be stratigraphy*). The in situ produced ^{10}Be becomes increasingly important for exposure dating and rates of geomorphic processes. For that purpose, ^{10}Be analysis is often combined with a determination of the cosmogenic ^{26}Al (^{10}Be – ^{26}Al method) and ^{36}Cl.

5.3.1 Methodological Basis

The average geochemical abundance of beryllium in the Earth's crust is 2.5 μg g^{-1}. Natural beryllium consists – besides the cosmogenic isotopes ^{7}Be ($t_{1/2}$ = 53.2 d) and ^{10}Be – only of the stable ^{9}Be. ^{10}Be is radioactive and decays to ^{10}B with $t_{1/2}$ = 1.51 Ma under β^{-}-emission. In the atmosphere ^{10}Be is produced mainly by neutron-induced spallation of nitrogen and oxygen.

Adsorbed to minute particles, ^{10}Be resides in the atmosphere for about 1 a before it is carried together with rain and snow to the surface and into sediments. Rainwater contains 10^3-10^4 ^{10}Be atoms g^{-1}. The global deposition rate is $\sim 1.2 \times 10^6$ ^{10}Be atoms cm^{-2} a^{-1}. Two-thirds of the atmospheric ^{10}Be is carried into the ocean, where it is bound to solid particles. After a residence time of ~1 ka in the water column, ^{10}Be is deposited in deep-sea sediments. The remaining one third of the ^{10}Be stays on the continental surface where it is taken up by soils and fine-grained sediments.

In oceanic sediments the ^{10}Be deposition rate varies strongly with the marine environment (Monaghan et al. 1985/86; Lao et al. 1992). Assuming a temporally constant ^{10}Be initial amount N_0 at deposition, the deposition age t of deep-sea sediments can be determined from the amount of ^{10}Be still present N, according to Eq. (2). Since N_0 obviously depends on the sedimentation rate r, it is convenient to plot the decrease in the ratio N/N_0 logarithmically against the depth d of a sediment core

$$\ln(N/N_0) = -(\lambda/r) \times d \qquad (40)$$

The slope λ/r of the straight line represents, inversely, the sedimentation rate r in vertical profiles (Fig. 40). ^{10}Be dating according to this simple model is often problematic, because N_0 varies with time. The ^{10}Be produc-

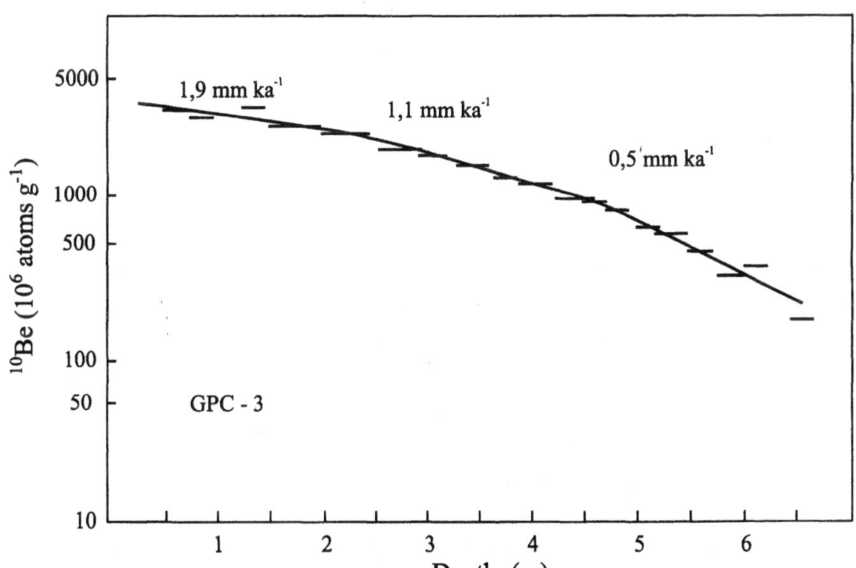

Fig. 40. ^{10}Be depth profile of sediment core GPC-3 from the northern Pacific Ocean (after Mangini 1986). The *slope* yields the sedimentation rate in this plot of logarithmic ^{10}Be content versus the linear depth axis

tion rate depends on the flux of cosmic rays and its modulation by the changing magnetic fields of sun and Earth. Owing to the short atmospheric residence time, any such ^{10}Be variations are quickly passed on to the terrestrial and marine reservoirs (Lao et al. 1992). In addition, the climate affects the ^{10}Be deposition. Furthermore, the initial beryllium content depends on lithologic, sedimentologic and pedologic properties. Another problem is the contamination by detrital ^{10}Be derived from continental denudation. Despite these difficulties there a numerous successful studies on sedimentation rates of deep-sea sediments and growth rates of manganese nodules. For terrestrial deposits the ^{10}Be method is more complex. Their sedimentation rate varies strongly and may even be interrupted by erosion. Detrital ^{10}Be contamination is omnipresent. There are attempts to date soil formation on stable surfaces under the presupposition that the ^{10}Be concentration increases with time due to the steady atmospheric supply and ultimately reaches radioactive equilibrium. To what extent this model is realistic, is difficult to judge. More promising are efforts to utilize the ^{10}Be in sediments and soils in order to study the related exogenic processes.

In the rock surface, cosmic rays produce ^{10}Be by spallation on oxygen and silicon with rates of several orders of magnitude lower than in the atmosphere. Contrary to the situation in which ^{10}Be originates from the atmosphere and is adsorbed to soils and sediments, this in situ produced ^{10}Be is not subjected to the exogenic cycle. It builds up in silicate minerals, in particular in quartz, within the upper few meters below the surface, and thus is well suited for exposure dating.

5.3.2 Practical Aspects

Because of various methodological problems ^{10}Be dating of sediments and soils is by no means an established procedure and thus commonly involves accompanying investigations of the method. For surface exposure dating the method is more reliable, although many unsolved questions remain. When sampling for exposure dating, the rock specimens are taken from a defined depth within the upper ~10 cm below uneroded, flat-lying surfaces so that no correction for cosmic ray exposure geometry is necessary. Because of the low levels of ^{10}Be concentration in minerals and sediments, the ^{10}Be detection by β-counting is completely replaced by AMS analysis. The AMS technique requires sample amounts of <1 g for deep-sea sediments, >10 mg for manganese nodules and loess, and ~10–20 g for quartz and olivine. The ^{10}Be is attached predominantly to the clay fraction in sediments. After sieving the sample is dissolved and the iron contamination is precipitated. Beryllium is extracted by ion exchange and introduced as BeO target into the spectrometer. The AMS detection limit is $\sim 5 \times 10^{-15}$ for the $^{10}Be/^{9}Be$ ratio or $\sim 5 \times 10^{5}$ ^{10}Be-atoms. The time necessary for AMS analysis of ^{10}Be is ~ 0.5 h with a precision around 2–5%. The analytical data are mostly given in ^{10}Be atoms g^{-1}.

5.3.3 Application

Soils. High ^{10}Be contents of up to 10^{9} atoms g^{-1} are found in soils and is explained by the following model. The soil evolution takes a long time during which the rock surface is exposed to chemical weathering. The ^{10}Be, which is brought in from the atmosphere, is preferentially adsorbed by the clay fraction of the soil. The clay fraction increases with progressing soil evolution, as does the ^{10}Be content until it reaches equilibrium due to radioactive decay and soil erosion. This straightforward model should enable the determination of the age and the erosion rate of soils if the local precipitation flux of ^{10}Be atoms is known (Monaghan et al. 1983; Pavich et al. 1986). Although ^{10}Be retention in soils seems to be larger than the ^{10}Be half-life, no reliable data as to age and erosion of soils have been obtained hitherto with the ^{10}Be method. This is mainly due to the open-system-behavior of ^{10}Be in soils, in particular its downward transport in soil profiles by acidic solutions. The complex exogenic system of beryllium is still too little known. Despite this limitation, the ^{10}Be data at least yield qualitative information on soil formation and erosion, as was demonstrated, for instance, in the North American Piedmont, where enormous amounts of soil were set in motion by the destructive farming from about 1700 (Brown 1987). This region is characterized by an increased ^{10}Be index which is the ratio of the number of ^{10}Be atoms leaving the basin relative to the number incident on it. In order to circumvent the problems introduced by the ^{10}Be mobility in soils, Barg et al. (1997) selected the authigenic mineral phases, such as clays and iron oxides, from the B-horizon for ^{10}Be analysis. It is assumed that these minerals form closed systems. The data obtained from several soil profiles in temperate and tropical climate regimes seem to validate the closed system ^{10}Be model.

Limnic and Fluvial Sediments. A wide range of ^{10}Be content from $<10^{7}$ to 10^{9} atoms g^{-1} has been observed in these sediments. This variation is caused by the different provenances of the ^{10}Be. Two ^{10}Be components are carried into sedimentary deposits: the detrital one originates from the soil erosion in the source area, and the atmospheric one comes with the precipitation. The detrital component bears information on the transportation balance, whereas the atmospheric one contains information on the age and the rate of deposition. In practice, however, it is hardly possible to discriminate between the two components (Morris 1991).

Loess. The ^{10}Be record in loess–paleosol sequences is a sensitive climate proxy and may thus be employed indirectly to date loess profiles, as was shown by Shen et al. (1992) in the Luochuan loess plateau, Shaanxi Province, China. A 55-m-high profile, which comprises the Pleistocene including the Matuyama–Brunhes boundary, was intensively sampled. Three g of each loess sample were processed for the AMS detection. Analyses on carbonate nodules from the bottom of the paleosol layers confirmed that carbonate dissolution does not contribute much to the mobility of ^{10}Be. The ^{10}Be concentration varied from 0.5 to 6.5×10^8 atoms g^{-1} over the profile, with low values in the loess and high values in the paleosol. It reflects the different sedimentation rates between the quickly accumulated, glacial loess and the interglacial paleosols which were exposed to the atmospheric ^{10}Be flux for a long time. Consequently, the ^{10}Be curve is climatically controlled. Its signature is very similar to that of the marine $^{18}O/^{16}O$ curve (Sect. 10.2.2). An excellent correlation is obtained after matching the main variations of both records by stretching and compressing the loess depth scale (Fig. 41). This has not only important implications for the global character of the Quaternary climate, it also allows to the transfer of the marine $^{18}O/^{16}O$ chronology to the continental loess profiles, probably with 5–10 ka accuracy. By working on additional loess–paleosol sequences in the Chinese Loess Plateau, Gu et al. (1996) extended the ^{10}Be record back to the last 5 Ma. In the loess profile of Luochuan, Beer et al. (1993) observed a correlation of the ^{10}Be content with the magnetic susceptibility, providing an additional stratigraphic feature.

Deep-Sea Sediments. The ^{10}Be which is carried with the precipitation into the ocean, resides there for ~1 ka – approximately the same time as is required for thorough oceanic mixing – before it arrives, attached to particles, at the sea floor. Besides this authigenic component detrital ^{10}Be from the continental soils may also be present in deep-sea sediments. The dating model requires a constant ^{10}Be flux into the deposits. This presupposition is commonly disturbed by changing lithology (mainly the clay content), sedimentation rates and bottom currents. The authigenic ^{10}Be fraction is the one relevant for dating. It is difficult to discriminate between the authigenic and detrital ^{10}Be components, and a NaOH leaching technique was proposed to separate them (Wang et al. 1996). These considerations are analogous to those already made in the context of the ^{230}Th excess method (Sect. 4.1.5). Both techniques are often jointly applied to deep-sea cores. If

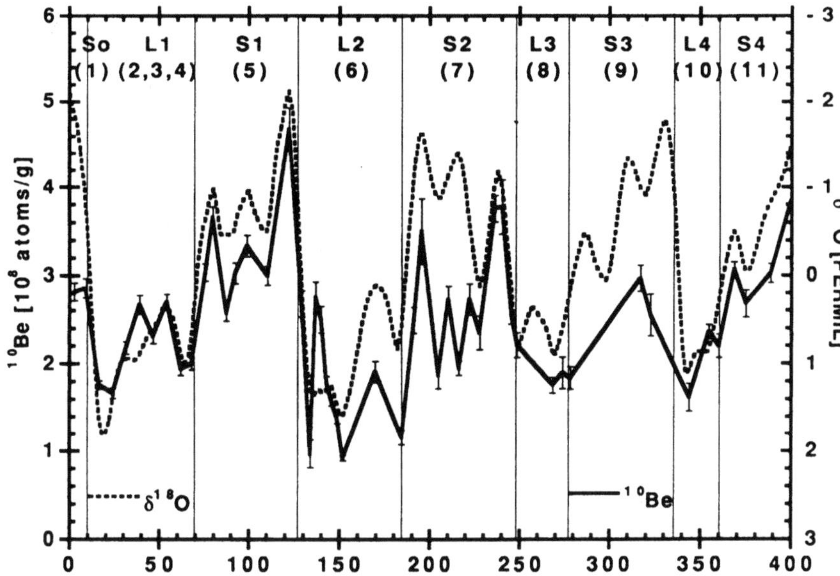

Fig. 41. ^{10}Be content (scale on the *left*) as function of deposition age (bottom scale; in ka) in a loess profile from Luochuan/Shaanxi in China). The ^{10}Be content changes from low values in the loess (L) to high ones in the paleosols (S) and correlates negatively with the marine δ^{18}O curve (scale on the *right*). This is suggestive of a climate-controlled ^{10}Be input. (Reprinted from Shen et al. 1992; copyright 1992, with kind permission from Elsevier Science Ltd., The Boulevard, Langford Lane, Kidlington OX5 IGB, UK)

the logarithmic ^{10}Be content is plotted against the linear depth of the core, the data points form a straight line, whose inverse slope gives the sedimentation rate. A changing slope indicates a changing sedimentation rate, as depicted in Fig. 40 for a North Pacific core (Mangini 1985). The faster sedimentation rate of 1.9 mm ka^{-1} during the past 1.3 Ma reflects the stronger detritus supply to the ocean caused by the glaciation of the northern hemisphere. The manganese nodules (Fig. 42) growing on the ocean bottom incorporate ^{10}Be from the seawater. This enables the determination of the age and the growth rate of the nodules (Mangini 1985). With ultrasensitive ^{10}Be detection by AMS, ages back to 15 Ma may be obtained. In arctic deep-sea cores Eisenhauer et al. (1990, 1995) observed climate-dependent ^{10}Be variations. The ^{10}Be contents in interglacial sediment sections are higher than those in glacial ones. The ^{10}Be signature correlates with the δ^{18}O as well as the ^{230}Th variation. Since arctic deep-sea sediments have only low carbonate contents and therefore are inappropriate for the δ^{18}O stratigraphy, the ^{10}Be variation (*^{10}Be stratigraphy*) is important for the chronology of such deposits.

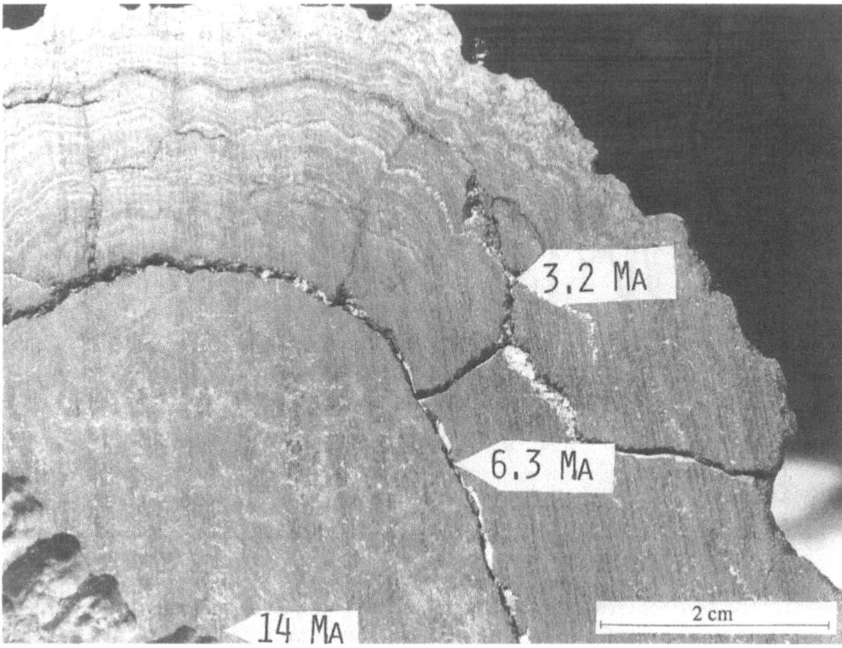

Fig. 42. Manganese nodules in the deep-sea incorporate ^{10}Be from the ocean water. Consequently, the age and growth rate are deducible from the ^{10}Be decay. The figure shows a cross section through a manganese nodule from the South Pacific from 5600 m depth. (After Segl et al. 1984)

Volcanites. The ^{10}Be produced in olivine in situ can be used for exposure dating of volcanic rocks. This was achieved by Shepard et al. (1995) for the Black Rock basaltic lava flow, Lunar Crater volcanic field, Nevada. For ^{10}Be analysis olivine phenocrysts 5–8 cm below the flat-lying original flow structures were sampled. The mean of three samples was 43.8 ± 11.4 ka, which is concordant with the ^{36}Cl whole-rock exposure age of 33.9 ± 6.7 ka, and is one order of magnitude younger than indicated by K–Ar dating. The latter may be erroneously high due to excess argon.

Atmospheric ^{10}Be can also be of interest for volcanites. In active subduction zones geologically young deep-sea sediments, together with the underlying oceanic crust, are transported into the upper mantle and partially melted. These melts may rise again to the Earth's surface and extrude as lava. Because the ^{10}Be content in the oceanic crust is much lower than in deep-sea sediments, the ^{10}Be in the lava can be utilized as a tracer isotope. Information on the mass-balance and the time scale of melting can be derived from the ^{10}Be content, as has been attempted by Sigmarsson et al. (1990) for volcanites of the Chilean Andes. However, Lao et al. (1992) pointed out that this approach is problematic without taking into account the inhomogeneous ^{10}Be precipitation in the Pacific.

Ice Cores. The cores drilled in ice sheets of Greenland and Antarctica cover the whole climate cycle of the last 150 ka. Their record exhibits strong ^{10}Be variation which correlates negatively with the ^{18}O content (Sect. 10.2.3). The ice core ^{10}Be curves correspond to each other, even on a global scale (Beer et al. 1987). During cold, dry periods the ^{10}Be accumulation in the ice is less diluted by snow, resulting in higher ^{10}Be contents. For instance, the ^{10}Be content increased by 40–100% during the "Little Ice Age" (17–18th century), caused by Maunder Minimum of solar activity (Beer et al. 1985). An even stronger increase (200–300%) was observed in the transition zone from the last Glacial to the Holocene ice ~11.5 ka ago (Yiou et al. 1985). In order to eliminate the strong, climatically controlled variations of the ^{10}Be input into the ice that disturb the dating, the ^{10}Be was combined with ^{36}Cl determinations as the *^{10}Be–^{36}Cl method* (effective half-life 370 ka), but without success (Elmore et al. 1987).

Quartz-Bearing Rocks. The cosmogenic ^{10}Be produced in situ can be used for surface exposure dating and erosion studies of various quartz-bearing rocks (Nishiizumi et al. 1986). Quartz forms a geochemically closed system so that the in situ ^{10}Be is not contaminated with atmospheric ^{10}Be. Because of low ^{10}Be contents in quartz 10–30 g are required. Frequently, the ^{10}Be analysis is combined with that of ^{26}Al (*^{10}Be–^{26}Al method*, Sect. 5.6). There is a wide spectrum of interesting geomorphic applications of the ^{10}Be method. One of them is dating the glacier melting by ^{10}Be analysis in glacially polished granite surfaces from the Sierra Nevada range, California (Nishiizumi et al. 1989). In such a case, the ^{10}Be clock is turned on when the covering ice that shielded from the cosmic rays is removed. The age of meteorite impacts can be derived by ^{10}Be exposure dating of ejecta blocks that were exhumed from greater depths to the surface during the impact, as demonstrated for the Meteor Crater, Arizona (Sects. 5.6.3 and 5.8.3). The ^{10}Be and ^{26}Al exposure ages of granitic boulders were used to estimate the timing of Sierra Nevada fan aggradation which, for its part, is related to earthquake recurrence (Bierman et al. 1995). Erosion rates of 0.2 and 20 m Ma^{-1} were determined from the ^{10}Be contents – combined with ^{26}Al analysis – in quartz from the Allan Hills, Antarctica, and Anza Borrego, California, respectively (Nishiizumi et al. 1986). The ^{10}Be–^{26}Al method revealed extremely low denudation rates of less than 0.7 m Ma^{-1} for southern Australian granitic domes (Bierman and Turner 1995; Sect. 5.6.3). Analyses of ^{10}Be in quartz from bedrock outcrops, soils, mass-wasting sites and fluvial sediments from the Icacos River basin, Puerto Rico, yielded a long-term average denudation rate of ~43 m Ma^{-1} (Brown et al. 1995). In an investigation of the history of Pleistocene lake levels, ^{10}Be dating was carried out on quartz separates from metamorphic quartzites and felsic rocks incorporated in a sequence of beach conglomerates ranging 3–160 m above sea level in Death Valley, California (Trull et al. 1995). The ^{10}Be ages obtained, between 17 and 135 ka, lack a systematic relationship with terrace altitudes, implying that, due to reworking, the ^{10}Be contents do not solely reflect accumulation at the present sample site. Thus the age data are upper limits for the exposure at the present location.

5.4 Radiocarbon (^{14}C)

Carbon is a basic compound of all organisms. One of its isotopes, ^{14}C (also known as *radiocarbon*), is radioactive, but is cosmogenically replenished in the atmosphere, so that an equilibrium level is maintained. From there it is transmitted by exchange to the biosphere and the hydrosphere. When the exchange is interrupted, the radiocarbon content declines steadily, which enables the dating of various organogenic remains, such as wood and bone, as well as of ice, groundwater and ocean water in the age range between 300 a and 40 to 50 ka. Regarding the chronological message of ^{14}C ages, it is often overlooked that a significant period of time may have elapsed between the dated time of formation of the organic material and its use by Man or its burial in sedimentary layers. In other words, the ^{14}C date usually represents a *terminus post quem*, i.e., an upper limit, for the event of actual interest.

The ^{14}C content is detected either by radioactive counting or by accelerator mass spectrometry. For the customary technique of radioactivity measurement the dating range stretches ca. 40 ka back depending on the available amount of sample and on the degree of contamination by young radiocarbon. For older samples the radioactivity cannot be discriminated from the background radiation. With special enrichment procedures ages back to 75 ka may be determined under favorable circumstances. Moreover, the radioactivity counting requires a relatively large amount of sample for precise analysis, which is often not available for archaeological objects. When the AMS technique of direct ^{14}C detection was introduced, it was originally hoped that the age range of ^{14}C dating would soon be extended to 75 ka ago. Although this goal is, in principle, attainable, the omnipresent inseparable ^{14}C contamination limits, in practice, AMS ^{14}C dating to the last 40–50 ka and under favorable conditions to 60 ka (Tuniz 1996). Important advantages of the AMS technique are the shorter time as well as the much lower amounts of sample necessary for analysis, so that even single cereal grains can be worked. Samples that contain carbon components of different ages can be analyzed selectively by the AMS technique, which has brought forth an essential progress in methods.

The possible occurrence of cosmogenic radiocarbon in nature was already discussed during the 1930s, but it was not until 1947 that its existence was experimentally proven (Anderson et al. 1947). The first ^{14}C ages reported by Arnold and Libby (1949) had the merit of a systematic scientific advance (Libby 1952), which was recognized by the award of the Nobel prize for chemistry in 1960. As the measuring precision improved during the 1950s and the body of data accumulated, it became evident that ^{14}C age values were commonly lower than independently derived ages, in particular those of historically dated objects of the Egyptian Old Kingdom. Because uncertainties in the Egyptian calendar could not be completely ruled out, an extensive program of ^{14}C measurements on dendrochronologically dated tree rings of ancient bristlecone pines in the White Mountains, California, was started, which confirmed the age disparity (de Vries 1958). The observation that the discrepancy took its course in an irregular manner led Suess (1965) to install a calibration procedure for ^{14}C dates. In the meantime, calibration curves based on annual tree rings exist for the past 12 ka and beyond – although with reduced accuracy – for the last 30 ka based on uranium-series dated corals (Stuiver et al. 1993).

Among the physical dating methods applied in archaeology, Quaternary geology and paleobotany, radiocarbon is the most known and by far the most common one. The geophysical and geochemical complications of the ^{14}C system, which have repeatedly caused confusion and misunderstanding during the past four decades, are in the meantime for the most part understood and controlled. The system's versatility and precision, coupled with the dendrochronological calibration, make the radiocarbon method indispensable for age determination. It has revolutionized prehistoric chronologies in archaeology (Renfrew 1976).

5.4.1 Methodological Basis

Recent natural carbon consists of three isotopes. These are the two stable isotopes ^{12}C (98.89 %) and ^{13}C (1.11 %), and in minor traces (ca. 10^{-12}) the radioactive ^{14}C, which decays ($t_{1/2}$ = 5730 a) under β^--emission. ^{14}C is predominantly produced in the stratosphere at 12–15 km altitude through the interaction of neutrons of secondary cosmic rays with nitrogen atoms, according to the nuclear reaction $^{14}N(n,p)^{14}C$. The average global ^{14}C production rate is 7.5 kg a^{-1}, but is subjected to considerable temporal and latitude-depending variations. The total ^{14}C present on Earth amounts to

~75 t. The in situ production in rocks of the Earth's surface is negligible (Zito et al. 1980).

Libby Model. Because of the balance between atmospheric cosmogenic production and radioactive decay a ^{14}C equilibrium level is maintained. Chemically, the ^{14}C behaves like the two other carbon isotopes. It combines with oxygen to carbon dioxide. It enters the lower atmosphere by diffusion as radioactive CO_2. Within 5 a ^{14}C becomes equally distributed throughout the atmosphere. The present CO_2 content of the atmosphere is around 0.03%. The radiocarbon moves into the biosphere within a few years through photosynthesis of plants and the food chain, and into the hydrosphere through carbonate formation, from where it gets into marine and continental sediments as $CaCO_3$. The carbon cycle through the various reservoirs is shown in Fig. 43. The ocean water contains most of the cycle carbon, that is 94%, whereas only 2% reside in the atmosphere. Provided that the ^{14}C mixing, the residence within and the exchange between the reservoirs are fast compared to the half-life, the same ^{14}C equilibrium concentration as in the atmosphere is found everywhere. These presuppositions do not hold true in each reservoir.

As long as the organisms participate in the cycle, their ^{14}C concentration maintains the equilibrium value. With death the organism becomes sepa-

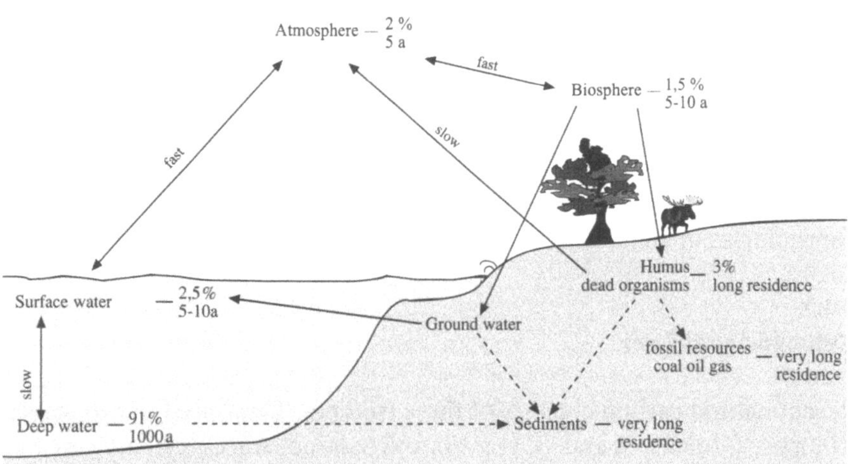

Fig. 43. Natural carbon cycle. A continuous exchange of carbon takes place between the large reservoirs of the atmosphere, hydrosphere and biosphere. The individual reservoirs contain different amounts (in %) of the total carbon participating in this cycle. Additionally, the mean residence times in the reservoirs are indicated. The lithosphere participates extremely slowly (compared to the half-life of ^{14}C) in this exchange, thus playing a subordinate role

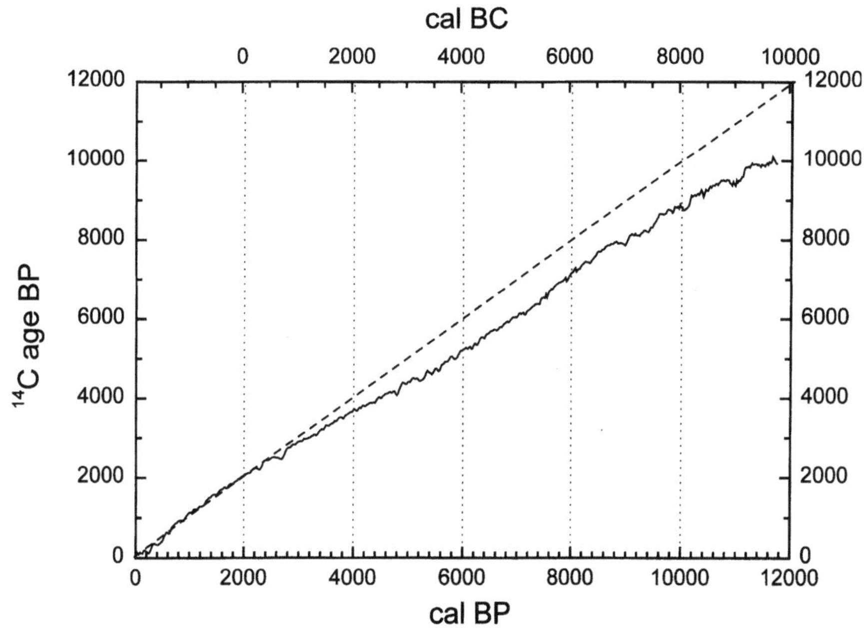

Fig. 44. Disparity between conventional ^{14}C years and calendar years measured on tree rings covering the last 12 ka (courtesy of B. Kromer). For this age range the ^{14}C ages are mostly too low, in fact up to ~ 2000 a. This difference is attributed to long-term variation of the initial $^{14}C/C$ ratio

rated from the cycle, so that the ^{14}C input is interrupted. From then on the initial $^{14}C_0$ declines due to radioactive decay. According to Libby's model the ^{14}C age t [a] is calculated from the remaining ^{14}C

$$t = 8033 \times \ln (^{14}C_0/^{14}C) \tag{41}$$

i.e., a ^{14}C decrease by 1% corresponds to an age increase of ~80 a.

The ^{14}C ages based on this model turned out to be too low, indicating that the simple model conditions of a temporal and spatial constant initial ^{14}C concentration are not valid (de Vries 1958). The temporal variation is recognized from the disparity between ^{14}C years and calendar years (Fig. 44). Because of the fast atmospheric mixing, the ^{14}C variation appears practically worldwide at the same time.

Temporal Variation of Initial Radiocarbon. One distinguishes ^{14}C variations of long duration from those of short duration as well as natural from anthropogenic ones. Traditionally it was believed that the long-term variations have an amplitude of ~10% and a period of ~20 ka. Tree-ring data show that the initial ^{14}C content decreased by ~10% within the past 11 ka,

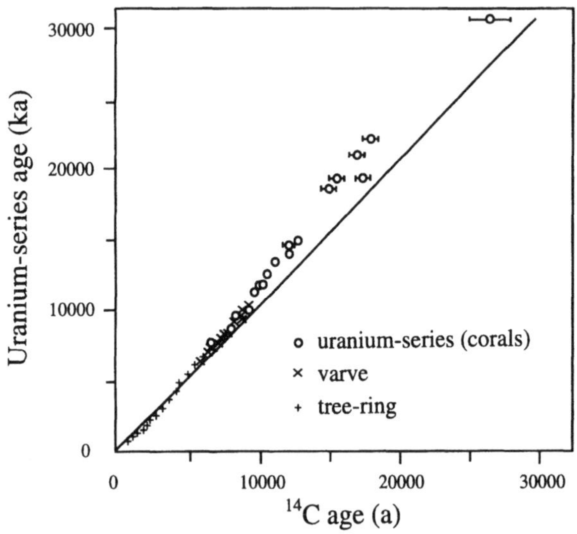

Fig. 45. Disparity between uranium-series and ^{14}C ages ($t_{1/2}$ 5730 a) of corals from Barbados for the age range 30–10 ka. The *crosses* in the lower age range symbolize dendro and varve ages for comparison. Provided that the uranium-series ages represent calendar years – which is proved for the Holocene by the correspondence with the dendro and varve years – the ^{14}C ages turn out to be too low, namely, up to 3500 a. This result bears crucial consequences for the temporal evolution of the initial $^{14}C/C$ ratio and the correction of ^{14}C ages in this age range. (Reprinted with permission from Nature; Bard et al. 1990b; copyright 1990, Macmillan Magazines Ltd.)

since a 10-ka-old sample is ~1 ka too low by its ^{14}C age. The apparent half-period of 10 ka simply was a consequence of the duration, for which dendro-dates were available. Such a simple periodic sinusoidal model for the long ^{14}C variation is now out of date. The comparison of AMS ^{14}C with TIMS uranium-series ages of Late Glacial corals (Sect. 4.3) shows that the gap between the ^{14}C years and solar years widens backward even more, namely to 3.5 ka for 20-ka-old corals (Fig. 45), so that the ^{14}C concentration of the atmosphere has been up to 40% higher than today (Bard et al. 1990b, 1992). A dramatic change of 15% decrease occurred around the end of the Younger Dryas (Edwards et al. 1993). Several causes have been discussed for the long-term ^{14}C variations. An essential role is commonly assigned to the Earth's magnetic field. The increase in the dipole moment results in stronger deflection of the cosmic rays and thus in reduced ^{14}C production. Also paleoclimatic circumstances are taken into account, in particular the increased meltwater input to the North Atlantic, which alters the oceanic circulation and thus affects the carbon balance between oceanic and atmospheric reservoirs. Such climate-controlled shifting releases "aged" carbon dioxide from deep ocean water to the atmosphere. As a consequence, the atmospheric ^{14}C concentration becomes diluted. This scenario of accelerated ocean ventilation is held responsible for the fast ^{14}C decline during the Younger Dryas (Goslar et al. 1995; Björk et al. 1996). Furthermore, primary oscillations of the cosmic ray flux are considered as a possible cause for the long-term ^{14}C changes.

Irregular cycles of 100–200 a duration and up to 2% amplitude appear in the plot of ^{14}C years against calendar years. These medium-term ^{14}C variations (*Suess wiggles*) are negatively correlated with solar activity changes, whereby the solar magnetic field modulates the cosmic ray flux. During high solar activity, the intensity of the cosmic rays and thus the ^{14}C production rate diminishes (*de Vries effect*). High solar activity is apparently accompanied by an amelioration of the climate (Suess 1979). Accordingly, the historically well-documented Maunder minimum of sunspot activity coincided with the"Little Ice Age" during the 17th/18th centuries and led to an increase in the atmospheric ^{14}C level by 2% (Fig. 46). Furthermore, an *11-year-cycle* in the ^{14}C production exists. The amplitude of this short-term variation is around 1.4% (Stuiver 1993) and is linked to the 11-year period of sunspot recurrence.

As a consequence of industrialization since the mid-19th century, large amounts of "dead" CO_2, i.e., without ^{14}C, were released by the combustion of fossil fuels, first coal and later petroleum also. This dilution of the atmospheric ^{14}C concentration by ~0.03% a^{-1} is known as the *Suess effect.* From 1850 until 1950 the ^{14}C concentration in the atmosphere decreased by 3%; implying that the ^{14}C age of a tree ring grown in 1950 would overestimate its real age by 240 a, if based on a pre-industrialization $^{14}C_0$ value. This impedes ^{14}C dating of samples formed after 1850.

The aboveground nuclear bomb tests, carried out between 1950 and 1964, temporarily nearly doubled the ^{14}C content in the atmosphere (bomb effect) and it is still ~20% above normal at present. Although it poses a danger of contamination for ^{14}C dating, the bomb ^{14}C spike can be utilized as an artificial tracer for meteorological and hydrological processes (Levin et al. 1992).

Spatial Variation of Initial Radiocarbon. Apart from the temporal changes there are spatial ^{14}C inhomogeneities in the materials and reservoirs participating in the carbon cycle. One distinguishes *isotope fractionation* and *reservoir effects.*

Isotopes belonging to the same element show identical chemical behavior, but have dissimilar kinetic reaction rates due to their different mass number. For most elements this phenomenon is insignificant, but this is not the case for the light ones, such as carbon. There the relative mass dif-

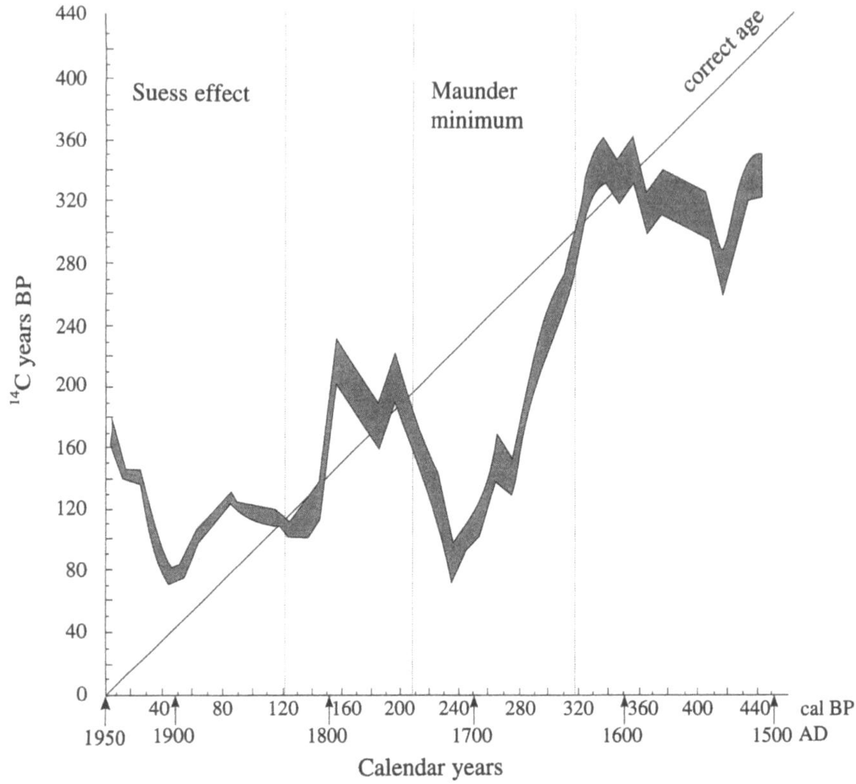

Fig. 46. Relationship between conventional ^{14}C ages (in ^{14}C years before 1950) and calendar years (before 1950), measured on tree rings (each 10 rings = 1 ^{14}C sample) of Douglas fir for the period 1500–1950. The *width* of the curve is twice the counting error in the measurements. In the case of age concordance the curve has to fall on the 1:1 *straight line*. The too high ^{14}C ages since 1830 are caused by industrial combustion of fossil carbon fuel. The too low ^{14}C ages between 1630 and 1750 are caused by the minimum of sun spot activity (Maunder minimum) with increased ^{14}C production. (Reprinted with permission from Nature; Stuiver 1978; copyright 1978, Macmillan Magazines Ltd.)

ferences (i. e., mass difference normalized to mass number) between the isotopes are sufficiently large to cause significant *isotope fractionation effects* in chemical reactions. Photosynthesis, for instance, enriches the light ^{12}C over the heavy ^{13}C, and the latter one in turn over the even heavier ^{14}C, so that the carbon in plants is isotopically lighter than in the atmosphere. Because the mass difference amounts to "one" from ^{12}C to ^{13}C, and to "two" from ^{12}C to ^{14}C, the isotope fractionation between ^{12}C and ^{14}C is about twice that between ^{12}C and ^{13}C. The isotope fractionation of a sample is expressed as δ-notation. For the pair $^{13}C/^{12}C$ the $\delta^{13}C$ value means the

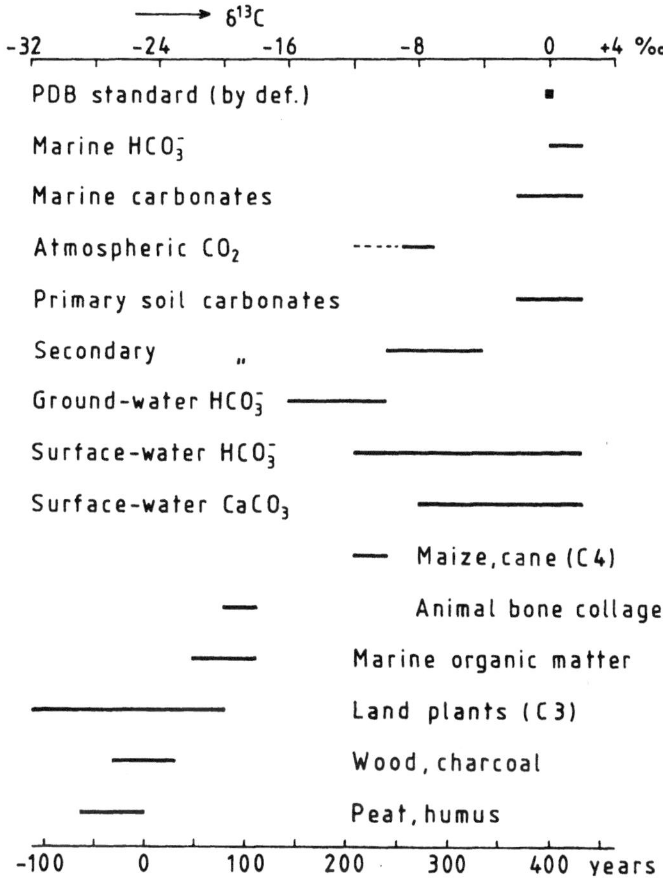

Fig. 47. Isotope fractionation of ^{13}C, given as $\delta^{13}C$ (in ‰, *upper scale*) in different reservoirs and materials normalized to the PDB standard. For the calculation of the conventional age the measured $\delta^{13}C$ value is converted to the value of wood (−25‰). The corresponding age correction is given on the *lower scale*. (After Mook and Waterbolk 1992)

deviation [‰] between the sample "P" and the standard "S" (belemnite PDB) and is calculated from

$$\delta^{13}C = 1000 \times [(^{13}C/^{12}C)_P - (^{13}C/^{12}C)_S]/(^{13}C/^{12}C)_S \tag{42}$$

By definition, $\delta^{13}C$ equals 0‰ for the PDB-standard. An overview of the typical $\delta^{13}C$ ranges in various reservoirs and materials is given in Fig. 47. Isotope fractionation implies that samples of the same age, but different $\delta^{13}C$ values, have different ^{14}C ages, with 16 a difference for 1‰ $\delta^{13}C$. The ^{14}C age apparently becomes younger with increasing $\delta^{13}C$. For this reason, age distortion by isotope fractionation requires correction. By international convention the (*conventional*) ^{14}C age refers to the $\delta^{13}C$ value of wood (−25‰). The $^{14}C_c$ content, corrected for isotope fractionation, is calculated from the measured content $^{14}C_m$ by

$$^{14}C_c = {}^{14}C_m \times [1 - 2(\delta^{13}C + 25)/1000)] \tag{43}$$

The factor 2 allows for the doubling of the isotope fractionation of ^{14}C as opposed to ^{13}C. To give an example, for a sample with a $\delta^{13}C$ value of −15‰ one derives $^{14}C_c = 0.980 \times {}^{14}C_m$ for the corrected content, i.e., such isotope fractionation would increase the ^{14}C age by 160 a, according to Eq. (41). For most materials, the age corrections that result from isotope fractionation are less than 80 a, but may be up to several 100 years, as in the case of marine limestones and organisms. Since the isotope fractionation of carbon depends on the reaction conditions, above all the temperature, $\delta^{13}C$ data are also utilized in paleoclimatic studies.

The *reservoir effect* deals with the isotopic variation of carbon within the reservoir, from which the organisms extract their carbon. Such spatial changes may have various causes. If the carbon stays within the same reservoir for a long time, without being quickly – compared to $t_{1/2}$ – mixed or exchanged with other reservoirs, the ^{14}C concentration declines ("aging" of carbon). Also, the admixture of "aged" carbon lowers the ^{14}C concentration. Reservoir effects result in an apparent increase in the ^{14}C age and are difficult to assess (Olsson 1979).

The ocean surface steadily exchanges carbon dioxide with the atmosphere, so that the surface water assumes the atmospheric $^{14}C/^{12}C$ ratio. In contrast, the deep oceanic circulation proceeds slowly within ~2 ka. The deep water is excluded from the atmospheric exchange and, consequently, its radiocarbon content declines. Deep ocean water exhibits ^{14}C ages up to 1.7 ka (*marine reservoir effect*). When such "aged" water re-emerges at the surface, it mixes with "young" surface water. Therefore, the waters of upwelling regions have apparent ^{14}C ages around 400 a. Since the mixing is controlled by local factors, such as coast shape, currents and winds, the apparent ^{14}C age of seawater varies from place to place and also with time (Stuiver and Braziunas 1993). When building their bodies and skeletons, the marine organisms take over the $^{14}C/^{12}C$ ratio of the seawater – apart from isotope fractionation during metabolism. Consequently, marine shells, foraminifers and corals may have ^{14}C ages which overestimate their true age. One possibility to assess this effect is to measure the ^{14}C content of recent organisms (but older than 1950 due to the bomb effect) from the same area. Although the reservoir effect (^{14}C age too high by ~400 a) and the isotope fractionation effect (^{14}C age too low, cf. Fig. 47) fortunately more or less compensate each other along many coasts, both phenomena need to be taken into consideration independently.

Another reservoir effect occurs between the atmospheres of the northern and southern hemispheres (*northern to southern hemisphere effect*). ^{14}C dating on tree rings of the same age showed that the ^{14}C age in the southern hemisphere are systematically higher by ~30 a (Lerman et al. 1970). The effect is attributed to the larger ocean surface of the southern hemisphere, so that the "recent" atmospheric CO_2 is stronger diluted by the "older" marine CO_2. The fast atmospheric exchange within each hemisphere is impeded at the equator by the prevailing winds blowing in opposite directions along the equator.

In carbonaceous ground and surface waters the *hard-water effect* is observed (Deevey et al. 1954). During migration through carbonaceous soils and aquifers, groundwater dissolves geologically "old" carbon which dilutes its "recent" organogenic carbon. The mixture of both carbon compounds increases the ^{14}C age of groundwaters. After being fed into rivers and lakes this ^{14}C-poor water is admixed in various portions with "young" surface water. Organisms, such as freshwater mollusks and plants, living in this environment therefore may have ^{14}C ages too old by several centuries. The hard-water effect also occurs in estuary regions. Because of its strong regional and temporal variability, the hard-water effect is difficult to correct for. "Dead" carbon may also be released during the decomposition of limestone by humic acids, and this reduces the ^{14}C concentration in the soil CO_2 and in calcareous sinters (Geyh 1970). The gas exhalation of active volcanoes contains CO_2 coming from the deeper crust and thus being free of ^{14}C. The admixture of this gas with the atmosphere reduces the atmospheric ^{14}C ratio so that plants growing near the vents may have excessive ^{14}C ages (Bruns et al. 1980).

As to age determination, all these temporal and spatial changes in ^{14}C concentration violate the assumption of a constant initial level $^{14}C_0$ as made by the Libby model. In order to apply ^{14}C dating correctly, one would have to know the true $^{14}C_0$ of each sample individually. Although spatial ^{14}C variations – in particular those caused by the reservoir effect – commonly cannot be evaluated quantitatively, they may be estimated qualitatively. More problematic are the temporal ^{14}C variations, since they are a priori unknown for the age which still is to be determined. In order to derive correct ages in spite of this dilemma, one proceeds in two steps and distinguishes two types of ^{14}C ages, the *conventional* and the *calibrated* ones. The calculation of the conventional age is subjected to defined model assumptions. It is reported in *^{14}C (or radiocarbon) years*. In the second step one attempts to convert the conventional age into a chronometrically correct age by applying a calibration procedure. The calibrated age is quoted in *calendar years*.

Conventional ^{14}C Age. The conventional ^{14}C age is calculated according to the following rules:

1. The year of reference is 1950 AD, expressed by the notation *BP* (before present) after the age value. If a ^{14}C age quotation is conventional it is identified by the symbol BP. For instance, the conventional ^{14}C age 1820 BP means that a sample has an age of 1820 ^{14}C years reckoned backward from, i.e., before 1950. The fact that the conventional "before present" does not mean the actual present day is frequently misunderstood. It is therefore recommended to apply the term BP only in the defined sense.
2. The initial $^{14}C_0$ activity (or ratio) is assumed to be temporally constant and is fixed to an NBS-oxalic-acid standard as primary standard, or secondary standards derived therefrom. The standards define the initial $^{14}C_0$ level of wood for time zero.
3. The calculation of the conventional age is based on the former "Libby half-life" of 5568 a, although it is known that it is too low by 3%. This value was chosen in order to facilitate the comparability with the formerly determined ^{14}C ages.
4. The isotope fractionation $\delta^{13}C$ is corrected for according to Eq. (43), i.e., it is based on $\delta^{13}C = -25‰$ (wood value).

It is obvious that conventional ^{14}C ages are systematically too low by 3% and that they do not take into account the temporal $^{14}C_0$ variations. The age errors introduced by this convention are made up for by calibration. The uncertainty quoted together with the conventional age is usually the 1 $\bar{\sigma}$ standard error of precision (Sect. 1.3).

Calibrated ^{14}C Age. Calibration is the conversion of conventional ^{14}C ages to calendar dates. For calibration, commonly accepted curves and programs are used which are based on precise ^{14}C measurements of dendrochronologically dated tree-ring sequences. The calibration curves are steadily improved and supplemented (Kromer et al. 1986; Stuiver and Kra 1986; Becker and Kromer 1990; Becker 1992; Kromer and Becker 1992; Stuiver et al. 1993). In order to distinguish calibrated ^{14}C ages from conventional ^{14}C ages (BP) they are characterized by the notations *cal BC* (calendar years BC), *cal AD* (calendar years AD) or *cal BP* (calendar years before 1950 AD). By the symbol *cal* the calibrated ^{14}C dates are set apart from historic dates which are quoted simply as BC or AD.

The calibration diagrams (Fig. 48) plot the dendrochronologically known calendar date along the x-axis. It is derived from wood that contains a series of consecutive tree rings (Sect. 10.1.2). Wood specimens that comprise a given number of rings (spanning commonly 20, 10 or single annual rings) are precisely ^{14}C dated, and the resulting conventional ^{14}C age is plot-

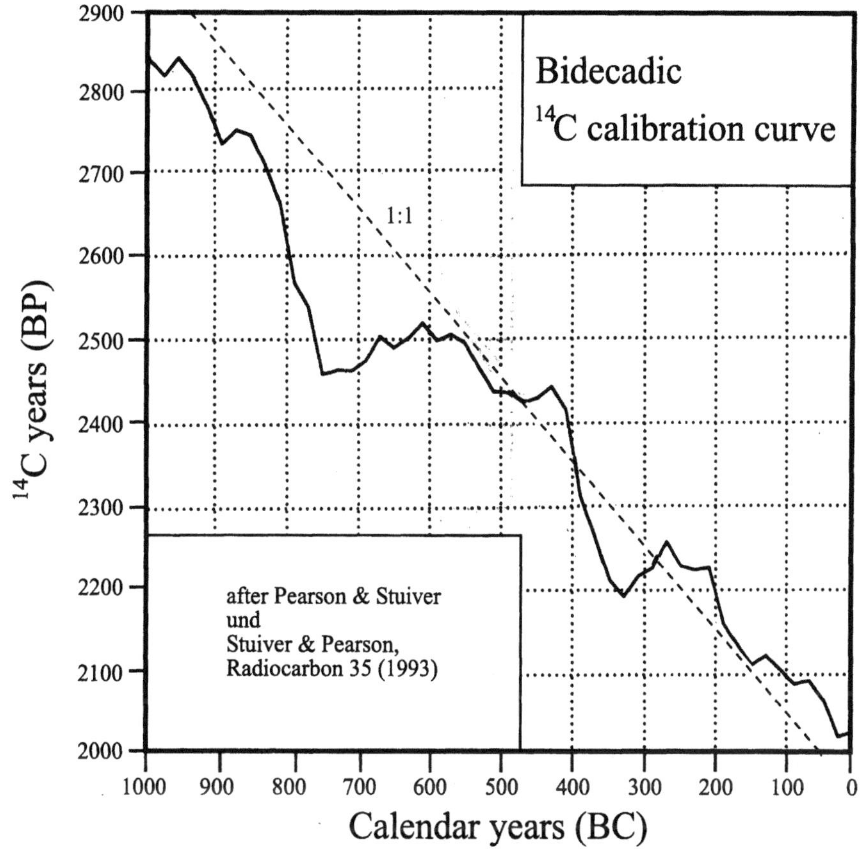

Fig. 48. Bidecadic calibration curve for conventional ^{14}C ages; section (1st millennium BC) of the high-precision curve ranging from 1950 AD to 6000 BC. The curve is based on dendrochronologically dated wood, where each ^{14}C sample was taken in form of 20 successive annual tree rings. The 1 $\bar{\sigma}_c$-error of the curve amounts on average to 12.9 ^{14}C years. In the age range 750–420 BC, the shape of the curve implies calibration ambiguities. The *dotted line* of reference marks age equality

ted along the y-axis. In the case of sections with more than one ring, the ^{14}C age is plotted over the mean calendar date of the section. In this manner, the continuous tree-ring chronology, nearly covering the past 12 ka, is dealt with. The x–y-data points are finally connected by a curve. The x-value is considered to be accurate to the year. The y-value bears an error (precision of the ^{14}C measurement) that is depicted in Fig. 46 as a band of width $1\bar{\sigma}_c$ instead of a thin line. This error averages ±13 a for the bidecadic calibration curve of the 1950 AD to 6000 BC interval (Stuiver and Pearson 1993). Calibration curves for the Holocene are based on various species of trees in different regions, essentially the Californian bristlecone pine

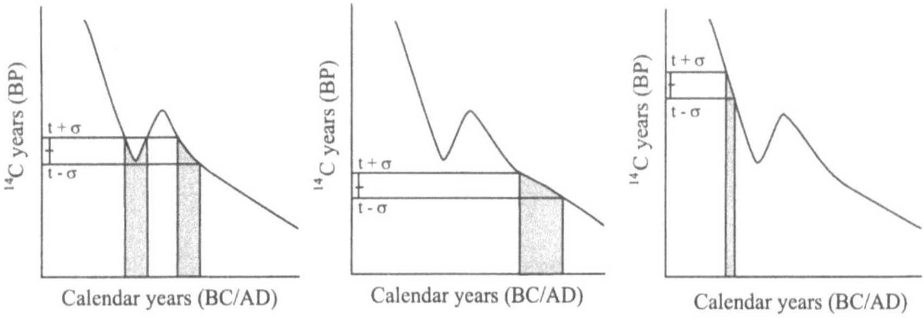

Fig. 49. Schematic representation of the intercept technique for calibration. The intercept points of $t + \sigma$ and $t - \sigma$ with the calibration curve are constructed. The range between the endpoints of intercept is taken as the confidence interval of the calibrated age. The points of intersection with the curve may yield one or several intervals of the calibrated age. (After Bowman 1990)

(*Pinus aristata*) and the European oak (*Quercus* sp.). The concordance among these curves – all derived from trees of the northern hemisphere – displays an important criterion for their general validity as well as their reliability. For the preboreal period, European pines (*Pinus*) are also employed (Becker and Kromer 1990). With respect to the resolution of the calibration curves, high-resolution curves, based on single annual rings (Stuiver 1993), need to be distinguished from those that rest on the combined analysis of several rings and are thus smoothed. Because of their smaller analytical uncertainty, bidecadic (e.g., Stuiver and Pearson 1993) and for short-lived (< 10 a) sample material also decadic calibration curves (e.g., Stuiver and Becker 1993) are usually applied.

All of these curves exhibit two important characteristics. In the first place, the conventional ^{14}C ages do not agree with the dendro-ages. The conventional ^{14}C ages are generally lower, with an increasing trend going back in time (Figs. 44 and 45). In the second place, this behavior is not systematic, but erratic. Because of the Suess wiggles, the ^{14}C ages may even decline when the calendar ages increase (Figs. 46 and 48). As a consequence of the former characteristic, the gap between the conventional ^{14}C and the calendar ages is, at the first approximation, different for different periods. For the last two millennia the conventional ^{14}C ages plot within ± 250 a of their calendar ages. For the preceding periods the conventional ^{14}C age is always lower than the calendar age: within the 2000 – 7300 cal BP interval the gap widens to 800 a and within the 8000 – 11,200 cal BP interval even to 1200 a. To move further back becomes increasingly difficult due to the climatically governed lack of oak and pine trees in northern mid-latitudes where the calibration curves were hitherto constructed. For earlier periods there is the prospect of using corals, whose age can be established independently by TIMS uranium-series dating, for the extension of the calibration curve. The ^{14}C ages of 12,330 a (^{230}Th age related to 1950 AD) old corals are already too low by 2130 a (Edwards et al. 1993). The difference amounts to 3.5 ka during the Glacial Maximum 20 ka ago (Fig. 45) and – according to other age evidence – possibly even to 5 ka, 35 ka ago (Vogel 1983; Bard et al. 1990b; Stuiver 1990) and 7 ka, 40 ka ago (Guyodo and Valet 1996). Contrary to these long-term trends, the phenomenon of the Suess wiggles cause principal difficulties for the calibration. In the calibration curve they cause result in reverse branches, where conventional ages decrease as the true ages increase, in contrast to the general trend with both types of age increasing. In such a case, the conversion of the conventional ^{14}C age to the calendar age results in multiple answers. In other words, a single conventional ^{14}C age corresponds to more than one possibility for the calibrated age, as illustrated in Fig. 49.

The calibration of the conventional age $t \pm \bar{\sigma}$ is conducted as follows: a straight horizontal line through t (y-axis) – if necessary, after reservoir correction – is constructed parallel to the x-axis. The intersection point of this line with the calibration curve yields the corresponding x-value in calibrated calendar years (Pearson 1987). However, this simple procedure does not take into account the associated error term $\pm \bar{\sigma}$. The irregular, mathematically undefinable shape of the calibration curve poses a problem for the transformation of the error. After conversion, the original Gaussian distribution of the conventional date is no longer Gaussian. Consequently, the calibrated age is not central within its error range which depends strongly on the shape of the respective part of the calibration curve. In view of this difficulty, basically two approaches exist: the *intercept* and the *probability* techniques (Bowman 1990). In the intercept technique (Fig. 49), first the error $\bar{\sigma}_c$ of the calibration curve is combined with the precision $\bar{\sigma}_r$ of the conventional age to the combined error $\bar{\sigma} = \sqrt{\bar{\sigma}_c^2 + \bar{\sigma}_r^2}$. The intersections between the upper and lower limits of the confidence interval, $t + \bar{\sigma}$ and $t - \bar{\sigma}$, and the calibration curve are taken as the endpoints of the probability range for the calibrated age, i.e., in the case of a $1\bar{\sigma}$-error bar the calibrated age lies with 68.3% probability within this interval, but not necessarily in its center. The information contained in the Gaussian distribution is lost in this procedure, so that the central value of the calibrated age is not more likely than any other value from the interval. The probability techniques (Fig. 50) attempt to make up for this shortcoming by transforming the Gaussian distribution into a probability distribution of the calibrated age. Presently, various computer-based procedures for it exist. The basic principle behind all of them is to group the Gaussian distribution into short intervals, to convert them separately, and to sum up the single results to a total probability histogram or curve. One variety of these

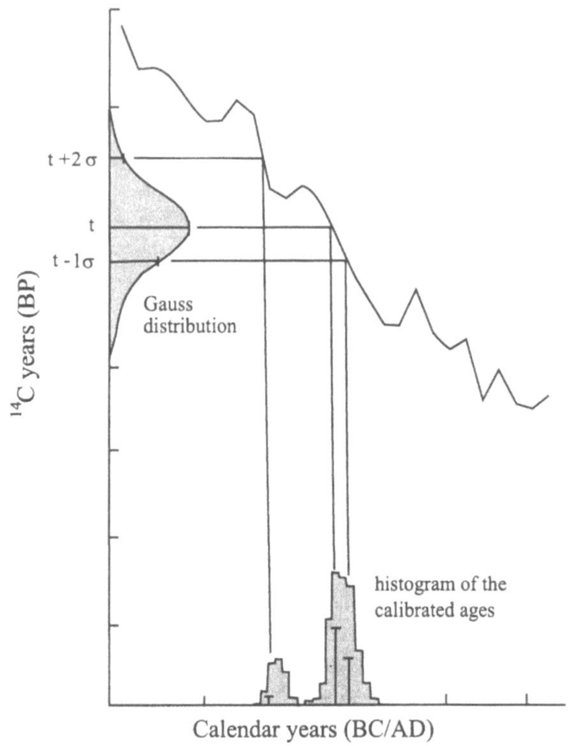

Fig. 50. Schematic representation of the probability technique for calibration. The Gaussian distribution of the conventional age is transformed at the calibration curve into a probability distribution (here as *histogram*) of the calibrated age. Various mathematical procedures are available for this technique. (After Bowman 1990)

techniques is the "2-D dispersion calibration" by Street at al. (1994) in which the corresponding Gaussian probability of the conventional date is taken for each calendar year through intersection with the calibration curve.

For a given conventional age $t \pm \bar{\sigma}$ the range of the calibrated age depends on the slope of the calibration curve: the steeper the slope, the narrower the range. Flat sections of the curve, in which the conventional age grows little with the calendar age, are less advantageous. There are even periods (*plateaus*), in which the conventional age does not grow at all with calendar age. These are predominantly the periods: 10,450–9250, 6900–6700, 4200–4000, 3350–3050, 2850–2600 and 800–400 cal BC. These periods cannot be resolved by ^{14}C dating.

An exceptional calibration technique is *wiggle-matching* (Pearson 1986). It is highly accurate (to ±10 a), but laborious. It is applied mostly to wood with at least 50 consecutive tree rings preserved. The underlying principle is the characteristic temporal variation of the conventional ^{14}C age that is recorded as fine structure (*wiggles*) in the calibration curve. In other words, the various sections of the calibration curve can be distinguished by

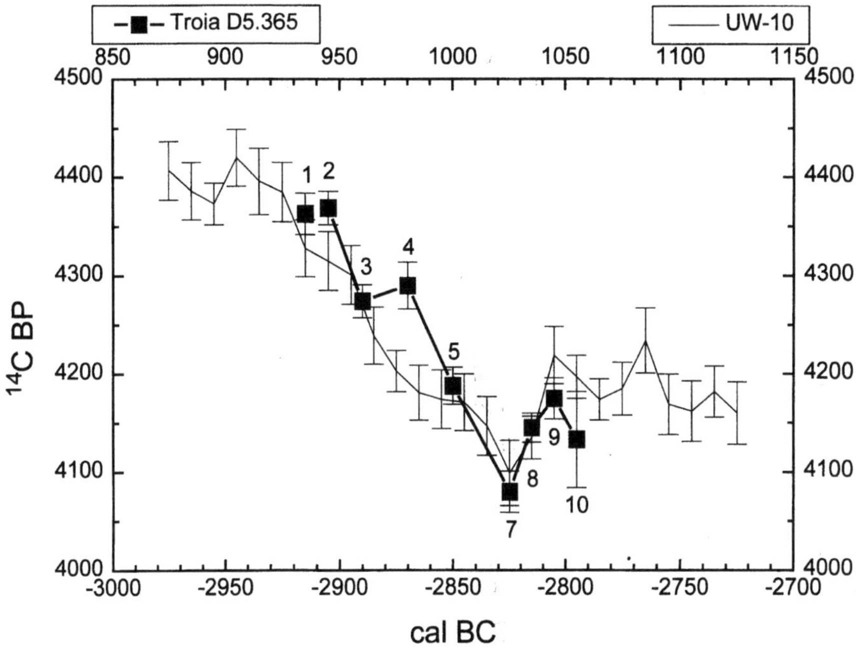

Fig. 51. Calibration by means of wiggle matching for a charcoal sample with preserved tree-ring pattern (arbitrary ring numbers; above scale) from the Early Bronze Age Troia I (after Korfmann and Kromer 1993). The wiggle pattern of ten conventional ^{14}C ages from the ring sequence was matched to a high-resolution master curve. For the last dated ring the date 2790 ± 15 BC is obtained. Owing to the 91 consecutive tree rings, the *terminus post quem* of the felling date is 2699 ± 15 BC. Because of the missing bark, the felling date is assumed to be ~30 years later

their "wiggle signature". For dating, the wiggle pattern of a sample is determined by measuring the ^{14}C age of different rings. The wiggles are matched (synchronized) with those of a high-resolution master curve and the sample is thus temporally fixed (Fig. 51). Annually layered sediments, the varves (Sect. 10.1.1), are also suited for wiggle-matching if they contain macrofossils to which AMS ^{14}C dating can be applied (Goslar et al. 1995). An analogous approach is *archaeological wiggle-matching* (Weninger 1986), in which the ^{14}C age pattern is derived from a stratified sequence of short-lived organic remains.

Contamination. As is the case for the other radiometric dating systems, the ^{14}C system also has to remain closed, i.e., carbon must neither enter nor leave the sample. The ^{14}C age is apparently increased by contamination with "dead" carbon and lowered by recent carbon. Dead carbon may be incorporated by geologically old carbonates, as already discussed in connection with the hard-water effect. Common sources of contamination with recent carbon are the presence of rootlets, humic acid infiltration and bioturbation. Inappropriate storing and processing of the sample may also introduce contaminant carbon. Old samples with low ^{14}C concentration are particular susceptible to contamination; for example, a contaminant fraction of 1% lowers the age by 10, 200 or 7000 a for samples aged 1, 10 or 40 ka, respectively (Geyh 1983; Mook and Streurman 1983). The smaller the authigenic ^{14}C amount and the older the sample, the larger the danger of contamination by modern carbon. For this reason, the applicability of ^{14}C dating toward high ages is limited by unavoidable contamination rather than by the instrumental capabilities of ^{14}C detection.

Maximum Age. The upper dating limit (*maximum age*) is reached, for detection by β-counting, when the ^{14}C activity $[^{14}C]$ of an old sample cannot be discriminated with sufficient statistical confidence (commonly double the standard deviation 2σ is applied) from the background $[^{14}C]_u$. The attainable maximum age varies between 35 and 65 ka (ca. ±4 ka) among laboratories owing to duration, background and equipment of counting. If the activity is no longer distinguishable from the background, i.e., $([^{14}C] - [^{14}C]_u) \leq 2\sigma$, then the age of the sample is equal or higher than the maximum age, which is annotated by the symbol > in front of the maximum-age value. For example, an age quoted as >47±4 ka (47±4 ka being the maximum age for a laboratory) implies that the sample is at least 43 ka old, and the laboratory's maximum age is the sample's minimum age. The AMS technique has the great potential to lower the instrumental background and thus to extend the maximum age, but is effectively limited by the background due to contamination.

5.4.2 Practical Aspects

The amount of sample required for ^{14}C dating depends on the carbon content, the conditions of preservation, the degree of contamination, and the technique of ^{14}C detection. For the β-counting, either in gas or liquid scintillation counters, 5–10 g of extracted carbon are needed. For miniature gas counters several 10 mg are sufficient. The AMS technique requires carbon in the mg range. Note that the quoted amounts refer to carbon and not to sample. The required amount of the latter one might be larger by a factor of 10 or so, depending on the carbon content.

Another aspect to take into consideration when sampling is the length of the period within which the sample was formed. As in the case of peat or wood which grow over tens to hundreds of years, the period may be larger than the precision of measurement. In such a case ^{14}C dating yields an average age of the sample that may not be related to the event of interest.

Since contamination with allogenic carbon impedes reliable age determination, samples with signs of corresponding influences, such as weathering, rootlets, humic acid infiltration, and secondary calcification should be possibly avoided. Such hazards, but also information on the moisture, the position and changes of the groundwater table as well as the water-hardness, together with a scaled sketch of the sample context should be recorded and reported to the laboratory. During packing and storing, any contamination with paper, textiles and other biogenic tissues must be avoided. Glass or aluminum containers are suited. Mold growth during extended storing in closed containers is prevented by drying (not heating!) the sample first. Dry, dark and cool storing conditions are optimal. Wet wood and peat samples should be brought in their natural state as soon as possible to the dating laboratory. It goes without saying that the sample must not be exposed to any organic solvents and adhesives. On the other hand, treatment with clean water does no harm.

In the laboratory any visible macro-remains, such as rootlets, twigs and insects are picked out of the sample. By chemical treatment one attempts to remove the carbon contamination, old calcium carbonate by diluted HCl or recent humic acids by NaOH, and to enrich the authigenic carbon component of the sample (Mook and Waterbolk 1985; Taylor 1987a; Hedges and van Klinken 1992). In the case of bone, for instance, the authigenic component suitable for dating is the protein collagen. The ^{14}C analysis of various purification fractions may give important insights in the kind and degree of contamination. The purified extracted carbon is burned to CO_2 which can directly be used for gas counting. The CO_2 is transformed to benzene for liquid scintillation and to graphite for AMS analysis.

For ^{14}C detection two different approaches exist: the β decay counting or the atomic counting by AMS. The amount of 1 g recent carbon contains 5.9×10^{10} ^{14}C atoms. After 50 ka 1.4×10^{8} atoms and after 100 ka only 3.2×10^{5} atoms are left over. The 5.9×10^{10} ^{14}C atoms produce 13.5 decays/min. Consequently, one has to count this sample for half a day (10,000 β-disintegrations) in order to attain a precision of 1%. In the case of the 50-ka-old sample, this precision would require already 8 months of counting. On the contrary, the AMS technique enables the ^{14}C detection down to 10^4 atoms and to $^{14}C/^{12}C$ ratios of 10^{-16} within about an hour.

The activity measurement is performed by gas proportional counting (Kromer and Münnich 1992) or by liquid scintillation counting (Polach 1992). For both techniques it is important to suppress the background radiation. Apart from cosmic rays, the natural radioactivity of uranium, thorium and potassium contributes essentially to the background. These elements may be found in the walls of buildings and in the instrumentation itself. The background can be reduced by selecting suitable building materials, by shielding the counter with old lead or steel, by placing anti-coincidence counters around the detector. In order to reduce the cosmic ray component further, laboratories are placed underground. For gas counting commonly few liters of carbon dioxide, occasionally also methane or ethylene, are used. The required duration of counting depends, of course, on the activity. Mostly several days are sufficient in order to register at least 10,000 counts. The doubling of the counting time or the sample size improves the precision of measurement only by the factor of $\sqrt{2}$. The application of mini-counters (Harbottle et al. 1979) reduces the sample size considerably, but requires correspondingly longer counting times up to several months. For liquid scintillation counting the carbon is transformed to benzene (several ml). A scintillator is added to the liquid and the light flashes, induced by the β-particles, are recorded by two photomultipliers (Otlet and Polach 1990). Both counting techniques make possible routinely age precision $\leq \pm$ 40 a and for younger samples even $\leq \pm$ 12 a.

By isotopic enrichment of ^{14}C the detection limit can be reduced and thus the dating range extended to ~75 ka (Grootes 1978; Stuiver et al. 1978). This technique, termed thermal diffusion, requires 10–150 g carbon. The efficiency of the ^{14}C enrichment is experimentally obtained from mass spectrometric determination of the simultaneous ^{18}O enrichment. In view of the fact that in 50–75 ka old samples the original content $^{14}C_0$ has declined to fractions of 10^{-3}–10^{-4}, the presence even of minute contamination with modern carbon limits the theoretically attainable age range.

The AMS technique has no problems with background radiation. However, during sample processing, some contamination with recent carbon cannot be avoided, which limits the age range in practice to 50–60 ka. The small sample amount (0.1–1 mg carbon) sufficient for AMS analysis allows one to apply the technique to precious and tiny samples. In particular, specific as well as different carbon fractions extracted from the same object can be dated. Since the ^{14}C detection by AMS is much faster than by β-counting – hours instead of days or weeks – AMS laboratories have higher dating capacity at comparable precision (Beukens 1992). This advantage must, however, be weighted against the higher costs of running an accelerator.

As to the quotation of ^{14}C ages, the laboratories release the conventional age value [^{14}C year BP] with the analytical uncertainty of $\pm 1\bar{\sigma}_r$ or $\pm 2\bar{\sigma}_r$, commonly also with the isotope fractionation. Many ages appear in the date list of the journal *Radiocarbon*. The laboratory identifies itself by a symbol and the sample by a number. The ^{14}C laboratory of the "Heidelberger Akademie der Wissenschaften", for instance, reports its data in the following scheme: HD-8844–8816 (charcoal) 1580 ± 35 BP (= $1\bar{\sigma}_r$), $\delta^{13}C = -23.99$ ‰. For the calibration of the conventional age, including the error assessment, various PC-based programs are available (see van der Plicht et al. 1990; Stuiver and Reimer 1993).

5.4.3 Application

The ^{14}C method is widely applicable with regard to materials as well as to problems in archaeology and Quaternary geology. With the introduction of the AMS technique, its attraction increased even further. A general point worth mentioning is the interpretation of ^{14}C ages. The age result does not necessarily relate to the *age* of the event to be dated. In the case of a biogenic sample embedded in a sediment, its ^{14}C age relates to the death of the organism, but not to the sedimentary deposition. In such a case, the meaningfulness of the age would depend on the question: is the age of the sample of interest or that of the context? Because death usually precedes burial, the ^{14}C age would define an upper limit, also called *terminus post quem*, for the deposition, without saying anything about the time difference between death and burial. On the other hand, the ^{14}C age of a downward burrowing animal or root would postdate the sediment, i.e., it would give a minimum age or *terminus ante quem*. This example demonstrates that one has clearly to distinguish between the analytical age result of a sample and the archaeological or geological event to be dated. Although this statement applies also to other dating methods, it needs to be kept in mind particularly for the high-resolving radiocarbon dating. Thoughtlessness on behalf of the consumer with the interpretative treatment of ^{14}C ages led frequently to avoidable confusion.

Wood and Charcoal. These are preferred materials for ^{14}C dating due to their generally low and easily removable contamination. Wood consists of cellulose, carbohydrate and lignin. Twigs and stems as well as roots of trees and shrubs are suitable for ^{14}C dating. In the case of stems it must be taken into account that trees are long-lived plants, up to several 100 years old. When growing, one ring is added to the stem each year. With the end of the growth season the newly formed annual ring ceases exchange with the atmospheric ^{14}C reservoir, so that the ^{14}C age of a single ring dates its specific year of growth. If interested in the year of planting, one has to sample the center of the stem, and if interested in the year of cutting, one samples the last ring directly below the bark. This increase in ^{14}C age from the center to the bark raises the *old-wood problem*. This phenomenon can cause considerable uncertainty, if an archaeological context should be ^{14}C dated by charcoal or a building by construction timber. Such date would provide a *terminus post quem* for the context. In the case of charcoal one must bear in mind that preferentially the inner old-wood core survives the burning, consequently its ^{14}C age may predate the fire and burial by a few 100 years (Warner 1990). Only in rare cases the last ring before felling can be identified. The old-wood effect can be circumvented by selecting the shorter-lived twiggy material which is recognized by the complete cross section with few rings. In order to gain information on the possible extent of age offset by the old-wood problem it is always helpful to identify the tree species of the wood or charcoal sample. Since, for practical reasons, several consecutive tree rings are commonly taken as one sample, its ^{14}C analysis yields an average, but not necessarily the mean age of the rings. Determined efforts have been undertaken to date consecutive ring sequences precisely by ^{14}C. Such data from dendro-dated samples allow the construction of reliable calibration curves (Fig. 44).

The chronology of the Aegean Late Bronze Age was re-examined by Manning and Weninger (1992) by means of ^{14}C dates on wood, but also on short-lived materials such as twigs and seeds from Pylos, Mycenae, Midea, Lefkandi, Asine and Kastanas. The dates obtained by probabilistic computer archaeological wiggle matching confirm the conventional archaeological chronology for the Late Helladic and Protogeometric periods (Fig. 52). Wiggle matching of timbers from Anatolian major archaeological sites to the ^{14}C calibration curve promises considerable improvement in the chronology of eastern Mediterranean prehistory (Kuniholm et al. 1996).

Interstadial wood of the early Wisconsin (= last Glacial in North America) was dated by isotope enrichment (Stuiver et al. 1978). A date of 74.7 (+2.7/−2.0) ka BP was found for the St. Pierre interstade. The ^{14}C dates of wood and peat, in combination with palynological studies, revealed a temperature history of the last Glacial in North America that in many aspects can be correlated with that of northwest Europe (Fig. 53). Glacier advances can be dated by ^{14}C on overridden tree trunks found in situ under moraines (Holzhauser 1984; Geyh et al. 1985). It is also possible to date strong earthquakes which are accompanied by coastal subsidence and by tidal mudflats covering former forests. The ^{14}C dating of stumps buried under tidal mud on the Pacific coast of Washington revealed that the last strong earthquake occurred 1680–1720 AD (Atwater et al. 1991).

The transition from the Middle to Late Paleolithic ~40 ka ago falls into the upper age range of the ^{14}C method. Particularly in the presence of contamination, it is advantageous in such a case to employ the AMS technique, as was successfully demonstrated by Valdes and Bischoff (1989) for the site El Castillo cave, Cantabria, Spain. Charcoal samples (62–233 mg)

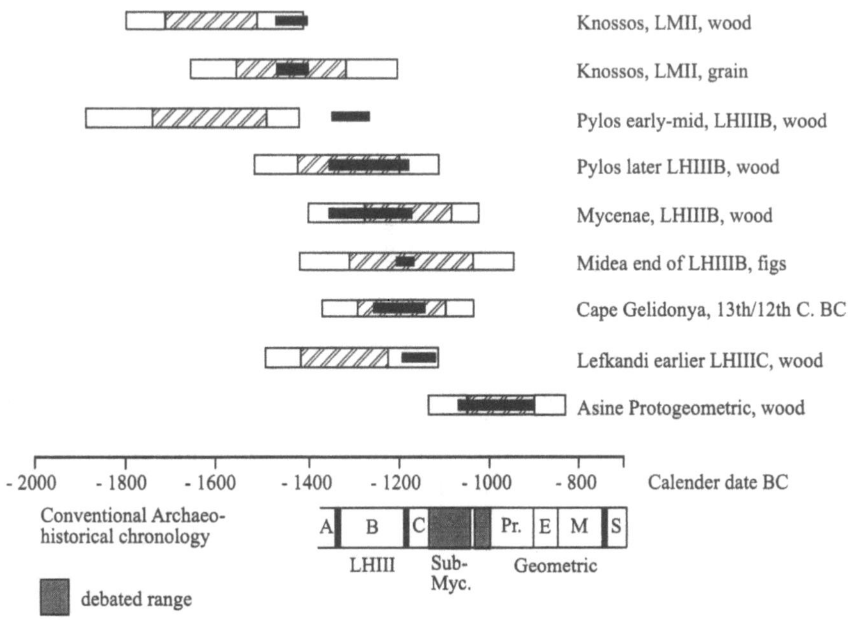

Fig. 52. Comparison between calibrated ^{14}C ages and archaeological dating of the Late Helladic and Protogeometric periods in the Aegean (after Manning and Weninger 1992). The *bars* represent 95% (*total length*) and 68% (*hatched length*) confidence intervals of the data sets. The archaeological ages (*black bars*) are essentially confirmed by the ^{14}C results. The old-wood effect for long-lived wood is clearly apparent in contrast to short-lived sample materials

were taken from different parts of the lowermost beds containing Aurignacian artifacts. After removing rootlets and treating with diluted HCl and NaOH, 5 mg each were analyzed by AMS. The mean ^{14}C age of the three samples amounted to 38.7 ± 1.9 ka BP, which is supported by additional AMS data (39.6 ± 1.5 ka BP, Hedges et al. 1994) and by ESR ages of enamel and dentine (36.2 ± 4.1 ka, Rink et al. 1996). This early date for the beginning of the Upper Paleolithic is identical to the AMS age (38.5 ± 1.0 ka BP) from the same cultural horizon at l'Arbreda cave, Catalonia, Spain, and significantly older than the earliest dates for Aurignacian industries north of the Pyrenees. It must be noted that these results rest on conventional ^{14}C ages and thus do not necessarily represent true ages because no calibration procedure exists for such an early period.

Seeds and Grains. The presence of seeds and grains in pre-ceramic, Neolithic layers testifies to the development of agriculture in the Near East. It is a long-standing goal to find out by ^{14}C dating the time frame when exactly the domestication of plants began and spread to other regions. Cereal

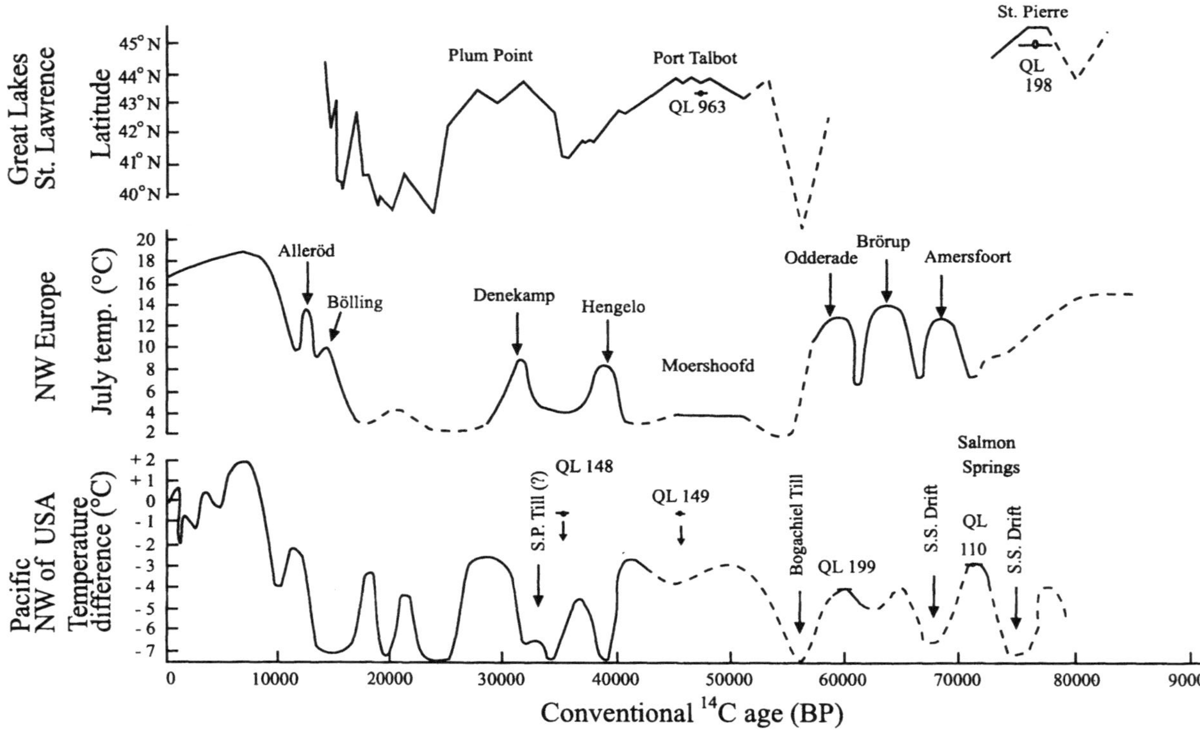

Fig. 53. ^{14}C chronologies for the last Glacial, based on wood and peat. The ages obtained by isotope enrichment date back to 75 ka. The northwest European and North American paleoclimate curves – established by palynological studies – correlate well with each other on the basis of the ^{14}C ages (Reprinted with permission from Science; Stuiver et al. 1978; copyright 1978, American Association for the Advancement of Science)

grains from the "Pre-Pottery Neolithic A" horizon at Jericho revealed ^{14}C ages of ~10 ka (Hopf 1969). Finds of grains with higher ^{14}C ages turned out to be undomesticated wild forms (Henry 1992). The AMS technique allows the analysis of single grains and thus the independent examination of their contextual consistence with the stratigraphic order (Wendorf 1987). In Egypt, a layer 17-ka-old contained grain kernels and date stones, but AMS dating revealed that these finds were in fact only a few 100 years old (Hedges and Gowlett 1986). Obviously, these bits were dispersed by bioturbation, such as the activity of rodents, worms, ants or Man, into deeper layers. The timing and direction of the neolithization in Europe have been reconstructed by means of a large ^{14}C data base of grains and other short-lived materials (Gläser 1991, 1996). Between ~6500 and ~6300 cal BC, the neolithization, coming from Anatolia, entered Europe and quickly proceeded northwest into the Carpathian Basin (Hungary), stagnating there for about 800 years, before it moved ~5500 cal BC, again quickly, further northwest to central Europe.

^{14}C age determination of grains and other short-lived samples plays an important role in dating the Late Bronze Age eruption of the Thera (Santorin) volcano. This date is of consequence for Bronze Age chronology all over the eastern Mediterranean. Archaeologically, the eruption is placed in the Late Minoan IA phase (LMIA). In conventional chronology based on the ceramic trade between the Aegean and Egyptian cultures, the LMIA period is placed in the first half of the 15th century BC. However, archaeological evidence in favor of an earlier placement, by about one century, has also been presented (Manning 1990). If anomalous tree rings ("frost rings") from North America as well as Ireland (La Marche and Hirschboek 1984; Baillie and Munro 1988) and acidity layers in a Greenland ice core (Hammer et al. 1987) are associated with a deterioration of the global climate and atmospheric pollution caused by the Thera eruption – which is plausible, but not yet proven – then the eruption occurred closely around 1630 BC, i.e., about one and a half centuries earlier than the date derived from conventional chronology. By ^{14}C dating of short-lived materials from the volcanic destruction horizon, which covers the Late Minoan settlement at Akrotiri on Thera island, an attempt was made to examine the different temporal emplacements of the Thera eruption independently. The calibrated ages range from the late 18th to the mid-16th centuries, with dominance in the 17th century BC. The 15th century drops out of this range, even if the confidence interval is doubled to $2\bar{\sigma}$ with 95% probability. This implies a revision of the Aegean Late Bronze Age chronology.

Pollen and Spores. Defined pollen zones are essential for the biostratigraphy and the vegetational history of the Quaternary and in particular the

Holocene. The application of AMS analysis to pollen enables the direct chronometric dating of these microfossils instead of the bulk sediment. The potential of this approach was demonstrated for pollen of defined tree species from early Holocene sediments of Mike Lake, North America (Brown et al. 1989). The pollen was concentrated from the sediment while removing carbon-containing contaminants. For an AMS analysis 200–500 spruce pollen grains or equivalent amounts of other species are sufficient. All AMS ^{14}C dates of pollen turned out to be younger than the ^{14}C dates of the corresponding bulk sediment organic fractions determined by β-counting. Furthermore, the pollen data showed better internal consistence and agreed with independent ages. In contrast to the more reliable pollen ^{14}C ages, bioturbation and contamination can introduce significant errors to bulk sediment ages. Pollen grains separated from wasp nests from northern Australian rock shelters were also analyzed by AMS ^{14}C (Roberts et al. 1997).

Phytoliths. These silica microfossils with plant-distinctive shapes are invaluable for reconstructing past environments. Silica phytoliths contain remnants (~1%) of the original plant material encapsulated within them. The silica seals the occluded organic material from the soil environment, thereby preventing allogenic contamination. The AMS technique has opened the possibility of analyzing this organic compound and thus of dating phytoliths directly. After separating the phytoliths from the fine silt fraction of the sediment, the organic material is chemically extracted from the phytoliths by a rigorous procedure (Mullholland and Prior 1993). AMS ^{14}C dates of phytoliths recovered from the Vegas site, Ecuador, confirmed the pre-ceramic age (pre-5500 BP) of the cultural layers. Dates between 7170 BP and 5780 BP were obtained for maize-bearing assemblages, proving the introduction of this domesticate to southwestern Ecuador during the late pre-ceramic period (Prior and Piperno 1996).

Peat and Sapropel. Organic sediments can be dated directly with ^{14}C. Reliable dating presupposes, however, that only contemporary organic components of the sediment are analyzed. Suitable sediments are peat, gyttja and sapropel, but also sediments with less organic material. Contamination with recent carbon may be caused by rootlets and humic acid infiltration. Age overestimates due to redeposited organic material and to calcium carbonates affected by hard-water are also known (Mathewes and Westgate 1980). The AMS offers the advantage of probing microsamples such as fossils selectively, instead of the bulk organic content (Törnqvist et al. 1990). Grootes (1978) applied the isotope enrichment technique to Early and Middle Weichselian peats of Dutch and North German profiles. In combination with palynological studies he was able to reconstruct the northwest European paleoclimate back to 75 ka ago (Fig. 53) and to date the three early Weichselian interstades Amersfoort, Brörup and Odderade in particular. In the profile Grand Pile, Vosges, which plays a key role in the reconstruction of the central European paleoclimate since the last Interglacial, Woillard and Mook (1982) dated pollen-bearing sediments (Sect. 10.2.4) back to 70 ka ago by ^{14}C isotope enrichment. An example for AMS dating of sediments with low contents of organic substance are the deposits from the Lake of Zürich. By means of ^{14}C dates, in combination with sedimentological studies, it was possible to reconstruct the late and post-Glacial development of the lake (Lister et al. 1990).

Paper and Textiles. When dating paper by ^{14}C attention must be paid to the fact that older materials (ragged clothes) are commonly added. Depending on the manufacture, the ^{14}C age of paper may be too high by up to several hundreds of years. Detailed studies on the production technique should precede the ^{14}C age determination (Burleigh and Baynes-Cope 1983). Plant fibers of textiles are excellently suitable for ^{14}C dating. Their raw material is short-lived and is worked within few years after harvest so that the ^{14}C date corresponds within the error limits to the textile manufacture. However, care must be taken regarding the different isotope fractionation effects for various plant species, necessitating the identification of the plant materials used in the textile manufacture and the determination of their $\delta^{13}C$ values. With the introduction of the AMS technique, the sample demand for ^{14}C dating became dramatically reduced, so that the ^{14}C method could for the first time be taken into consideration for precious objects, as for instance the Shroud of Turin. It is believed by many people that the shroud served to wrap Christ's body after descent from the cross. Historically, the Shroud of Turin can be traced with certainty only back to medieval France, a fact that again and again raised doubts about its authenticity. For ^{14}C analysis a strip ($\sim 10 \times 70$ mm) was cut from the 4.5-m-long and 1.1-m-wide shroud, and three specimens (sample 1), each ~ 50 mg in weight, were prepared from this strip and given to the three AMS laboratories involved (Damon et al. 1989). As control, three samples whose ages had been determined independently were also dated (sample 2: 12/13th centuries AD linen from Nubia; sample 3: early 2nd century AD linen from Egypt; sample 4: 1290–1310 AD threads of a cope from France). Because of possible contamination, all three laboratories subdivided the samples and subjected the pieces to several different mechanical and chemical cleaning procedures. The $\delta^{13}C$-corrected conventional ^{14}C ages are shown in Table 4. The measurements of the three laboratories agreed with each other within their $1\bar{\sigma}$-errors. The calibrated dates corresponded to the known ages for

Table 4. Conventional and calibrated ^{14}C ages of the Shroud of Turin (sample 1) and three control samples. (Damon et al. 1989)

AMS laboratory	Sample 1	Sample 2	Sample 3	Sample 4
Arizona	646 ± 31 a BP	927 ± 32 a BP	1995 ± 46 a BP	722 ± 43 a B.
Oxford	750 ± 30 a BP	940 ± 30 a BP	1980 ± 35 a BP	755 ± 30 a BP
Zürich	676 ± 24 a BP	941 ± 23 a BP	1940 ± 30 a BP	685 ± 34 a BP
Mean	691 ± 31 a BP	937 ± 16 a BP	1964 ± 20 a BP	724 ± 20 a BP
Calibrated date	1273–1288 cal AD	1032–1048 cal AD, 1089–1119 cal AD, 1142–1154 cal AD	11–64 cal AD	1268–1278 cal AD

the three control samples. A mean for conventional ages of 691 ± 31 a BP was obtained for the Shroud of Turin, which converts by calibration to 1273–1288 cal AD, and with 95% confidence to the range 1262–1384 cal AD. This result provides conclusive evidence that the linen of the Shroud of Turin is medieval (Damon et al. 1989). The reliability of these results was questioned by Kouznetsov et al. (1996) on the basis that fire-induced carboxylation of the Turin Shroud may have incorporated ^{14}C and thus the ^{14}C ages might be underestimates. However, this possibility was dismissed by Jull et al. (1996) as unsubstantiated and incorrect.

Bones and Antler. Bones were considered as problematic material for ^{14}C dating for a fairly long time. Their inorganic fraction (carbonate hydroxyapatite) – aided by the high porosity – tends to exchange ions with the groundwater and is, therefore, composed of a primary and a secondary component. Consequently, age determination is unreliable on the inorganic bone fraction. However, under favorable, dry conditions carbonate hydroxyapatite may yield reliable ^{14}C ages, as has been shown for human bones from Saharan tombs (Saliege et al. 1995). The absence of carbon exchange was demonstrated by the original $^{13}C/^{12}C$ ratio in the carbonate hydroxyapatite. Besides this main component, bones also contain a few percent of organically bound carbon which is much more resistant to exchange even in the presence of diagenetic alteration. The organic bone substance consists predominantly of various proteins, generally classed as *collagen*. The collagen is chemically extracted as acid-insoluble residue. Rather than analyzing the bulk collagen it is advantageous to separate different collagen fractions and to date them by AMS (Taylor 1987b; Hedges and van Klinken 1992). An example of this approach is the dating of a late Glacial mammoth bone from Domebo (Stafford et al. 1987) of which 15 different chromatographically separated, organic fractions were analyzed (Fig. 54). All fractions gave ^{14}C ages

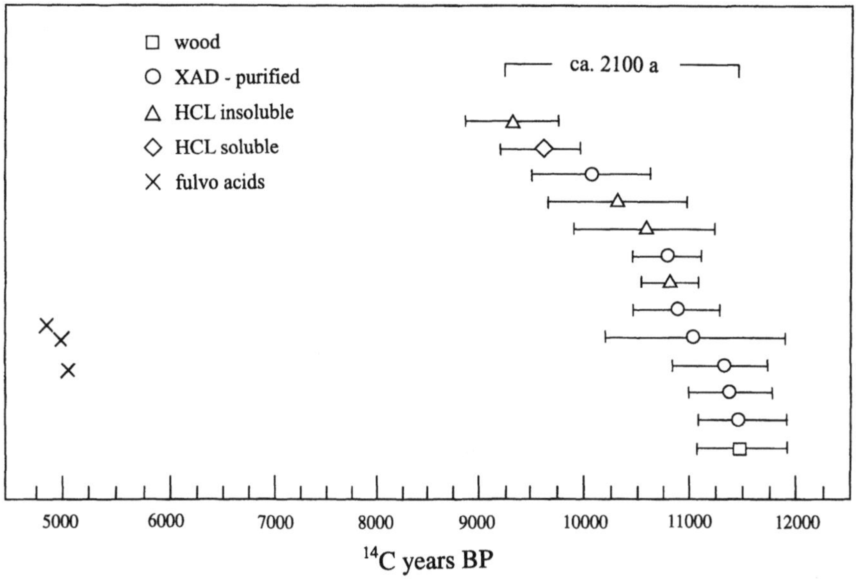

Fig. 54. AMS-^{14}C ages of different collagen fractions of a late Glacial mammoth bone from Domebo (after Stafford et al. 1987). The ages are up to 2 ka younger than the age of an associated wood sample. The age lowering is attributable to contamination from 5-ka-old humic acids (fulvo acids)

which are more or less younger than a wood sample (11,490 ± 450 a BP) from the same context. The apparent age lowering is caused by contamination with humic acids which themselves have ^{14}C ages around 5000 a. Sample weights of a few grams of bone are sufficient for the AMS technique. Also, Upper Paleolithic and Mesolithic artifacts made of deer antler were successfully dated by AMS (Smith and Bonsall 1990). AMS dating of bone also contributed essentially to the solution of the long-standing enigma when Man first entered the North American continent. Previous ^{14}C dating by means of β-counting as well as uranium-series and racemization dates of human bones had revealed an age range between 17 and 70 ka. However, AMS dating on extracts of specific amino acids from the same bones gave results which were 7 ka at the oldest (Taylor 1991). By ^{14}C charcoal dating on Alaska occupation sites, paleo-Indians can presently be traced back to 13–13.5 ka cal BP (Kunz and Reanier 1994).

Bone ^{14}C dating in combination with dendrochronology at Early Bronze Age burial places in Singen and the Neckar region, Germany, caused a shift in the estimated date of the beginning of the Early Bronze Age in central Europe (Becker et al. 1989). With the archaeological-historical method, this beginning traditionally had been placed at the turn of the 18th/17th

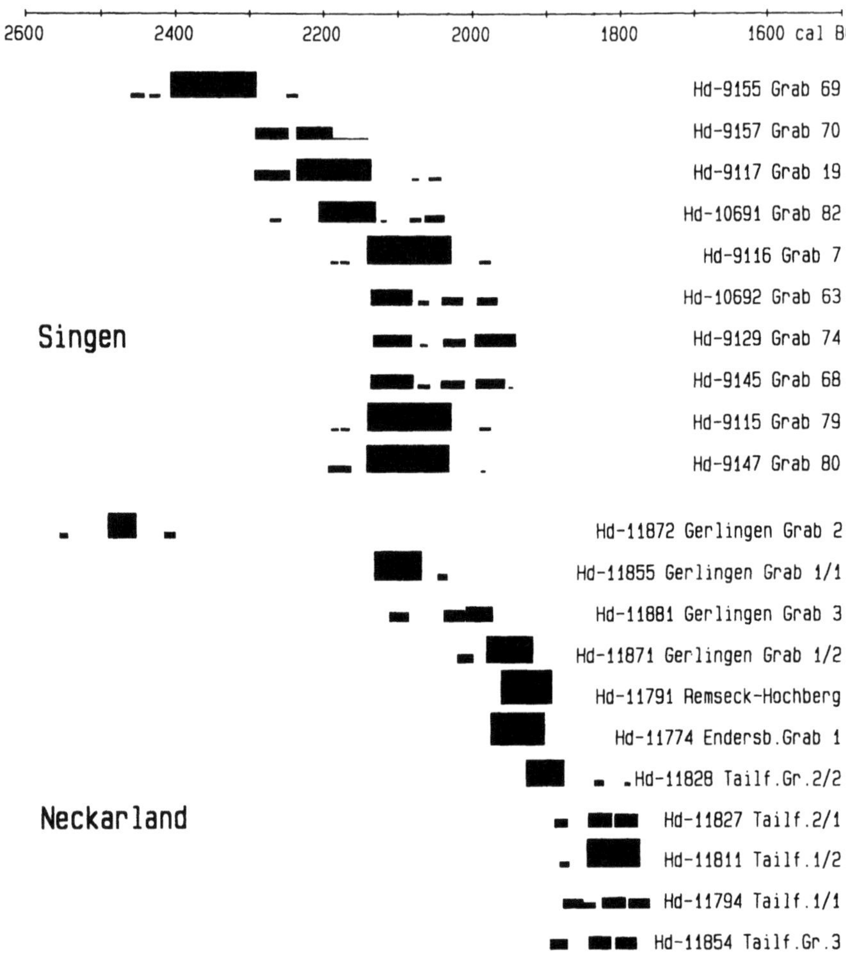

Fig. 55. Calibrated ^{14}C dates of bones from Early Bronze Age burial places in southwestern Germany. Each *horizontal block* represents a sample. The *height* of a block represents the probability of the age value for the age range. (After Becker et al. 1989)

centuries BC. Collagen was isolated from 22 bones with sufficiently high collagen fractions (8.5–17.4%). The $1\bar{\sigma}$-precision of calibrated ages for the Singen samples amounted to ± 35–40 a uncertainty, whereas those for the Neckar region samples with double the size and counting time to ± 25–35 a. The calibrated dates are presented in Fig. 55. The oldest dates range in the 23rd/24th centuries, but the majority were within 2100–2000 BC (Singen) and 1950–1750 BC (Neckar region). The ^{14}C dates of Singen push the beginning of the Early Bronze Age backward to the 23rd century BC which has important implications for the development of tin bronze technology in central Europe. As another consequence of these dates, with the duration of 700 a, the Early Bronze Age in central Europe was considerably longer than hitherto assumed. A further example of dating bones by ^{14}C is the Middle Neolithic site at Künzing-Unternberg, Germany (Petrasch and Kromer 1989), where numerous animal bones were found in the sedimentary filling of ditches < 4 m deep. ^{14}C dating was applied in order to obtain information about the duration and mechanism of the ditch filling. Six bones each were collected from the base and the top layers of the sedimentary filling. The collagen content of the bones was relatively high (4 and 11%). Since small differences in age were expected, high precision measurements with double sample size and extended β-counting (1 week) were performed, resulting typically in ± 35 a (1σ-precision). The sample groups differed only slightly in their conventional ages: 5910–5840 a BP for the lower and 5850–5650 a BP for the upper group. After calibration and statistical analysis the data revealed a time-span of 60 a between 4840 and 4780 BC, within which the ditch had been used and steadily renewed. Finally, the ditch was abandoned and slowly (0.3 cm a^{-1}) filled up with colluvial sediments between 4740 and 4590 BC.

Soils. Paleosols contain carbon in various fractions, predominantly as allogenic carbonate, as rootlets, both of which have to be removed, and as authigenic organic compounds. These last consist of NaOH-soluble humic acids and NaOH-insoluble micro-residues; both of them can be used for ^{14}C dating. The ^{14}C age derived from humic acids cannot be taken directly as the mean age of the soil formation due to the complexity of pedogenesis. The interpretation requires some kind of pedological model. Furthermore, the contamination of older soils by descending recent humic acids disturbs the dating (Scharpenseel and Schiffmann 1977; Head et al. 1988). For the pedogenetical interpretation of the age results it is helpful if ^{14}C data of the NaOH-soluble as well as the insoluble fractions exist and can be compared to each other (Geyh et al. 1985). Soil carbonates also raise difficulties due to their complex genesis, and their ^{14}C age data may not be related to the time of their actual formation or that of the soil (Netterberg 1978).

Limnic Sediments. Limnic sediments form an important archive for the climatic fluctuations of the past and thus require reliable chronometric dates. If lake sediments contain organic matter ^{14}C dating can be directly applied, as already discussed in connection with sapropels. Macrofossils from the sediments, such as nutlets, fruit scales and leaves, are excellent material for AMS ^{14}C dating. Such fossils are short-lived and free of the hard-water effect. The ^{14}C ages of macrofossils collected from varved sediments in lake Gosciaz, Poland, enabled wiggle matching with the German oak chronology. Using this approach, the Younger Dryas/Preboreal boundary was dated at 11,440 ± 120 a cal BP (Goslar et al. 1995). Analogously, Björk et al. (1996) placed this boundary at 11,450 – 11,390 ± 80 a cal BP (revised to 11,500 ± 20 a cal BP; Spurk et al. 1998) using varved sediments from southern Sweden.

An attempt has also been made to apply ^{14}C dating to the carbonate content of limnic sediments. In a study on the Pleistocene Lake Lahontan, Nevada, Lin et al. (1996) demonstrated the specific difficulties of this approach. Samples of tufa, probably precipitated by algae, were collected from former, elevated shorelines. Both the $^{230}Th/^{234}U$ isochron (Sect. 4.1.1) and the AMS ^{14}C techniques were applied. The calibrated, but not reservoir-corrected, ^{14}C ages ranged between 12 and 20 ka, and were generally lower than the corresponding $^{230}Th/^{234}U$ ages by up to 2.3 ka. The hard-water effect of a Pleistocene lake-water reservoir is difficult to assess, and if corrected for, it would even increase the discrepancy between the two dating methods. Modern carbon contamination and/or the deposition of secondary calcites are likely to be responsible for part of the discrepancies.

Calcareous Cave Deposits and Travertine. Since secondary calcareous sinter recruits its carbonate content from the groundwater, methodological analogies with the ^{14}C dating of groundwater exist. As in the case of groundwater, only a certain fraction of carbon comes from the atmosphere and biosphere. In accordance with the carbonate/bicarbonate chemistry of groundwater, the other fraction comes from geologically old limestone (cf. Eq. 44). Consequently, calcareous sinters already have at the moment of their precipitation a reservoir age that may amount up to several 1000 years, depending on the ratio in which both carbon components are mixed. Of the various cave sinters stalagmites are the best suited for ^{14}C dating because they grow in defined layers upward and not radially, as stalactites do (Fig. 7). By means of ^{14}C measurements, the growth rates of Holocene stalagmites turned out to be in the order of 0.1 mm a^{-1}, but were significantly lower in the preceding interstades (Geyh and Franke 1970). Isotopic exchange with the moisture, recrystallization as well as secondary carbonate precipitation in pores may lead to ^{14}C ages that underestimate the true age (Geyh and Hennig 1986). When dating travertine, one also has to take into account the complication that the groundwater may already have a significantly high ^{14}C age at the moment of carbonate precipitation. During travertine formation, algae play an important role (Srdoc et al. 1989). Another approach to dating travertine is to utilize its organic component. Microbial mats which grow in the travertine consist mainly of bacteria, diatoms, algae and bryophytes as well as of trapped pollen and spores. This novel approach was successfully tested on travertine encrusted in Precolumbian canals of the Tehuacan valley of Pueblo, Mexico (Winsborough et al. 1996).

Mollusk Shells. Mollusk shells consist of an organic matrix (conchiolin) in which calcium carbonate is incorporated as the main component. Most ^{14}C dating of shells use this inorganic component. The initial $^{14}C_0$ content in the carbonate of marine mollusks is – apart from isotope fraction effects – governed by that of the seawater at the time of the shell growth. Compact calcareous shells with well-preserved texture are optimally suited for ^{14}C dating. Shells with signs of recrystallization and groundwater exchange reactions must be avoided.

^{14}C analysis is commonly applied to marine shells from fossil, elevated shorelines and beach terraces with the aim of tracing sea level changes or reconstructing the uplift pattern. One of these studies was performed on shorelines along the Woodfiord, Spitsbergen (Brückner and Halfar 1994). The ^{14}C dates of shells from shorelines 2–39 m above sea level revealed – after subtracting 400 a for the reservoir effect and calibration – that the glacio-isostatic rebound occurred during the final phase of the Weichselian and in the early Holocene. ^{14}C dates of marine mollusk shells are also of archaeological interest. Prehistoric coast settlers left numerous shell middens. Apart from their archaeological significance, middens are also sensitive indicators of sea level and climate changes due to their position close to the shoreline. ^{14}C dates of shell carbonate suffer from the marine reservoir effect and from secondary exchange reactions. A systematic age difference between the ^{14}C dates of the carbonate fraction and accompanying charcoal has been observed for prehistoric middens in Oman (Uerpmann 1990). After reservoir correction (–800 a) and taking isotopic fraction into account, the ^{14}C dates place the coastal settlement activities in Oman in the Neolithic and Bronze Age periods. The ^{14}C dating of non-marine mollusk shells, predominantly land and freshwater snails, is often problematic, since carbon from old limestone is taken in with the food. Depending on their environment and nutrition, recent freshwater snails have reservoir ages varying from several hundreds to a thousand years (Evin et al. 1980).

Eggshell. Avian eggshells contain carbon in the carbonate matrix as well as in the organic matter. For dating purposes the organic component is more straightforward than the carbonate since its proteins are directly derived from the biosphere. Furthermore, it obeys more likely the closed system assumption. However, the carbonate component also seems to be suitable for ^{14}C dating. Feeding experiments have shown that its ^{14}C content reflects that of feed and atmosphere (Long et al. 1983). Ingestion of "old" carbonate seems to be negligible. Isotopic fractionation effects due to pedological and biochemical processes need to be taken into account.

Corals and Foraminifers. Fossil corals are important indicators of sea level fluctuations. For dating their calcareous skeleton, the ^{14}C AMS method is well suited. The marine reservoir effect of the surface water is taken into consideration usually by subtracting 400 a from the conventional date. By dating over 100-m-deep drill-hole samples from the Barbados reef, the sea level rise by a maximal 131 m since 18,200 a BP was reconstructed (Fairbanks 1989). Parallel mass spectrometric $^{230}Th/^{234}U$ dates on the same samples turned out to be up to 3500 a older. The comparison of both ages (Fig. 45) was an attempt to extend the ^{14}C calibration curve beyond the dendro-based section (last 12 ka) back to 20 ka ago (Bard et al. 1990b, 1992). Analogous studies were carried out on reefs of the Abrolhos Islands, Australia (Eisenhauer et al. 1993), and the Huon Peninsula, Papua New Guinea (Fig. 29; Edwards et al. 1993).

In connection with oxygen isotope analyses and micropaleontological as well as sedimentological studies ^{14}C AMS dating on monospecies foraminifers allows the reconstruction of the late Glacial and Holocene paleoclimate (Bond et al. 1992; Duplessy et al. 1992). With the goal of obtaining information on the causal connection between the deep oceanic circulation and the paleoclimate during the past 20 ka, planktonic and benthonic foraminifers were analyzed with the AMS technique (Andree et al. 1984a). The sample size (8 mg) required 30–1000 single shells (depending on the species), which were hand-picked from the deep-sea sediment core. Bioturbation limited the time resolution of the age data in this study.

Rock Varnish. Rock surfaces, which are exposed to the atmosphere for a long time under arid or semiarid climatic conditions, are coated with a thin patina, known as rock or desert varnish. Rock varnish consists predominantly of iron-manganese oxides and clay minerals. An attempt to date the varnish – in addition to that made by using chemical cation analysis (Sect. 8.6) – was made using ^{14}C. Rock varnish, at least that occurring in the Californian Mojave Desert, contains some ($<2\%$) organic matter, which allows the application of the AMS technique. It is presupposed that the organic carbon in the innermost, oldest layer of the varnish stems from the time that the varnish began to form, and it was not exchanged since then with that of the biosphere or atmosphere. Owing to its shallowness (mostly $<100\ \mu m$), large surface areas (several $10^3\ cm^2$) are necessary for sampling, even for the sensitive AMS-detection method (Dorn et al. 1986, 1989). After laborious mechanical separation of the very bottom layer of varnish from the rock surface, the varnish is chemically processed in order to extract the organic carbon. The feasibility of varnish dating was demonstrated in the deserts of Arizona and California, where ^{14}C ages between 175 ± 45 and 42,900 ± 3200 a BP were found (Dorn et al. 1986, 1989). Such results are useful, on the one hand, for the morphogenesis of arid regions and, on the other hand, enable the calibration of the cation dating method. Weathering crusts on moraine boulders in the semiarid eastern Sierra Nevada were also successfully dated (Dorn et al. 1987). AMS-^{14}C dating of the Tioga moraine gave 19,050 ± 420 and 13,170 ± 200 a BP. The possibility that these ages are somewhat too young, due to contamination with young organic carbon, cannot be excluded. The problem of contamination has also been encountered in a study of paleo-Indian petroglyphs in the lower Colorado River area of southwestern North America (von Werlhof et al. 1995).

Mortar. Mortar is produced from limestone $CaCO_3$, whereby the CO_2 is driven out by burning and quicklime CaO remains; this is then slaked with water to $Ca(OH)_2$. The slaked lime sets with the atmospheric CO_2, thereby changing back to $CaCO_3$, the mortar. This reaction is the decisive step for ^{14}C dating, because young atmospheric carbon is bound. In a mortar formed according to this simple model, the total content of inorganic carbon originates from the atmospheric carbon dioxide. If there is no further carbon exchange with the atmosphere after setting, the ^{14}C age of the mortar dates the construction in which the lime was used. However, there are two spots where this dating model can be seriously disturbed. First, the burning may have been incomplete, so that unburned carbonate with old carbon is inherited by the mortar, thus causing the ^{14}C ages to be too high, as observed for instance in the mortar floor of the aceramic Neolithic cult building at Nevali Cori, Turkey (Reller et al. 1992). Secondly, materials which contain old carbonate may be added during mortar production. Therefore it is not surprising that ^{14}C ages of mortar are frequently unreliable. Furthermore, depending on the length of the diffusion path which the CO_2 takes through the wall, various degrees of isotopic fractionation occur, requiring the determination of the $\delta^{13}C$ value. An example of mortar dating is the study of the Franciscan convent Kökar on Aland Island, Finland (Gustavsson et al. 1990). Three mortar samples each were collected from the refectory and the south house. The $\delta^{13}C$ values range from 15.0 to 21.2‰. The ^{14}C results reveal construction phases of 1225–1270 and 1260–1280 cal AD, respectively. These dates are 100–150 a earlier than suggested by the oldest archaeological finds. Charcoal inclusions in mortar can also be used for dating buildings as has been demonstrated by Berger (1990) for early medieval Irish churches.

Stone Tools and Rock Paintings. When stone tools get into contact with organic material during production or use, traces of organic material commonly remain on the rock surface. The organic traces comprise animal residues, such as blood, muscle fibers, hair, feathers and fatty acids, as well as plant residues, mainly resin, starch and fibers. They are quite resistant to secondary alteration. Blood residues, for instance, may be preserved over 10^5-10^6 a. The manufacture or use of the tools can be directly dated by ^{14}C. Because of the minute size of such remainders, only the AMS technique is suitable (Nelson et al. 1986).

Paleolithic rock paintings commonly contain organic material, such as charcoal, carbonized plant matter, pigments, plant fibers, blood, fatty acids and beeswax (Nelson et al. 1995), and this enables ^{14}C dating. The rock art at Laurie Creek, North Australia, that contains human blood in the pigment, yielded a ^{14}C age of 20,320 (+3100/−2300) a BP (Loy 1991). The recently discovered rock paintings at Chauvet-Pont d'Arc, France, gave ^{14}C ages ~31,000 a BP, using microsamples of charcoal from the paintings (Clottes et al. 1995); these paintings thus belong to the earliest examples of prehistoric rock art so far discovered. Tuniz (1996) reports successful AMS ^{14}C dating by laser combustion of charcoal-containing single consecutive laminae, separated from a 2-mm-thick rock surface accretion that covers rock art in North Australia. In this case, a decreasing age gradient from the inner toward the outer laminae was observed.

Ceramics. Prehistoric chronologies rest largely on the typological development of ceramics. For the temporal placement of such an established relative framework, chronometric age determination on the ceramic material directly is desirable. Apart from the TL method (Sect. 7.1.3) the ^{14}C method can also be utilized for that purpose. Because of the tiny amounts of organic carbon in ceramics, this ^{14}C dating only became practicable with the introduction of the AMS technique, which requires only milligram carbon samples. However, in rare cases, β-counting has also been used (Bollong et al. 1993). Suitable materials are burned food remains in cooking vessels, organic glues on fractures, inclusions of plant chaff and seeds. The validity of this approach has been successfully demonstrated for Neolithic and Iron Age pottery of independently determined ages from various Swiss archaeological sites (Bill et al. 1984).

Metallurgical Slag. Slags are often the only remains of ancient smelting activities and present an important source of information for the development of metallurgy. However, their dating, neglecting the rare cases when they are found in stratigraphically safe positions, commonly raises difficulties. Charcoal, which was used as fuel and thus frequently occurs on smelting places, must often be excluded as a dating object due to its unclear context association. Smelting and reworking on the same spot frequently went on for centuries. This problem is circumvented if tiny inclusions of charcoal are found within single pieces of slag to which the AMS can be applied. For instance, a 0.4-mg-small charcoal inclusion in a Late Bronze Age copper slag from Trentino, Italy, yielded the archaeologically reasonable age of 2780 ± 330 a BP (Bill et al. 1984).

Groundwater. Age determination of groundwater belongs to the most problematic applications of ^{14}C. The carbon content in groundwater is a mixture of two components: the fraction which originates from the biosphere and atmosphere reservoirs, which carries the ^{14}C, and the fraction which originates from geological carbonates and bears no or very little ^{14}C. Depending on the mixing ratio of both components, recent groundwaters exhibit various $^{14}C_0$ concentrations. Furthermore, a groundwater body does not behave as a geochemically closed system. It undergoes steady exchange with the minerals of the aquifer and may also mix with other water bodies. Despite these complications, ^{14}C dating of groundwater yields hydrogeological meaningful results in many cases. However, ^{14}C analysis on groundwater does not primarily aim at its actual dating, but rather at its genesis and hydrodynamics. This presupposes hydrological model constraints (Fontes 1992; Geyh 1992).

In groundwater the carbon is present as free carbonic acid and as calcium bicarbonate (Fig. 56). Rainwater infiltrates the topsoil and combines there with the biogenic CO_2 to carbonic acid. The carbonic acid dissolves fossil carbonate in the ground according to the equilibrium reaction

$$CO_2 + CaCO_3 + H_2O \leftrightarrow Ca(HCO_3)_2 \quad (44)$$

The old carbonate component, whose ^{14}C content has already more or less or even completely decayed, dilutes the ^{14}C content of the young biogenic component. Therefore recent groundwater already has a ^{14}C reservoir age up to several 1000 years. Isotopic exchange between the ^{14}C-containing groundwater and ^{14}C-free limestone in the aquifer increases the apparent ^{14}C age even further. The extent of such secondary reactions can be assessed by accompanying ^{13}C analysis (Münnich 1968) since the $\delta^{13}C$ values of plants and limestone differ from each other (Fig. 47).

In an extensive investigation of Sahara groundwaters by Sonntag (1980), it was shown that the deep, often artesian water was formed during a long humid phase 50–20 ka ago (Fig. 57). In this Pluvial period, which corresponds to the last Glacial, the Sahara received sufficient winter rain with the west drift as is evident from accompanying δD and $\delta^{18}O$ measurements. These data reveal that due to the continental effect the rain becomes iso-

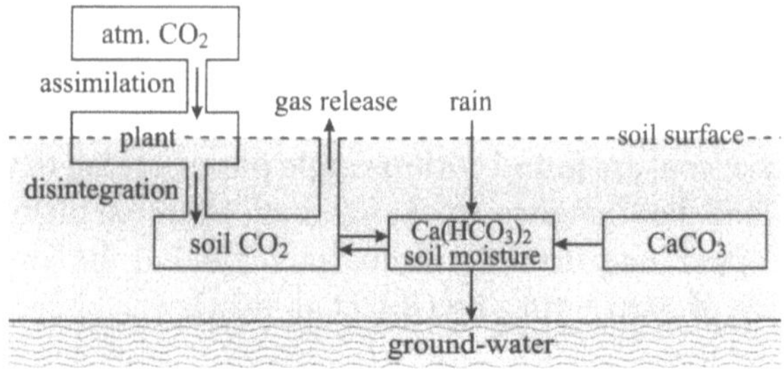

Fig. 56. Schematic pathway of carbon from the atmosphere to the groundwater. (After Münnich 1968)

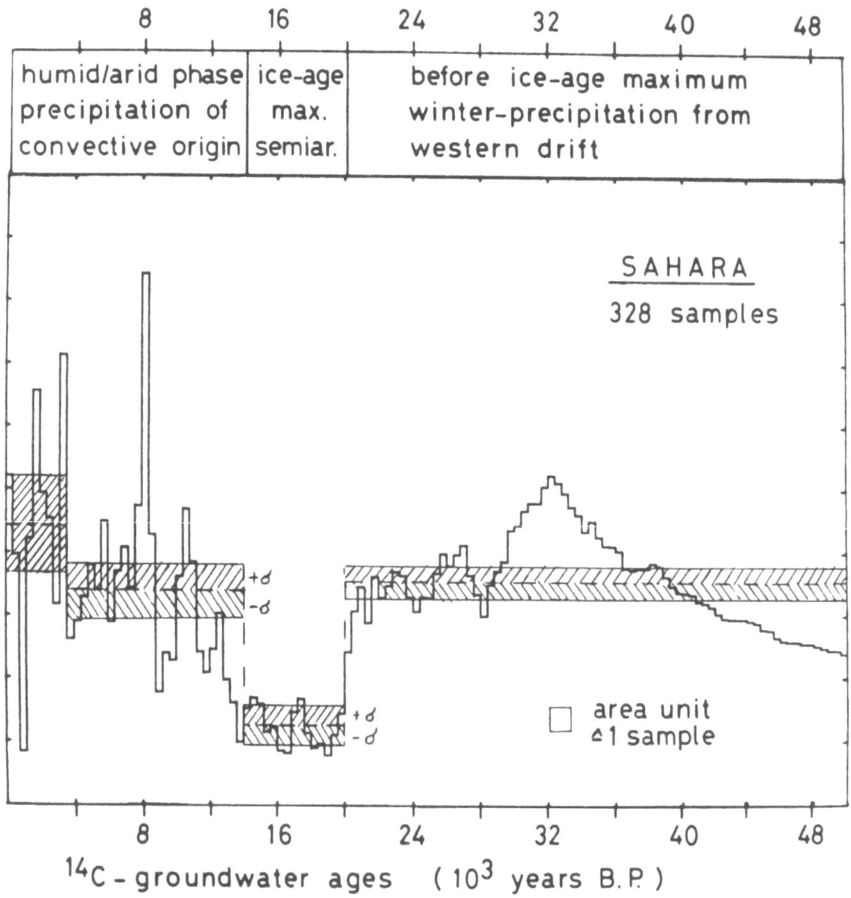

Fig. 57. Frequency distribution of the ^{14}C ages of artesian groundwater from the Sahara (after Sonntag 1980). The distribution reflects the climatic development of the last Glacial and the Holocene in this region. The renewal of the groundwater strongly decreased 20–14 ka ago because of increasing aridity

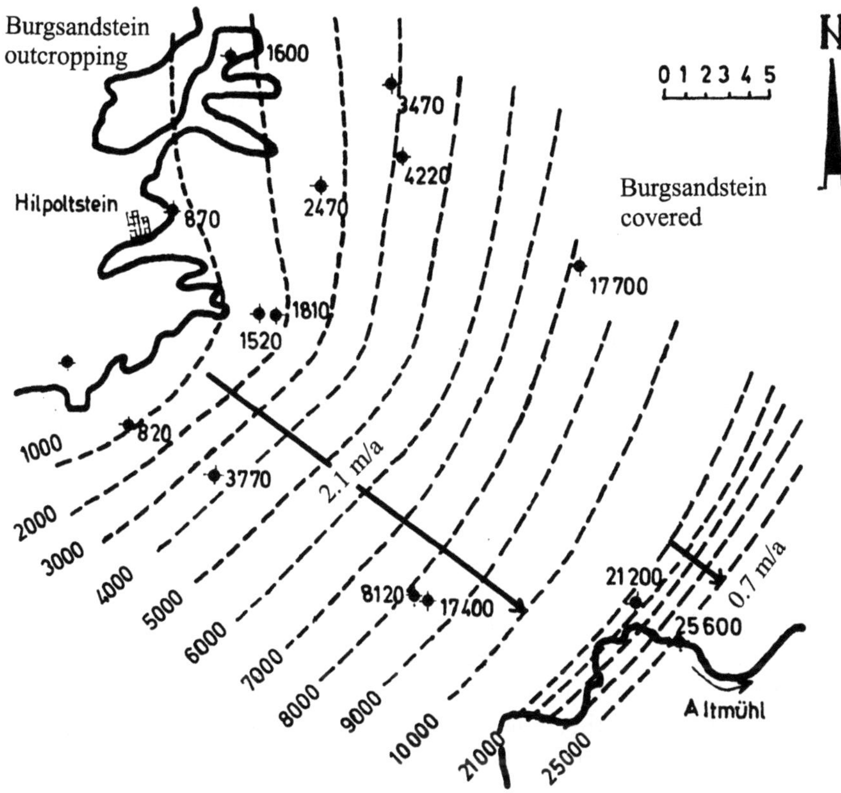

Fig. 58. ^{14}C age isochron of the groundwater in a Mesozoic sandstone aquifer (Burgsandstein), Frankenalb, southern Germany (after Andres and Geyh 1972). The age hiatus at 20–11 ka is caused by the inhibition of groundwater renewal under the permafrost conditions during the glacial maximum. During the Holocene (< 11.5 ka) a continuous groundwater renewal took place, roughly three times as fast as 25 – 20 ka ago (scale in km)

topically lighter eastward with increasing distance from the Atlantic coast. In the time span 20 – 14 ka, during the climax of the last Glacial, the groundwater formation declined strongly due to the arid climate. Also, in the periglacial regions of central Europe, the groundwater formation receded – although for different reasons – at the same time (20 – 11 ka ago), as shown in a study of the Keuper sandstone aquifer in the southern Frankenalb, Germany (Andres and Geyh 1972). The ^{14}C age distribution (Fig. 58) shows that during the Glacial maximum, when permafrost sealed the land surface, groundwater renewal was largely cut off.

Ice Cores. During the growth of glacier ice, air is incorporated with the accumulating snow and dust. With progressing diagenesis the air becomes firmly enclosed in small vesicles and thus isolated from further exchange with the atmosphere. The carbon dioxide contained in the enclosed air can be utilized for ^{14}C dating. It is expected that the ^{14}C method dates the ice formation at when the system became closed. Contaminating carbon may be introduced into the ice as terrestrial carbonate dust or as organic matter. Since 1 kg ice contains only ~15 µg carbon as CO_2, about 1 t of ice needs to be collected for CO_2 extraction for ^{14}C detection by activity measurement, even if highly sensitive counting equipment is employed (Oeschger et al. 1966). In contrast, 120–140 µg carbon is enough for AMS analysis, which means that only 10 kg of ice is required. The Greenland ice core Dye 3 was dated using this approach. After calibration the ^{14}C ages agreed well with those derived from annual ice-layer counting (Andree et al. 1984b).

5.5 Neon-21

The isotope ^{21}Ne of the noble gas neon is produced in situ in the rock surface by cosmic rays. Like cosmogenic ^{3}He, ^{21}Ne – in opposition to the other cosmogenic nuclides utilized for dating – is stable, so that both methods are similar to each other and are frequently applied together. The ^{21}Ne content grows proportionally to the duration of irradiation. Thus, for known ^{21}Ne production rates, the exposure time can be inferred from the cosmogenic ^{21}Ne amount. Surface exposure ages (1 ka up to several Ma) as well as erosion rates of basalts and quartz-rich rocks can be determined with the ^{21}Ne method. Cosmogenic ^{21}Ne in terrestrial rocks was first detected by Marti and Craig (1987). Meanwhile the feasibility and potential of the ^{21}Ne method have been convincingly demonstrated. Insufficient knowledge of the cosmogenic neon production rate still poses a major obstacle.

5.5.1 Methodological Basis

On Earth the noble gas neon occurs in the atmosphere and in rocks. It consists of the isotopes ^{20}Ne, ^{21}Ne and ^{22}Ne with atmospheric abundances of 90.5, 0.3 and 9.2%, respectively. All three isotopes are stable. Neon was already present as primordial neon at the formation of the Earth. Additionally, all three isotopes are formed by the interaction of cosmic rays with rocks close to the Earth's surface. Such rocks may contain neon which is a mixture of three components, namely, cosmogenic and atmospheric neon as well as primordial neon originating from the Earth's mantle. Nucleogenic ^{21}Ne, formed by (α, n) reaction in O and F, has also been observed in quartz (Niedermann et al. 1994). Since these components differ in their isotopic ratio, the isotopic composition of the neon mixture varies accordingly.

The most important in situ production process in surficial (< 2-m-deep) rocks is neutron-induced spallation of heavy nuclei, such as O, Si, Mg and Fe. Since these elements are common in quartz, olivine and clinopyroxene, quartzite and mafic rocks, such as basalt, are of major interest for cosmogenic ^{21}Ne detection (Graf et al. 1991). The production rate of cosmogenic ^{21}Ne is derived through that of ^{3}He (Sect. 5.2.1) and the $^{3}He/^{21}Ne$ production ratios. The latter one depends strongly on the chemical composition of the target material, e.g., 4.5 for quartz and 3.5 for olivine (Staudacher and Allegre 1991, 1993). Also, the effects of shielding by overlying rocks, of the geomagnetic latitude and the elevation need to be taken into consideration. An unsolved problem – in particular for low radiation ages less than 10 ka – is the temporal variation of the ^{21}Ne production. As in the case of the other cosmogenic nuclides, erosion and surface coverings also effect variations in the actual production rates.

Since only the cosmogenic $^{21}Ne_c$ correlates with the exposure time, the measured neon needs to be split up into its single components by means of their known isotopic composition (Staudacher and Allegre 1991). Cosmogenic neon has $(^{20}Ne/^{22}Ne)_c$ and $(^{21}Ne/^{22}Ne)_c$ ratios of 0.84 and 0.96, respectively, which is known from meteorites. Atmospheric neon has respective ratios of 9.8 and 0.03 and the upper mantle component (MORB-neon) 13.4 and 0.07. In near-surface quartz one finds a mixture of cosmogenic and atmospheric neon, whereas in basalts, mantle neon also occurs.

5.5.2 Practical Aspects

For dating basalt flows ~1-kg samples are taken from shallow (few cm) and defined depths below the surface. As already mentioned in connection with the ^{3}He method, it is crucial that the surface is not eroded, which is indicated by the presence of well-preserved flow structures and chilling rinds. From the ground basalt 0.5–1.6 mm-sized olivines and clinopyroxenes are separated magnetically and finally by hand-picking. After cleaning the grains are fused and the released neon analyzed by mass spectrometry. The neon measurement can be performed jointly with that of helium. With regard to the retention characteristics, in particular for quartz, the hazard of diffusive loss of neon is less than that of helium (Graf et al. 1991).

5.5.3 Application

Basalt. For dating basalts by ^{21}Ne, olivine separates are utilized, as accomplished by Staudacher and Allegre (1993) for basalts from Réunion Island. The Piton de la Fournaise volcano experienced four eruptions and three intermittent caldera collapses. Samples were collected from basalt flows of the last two eruptions and the caldera wall of the last collapse at ~2330 m altitude. The samples come from the upper most 10 cm below the surface. The $^{21}Ne_C$ production rate for basaltic olivine at 2330 m elevation and 21°S geographic latitude (Réunion) amounts to 169 atoms g^{-1} a^{-1} (Sarda et al. 1993). In the case of the caldera olivine, the production rate is geometrically reduced due to the steep dip angle (75°) of the wall. The ^{21}Ne ages for the eruption phases are 62 ± 4 and 3.4 ± 1 ka, respectively. The age of the younger event is supported by ^{14}C dates. The intermittent caldera collapse happened 23.8 ± 2 ka ago. This age is a minimum age for the collapse because rock slides may have occurred on this steep wall and thus exposed a fresh surface. The ^{21}Ne ages agree well with ^{3}He ages determined on the same olivine separates. For one of the basalt flows a systematic age difference between the K–Ar and the ^{21}Ne ages was observed, where the ^{21}Ne age was lower by 13%. According to Sarda et al. (1993), this difference is caused by an average erosion rate of 3.5 ± 1.7 µm a^{-1}.

Moraines. ^{3}He and ^{21}Ne exposure ages of 1.35 ± 0.07 Ma and 2.21 ± 0.12 Ma, respectively, were obtained on quartz separates of an quartzitic moraine boulder from the Arena Valley in the Transantarctic Mountains (Staudacher and Allegre 1991). The ^{21}Ne age agrees with ^{10}Be as well as ^{26}Al exposure ages of the same quartzite. It dates the formation of the moraine. The age underestimate by ^{3}He is attributed to diffusive loss of helium. The ratios $^{21}Ne/^{10}Be$ and $^{21}Ne/^{26}Al$ might also be of interest for reconstructing the erosion history of quartz-containing rocks (Graf et al. 1991). A ^{21}Ne production rate of 21 atoms g^{-1} a^{-1} in quartz (at sea level) was determined, using quartz separates of Sierra Nevada rocks which were brought to the surface by glacial scouring during the Tioga period 11 ka ago (Niedermann et al. 1994).

5.6 Aluminum-26

Cosmogenic ^{26}Al is produced in the atmosphere as well as in situ at the rock surface. The atmospheric ^{26}Al becomes adsorbed to aerosols and, together with them, is washed out by the rain to the Earth, where it is incorporated into sediments. There the decaying ^{26}Al content is a measure for the time since deposition. In contrast, the in situ ^{26}Al in surface rocks builds up with time and thus allows the dating of the surface exposure and the study of erosion rates. The ^{26}Al method is applicable in the 1 ka^{-1} to Ma range. Its beginning goes back to the 1960s (Lal and Peters 1967; Tanaka et al. 1968). The highly sensitive AMS detection technique has much increased the potential of the method (Raisbeck et al. 1983). It is applied to dating ice cores, deep-sea sediments and quartz-containing surface rocks. The ^{26}Al method bears strong analogies to the ^{10}Be method. Difficulties caused by the initial ^{26}Al content and temporal variations in the production rate are circumvented by the combined analyses with ^{10}Be, known as the *^{10}Be–^{26}Al method.*

5.6.1 Methodological Basis

Aluminum is with 8.1% – after oxygen and silicon – one of the most abundant elements in the Earth's crust. Natural aluminum consists – apart from cosmogenic ^{26}Al – only of the stable isotope ^{27}Al. ^{26}Al is radioactive and decays with $t_{1/2} = 705$ ka under β^+ and γ emission to ^{26}Mg.

In the atmosphere ^{26}Al is predominantly produced by spallation of argon nuclei. The production rate depends strongly on the geographic latitude, due to the geomagnetic shielding of cosmic rays, and thus has changed with the temporal variation of the dipole field. The geochemical behavior of ^{26}Al in the exogenic cycle is similar to ^{10}Be since both are particle-active. ^{26}Al is transported with the precipitation down to the surface of the Earth and hence to marine and continental sediments. The ^{26}Al content in sediments, in which ^{26}Al preferentially is attached to the fine-grained fraction, is $(0.5-50) \times 10^9$ ^{26}Al atoms g^{-1}. The ratio ($^{26}Al/^{10}Be$) of the atmospheric production rates of ^{26}Al and ^{10}Be is, despite their temporal and spatial variations, nearly constant and amounts to $\sim 4 \times 10^{-3}$. Assuming a temporally constant initial $^{26}Al_0$ concentration during deposition, the sedimentation ages t of the various layers within a deep-sea core can be derived from the ^{26}Al still present (Eq. 2). The dating of deep-sea cores requires – analogously to ^{10}Be – constant sedimentation rates, which manifest themselves as straight lines in plots with logarithmic ^{26}Al against linear depth scales (Eq. 40). The temporal variation of $^{26}Al_0$ impedes the utilization of atmospheric ^{26}Al for sediment dating. Since such variation affects ^{10}Be in a similar manner, the combined application of both nuclides is advantageous (^{10}Be–^{26}Al method).

The in situ ^{26}Al is produced by muon-capture and spallation of silicon nuclei within the upper meters of the rock surface. The production rates are lower than in the atmosphere by several orders of magnitude. The in situ ^{26}Al can be utilized for exposure age as well as erosion rate determination of surfaces, according to Eqs. (31) and (34). The analysis of ^{26}Al is commonly combined with that of ^{10}Be in order to eliminate the disturbing influence of the temporal and spatial variations in the production rate (due to latitude, elevation and mass cover). A value of 6.1 was reported for the ratio of the ^{10}Be to ^{26}Al production rates in quartz (Nishiizumi et al. 1989). Once the rock is removed from further cosmogenic production, this ratio decays with an effective half-life of 1.3 Ma, which enables its application in the 10 ka to 10 Ma age range.

5.6.2 Practical Aspects

Just as for ^{10}Be, the AMS technique has largely replaced β-counting for ^{26}Al detection. Sample sizes of a few grams are sufficient for AMS ^{26}Al analysis. The aluminum is chemically extracted from the sample. Since $^{26}Al/^{27}Al$ ratios as low as 10^{-15} need to be precisely measured, the AMS analysis is very demanding. The attainable precision is 5–10%. However, the high aluminum (^{27}Al) contents in many minerals and rocks limit the utilization of the ^{26}Al method. Therefore minerals very low in aluminum, and of course rich in silicon, such as quartz, are optimal.

5.6.3 Application

Deep-Sea Sediments. Because of the temporal variation of its initial concentration, ^{26}Al cannot be directly applied for dating of deep-sea sediments. In order to circumvent this difficulty, Reyss et al. (1976) combined the ^{26}Al measurement with that of ^{10}Be. For a manganese nodule, Guichard et al. (1978) determined a growth rate of 2.3 ± 1.0 mm Ma^{-1} by ^{26}Al analysis, a figure which is in accordance with ^{10}Be data. Wang et al. (1996) extracted the authigenic aluminum and beryllium from opal-rich sediments of the North Pacific by NaOH leaching, allowing the detection of the cosmogenic ^{26}Al and ^{10}Be more effectively.

Quartz-Bearing Rocks. Quartz is geologically abundant, is chemically pure with aluminum contents around 10^2 µg g^{-1}, and is resistant against weathering and thus is not contaminated with atmospheric ^{26}Al. These characteristics are advantageous presuppositions for ^{26}Al dating. The analysis of ^{26}Al in quartz is commonly applied together with that of ^{10}Be (Sect. 5.3.3) as the ^{10}Be–^{26}Al method. Exposure ages of rocks which were

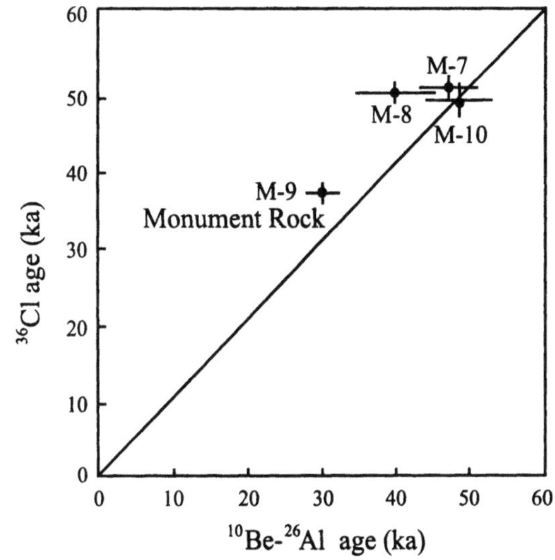

Fig. 59. ^{36}Cl and ^{10}Be–^{26}Al exposure ages of ejecta blocks at Meteor Crater, Arizona. Both methods yield concordant ages for the exhumation due to the meteorite impact (49.7 ± 0.85 ka and 49.2 ± 1.7 ka, respectively). The younger age (*M-9*) is attributed to surface erosion induced by weathering. (Reprinted from Nishiizumi et al. 1991a; copyright 1991, with kind permission from Elsevier Science Ltd., The Boulevard, Langford Lane, Kidlington OX5 IGB, UK)

instantaneously brought from the depth to the surface by processes such as glacier activity, volcanic eruption or meteorite impact, can be determined. For quartz separated from the impact ejecta of Meteor Crater, Arizona, a ^{10}Be – ^{26}Al age of 49.2 ± 1.7 ka was determined (Nishiizumi et al. 1991), which agrees well with ^{36}Cl (Fig. 59) and TL dates (Sects. 5.8.3 and 7.1.3). An example of ^{26}Al analysis of glacially polished granites in the Sierra Nevada was given by Nishiizumi et al. (1989). In that case the ^{26}Al exposure age dates the moment when the ice cover melted. From the ^{26}Al and ^{10}Be contents in quartz fractions of rock samples from southern Australian inselbergs extremely low mean erosion rates of less than 0.7 ± 0.1 m Ma^{-1} over at least 0.5 Ma were obtained, proving the antiquity and geomorphic development of these enigmatic landforms (Bierman and Turner 1995). As already mentioned in connection with ^{10}Be, ^{10}Be – ^{26}Al exposure ages of granitic boulders of an offset debris flow fan revealed an earthquake recurrence interval of 5.8 – 8.0 ka for the Owens Valley fault in California (Bierman et al. 1995). However, significant excess of ^{26}Al – due to nucleogenic ^{26}Al production – relative to ^{10}Be may preclude exposure age determination as has been pointed out by Brook et al. (1995).

5.7 Silicon-32

The radioactive silicon-32 is produced in the atmosphere by cosmic ray. It falls to the Earth's surface with the precipitation. The half-life of 140 a allows dating in the 50 to 500 a age range and bridges, in principle, the gap between the ^{3}H and ^{14}C methods. The beginning of ^{32}Si dating traces back to Lal et al. (1960). ^{32}Si dating is applied to oceanic sediments and glacier ice, occasionally also to groundwater and ocean sediments, but uncertainties about the value of the half-life, the complexity of silicon chemistry and the very low natural $^{32}Si/Si$ ratios have limited these applications.

5.7.1 Methodological Basis

Silicon is the second most abundant element in the Earth's crust and consists – apart from cosmogenic ^{32}Si – of the three isotopes ^{28}Si (92.2%), ^{29}Si (4.7%) and ^{30}Si (3.1%). ^{32}Si is radioactive and decays under low energy β^- ($\approx$100 keV) emission to the radioactive nuclide ^{32}P (β^- emitter of 1.7 MeV). For many years the half-life of ^{32}Si was only roughly known with estimates between 100 and 700 a, but a more accurate value of 140 ± 6 a is now available (Morgenstern et al. 1996).

In nature ^{32}Si is produced in the upper atmosphere by cosmic ray spallation of argon according to the reaction $^{40}Ar(n,p\,2\alpha)^{32}Si$. It reaches the ground dissolved in the precipitation as silicic acid or attached to aerosols. By its incorporation in the hydrologic cycle ^{32}Si enters the glacier ice, freshwater lakes, groundwater as well as the seawater. There are considerable temporal and spatial variations in the initial ^{32}Si concentration (Morgenstern et al. 1996). There are seasonal variations in the precipitation. Because of the latitude effect, the highest concentrations are found in Antarctica. Snow shows much lower concentrations than rain. Variations with solar activity and nuclear weapon tests were also observed. The annual mean of the ^{32}Si activity concentration in central European precipitation was estimated as 3.0 mBq m^{-3} (Franke et al. 1986).

In closed systems, most nearly realized in glacial ice, the initially present ^{32}Si gradually decays so that its remaining concentration dates the time elapsed since deposition. In contrast, water bodies rarely behave as closed systems, and may be mixed. Furthermore, the complex geochemistry of silicon and exchange processes impede the use of such simple ^{32}Si dating models.

5.7.2
Practical Aspects

About 1 ton of water or ice is necessary for ^{32}Si analysis by activity counting. The extraction of the silicon from such large sample amounts involves a laborious chemical procedure (Franke et al. 1986). For marine sediments where the ^{32}Si is situated in the biogenic silica component, ca. 600 g of sample is required. The activity of ^{32}Si is determined indirectly by β-counting of its radioactive daughter ^{32}P, which after 2–3 months is already in radioactive equilibrium with ^{32}Si. The extraction procedure of phosphorus was described by Delmas (1989). By means of the AMS technique, which allows the detection of $^{32}Si/Si$ ratios as low as 4×10^{-15} (Heinermeier et al. 1987; Morgenstern et al. 1996), the sample amounts can be much reduced.

5.7.3
Application

Glacier Ice. The ^{32}Si is deposited with the snow and thus becomes incorporated in the ice. If the initial ^{32}Si concentration is known and the ice stays a closed system, the age of the ice can be calculated directly. From the decreasing ^{32}Si activity down the glacier Hintereisferner in Austria, Clausen et al. (1967) were able to obtain an average flow rate of 6.6 m a^{-1}. This rate was calculated with $t_{1/2} = 500$ a; and for $t_{1/2} = 140$ a it would amount to 23.6 m a^{-1}. The Greenland ice cores Dye 3 and Camp Century were also analyzed by ^{32}Si (Clausen 1973). From the decreasing ^{32}Si activity with increasing depth and from independent age information, an attempt was made to derive the ^{32}Si half-life resulting in values between 200 and 300 a.

Groundwater. The ^{32}Si is carried with the surface water into the groundwater. If isolated from further exchange the gradually decreasing ^{32}Si concentration should yield the age of the groundwater formation. Groundwaters with decreasing ^{32}Si activity down the flow have been reported from India (Lal et al. 1970). The hydrologic interpretation of ^{32}Si data is impeded by the uncertainties about the value of the initial ^{32}Si activity, the mixing of water bodies, exchange processes with the aquifer and biologic influences. A significant reduction of the ^{32}Si concentration by geochemical processes was observed for karst waters (Morgenstern et al. 1995).

Deep-Sea Sediments. Diatoms and radiolarians incorporate silicon, dissolved in the ocean water, into their silicic skeletons. After death they sink to the ocean bottom where they are imbedded in the sediment. Since no further exchange with the seawater takes place, the ^{32}Si activity begins to decline. There have been attempts to utilize this phenomenon to derive the sedimentation rate of deep-sea deposits, in particular in connection with ^{230}Th (Kharkar et al. 1969) and ^{210}Pb dating (De Master and Cochran 1982). The coincidence that the ^{32}Si half-life is in the order of magnitude of the oceanic circulation renders this isotope in principle suited for studying the marine circulation pattern (Lal et al. 1976).

5.8 Chlorine-36

The radioactive ^{36}Cl is a cosmogenic nuclide which is produced in the atmosphere as well as in situ in the rock surface. The atmospheric cosmogenic ^{36}Cl reaches the ground quickly with the precipitation, where it enters the hydrologic cycle. It can be used for dating water and ice. The in situ cosmogenic ^{36}Cl enables the dating of surface exposure and the assessment of the denudation rate. The half-life of 301 ka allows applications in the 10^3 to 10^6 a age range, for calcite even down to 10^2 a. For karst surfaces denudation rates from 1 µm a^{-1} to 1 mm a^{-1} can be determined. The beginning of the ^{36}Cl method traces back to the 1950s (Davis and Schaeffer 1955). The highly sensitive AMS technique has widened the geologic potential of the method with the consequence that the investigation of ^{36}Cl in rocks has intensified since the mid-1980s (Phillips et al. 1986b).

5.8.1 Methodological Basis

Chlorine occurs in the Earth's crust with 180 µg g^{-1} on average. Natural chlorine consists – apart from the cosmogenic isotope ^{36}Cl – of the stable isotopes ^{35}Cl (75.77%) and ^{37}Cl (24.23%). ^{36}Cl is radioactive and decays under emission of an electron to ^{36}Ar. The atmospheric production of cosmogenic ^{36}Cl mainly follows the reaction $^{36}Ar(n, p)^{36}Cl$. After a mean residence time of 2 weeks the cosmogenic ^{36}Cl reaches the ground both by dry and wet deposition, bonded to aerosols and dissolved in the precipitation. This relatively short duration is too small for global atmospheric mixing, so that latitude effects of the ^{36}Cl production are reflected in the precipitation. In addition, a continental effect exists: the ^{36}Cl content of the precipitation increases from the coast to the continental interior, caused by the decreasing dilution with ^{36}Cl-free chlorine from the ocean spray. Dissolved ^{36}Cl is carried with the precipitation into the groundwater (Bentley et al. 1986). If the initial ^{36}Cl concentration is known and the systems remain closed, the formation age of groundwater and glacier ice can be determined from the left over ^{36}Cl. However, owing to its strong regional variation, the initial ^{36}Cl concentration needs to be determined on younger groundwaters of the region. Also, temporal variation of the initial ^{36}Cl concentration needs to be taken into account. Studies on fossil rat urine revealed that the initial $^{36}Cl/Cl$ ratio may have dropped by 50% since 40 ka ago (Plummer et al. 1997). Attention must be paid to the danger of contamination by ^{36}Cl originating from nuclear weapon tests during the late 1950s and early 1960s. The dating of groundwater may also be disturbed by a nucleogenic ^{36}Cl component produced in the aquifer by neutron capture of ^{35}Cl, whereby radiogenic (α, n)-reactions supply the neutrons.

Cosmogenic ^{36}Cl is produced in the uppermost few meters of the rock surface predominantly by spallation of ^{40}Ca and ^{39}K, whereas muon capture by ^{40}Ca dominates below, down to ~100 m. The in situ ^{36}Cl production rates are influenced by pronounced latitude and altitude effects (Zreda et al. 1991). Production rates of ^{36}Cl in calcium feldspar (at sea level, 38.9° latitude) were 48.8 ± 3.4 atoms (g Ca)$^{-1}$ a^{-1} by spallation and 4.8 ± 1.2 atoms (g Ca)$^{-1}$ a^{-1} by muon capture (Stone et al. 1996). The growth of the ^{36}Cl content up to the equilibrium level (Fig. 36) in surface rocks can be used for exposure dating according to Eq. (31). Particularly suited are geologic phenomena by which previously shielded rocks become suddenly exposed to cosmic rays, such as meteorite impact, volcanism and glacial activity. For systems already in equilibrium, erosion rates can be assessed. Besides the cosmogenic ^{36}Cl, a small nucleogenic ^{36}Cl background may be present and taken into account as already mentioned above.

5.8.2 Practical Aspects

When sampling rocks for exposure dating, it is important that the samples come from a position of defined and constant depth (<1 m). Such surfaces should be uneroded. Known rates of erosion can be accounted for. In order to avoid shielding effects by soil, snow or volcanic ash covers, slightly dipping faces may be advantageous. Because of their high calcium content, calcite and calcium feldspars are particularly suited for ^{36}Cl dating.

The ^{36}Cl analysis is performed by AMS. The AMS detection requires ~10 mg Cl, which corresponds to amounts of 10–100 g for rock samples or of several liters for water samples. Meteoric ^{36}Cl contamination is removed by washing the crushed rock samples. After several chemical purification steps the extracted chlorine is precipitated as AgCl for AMS measurement (Phillips et al. 1986b; Zreda et al. 1991). A favorable presupposition for the ^{36}Cl analysis by AMS – for instance in comparison to ^{26}Al – is the low geochemical abundance of chlorine and relatively high $^{36}Cl/Cl$ ratios. The $^{36}Cl/Cl$ detection limit is ~10^{-15}, and the attainable precision is around ±10% at levels of ~5×10^4 atoms g^{-1} (Stone et al. 1996).

5.8.3 Application

Moraines. Glaciers excavate rocks from the underground and deposit them in moraines, exposing previously shielded rock surfaces to cosmic rays. Also, the retreat of ice during melting freshly exposes glacially abraded faces to cosmic rays. Thus the ^{36}Cl clock begins to run. Consequently, both the glacier advance, using the moraine blocks, and the glacier retreat, using the abraded face, can be dated by ^{36}Cl. By dating moraines a chronology for the glacial deposits of the Bloody Canyon, Sierra Nevada, in California was obtained (Phillips et al. 1990). There, a glacial sequence of several morphologically distinct moraines is preserved. The samples were collected from the top 5 cm of the largest boulders along the crest of each of the moraines. In that position the boulders are least likely to be affected by erosional redeposition and snow covering. The ^{36}Cl ages revealed that episodes of glaciation occurred at about 200, 145, 115, 65, 24 and 21 ka ago. These glaciations correlate with the peaks of global ice volume inferred from the marine $^{18}O/^{16}O$ record (Fig. 60). The ^{36}Cl dating of moraines (Phillips et al. 1996) also revealed that the late Pleistocene Sierra Nevada glaciers advanced 49 ± 2, 31 ± 1, 25 ± 1, 19 ± 1, and 16 ± 1 ka ago in association with the expulsion of large numbers of icebergs from the ice caps surrounding the North Atlantic, known as Heinrich events. In this study a rock-surface erosion of 5 mm ka^{-1} and the progressive exhumation of moraine blocks by soil erosion was taken into account for the age evaluation. The deglaciation

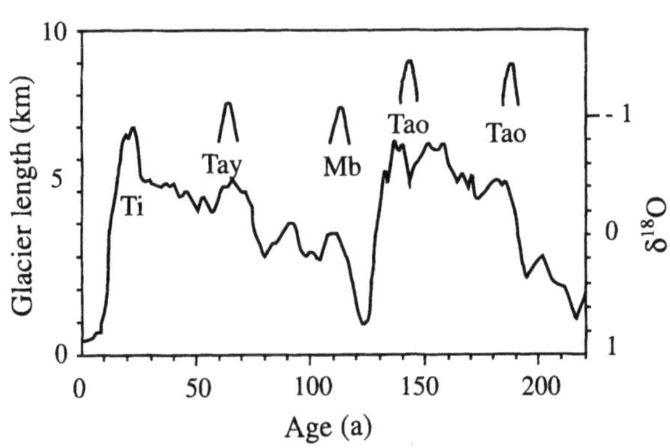

Fig. 60. ^{36}Cl exposure ages of the glacial advances. *Mb* Mono Basin; *Tao* Older Tahoe; *Tay* Upper Tahoe; and *Ti* Tioga of the Bloody Canyon glacier in the Sierra Nevada, California. The ^{36}Cl exposure ages of the moraine blocks of five advances – plotted versus the respective glacier length – correlate with the global ice maxima of the marine $\delta^{18}O$ record. (Reprinted with permission from Science; Phillips et al. 1990; copyright 1990, American Association for the Advancement of Science)

at Tioga Pass was dated by collecting glacially abraded marbles (Stone et al. 1996). From the glacial polish on the adjacent quartzite an erosion of 1–1.5 cm for the marble was inferred. Therefore an erosion rate of 1 μm a^{-1} was incorporated into the age calculation. The average ^{36}Cl age of three surfaces is 14.9 ± 0.5 ka. Since no allowance for shielding by winter snow was made, the date might slightly underestimate the true retreat of the ice.

Lava Flows. During volcanic eruptions lava extrudes on the surface, solidifies and thus is freshly exposed to cosmic rays. The growth of ^{36}Cl allows the dating of the eruption. Phillips et al. (1986b) measured the $^{36}Cl/Cl$ ratio in lava flows of known K–Ar ages from western North America. It turned out that the ratio grows with age in accordance with the model assumption (Fig. 61). Similar studies were performed on young lava flows from Mauna Kea, Hawaii (Zreda et al. 1991). ^{36}Cl dating was applied to 11 basalt samples, taken less than 5 cm below the surface from lava flows and volcanic bombs,

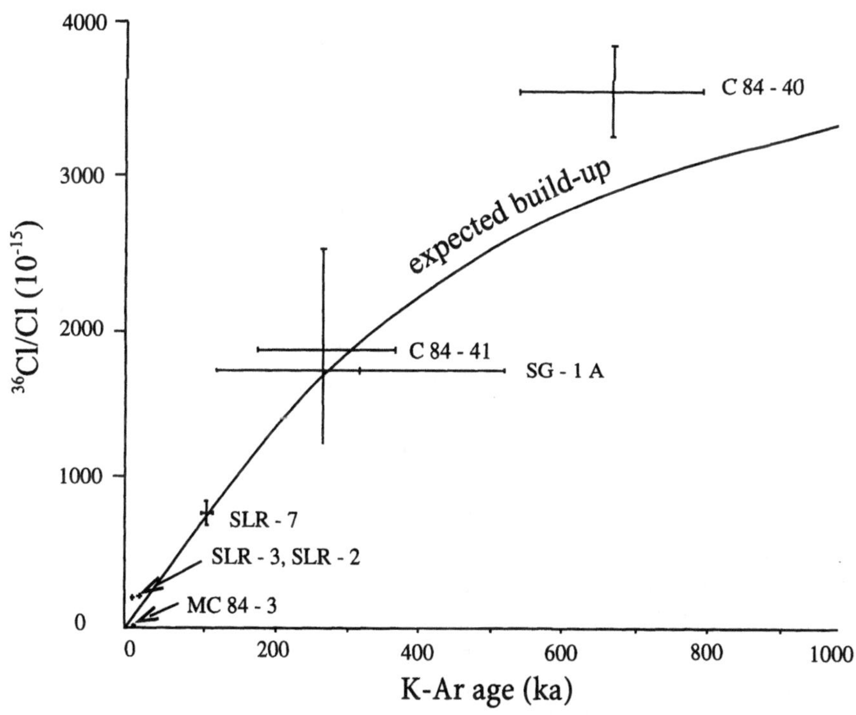

Fig. 61. The in situ produced ^{36}Cl in the surface of lava flows correlates with the K–Ar age. The analytical data confirm the exponential character of the ^{36}Cl growth as deduced from model assumptions (Fig. 36). (Reprinted with permission from Science; Phillips et al. 1986b; copyright 1986, American Association for the Advancement of Science)

from the Lathrop Wells volcanic center, Nevada; the result obtained was 81 ± 7.9 ka (Zreda et al. 1993). The lava samples collected from the tops of pressure ridges required age correction for surface geometry. Four samples from the Black Rock basaltic lava flow, Lunar Crater volcanic field, Nevada, yielded a ^{36}Cl exposure age of 33.9 ± 6.7 ka (Shepard et al. 1995), which agrees with the ^{10}Be date (Sect. 5.3). ^{36}Cl exposure data, ranging from 120 to 560 ka, for a 600-ka-old (K–Ar) flow north of the Lunar Crater maar, are consistent with an average erosion rate of ~3 mm ka^{-1}.

Impactites. During impact of giant meteorites large masses of rocks are exhumed and deposited as ejecta on the surface where they are exposed to cosmic rays. Therefore the accumulated cosmogenic ^{36}Cl in the ejecta material as well as in the exposed original crater walls should date the impact event. ^{36}Cl analyses on four dolomite boulders ejected from the Meteor Crater, Arizona, gave a mean age of 49.7 ± 0.85 ka (Fig. 59) which is in agreement with the $^{26}Al/^{10}Be$ age of 49.2 ± 1.7 ka (Sect. 5.6.3), the TL age of 49 ± 3 ka (Sect. 7.1.3) and which is further supported by undetectably low ^{14}C in the oldest rock varnish sample (Phillips et al. 1991).

Ice Cores. Because of its half-life ^{36}Cl is potentially suited for dating ice from the early glacial cycles (Dansgaard 1981). Atmospheric cosmogenic ^{36}Cl enters the ice together with the snow. Ice forms a closed system for chlorine. For known initial $^{36}Cl/Cl$ ratio, the decay of ^{36}Cl allows the dating of the formation of the ice. A major difficulty is caused by the temporal and regional variation of the initial value. In order to reduce this uncertainty, the analysis of ^{36}Cl was combined with that of ^{10}Be, as the *^{10}Be–^{36}Cl method* (with an effective half-life of 370 ka), with the assumption that the initial concentrations of both nuclides fluctuate in a similar pattern so that their ratio stays constant. However, attempts to test this method on the Greenland Camp Century ice core were without success (Elmore et al. 1987).

Groundwater. The cosmogenic nuclide ^{36}Cl is suited for groundwater dating due to its large atmospheric input and its solubility in water. It covers an age range up to 2 Ma beyond that of the ^{14}C method. According to the dating model of the closed system, the $^{36}Cl/Cl$ ratio should decrease with increasing age of the groundwater, i.e., down the groundwater stream. A smooth $^{36}Cl/Cl$ decrease with distance down-gradient was observed for groundwater in the Milk River aquifer, Canada. The ^{36}Cl ages increase gradually up to 2 Ma indicating very slow flow (Phillips et al. 1986 a). However, mixing of water bodies and geochemical exchange with the aquifer commonly limit the quantitative interpretation of the data (Fabryka-Martin et al. 1987). Nevertheless, ^{36}Cl is an important hydrologic tracer (Yechieli et al. 1996). In granites the dating of groundwater may be disturbed by nucleogenic ^{36}Cl contamination. The bomb ^{36}Cl of late 1950s and early 1960s is utilized as a temporally defined tracer in hydrologic systems (Bentley 1982).

Evaporites. Salt lakes are important archives for paleoclimate proxies, since their water level, salinity and sedimentary deposits change sensitively with climatic fluctuations. The hydrologic input of cosmogenic ^{36}Cl into the deposits enables the dating of the sediments. The sources of ^{36}Cl comprise both atmospheric and in situ components. The latter are derived from eroded near-surface rocks and soils. The ^{36}Cl method was tested on a sediment core from Searles Lake, California (Phillips et al. 1983). The $^{36}Cl/Cl$ ratio of five halite samples decreased with increasing depth. Under the assumption of an initial $^{36}Cl/Cl$ value of 56×10^{-15}, ages between 10 and 922 ka were obtained, which agree with ^{14}C and ^{230}Th ages and with the magnetostratigraphy. However, the assumption of a temporally constant initial $^{36}Cl/Cl$ ratio is problematic, as studies in the Qaidam basin, China, have shown (Phillips et al. 1993). Because of the high water solubility of chlorides, the dating of evaporite by ^{36}Cl is possible only in arid regions, where the original salt deposits are not remobilized.

5.9 Argon-39

Cosmogenic ^{39}Ar is produced in the atmosphere and is passed on through precipitation to surface waters and ice. If enclosed in ice or surface water and excluded from further exchange, ^{39}Ar dates their last gas exchange with the atmosphere. The half-life allows dating in the range of a few decades to 1000 a and bridges the age gap between the ^{3}H and ^{14}C methods. The methodological development of ^{39}Ar dating of terrestrial samples traces back to the 1960s (Loosli and Oeschger 1968). The ^{39}Ar method is mainly used for dating glacial ice, ocean water and groundwater.

5.9.1 Methodological Basis

The noble gas argon is the third most abundant gas in the atmosphere and is composed – apart from the cosmogenic ^{39}Ar – of the three isotopes ^{36}Ar, ^{38}Ar and ^{40}Ar (Sect. 3.1.1). ^{39}Ar is radioactive and decays with a half-life of 269 a under β^--emission to ^{39}K. In nature cosmogenic ^{39}Ar is produced in the stratosphere predominantly by neutron capture on ^{40}Ar according to the reaction $^{40}Ar(n,2n)^{39}Ar$. Other sources of ^{39}Ar in the atmosphere can be neglected. As a noble gas argon does not enter chemical compounds. Because of rapid atmospheric mixing, a constant $^{39}Ar/Ar$ equilibrium concentration exists. Variations in the atmospheric $^{39}Ar/Ar$ ratio of up to 7% may have occurred over the last 1000 a but this is practically negligible for dating purposes (Loosli 1983). Because of the fast argon exchange with precipitation and surface water, the same $^{39}Ar/Ar$ equilibrium ratio prevails in all these reservoirs. In groundwater a nucleogenic ^{39}Ar component can also occur that is formed in rocks through the capture of (α, n)-neutrons according to the reaction $^{39}K(n,p)^{39}Ar$ (Forster et al. 1992). The argon is enclosed in snow and ice or dissolved in equilibrated surface ocean water and groundwater. When it becomes isolated from further exchange with the atmosphere, ^{39}Ar declines and thus may serve as a radiometric clock. If the systems stay closed the ^{39}Ar age dates the formation of ice and water bodies (Eq. 26). However, in reality, the hydrologic cycles rarely behave in such a simple manner.

5.9.2 Practical Aspects

The argon dissolved in the water or occluded in the ice samples is extracted. It is separated from oxygen, nitrogen and other contaminants. For ^{39}Ar measurement purified argon amounts of 0.3–2 l are needed, which requires sample sizes of several m^3 of ice or water. The ^{39}Ar content is determined by its β^- activity in gas proportional counters. Since the specific ^{39}Ar activity is very low (~70 times lower than the modern specific ^{14}C activity), the laboratory background must be extremely low. An argon sample of 2 l requires a counting duration of up to a few weeks. Contamination with recent air must be avoided. Such contamination can be recognized by the presence of technogenic ^{85}Kr and, if necessary, corrected for (Loosli 1983).

5.9.3 Application

Glacier Ice. During the formation of glacial ice, air and therewith radioactive ^{39}Ar is occluded. After the deposition of snow this air is still in exchange with the atmosphere. With growing snow cover the snow is compacted and the interconnecting voids form into disconnected pores filled by gas bubbles which thus are entrapped in the ice. At this moment the ^{39}Ar system becomes isolated from the atmosphere and the radiometric clock starts to run. Therefore the ^{39}Ar age does not date the moment of snow deposition, but that of compaction and air enclosure. For the Greenland ice core Crête, Loosli (1983) was able to show that the final closure of the ^{39}Ar system took place at a depth of 70 m below the glacier surface (Fig. 62). Be-

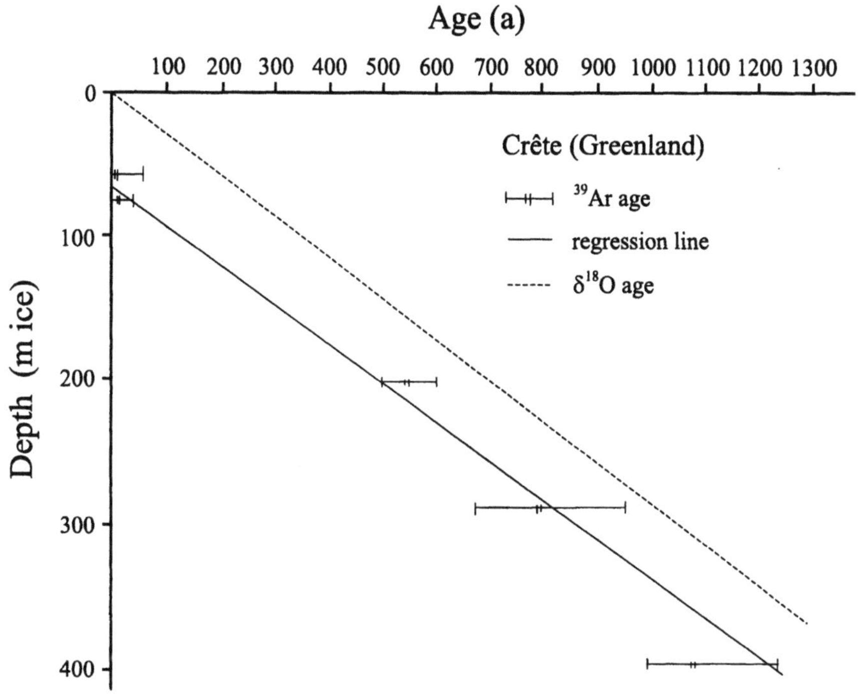

Fig. 62. ^{39}Ar age–depth-profile of the ice core Crête, Greenland. For comparison, the *dotted line* shows the $\delta^{18}O$ ages. The systematic age shift between both methods is due to the closure of air bubbles at 70 m ice depth 250 years after snow deposition. (Reprinted from Loosli 1983; copyright 1983, with kind permission from Elsevier Science Ltd., The Boulevard, Langford Lane, Kidlington OX5 IGB, UK)

cause of this phenomenon, a systematic shift of 250 a between the ^{39}Ar ages and the independently determined ages dating the snow deposition was observed in this ice core for the last 1000 a.

Ocean Water. The oceanic circulation pattern strongly influences the climate. Since the ^{39}Ar half-life is similar to the period of the circulation (1 ka), this nuclide is particularly suited for oceanographic studies. A further advantage is that cosmogenic ^{39}Ar – in contrast to ^{3}H and ^{14}C – is not disturbed by an anthropogenic component. At the ocean surface the water undergoes gaseous exchanges with the atmosphere whereby recent ^{39}Ar enters the water. When the water body sinks, the exchange with the atmosphere is interrupted and the ^{39}Ar activity decreases. Ocean waters upwelling from the depths, as for instance the seawater in the Antarctic McMurdo Sound, exhibit significantly lowered ^{39}Ar activities. Additional information on the oceanic circulation pattern is gained if the ^{39}Ar analysis is combined with that of ^{3}H, ^{14}C and ^{85}Kr (Smethie et al. 1986).

Groundwater. The surface water infiltrating the ground carries dissolved recent ^{39}Ar to the groundwater. In confined aquifers the groundwater is excluded from atmospheric exchange, so that the its ^{39}Ar activity declines. From the remaining activity the time since the water body became isolated from exchange can be determined, provided the systems stays closed and no mixing with other waters occurs. However, these model conditions are rarely realized in nature. ^{39}Ar activities decreasing downstream were observed in groundwaters in the Frankenalb, Germany (Loosli 1983). When compared with ^{14}C ages of the same groundwater samples, the ^{39}Ar age turned out to be lower. The discrepancy cannot be explained only by assuming the mixing of water bodies. The most probable reason for the falsification of the ^{39}Ar ages is the nucleogenic subsurface production of ^{39}Ar which is picked up by the groundwater, in particular if potassium-rich rocks such as granite are present in the hydrologic system (Forster et al. 1992). In any case ^{39}Ar dating yields a minimum age for the groundwater formation, whereas ^{14}C dating tends to give upper age limits (Sect. 5.4.3).

5.10 Calcium-41

Cosmogenic ^{41}Ca is produced in situ in the uppermost few meters of the Earth's surface. From there it reaches the groundwater and the biosphere. If calcium-containing material is removed from the surface zone to deeper zones shielded from cosmic rays, the ^{41}Ca concentration declines and the radiometric age can be inferred from the remaining ^{41}Ca. Bones, calcareous sinter and concretions should be datable. The ^{41}Ca half-life of ~103 ka enables application in the 20–500 ka age range. The method was proposed by Raisbeck and Yiou (1979), but has not yet developed into a method of practical use. Major problems, such as the ^{41}Ca detection, the wide variation of the initial $^{41}Ca/Ca$ ratio and the geochemical behavior of the ^{41}Ca system, have still to be overcome (Fink et al. 1990).

5.10.1 Methodological Basis

Calcium is with 3.6 % the fifth most abundant element in the Earth's crust. Natural calcium consists of six isotopes: ^{40}Ca (96.94 %), ^{42}Ca (0.65 %), ^{43}Ca (0.14 %), ^{44}Ca (2.08 %), ^{46}Ca (0.0033 %) and ^{48}Ca (0.185 %). Calcium is geochemically mobile, that is, it is easily exchanged between rock carbonate and the groundwater depending on the pH conditions and the temperature. Dissolution and precipitation of calcium carbonate are common in the groundwater zone. Biochemically calcium is a major constituent of bones and skeletons.

Cosmogenic ^{41}Ca is primarily produced by thermal neutron capture of ^{40}Ca according to the reaction $^{40}Ca(n, \gamma)^{41}Ca$ in near-surface, calcium-bearing rocks (Fig. 63). ^{41}Ca decays under electron capture to ^{41}K. In the near-surface zone, the $^{41}Ca/^{40}Ca$ ratio reaches a steady state between production and decay after ~500 ka. At an average neutron flux of 3×10^{-3} neutrons cm^{-2} s^{-1}, the $^{41}Ca/^{40}Ca$ equilibrium level amounts to ~10^{-14} in the uppermost meter.

The cosmogenic ^{41}Ca participates in the geochemical and biochemical calcium cycle and thus enters calcium carbonate precipitations and bones. As long as these processes take place near the surface, the equilibrium $^{41}Ca/^{40}Ca$ ratio should be maintained. Not until the removal of the calcium-bearing material from the near-surface zone to a greater depth (> 3 m) does

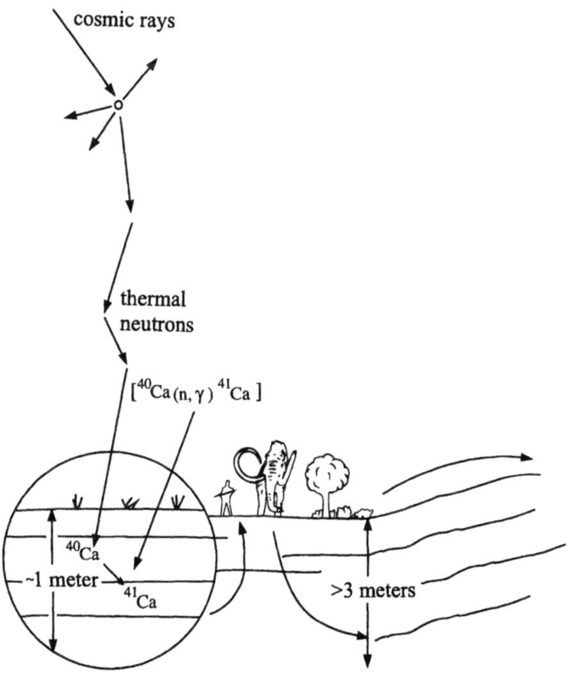

Fig. 63. Dating model of the ^{41}Ca method (after Taylor et al. 1989). Cosmogenic ^{41}Ca is built up by neutron capture of ^{40}Ca in the uppermost meters of the rock surface until an equilibrium between production and radioactive decay rate is attained. The ^{41}Ca participates in the geochemical and biochemical calcium cycle, and thus enters secondary carbonates and bones. With the shielding (> 3 m below the surface) from cosmic rays the ^{41}Ca clock commences to run

the $^{41}Ca/^{40}Ca$ ratio begin to decline. The amount of ^{41}Ca remaining should make it possible to date the event when the material became shielded from the cosmic rays, such as sedimentary burial of bones, formation of cave sinters as well as the deposit of bones and limestone fragments in caves. However, this would presuppose the knowledge of the initial $^{41}Ca/^{40}Ca$ ratio and a closed $^{41}Ca/^{40}Ca$ system after the shielding event. To what extent these conditions are true is still an open question, crucial for the ultimate suitability of the ^{41}Ca dating method. Also, the accurate value of the ^{41}Ca half-life is still uncertain. Besides dating events of deposition, surface exposure dating should also be possible with the ^{41}Ca clock. When calcium-rich rocks suddenly come from the cosmic-ray-protected depth to the surface, cosmogenic ^{41}Ca starts to build up, allowing exposure dating up to few 100 ka. In karstic and Ca-feldspar-bearing terranes, the ^{41}Ca data should also allow the estimation of denudation rates (Fink et al. 1990).

5.10.2 Application

Bones, Cave Sinter and Calcium Carbonate Concretions. Only when the AMS technique became available was it possible to detect sensitively $^{41}Ca/^{40}Ca$ ratios as low as 10^{-14} to 10^{-15} (Raisbeck and Yiou 1980). First, $^{41}Ca/^{40}Ca$ measurements of recent bones and limestones from the surface and from 11 m depth were achieved, whereby the extracted calcium had to be isotopically pre-enriched (Taylor et al. 1989). The overall $^{41}Ca/^{40}Ca$ spread for modern bones is two orders of magnitude, and even up to a factor of seven among different species at a single site (Fink et al. 1990). This high variability is probably caused by local environmental factors and will affect the zero-age of the ^{41}Ca clock. Further methodological studies on ^{41}Ca detection and the calcium cycle in nature are necessary in order to develop the practical use of ^{41}Ca for dating Pleistocene bones, calcareous sinters and concretions.

5.11 Krypton-81

Cosmogenic ^{81}Kr is produced in the atmosphere. It is carried by the precipitation to surface waters. If enclosed in groundwater and glacier ice and excluded from further exchange, the decaying ^{81}Kr allows the dating of these materials. Because of its half-life of 210 ka, the datable age range is 50 ka to 1 Ma. The development of ^{81}Kr dating for terrestrial samples started, as the related ^{39}Ar method, in the 1960s (Loosli and Oeschger 1968). The other radioactive krypton isotope, ^{85}Kr with $t_{1/2} = 10.8$ a, is released from nuclear plants from where it escapes into the atmosphere, and it is also formed nucleogenically to a minor extent in underground rocks (Loosli et al. 1989). ^{85}Kr may also be used for dating purposes, mainly in connection with hydrologic studies, in which it supplements ^{3}H data (Salvamoser 1982).

5.11.1 Methodological Basis

The noble gas krypton occurs as a trace element (10^{-9}) in the atmosphere. It consists – apart from the radioactive isotopes ^{81}Kr and ^{85}Kr – of the stable isotopes ^{78}Kr (0.4%), ^{80}Kr (2.2%), ^{82}Kr (11.6%), ^{83}Kr (11.5%), ^{84}Kr (57%) and ^{86}Kr (17.3%). ^{81}Kr is radioactive and decays under electron capture to ^{81}Br. It is cosmogenically produced in the atmosphere by neutron capture and spallation on the stable krypton isotopes. As a noble gas krypton is chemically inactive. The fast mixing of the atmosphere homogenizes the cosmogenic krypton, so that in the atmosphere a steady state ^{81}Kr/Kr ratio of 5×10^{-13} is established between cosmogenic production and radioactive decay. The precipitation transfers this equilibrium ratio to the recent groundwater, which contains only 1300 ^{81}Kr atoms per liter at a solubility of 9.2×10^{-5} cm^3 krypton per liter water (Lehmann et al. 1985).

When the exchange with the atmosphere is interrupted, the enclosed ^{81}Kr content declines and the moment of gas enclosure can be inferred from the remaining ^{81}Kr, provided the ^{81}Kr system stays closed (Eq. 26). The age calculated under such a model assumption should date the formation of glacier ice or groundwater. Temporal variations in the production rate of ^{81}Kr, which occur within periods of $<10^5$ a, are without consequence for dating since they are short compared to the half-life and balanced by the atmosphere.

5.11.2 Practical Aspects

Krypton is extracted from several liters of water or ice together with the other noble gases and separated from oxygen, nitrogen and other contaminants. The separation of krypton from the other noble gases is performed by gas chromatography. After isotopic enrichment ^{81}Kr is analyzed by resonance ionization spectroscopy (Chen et al. 1984; Lehmann et al. 1985). Contamination with recent air must be avoided, but can be discovered by the presence of ^{85}Kr (Loosli 1983).

5.11.3 Application

Glacier Ice. The ^{81}Kr method has an interesting potential for dating polar ice cores. This method allows to study the history of deposition over the middle Pleistocene period back to several 100 ka.

Groundwater. The surface water carries dissolved ^{81}Kr together with the air into the groundwater. As long as the groundwater is in contact with the atmosphere, it maintains the ^{81}Kr equilibrium concentration. In confined aquifers the water is cut off from atmospheric exchange, so that the ^{81}Kr concentration gradually declines. Because of its long half-life, ^{81}Kr dating is only applicable to rather old groundwaters. The feasibility of ^{81}Kr dating was demonstrated by Lehmann et al. (1985) for groundwater in a sandstone aquifer near Zürich, Switzerland.

6 Particle Tracks

There are two dating methods, the *fission track* and the *α-recoil track*, which are based on nuclear particle tracks. During nuclear fission and α-decay heavy nuclear particles acquire high kinetic energies. As a result of interactions with the atoms of solid state detectors, they leave along their path a trail of radiation damage: the latent particle track. In this way single decay events can be detected. Many minerals and glasses are able to record particle tracks. Fission tracks are formed during the spontaneous nuclear fission of ^{238}U. The two heavy fission fragments leave a fission track of 10–20 μm length. The α-recoil tracks are formed during the α-decay of uranium and thorium as well as of their radioactive decay products. The nucleus left over by the α-particle recoils by a few 10^{-2} μm. The diameter of latent particle tracks amounts to some 10^{-3} μm.

The particle tracks are directly visible under the transmission electron microscope. By etching, the diameter of the latent tracks is increased 1000-fold at the expense of the undamaged material, to a size which is visible under an optical microscope. The etching behavior depends on the composition and structure of the solid, the properties of the track as well as the etching conditions (Wagner and Van den haute 1992). A crucial phenomenon for the quantitative aspects of dating is the etching efficiency, because not all of the tracks which intersect a surface are revealed as etched, visible tracks. The etchant attacks not only the track, but also the undamaged material and thus lowers the etched face parallel to itself. As a result, tracks inclined at less than a critical angle to the surface are not revealed. The critical angle depends on the detector and on the etching conditions. Also, new tracks are exposed with progressive etching of the detector surface, and overetched tracks may be lost. The density of etched tracks per unit area of surface, therefore, also depends on the etching time (Fig. 64) and well-defined, optimal etching conditions are a necessity for accurate age determinations.

Track Accumulation Age. Under the presupposition that all tracks have been preserved since the formation of a sample, i. e., that the system stays

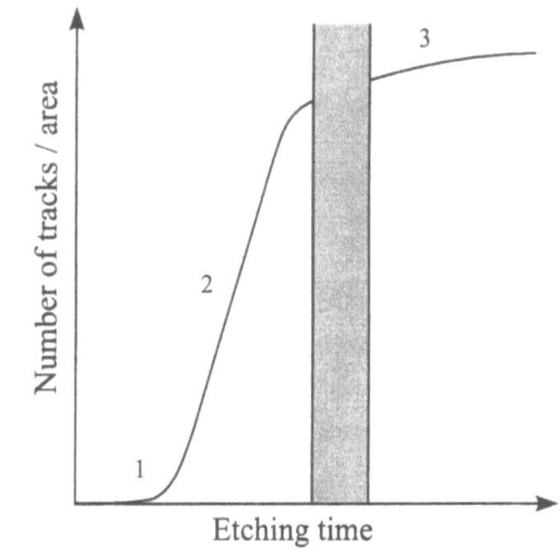

Fig. 64. Fission track density in crystals as a function of etching time. (After Wagner and Van den haute 1992). *1* No visible tracks are developed. *2* Revelation of surface tracks. *3* Revelation of added volume tracks; a balance between overetched and newly etched tracks is established resulting in a plateau. Optimal etching conditions are met in the *hatched area*

closed, the total number of accumulated tracks is a measure for the sample's age. In the scheme of the radiometric clock in which the radioactive parent disintegrates to the radiogenic daughter nuclide, the particle tracks represent the daughter. The age value dates the mineral formation or the last heating with complete erasure of the tracks previously present. The track number is measured as areal track density [cm^{-2}]. For counting the µm-sized fission track etch pits a highly magnifying petrographic microscope is suitable (Fig. 65). For α-recoil track counting a phase-/interference-contrast or electron scanning microscope is required.

The presupposition for dating – that all tracks, once they are formed, are preserved – is often not realized, so that the age becomes too low with respect to the sample's formation. One observes partial or even complete loss of tracks. The fading of tracks occurs essentially at elevated temperatures, a phenomenon known as track annealing. The various track recording materials differ much in their annealing properties. The size of the fission tracks is measured in order to examine whether they are partially annealed. This is effectively achieved with image analyzing systems (Fig. 65). If all previous tracks are annealed the track age dates the moment of heating.

Particle track dating requires the knowledge of the concentration of the parent nuclide; this is uranium in the case of fission tracks and uranium plus thorium in the case of α-recoil tracks. In principle, these elements could be analyzed explicitly, but this would involve additional systematic error sources to the dating procedure. It is advantageous to measure uranium by induced ^{235}U fission tracks after irradiation with thermal neutrons, and thorium with induced ^{232}Th fission tracks after irradiation with fast neutrons. This means a double irradiation in the case of α-recoil track

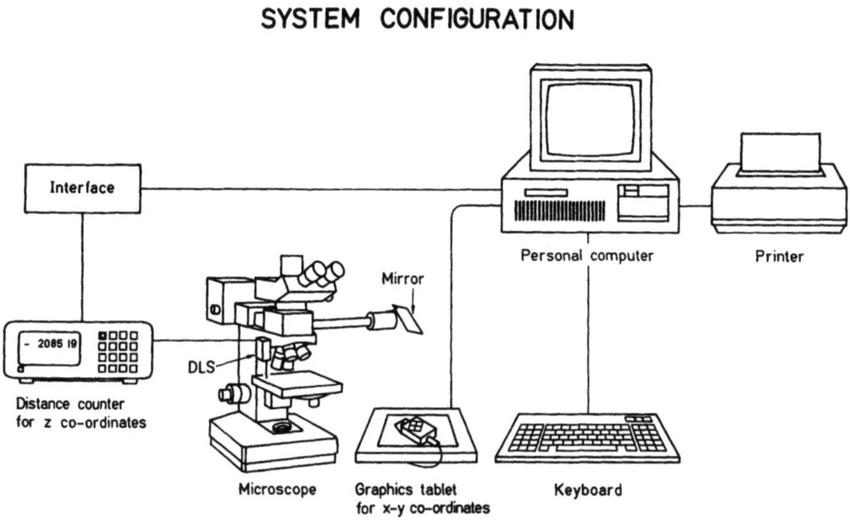

Fig. 65. Instrumentation required for measuring particle tracks. In addition to particle track counting, the track sizes are also measured to obtain information on the degree of track annealing

dating, first with thermal neutrons for uranium and then with fast neutrons for thorium (Wagner 1980 a). The quantitative analysis is achieved through reference materials, usually glasses, added during neutron irradiation. The detection of uranium and thorium by means of induced nuclear fission has the further advantage that information on their microscopic distribution is also obtained exactly for the region where the fossil tracks were counted. This microanalysis enables the dating of single mineral grains.

6.1 Fission Tracks

The fission track method is based on the spontaneous fission of uranium which is a rare type of natural radioactivity. Fission track dating began with the discovery that the heavy fragments of nuclear fission leave etchable, microscopically visible *fission tracks* in minerals and glass (Price and Walker 1962, 1963). Since the single fission events are detected by means of the tracks, the fission track method is applicable also to the relatively short time span of the past 10^6 a in spite of the long half-life for the spontaneous fission of ^{238}U (8.2×10^{15} a). The practicability of fission track dating in this young age range requires a sufficiently high uranium concentration. If an areal density of 10 tracks cm^{-2} is presupposed as minimum, the uranium content must be at least 1 $\mu g\ g^{-1}$, in order to determine

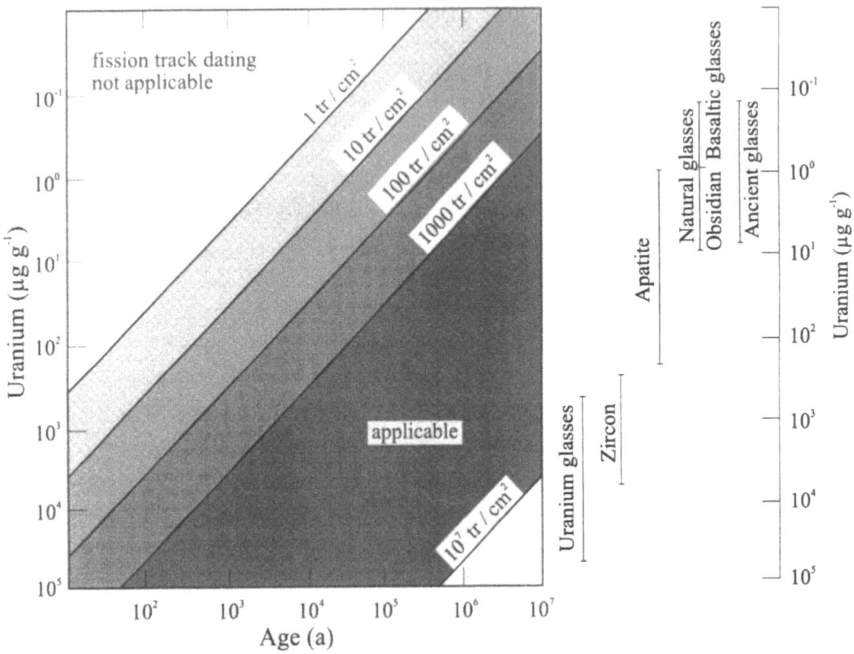

Fig. 66. Relationship between the age range datable by fission tracks and the uranium content of the sample. Counting track densities $< 10^2$ cm^{-2} is labor intensive, and densities $> 10^7$ cm^{-2} are no longer resolvable in optical microscopic counting. (After Wagner 1976)

an age of 10^5 a precisely. For lower ages higher uranium concentrations are needed and under favorable circumstances the lower dating limit even reaches the archaeological age range of few ka. The connection between fission track density, uranium content and age is shown in Fig. 66.

The principles and application of fission track dating were described in detail by Wagner and Van den haute (1992). Zircon is the mineral most frequently used in the young age range, due to its high uranium content. Accessory zircon occurs in volcanic rocks. Of particular interest are volcanic ashes which are intercalated in sediments so that Quaternary sequences – among them those with hominid remains and Paleolithic implements – are datable by fission tracks. Volcanic extrusive rocks may heat rock contacts up to temperatures at which the fission tracks in zircon, titanite and apatite anneal, and thus the fission track clock is reset. Analogously the firing of ceramics and the heating of stones by Man can be dated, provided these objects contain uranium-rich minerals. Volcanic glasses, such as obsidian and pumice, are also frequently used for fission track dating. Obsidian implements were often heated, which is important for archaeological applications. In addition to naturally occurring glasses, man-made glass can be dated with fission tracks. Because of their recent origin their uranium contents must be extraordinarily high, as is actually the case in the uranium glasses of the 19th and 20th centuries.

The precision of the fission track age of young samples is governed by the number of counted fission tracks, which in turn depends on the areal track density and the available counting area. The microscopic counting of an area of 1 cm^2 requires several hours. In order to attain 10 % precision one has to count at least 100 tracks (Sect. 1.3). This goal may be unrealistic due to the limited size of the sample, especially in case of mineral grains and glass fragments.

6.1.1
Methodological Basis

Natural uranium consists of the isotopes ^{238}U (99.3 %) and ^{235}U (0.7 %) in a known abundance ratio. Both isotopes decay by α-emission and are the parent nuclides of long decay chains (Fig. 22). The isotope ^{238}U decays – apart from its α-activity – also by spontaneous nuclear fission. The decay rate of the spontaneous fission is 10^6 times less than that of the α-decay. During fission, the uranium nucleus splits into two unequally heavy fission fragments and 2–3 neutrons (Fig. 3). Each fission event releases energy of ~200 MeV, which appears primarily as kinetic energy of the fission fragments which are expelled in opposite directions. These fragments lose their energy by ionization and leave a zone of radiation damage along their paths. Both zones form together a straight fission track of 10–20 µm length and several 10^{-3} µm diameter.

The latent fission tracks are enlarged by etching until they become visible under an optical microscope (Fig. 67), which is decisive for the feasibility of fission track dating. In order to be etchable the track, confined within a solid, needs to be intersected by a surface, such as polished faces or cleavage planes, so that the etchant has access to the radiation-damaged material (Fig. 68). Such fission tracks are termed *surface tracks*. As a result of surface preparation only a fraction of their total length is still present in the solid. If the total length should be etched one has to search for *volume tracks* still confined in the solid. Their etching requires host channels, such as surface tracks or cleavage planes, through which the etchant has access.

Fission Track Age. If all the fission tracks are preserved, their number reflects the age of the sample. Since the track number depends also on the uranium concentration, this needs to be known. For uranium analysis the

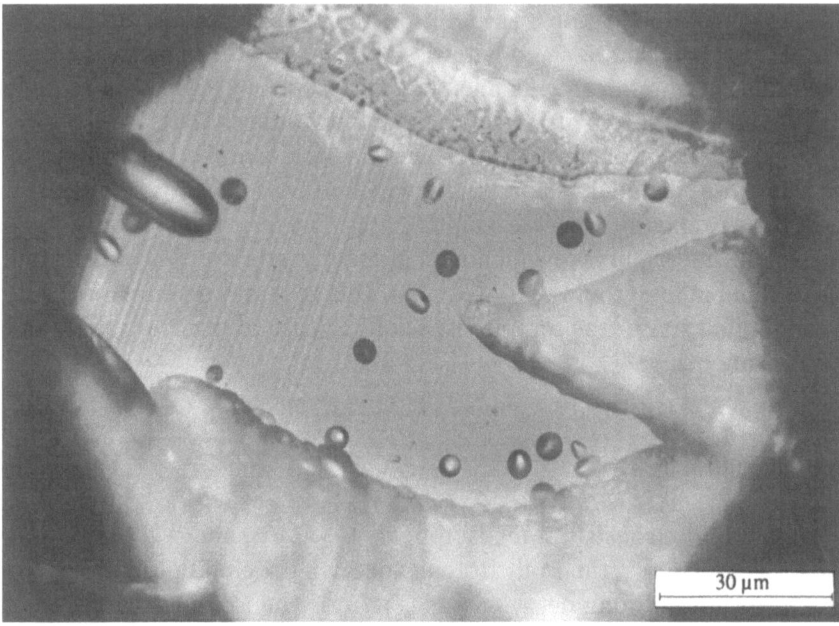

Fig. 67. Glass fragment (etched for 110 s at 23 °C in 24% HF) from Banks Island Tuff, Canada, with preannealed (90 d at 100 °C) fission tracks. (Courtesy of J. Westgate)

phenomenon that thermal neutrons induce fission of ^{235}U is exploited. This induced ^{235}U fission must be clearly distinguished from the spontaneous ^{238}U fission. Both types of fission result in fission tracks which are physically identical. The number of the induced ^{235}U fission tracks depends on the uranium content and the neutron fluence.

The procedure of fission track dating essentially involves the counting of spontaneous ^{238}U fission tracks before and induced ^{235}U fission tracks after a neutron irradiation. The age t [a] is calculated according to

$$t = 1/\lambda_d \times \ln[1 + (\varrho_s/\varrho_i) \times (\lambda_d/\lambda_f) \times (\phi \times I \times \sigma)] \tag{45}$$

whereby ϱ_s and ϱ_i are the spontaneous and induced fission track areal densities [cm^{-2}], respectively, ϕ is the thermal neutron fluence [cm^{-2}], λ_d and λ_f are the constants of the total decay [1.55125×10^{-10} a^{-1}] and the spontaneous fission [8.46×10^{-17} a^{-1}] of ^{238}U, I is the $^{235}U/^{238}U$ ratio [7.2527×10^{-3}] and σ is the ^{235}U fission cross section for thermal neutrons [580.2×10^{-24} cm^2].

There are several alternative procedures when carrying out fission track dating. One has to distinguish between the physically independent age measurement and the calibration by means of age standards. One can

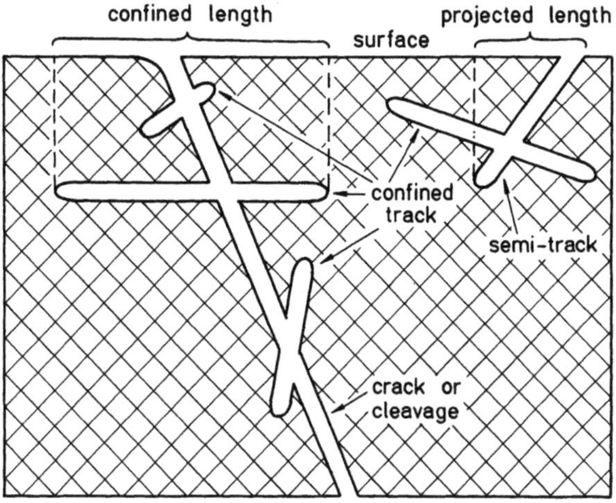

Fig. 68. Schematic representation of etched surface tracks and volume tracks. Surface tracks are intersected by the polished surface thus providing direct access to the etchant. Volume tracks – representing the total track length – however, are only accessible for the etchant through host tracks or cleavage planes

choose between the population technique and the external detector technique as well as some other technical variants. The selection of the optimal technique is governed by the type and size of the sample, and the homogeneity of the uranium distribution (Gleadow 1981). The independent age determination is performed according to Eq. (45). The technique employing age standards and reference glasses is known as *ζ-calibration* (Hurford and Green 1983). In the *population technique* the sample is divided into two about equal aliquots. The spontaneous fission tracks are etched and counted in the first aliquot while the induced fission tracks – after fully annealing the spontaneous tracks and neutron irradiation – in the second aliquot. This technique requires homogeneous uranium distribution. It has the advantage that both types of fission tracks are recorded in the same material and under identical experimental conditions. In the *external detector technique* the spontaneous and induced fission tracks are recorded in different materials. After etching the spontaneous fission tracks on a sample face, the face is covered for neutron irradiation with an external detector which records the induced fission tracks. Suitable detector materials are uranium-free muscovite sheets and polycarbonate foils. After irradiation, the detector is removed and etched. The spontaneous tracks are counted on the sample face while the induced tracks are counted on the detector face. The external detector technique offers the advantage that spontaneous and induced fission tracks originate from the same micro-region within the specimen, so that samples with inhomogeneous uranium distribution and single grains can be dated. Because the different etching efficiencies and registration geometries in sample and detector are difficult to quantify, the external detector technique needs calibration with age standards.

Track Annealing. Latent fission tracks are – as are other types of radiation damage – in a metastable stage. They gradually fade over time. The fading is much accelerated at elevated temperatures, and the process is known as track annealing. Since track annealing reduces the apparent age, it is of fundamental importance for the fission track dating system. The fading phenomenon can be studied in laboratory experiments. Minerals and glasses of a given fission track density are subjected to various temperatures for various lengths of time. From the degree of observed track loss the kinetic laws of annealing are derived and extrapolated to the long-term geological conditions. Such studies revealed that the fission track stability is relatively high in titanite and zircon, but low in apatite. For glasses the stability decreases with the silica content (Table 5).

Track annealing is a gradual process. There is a wide temperature range in which partial annealing occurs. Complete track loss resets the fission track clock which enables the dating of the heating event. Partial annealing, however, results in more or less lowered fission track ages that have no chronological meaning. It is a great advantage of the fission track method that partially annealed fission tracks can be recognized by their reduced size. The track size is measured as length or as diameter of the etched track in minerals and glasses, respectively. The experimentally established correlation between the reduction of track density and track size with pro-

Table 5. Annealing characteristics of fission tracks: the temperatures (in °C) are given for various durations at which the track density is reduced to half of its original value

Material	1 h	1 a	10^2 a	10^4 a	10^6 a	Reference
Titanite	610	490	440	390	350	Naeser and Faul (1969)
	600	480	430	380	340	Nagpaul et al. (1974)
Zircon	700	550	480	420	370	Fleischer et al. (1965)
	700	520	450	390	340	Krishnaswami et al. (1974)
Apatite	336	220	174	138	108	Wagner (1968)
	330	220	175	140	110	Naeser and Faul (1969)
	323	227	186	153	123	Watt and Durrani (1985)
Obsidian	390	245	190	145	110	Suzuki (1970)
	450	260	210	160	115	Suzuki (1973)
Basaltic glass	275	200	160	130	100	Aumento (1969)
	170	90	60	30	10	Macdougall (1976)
	235	124	84	51	24	Storzer and Selo (1978)
Australite	360	210	160	120	88	Storzer and Wagner (1969)
Uranium glass	220	140	110			Wagner et al. (1975)

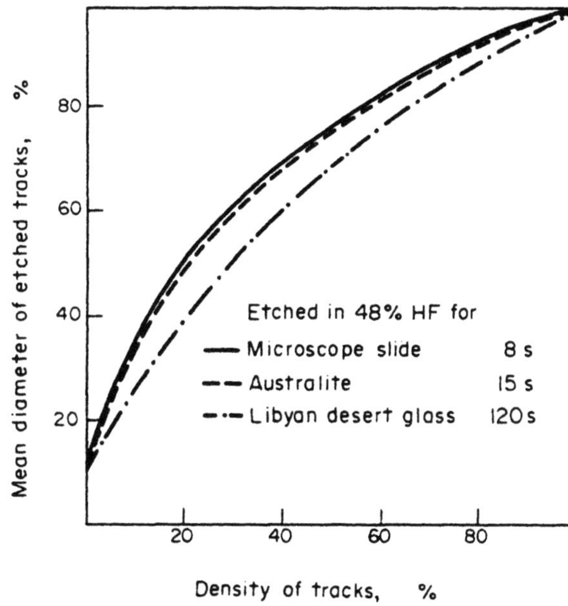

Fig. 69. Experimentally established correlation between track density and track sizes during progressive annealing for different glasses. Such curves are used for the correction of lowered fission track ages. (Gentner et al. 1969)

gressing annealing (Fig. 69) can be utilized for correcting lowered fission track ages. In the *track size correction procedure* (Storzer and Wagner 1969) the sizes of spontaneous and induced fission tracks are compared to each other and any size difference is accounted for. The alternative *plateau correction procedure* (Storzer and Poupeau 1973) is based on the observation that partially annealed, spontaneous fission tracks are more resistant against further annealing than are the induced ones. The progressive annealing is continued until both types of tracks exhibit the same annealing behavior. This is indicated by the appearance of a plateau in the apparent fission track age which is taken as the corrected age (Fig. 70).

6.1.2 Practical Aspects

Due to the low track densities expected for archaeological and Quaternary geological samples, the collected specimen must be sufficient in size to allow for about cm^2-sized areas for track counting. This demand usually presents no difficulties with obsidian. However, glass shards from volcanic ash and vesicular pumice are often below 1 mm in size. Accessory minerals are even smaller, commonly down to 0.1 mm which is the lower limit that is acceptable. In such cases a sufficient amount of sample for mineral separation needs to be collected. For fission track analysis, contaminants in

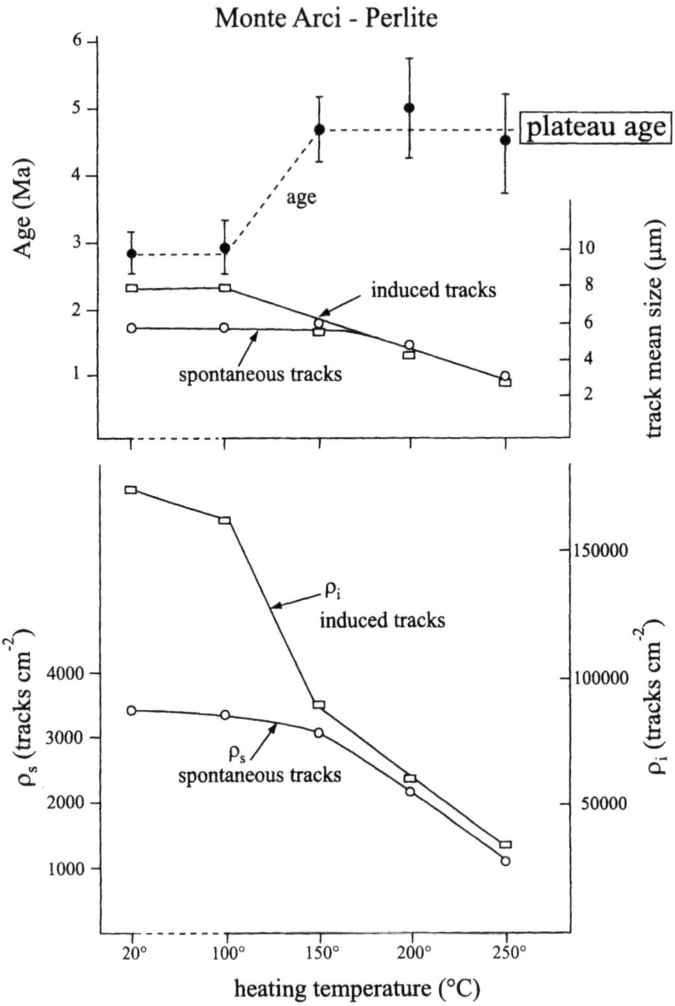

Fig. 70. Plateau correction for perlite from Monte Arci, Sardinia. Two subsamples, one with spontaneous (ϱ_s) and one with induced (ϱ_i) fission tracks are subjected to progressive annealing. After each step the track densities (*lower diagram*) and track diameters (*upper diagram*) are measured and the age (*upper diagram*) is calculated. The age value reaches a plateau, when both types of tracks exhibit the same annealing behavior. The plateau is taken as the correct fission track age of 4.67 ± 0.45 Ma. (Reprinted from Arias et al. 1981; copyright 1981, with kind permission from Elsevier Science Ltd., The Boulevard, Langford Lane, Kidlington OX5 IGB, UK)

Table 6. Commonly used etching conditions

Material	Face	Etchant	Temperature (°C)	Duration
Apatite	Prism	5% HNO_3	20	1 min
Titanite		50 N NaOH	120	30 min – 1 h
		$HF:2HCl:3HNO_3:6H_2O$	20	1 – 25 min
Zircon	Prism	KOH:NaOH eutect.	220	4 – 100 h
	Prism	6KOH:14NaOH:1LiOH	200	2.5 – 4.5 h
	Prism	$HF:H_2SO_4$	165	1 – 10 h
Obsidian		16% HF	23	5 – 7 min

grain separates are not as critical as for other dating methods, since the single grains are microscopically screened so that any contaminating grain can easily be rejected. When applying the single grain technique a minimum of 15 crystals, preferably 20, should be counted in case of zircons separated from tephra (Seward and Kohn 1997). Statistical treatment such as the radial-plot, chi-square statistics and the variance help to indicate the presence of detrital contaminants. The samples are embedded in epoxy, ground and polished. A well thermalized and homogenized reactor position is necessary for neutron irradiation (De Corte et al. 1991). Only few research reactors offer such facilities. The etching of the spontaneous and induced fission tracks needs to be carried out under defined conditions (Table 6).

Which of the alternative dating procedures mentioned should be applied for a given, young sample? The population technique is usually adequate for glass and apatite with homogeneous uranium distribution. The external detector technique is recommended for zircon and titanite due to the dangers of uranium inhomogeneity and grains of detrital origin. This technique requires age standards, for which the Fish Canyon Tuff zircon together with reference glasses are commonly used (Hurford 1990). In the population technique the neutron fluence is recorded with calibrated reference glasses or metal monitors (Van den haute et al. 1988).

In order to allow the judgment of the analytical quality of published data, the numerical presentation must include technical information (Naeser et al. 1979; Hurford 1990). The error evaluation usually represents only the precision. Major systematic errors may be caused by the neutron fluence measurement, the selection of the fission decay constant and the age calibration (Wagner and Van den haute 1992).

6.1.3 Application

Either the formation or a secondary heating event is dated by fission tracks. This potential is illustrated in Fig. 71. Formation dating requires that all fission tracks are preserved in the sample since its formation. For dating thermal overprints the sample must have been sufficiently heated in the past so that all previous tracks were erased. The temperatures necessary for complete annealing depend on the sample material and duration of heating (Table 5). Incomplete annealing results in mixed ages which lie apparently between the moments of formation and heating and which have no direct meaning. Partial annealing is recognized by size analysis of

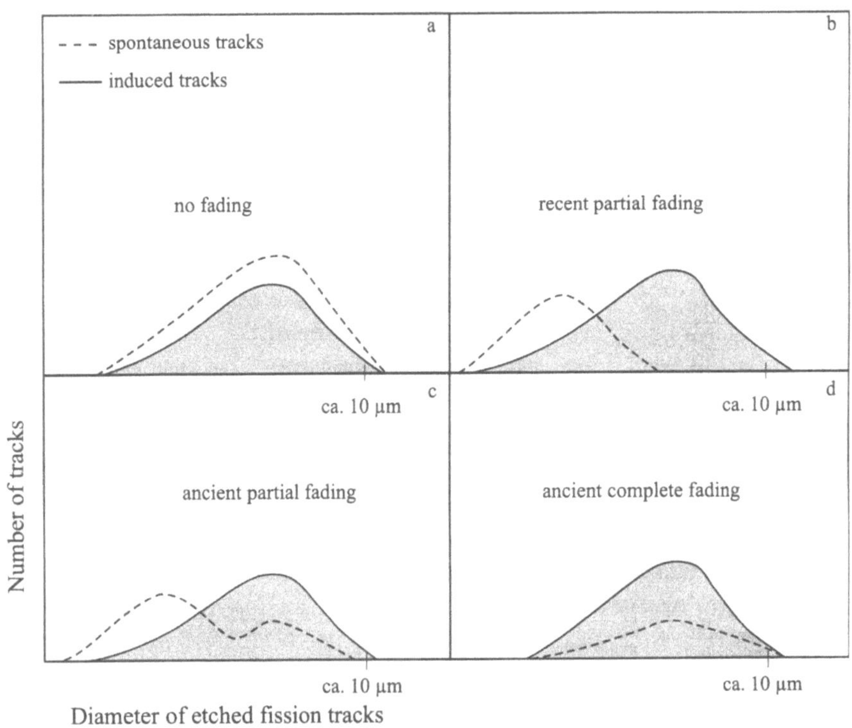

Fig. 71 a–d. The interpretation of fission track ages depends on the thermal history deducible from track size measurements. Four cases are schematically displayed: **a** both track types possess identical size distributions; the fission track age dates the formation. **b, c** The spontaneous tracks are smaller than the induced ones, indicating recent or ancient partial annealing; the fission track ages are mixed ages. **d** There are few spontaneous tracks, whose size distribution equals that of the induced tracks; the fission track age dates a secondary heating event through which all previous tracks were completely erased. (Wagner 1978)

the fission tracks, which makes fission track dating into a reliable tool for chronometry. The restricted application which fission track dating has had, so far, in archaeology and Quaternary geology is probably the consequence of the laborious and tedious procedure of track counting.

Basalts. The presence of zircon in basalts and tuffs enables the dating of volcanic eruptions. Fission track dating may also be applicable to sufficiently heated xenoliths included in volcanic rocks. These foreign inclusions originate from the crystalline basement and may contain zircon, titanite and apatite. Such zircon inclusions occur in the Eifel region, Germany. A 3.5-mm-sized, dark-red zircon crystal with 1040 $\mu g\ g^{-1}$ uranium and a 10-mm-long, needle-like zircon with 182 $\mu g\ g^{-1}$ uranium from the upper lava flow at Niedermendig yielded a mean age of 158 ± 10 ka. Two several mm-sized zircons with 455 and 356 $\mu g\ g^{-1}$ uranium from selbergite tuff from Wehr revealed 446 ± 39 ka (Wagner and Storzer 1970). These ages date Middle Pleistocene volcanic events and can be linked by tephrochronology with important Paleolithic sites within this region.

Volcanic Glasses. Its homogeneous and compact texture as well as high uranium concentrations up to 20 $\mu g\ g^{-1}$ render obsidian excellently suited as a material for fission track dating. However, the frequent presence of tiny microlites and vesicles within the glass matrix may disturb the results. During etching they develop pits which resemble the fission track etch pits and may be mistaken for them. Track counting becomes especially difficult at high densities of such "spurious tracks". Another difficulty is raised by the frequent phenomenon of natural track fading. As already discussed, it is detected by track size analysis and, if present, demands age correction. There are many examples of successful applications of fission track dating to Quaternary obsidians, among them those from the Mediterranean and Anatolia (Bigazzi et al. 1971; Wagner et al 1976; Bigazzi 1996). For instance, the obsidian flow (13.8 $\mu g\ g^{-1}$ U) at Gabelotto on Lipari island, Italy, gave an age of 8.6 ± 1.5 ka, while the apparent age of 6.4 ka had to be corrected due to the reduced spontaneous fission track diameter. Such minor degrees of partial track fading do not necessarily indicate distinct thermal events. They may be caused by ambient surface temperatures. Occasionally xenolithic inclusions in volcanites bear fritted glassy rinds which can be dated, as for example at Kalem in the Eifel region, Germany. For glass shards of 3–5 mm size a fission track age of 420±60 ka was obtained after plateau correction.

Tephra. Fission tracks contribute much to the dating of volcanic ashes. Volcanic glass shards, zircon and to a lesser degree apatite and titanite are used. Zircon has the advantage over glass that it is much more resistant against track fading. However, when relying on heavy minerals, the problem of various provenance needs to be taken into consideration since ash layers, as a sedimentary deposit, may contain detrital components. The discrimination of the primary volcanic grains from those of detrital provenance is achieved – apart from mineralogical criteria – by single grain fission track dating and their statistical treatment (Seward and Kohn 1997).

The potential and limitation of the tephrochronological application of zircon fission track ages can be readily demonstrated by means of the KBS Tuff, East Turkana, Kenya. The Plio/Pleistocene sedimentary sequence of the Koobi Fora formation contains several tuff horizons. The tuff primarily consists of glass fragments and pumice cobbles and shows signs of redeposition. Of particular interest is the KBS Tuff, because it embodies stone implements and is intercalated in hominid bearing layers. K–Ar data on the KBS Tuff raised a controversy in the 1970s between supporters of a long chronology (2.61 ± 0.26 Ma, Fitch and Miller 1970) and those of a short chronology (1.82 ± 0.04 Ma, Curtis et al. 1975). Fission track dating on zircon was carried out with the aim of undertaking an independent check. At first, the fission track results of 2.44 ± 0.08 Ma (Hurford et al. 1976) on zircon crystals, which were separated from the pumice cobbles, seemed to support the high K–Ar age. The lasting reluctance to accept such high value led to a renewed attempt of fission track dating, again on zircon from the pumice, this time with the result of 1.87 ± 0.04 Ma (Gleadow 1980) in accordance with the low K–Ar age (1.88 ± 0.02 Ma, McDougall 1985). Besides methodological aspects, the main reasons for the previous fission track overestimate of the KBS Tuff are detrital, old zircon grains and "spurious tracks" in form of needle-like inclusions within the zircon crystals.

For the first appearance of the genus *Homo* outside Africa, the discovery of stone artifacts near Riwat, northern Pakistan, are of great interest (Denell et al. 1988). Several tools made of quartzite were found in a conglomerate of the Neogene and Quaternary Siwalik Group. The artifact-bearing layer is inclined by folding and discordantly overlain by other sediments which contain a volcanic ash. Zircons separated form this tuff gave a fission track age of 1.60 ± 0.18 Ma (Johnson et al. 1982). This date represents a terminus ante quem for the artifacts. Rendell et al. (1987) infer, based on paleomagnetic measurements as well as on the tectonic and stratigraphic position of the artifact horizon, even an age of 2.0 ± 0.2 Ma, which implies a more or less simultaneous presence of *Homo habilis* in East Africa and South Asia.

In contrast to the heavy mineral zircon, glassy components in ash layers are mostly of primary volcanic provenance. The omnipresent track annealing in glasses always demands size analysis of fission tracks, whereby the long axes of the elliptical to rounded fission track etch pits are recorded, and corresponding age correction. Other difficulties are caused by the tiny size of the glass shards and by vesicles and microlites. The size of the shards must be at least 100 μm (Fig. 67). Corrected fission track ages on glass fragments from tephra contributed much to the Quaternary chronostratigraphy. The Valle Ricca Tuff in Latium, Italy, which is stratigraphically situated at the Plio/Pleistocene boundary in sandy clay, yielded, size-corrected, 2.13 ± 0.27 Ma and, plateau-corrected, 2.03 ± 0.26 Ma (Arias et al. 1981). An important glacial marker horizon in Alaska is the Old Crow Tephra, which is widely spread over 1000 km. Plateau-correction revealed 149 ± 13 ka, which means that this tuff must be placed in the penultimate Glacial and not, as hitherto assumed, in the last Glacial (Westgate 1988).

Deep-Sea Volcanites. Magmas that extrude at the ocean bottom along the mid-ocean ridges are chilled by contact with the cool ocean water, leaving a glassy rind on the pillow lavas. These rims can be dated by fission tracks (Fleischer et al. 1968). Their basaltic chemical composition implies low uranium concentration of less than 1 $\mu g\ g^{-1}$ and low thermal stability of the tracks. Even at the temperature of 4 °C, prevailing at the sea-bottom, annealing of fission tracks is observed, as is evident from track size measurements (Selo and Storzer 1981), so that fission track age correction may be necessary. Because of the low uranium content often only a few spontaneous fission tracks are found per cm^2. Despite this restriction fission track ages of less than 10 ka were determined on deep-sea glasses from the Mid-Atlantic Ridge (Storzer and Selo 1976). The ages increase with distance from the ridge axis and reveal an episodic and asymmetric spreading behavior (Fig. 72). The resulting sea floor spreading rates correspond to those derived from paleomagnetic studies.

Impact Glasses. Tektites and meteorite crater glasses are well suited for fission track dating. They belong to the first materials to which this method was applied (Fleischer and Price 1964). Crater glasses are generally inhomogeneous and often contain unmelted rock relics, which impedes fission track dating. Like volcanic glasses, impact glasses often suffer from track annealing, as was first realized in australites (Fig. 73). Numerous young impact craters were dated by fission tracks, among them the Henbury crater in Australia with 4.2 ± 1.9 ka, the Bosumtwi crater in Ghana – after age correction – with 1.04 ± 0.11 Ma (Storzer and Wagner 1977; Koeberl et al. 1997) and the Saltpan crater in South Africa with 220 ± 52 ka (Koeberl et al. 1994). Also in the Quaternary period falls the formation of the South-east-Asian-Australian tektites 700±100 ka and the Ivory Coast tektites

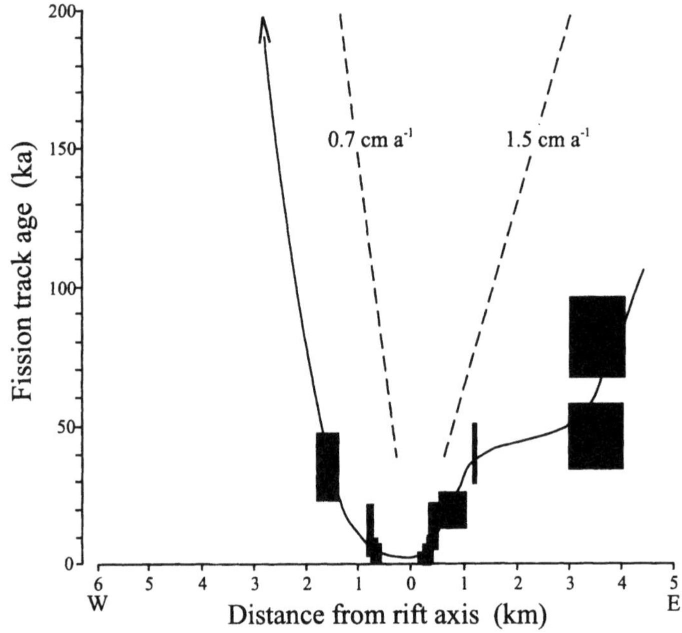

Fig. 72. Fission track ages of basaltic glasses from the Mid-Atlantic Ridge at 37°N. The ages are plotted against the distance of the sampling site from the ridge axis. The data reveal an asymmetric, episodic spreading behavior of the ocean floor. (After Storzer and Selo 1976)

Fig. 73. Size distribution of spontaneous and induced fission tracks in an australite. The reduced size of the spontaneous tracks is indicative of track annealing. (Storzer and Wagner 1969)

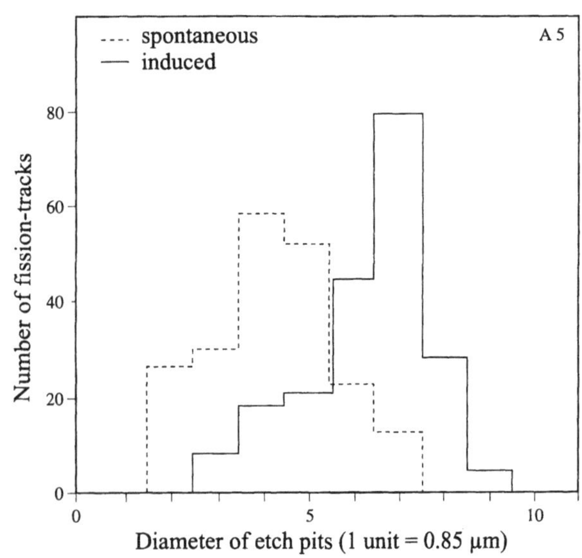

1.01 ± 0.10 Ma ago (Gentner et al. 1969). These tektite strewnfields extend into the adjacent oceans, where they appear as microtektites in deep-sea sediments. The temporal correlation between marine microtektites and corresponding continental tektites was established by fission track dating (Gentner et al. 1970). The concordance of the fission track age of the Bosumtwi crater with that of the Ivory Coast tektites established important evidence for the terrestrial origin of tektites.

Pseudotachylite. The vitreous phase of pseudotachylites, if sufficiently free of vesicular and mineral inclusions, can be dated by fission tracks. The method was applied to the pumice-like köfelsite from the post-Glacial landslide near Köfels, Austria. The 40–90 µm sized glass fraction yielded an age of 8 ± 6 ka (Storzer et al. 1971) which agrees with the ^{14}C age of 8710 ± 150 a BP determined on a tree covered by the slide. The large error of the fission track age is caused by the extremely low count number, namely two, of spontaneous fission tracks. A better precision is expected for older pseudotachylites.

Heated Obsidian Artifacts and Stones. Fission track analyses of obsidian implements reveal that during manufacture the raw material was subjected to heat treatment, for some unknown reason. The temperatures reached up to 400 °C resulting in partial or complete loss of fission tracks. Also the tools may have been exposed to elevated temperatures during use. Annealing of fission tracks was observed in obsidian artifacts from various cultures. At the Ecuadorian site El Inga near Quito a large quantity of obsidian flakes and tool fragments were found. Three flakes, made of obsidian from the same source with 11 µg g^{-1} uranium, gave fission track ages of 1.1 ± 0.07 Ma, 150 ± 11 ka and 2.06 ± 0.27 ka (Miller and Wagner 1981). Track size analysis in the three individual specimens revealed moderate, strong and no reduction, respectively, indicating that the fission tracks were partially, but differently, annealed in the first two specimens, but were completely annealed in the third specimen. After correction of the partially lowered data a fission track age of 1.72 ± 0.09 Ma was obtained which dates the geological obsidian source. The tools were manufactured from this raw material 2060 ± 270 a ago while the material was subjected to various temperatures up to 400 °C.

Baked stones and soils as well as ashes at prehistoric fire places and burning horizons may have been heated sufficiently to reset the fission track record in apatite, zircon and titanite. Such a case was encountered at the discovery site of Peking Man, *Homo erectus pekinensis*, at Zhoukoudian near Peking. Several hundred grains of titanite in the size range of 50–300 µm were separated from ashes of layers 10 and 4. The length distribution of the fission tracks was utilized as the criterion for discriminating between partial and complete track annealing of the titanite grains by the fire. Altogether, 100 grains showed complete resetting and gave mean ages of 462 ± 45 ka for layer 10 and 306 ± 56 ka for layer 4 (Guo et al. 1991).

Artificial Glasses. Due to their young age, low uranium content and consequently very low track densities, most artificial glasses, although in principle suited, are hardly recommendable for fission track dating. This is demonstrated by a glass shard recovered from the mortar of a Roman bath at Chassenon, France. Its uranium concentration was 3 $\mu g\ g^{-1}$. After 100 h of track counting and repeated repolishing 25 cm^2 were scanned and a total of 29 spontaneous fission tracks were found. The calculated fission track date of 150 AD ± 350 a agrees well with the historical date, but the experimental effort was not worthwhile in view of the analytical uncertainty and is rewarding only when independent age information is not available. The situation is different for the uranium rich glasses, so-called *uranium-glasses*, which have been manufactured since the mid-19th century, first in Bohemia and later also in other parts of the world (Fig. 12). Their high uranium concentration of up to several percent enables them to be dated precisely (Brill 1965).

6.2 Alpha Recoil Tracks

Unlike the fission track method, which utilizes the etchable tracks of the fission, the α-recoil track method is based on the etchable tracks of the α-decay. When attempting to reveal spontaneous fission tracks in mica by etching, Huang and Walker (1967) observed a background of numerous shallow etch pits. The fact that their number correlates with that of fission tracks implied, apart from other criteria, their interpretation as recoil tracks of the α-decay. This opened the possibility of employing α-recoil tracks also for dating. Since α-disintegration is much more frequent than spontaneous fission of uranium, dating by α-recoil tracks should be much more sensitive, allowing the dating of samples as young as a few hundred years. Despite its suitability in principle for the archaeological age range this method is still in an incipient stage of development and still lacks a reliable physical model. So far the revelation of etchable α-recoil tracks is restricted to mica. The validity of α-recoil track dating was demonstrated for mica inclusions in ceramics and young volcanic rocks. Other potential objects for dating are mica-containing heated stones and fire places. The methodological problems entail that the successful application of α-recoil track dating requires calibration by materials of known age.

6.2.1 Methodological Basis

During α-decay of heavy atomic nuclei, energy of several MeV is set free that appears predominantly as kinetic energy of the ejected α-particles. The remaining heavy nucleus suffers recoil according to the principle of momentum conservation. The energy ratio of the recoiling nucleus and the α-particle is inversely proportional to their mass ratio and thus the energy of the α-recoil nuclei within the uranium and thorium decay chains is ~2% of the total decay energy, i.e., several 10–100 keV per recoil. In solids the recoil nuclei have ranges of a few 10^{-2} µm. They lose their energy along their path by interaction with the lattice atoms, predominantly by atomic collision, and thus damage the structure of the solid. During etching shallow etch pits develop at the damaged zones, the *α-recoil tracks*, which are visible with optical phase-contrast (Fig. 74) or scanning electron microscopy. Unlike recoil nuclei, α-particles themselves leave no etchable tracks. Despite intensive efforts, α-recoil tracks were observed only on cleavage planes in mica – a phenomenon which is probably linked to the relatively low etching rate perpendicular to these faces.

Important natural α-emitters are ^{238}U, ^{235}U and ^{232}Th as well as some of their radioactive daughter nuclides in the decay chains (Fig. 22). Altogether 8 α-particles are emitted within the series beginning with ^{238}U and ending with ^{206}Pb, 7 α-particles are emitted within the series from ^{235}U to ^{207}Pb and 6 α-particles are emitted within the series from ^{232}Th to ^{208}Pb.

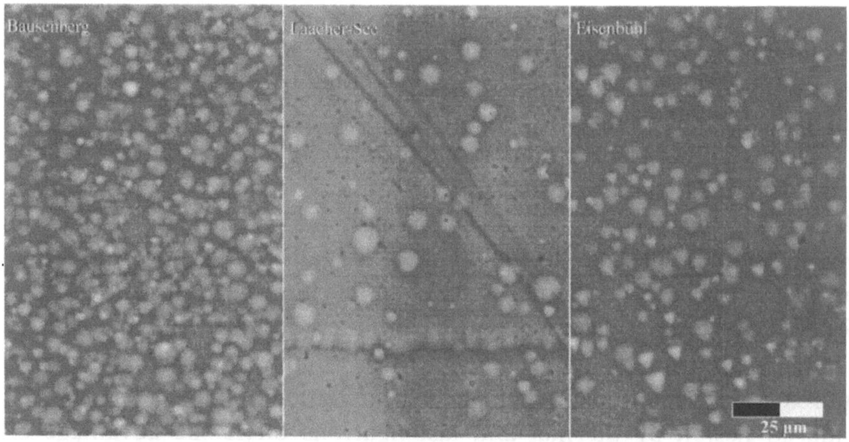

Fig. 74. Etched (10 min in 40% hydrofluoric acid at room-temperature) α-recoil tracks in Quaternary biotites from Bausenberg, Laacher See (both in the Eifel volcanic field) and Eisenbühl (western Czech Republic). (Courtesy of K. Gögen)

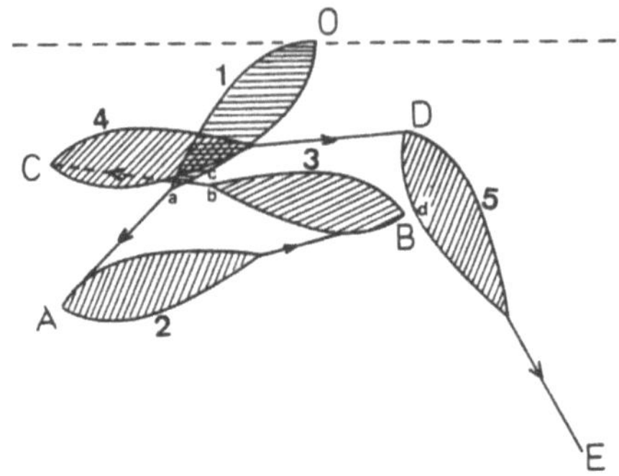

Fig. 75. Path of recoil nucleus from position O to E as consequence of 5 successive α disintegrations. (Reprinted from Hashemi-Nezhad and Durrani 1981; copyright 1981, with kind permission from Elsevier Science Ltd., The Boulevard, Langford Lane, Kidlington OX5 IGB, UK)

Within each of these decay cascades the remaining nucleus suffers repeated recoils by ~0.02 μm per recoil. Since the directions of the single recoils are arbitrary, the recoiling nucleus follows a zigzag path (Fig. 75). Instead of a linear track a cluster of radiation damage of several 10^{-2} μm diameter is produced, which represents one "α-recoil track". For the quantitative dating model it is essential to know whether the existence of one single α-recoil is already sufficient to produce one etchable track or whether several consecutive α-recoils are required. The observation by Hashemi-Nezhad and Durrani (1981) suggests that a single recoil is sufficient, at least in biotite. Another difficulty is the danger that nuclides within the decay chain are lost, in particular radon by diffusion, so that the later α-recoils occur at a different site from the previous ones, in other words that more than one track is formed for each uranium disintegration.

The α-recoil tracks accumulate with time so that their number becomes a measure for the age of a sample. The connection between age and areal density of α-recoil tracks is more complicated than for fission tracks. In young samples with ages in the range of the half-lives of the α-decays, in particular 75 ka for ^{230}Th, radioactive equilibrium may not yet be established, i.e., the α-activity and the track production rate are still increasing. The concentration of the parent nuclides uranium and thorium is required for α-recoil track dating. The determination of these elements with induced fission by thermal and fast neutrons offers the advantage of obtaining information on their microscopic distribution also.

Without the thorium contribution to α-recoil tracks one would expect a fixed abundance ratio between the α-recoil tracks and the fission tracks.

The ratio of ~3500, observed by Huang and Walker (1967) in micas of various ages, corresponds roughly to the theoretical prediction. When assessing this ratio one needs to keep in mind that the α-decay of uranium certainly is 2×10^6 times more frequent than the spontaneous fission, but the lengths of the α-recoil tracks and, therewith, their probabilities of intersection by the etched plane are 10^3 times lower than for fission tracks. Like fission tracks, α-recoil tracks fade. There are only few data available on the stability of α-recoil tracks in mica at elevated temperatures. Hashemi-Nezhad and Durrani (1981) report that the duration for complete annealing of α-recoil tracks in biotite is at least one order of magnitude shorter than for fission tracks.

6.2.2 Practical Aspects

Mica, the only mineral to which α-recoil track dating seems applicable, is a common compound in various rocks and artifacts. Restriction to mica is an experimental facilitation for the practical performance of α-recoil track dating. Since the tracks are revealed on cleavage planes, grinding and polishing are omitted. Due to the high track density less than mm-sized mica flakes are sufficient, and single flakes can be dated. It is important to analyze only the interior faces of the mica "booklets", because uranium leakage and deposition by migrating fluids are thereby minimized. A suitable etchant is concentrated hydrofluoric acid at room temperature, for muscovite ca. 2 h and biotite ca. 1 min. Since it was found that the track density for a given biotite depends strongly on the etching time, it is important to obey defined experimental criteria (Gögen and Wagner 1998, unpubl.). The very shallow α-recoil tracks require a phase-contrast or interference-contrast facility or a scanning electron microscope for visualization. When counting the α-recoil tracks in the mica flakes it is desirable to keep away from fractures and margins because such sites often bear signs of secondary alteration, such as increased uranium concentration. For uranium and thorium analyses a double irradiation technique with thermal and fast neutrons is optimal, because it allows the determination of the distribution of these elements on the same faces on which the α-recoil tracks were counted (Wagner 1980 a).

6.2.3 Application

Ceramics. Attempts were undertaken to date mm-sized muscovite inclusions in pre-Columbian pottery (Garrison et al. 1978; Wolfman and Rolniak

1978). The objects studied originate from various Paleo-Indian cultures of the past two millennia, and the ages were independently known for most of them. It was assumed that the mica temper was sufficiently annealed during ceramic manufacture to allow for complete annealing of all previously stored α-recoil tracks, thus resetting the clock. After separating and etching the mica plates, a linear correlation between the areal density of the α-recoil tracks, especially after normalization to uranium, and the known age was obtained (Fig. 76). The correlation probably would have been even better if thorium had been taken into account also. The linear age dependence of the α-recoil track is an encouraging result for the further development of the method, at least in the young age range up to several 10^3 a.

Volcanites. Biotite is a common constituent in young volcanic rocks. Of particular significance is its presence in tephra layers which, in their turn, are of great interest as stratigraphic marker horizons. In order to explore this potential, biotite was separated from lower and middle Pleistocene

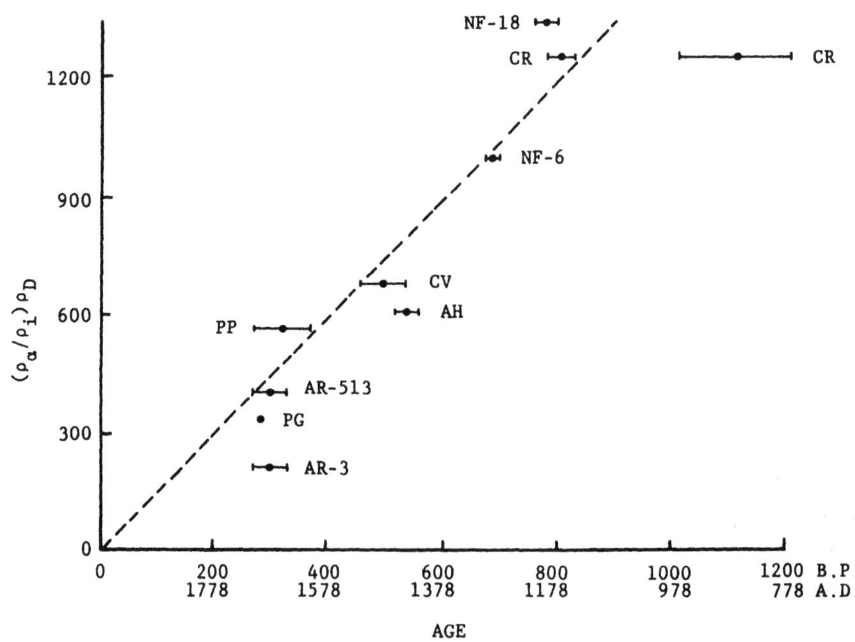

Fig. 76. Correlation between the α-recoil track density in muscovite temper and the known age of various ceramics (Wolfman and Rolniak 1978). The firing of the ceramic resulted in annealing of the pre-existing α-recoil tracks thus resetting the clock. The areal density of the α-recoil tracks is normalized to the induced fission track density, i. e., the uranium content

volcanites of the Eifel region, Germany (Unger 1993). The mm-sized mica plates were counted under the phase-contrast microscope. The α-recoil track density was normalized to the uranium content determined by induced fission tracks on the same faces. A linear correlation between the α-recoil track density and the known age of the rock samples was observed in the wide age range of 10 ka – 0.5 Ma. The validity of this linearity was confirmed in the meanwhile by additional samples from the Eifel region (Fig. 77). Even single biotite flakes as tiny as 300 µm, separated from a tephra layer, exhibited a sufficiently high track number for precise analysis. At present, it appears that the method is still too complex to evaluate physically independent age data. However, the confirmed linear relationship between age and track abundance is a promising basis for successfully developing α-recoil track dating as a technique relying on age calibration.

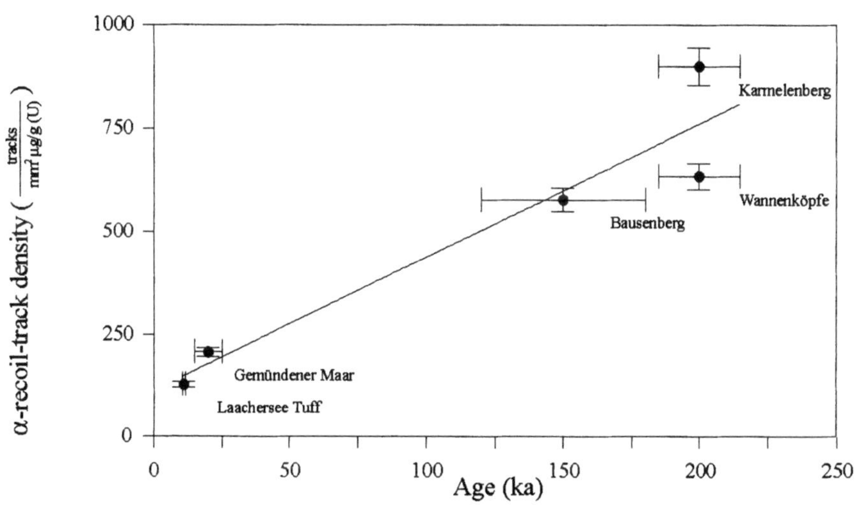

Fig. 77. Correlation between the areal density of α-recoil tracks in biotite – normalized to a uranium concentration of 1 µg g^{-1} – and the independently known rock ages of the volcanic Eifel region, Germany. (Courtesy of K. Gögen)

7 Radiation Dosimetry

The three dosimetric dating methods *thermoluminescence* (*TL*), *optically stimulated luminescence* (*OSL*) and *electron spin resonance* (*ESR*) are based on the time-dependent accumulation of radiation damage in minerals. Hence they have many features in common. What basically distinguishes them from each other is the physical phenomenon which is exploited to detect the radiation damage. Each of these phenomena has its specific potential for dating.

A low level of ionizing radiation, which originates from natural radioactivity and cosmic radiation, is omnipresent in nature. The interaction between this radiation and the atoms of minerals results in a gradually increasing radiation damage. The intensity of the radiation damage is a measure of the natural radiation dose *ND*, which a mineral has received ever since it was formed or its system was reset for the last time. The mineral is used as a natural radiation dosimeter. Once one has "read" the natural dose by means of a TL, OSL or ESR measurement and knows the natural dose rate *NDR* – i.e., the dose per unit time – the age *t* is obtained according the equation

$$t = \mathrm{ND/NDR} \tag{46}$$

In physics the term dose *D*, strictly speaking energy dose, comprises absorbed radiation energy per unit mass. Its unit is *Gy* (formerly in use was the unit *rad* which is still frequently met, with 100 rad = 1 Gy). The dose rate is measured in $Gy\ s^{-1}$. With respect to natural dose rates it is more convenient to use the $mGy\ a^{-1}$ or – what is equivalent – $Gy\ ka^{-1}$.

The age range covered by the radiation dosimetric methods comprises 10^2 to 10^6 a. The lower limit is essentially determined by the detection sensitivity of the TL, OSL or ESR signals and the dose rate, whereas the upper limit depends predominantly on the saturation and stability characteristics of the signals.

First some introductory remarks are given on the phenomenon of radiation damage and on the terms "natural dose" and "dose rate", which are fundamental to all three radiation dosimetric methods of dating. The particularities of the dating methods are then dealt with separately in the sections covering the individual methods.

Radiation Damage. The radiation damage to which luminescence and electron spin resonance phenomena are linked implies electrically non-conducting solids containing lattice defects which disturb the regular order of the crystal lattice. Implicit in the real crystal structure are defects on the microscopic and atomic scale. Hence the real crystal shows smaller or larger deviations from the perfectly ordered crystal structure. Primary defects are already acquired during mineral formation, whereas secondary defects are manifested at a later stage.

In this context only the atomic lattice defects are of concern. Three categories of atomic lattice defects can be distinguished: vacancies, impurities and Schottky-Frenkel type defects. Vacancies are non-occupied lattice positions. Chemical impurities may either substitute for regularly positioned atoms or occupy interstitial lattice positions. Schottky-Frenkel type defects refer to a defect pair consisting of a vacancy and a corresponding interstitial atom. In contrast to regularly occupied lattice positions, atomic defects are commonly characterized by positive or negative charge deficits, which act as traps on free charges diffusing through the lattice. The defect concentration of quartz and feldspars ranges around 10^{-8} to 10^{-7}.

In nature, crystals are exposed to more or less strong ionizing radiation. When radiation, consisting predominantly of energetic α- and β-particles as well as photons (γ-rays), interacts with the atoms of a solid, radiation energy is transferred to the atoms. The processes involved include ionization and scattering as well as Compton, photoelectric and pair-production effects, all of which finally result in freed electrons and heat. The electrons (negative charge) freed from the atoms leave behind an absence of electrons, referred to as holes (positive charge). These free charges move through the crystal lattice. The holes are very quickly trapped at hole traps which are lattice defects with positive charge deficits. The electrons wander through the crystal until they either encounter a trapped hole and recombine or become trapped at electron traps which are lattice defects with negative charge deficits (Fig. 78). Defects which are occupied by charges are called centers. The phenomena of TL, OSL and ESR are caused by these centers, however, the same centers do not need to be responsible for these different phenomena. The question still remains largely open whether distinct TL, OSL and ESR signals can be related to identical centers. With longer irradiation times the concentrations of the different centers increase, until a stage of saturation is reached at which all available defects are occupied. Assuming a constant, finite number of defects an

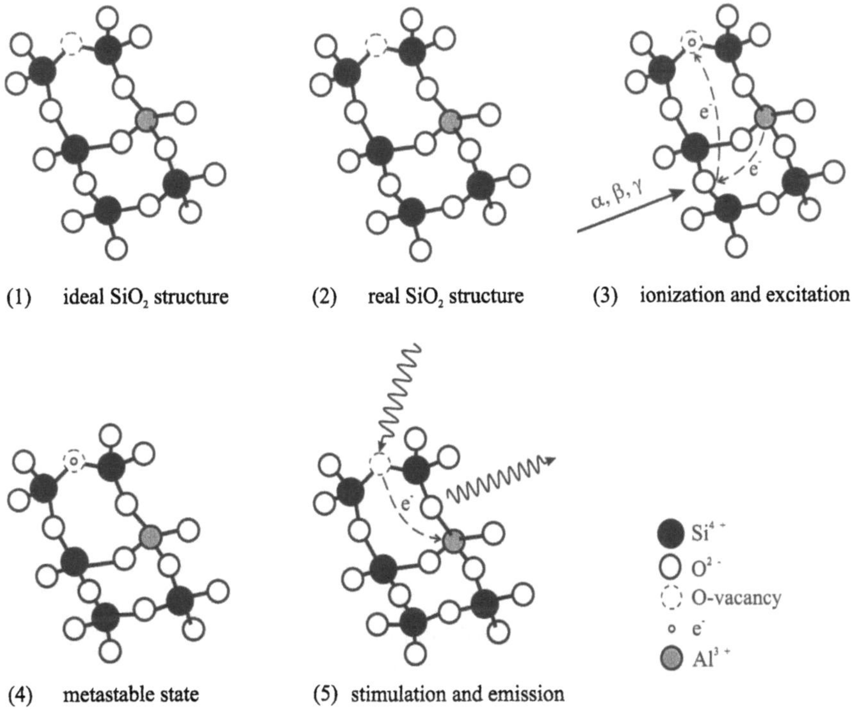

Fig. 78. Luminescence mechanism, shown schematically for quartz structure. Atomic defects – depicted as an O^{2-} vacancy and an Al^{3+} substituting Si^{4+} – are associated with charge deficits which serve as traps for freed charges. Ionization generates electrons and electron holes which can wander through the crystal lattice until they are trapped, the electron at the O^{2-} vacancy and the hole at the Al^{3+} defect. In the excited stage the system is metastable. Thermal or optical stimulation releases the trapped electrons which recombine with the Al center under luminescence emission

exponential saturation function results. The intensity I of a TL, OSL or ESR signal, all of which are proportional to the number of centers, increases as a function of the irradiation time t at dose rate DR:

$$I = S \times DR \times \tau_1 \times (1 - e^{-t/\tau_1}) \qquad (47)$$

S stands for the sensitivity with which I grows per unit dose D ($S = I/D$) and τ_1 is the time required for the filling of $(1 - e^{-1}) = 63.2\,\%$ of the available traps. At the signal intensity I_S (Fig. 79) saturation is reached. In this context it is worthwhile to call attention to the phenomenon that during irradiation not only does the occupation of primary defects take place, but secondary defects are also created. The latter disturb the exponential course of the function. The creation of such secondary defects is presumably caused mainly by α-recoil (Sect. 6.2).

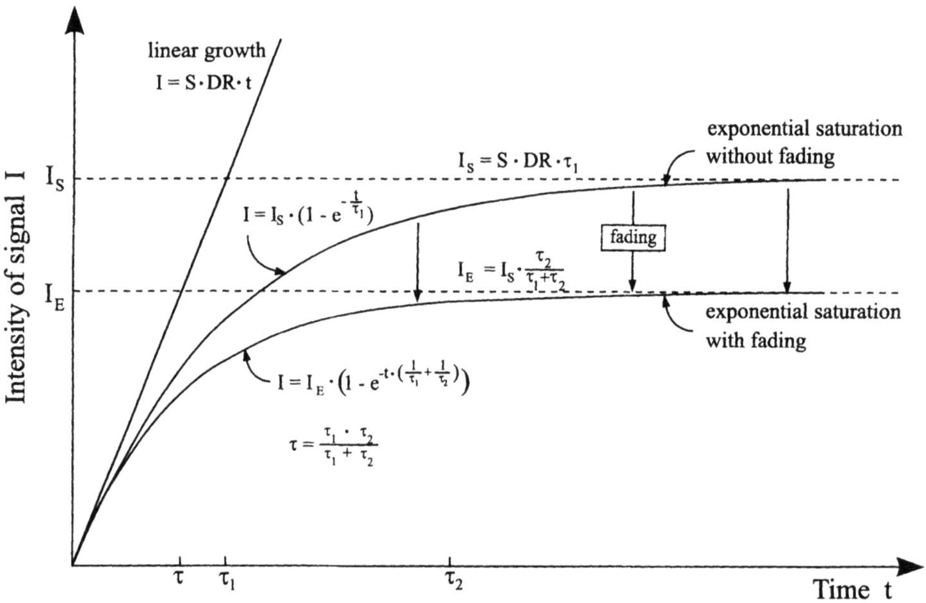

Fig. 79. Linear growth and exponential saturation of signal I (TL, OSL or ESR) with irradiation time t for given sensitivity S and dose rate DR. In the saturation function the signal converges to the saturation value I_s. Concurrent saturation and thermal fading (with the signal's mean lifetime τ_2) results in the growth of the signal I with the apparent mean lifetime τ, and finally attains the thermal equilibrium at the intensity I_E

The sensitivity S is the same for β- and γ-radiation but different for α-radiation. The higher ionization rate per unit length along an α-path produces an excess of free charges and therefore the probability that these charges find unoccupied defects is smaller than in the case of β- or γ-radiation. The sensitivity ratio S_α/S_β, the so-called a-value, is typically around 0.1.

The concentration of centers is determined by TL, OSL and/or ESR measurements. To evoke the emission of TL and OSL, additional thermal or optical stimulation energy, respectively, has to be supplied to the crystal. During the luminescence measurement, the traps are emptied and thus the affected centers are destroyed. The situation is different for ESR, where the ESR centers are determined in situ by irradiating with microwave energy, and the concentration of the centers is not altered during the measurement. There are methodological advantages inherent in this latter feature.

Natural Dose. From the intensity of the radiation-induced TL, OSL or ESR signals, the natural radiation dose *ND*, which a sample has received ever since its formation or resetting, needs to be evaluated. For this the sample

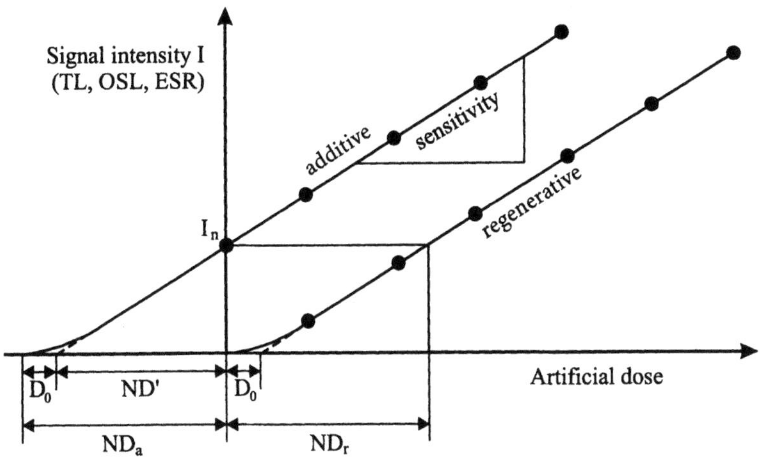

Fig. 80. Determination of the natural dose ND from the signal I (TL, OSL or ESR). The additive technique requires a further artificial irradiation in addition to the natural signal I_n. The natural dose ND_a is obtained from backward extrapolation of the growth curve to the intersection ND' with the dose axis, and, if necessary, i.e., in case of supralinearity in the lower dose range, the intercept correction D_0 must be added. In the regenerative technique the signal I_n is thermally or optically erased and subsequently regenerated by artificial irradiation. The dose required to regenerate I_n gives the natural dose ND_r. ND_r equals ND_a under the supposition that no sensitivity change has occurred

is irradiated with certain doses being either added to their natural dose or regenerated after artificial resetting. The particular signals are measured and an additive or a regenerative growth curve is constructed (Fig. 80). For γ-irradiation calibrated ^{60}Co or ^{137}Cs, for β-irradiation ^{90}Sr, and for α-irradiation ^{241}Am sources are commonly in use.

From an additive growth curve *ND* is evaluated through backward extrapolation of the growth curve onto the dose axis. This requires, however, the knowledge of the mathematical function of the signal growth with increasing radiation dose. It is desirable that signal versus dose should be a linear relation, however, the exponential growth curve shows a quasi-linear increase only in the beginning and then, as the concentration of unoccupied defects decreases, the curve asymptotically approximates to a saturation value I_s. Unfortunately, most observed growth curves are mathematically more complex and physically scarcely explainable. In the lower dose range supralinearity commonly is observed. Tentative physical explanations for this phenomenon comprise dose rate effects, creation of new defects and charge-entrapping competition at different traps. Through a given set of data mathematically different regression curves (linear, exponential or polynomial) can be fitted and consequently

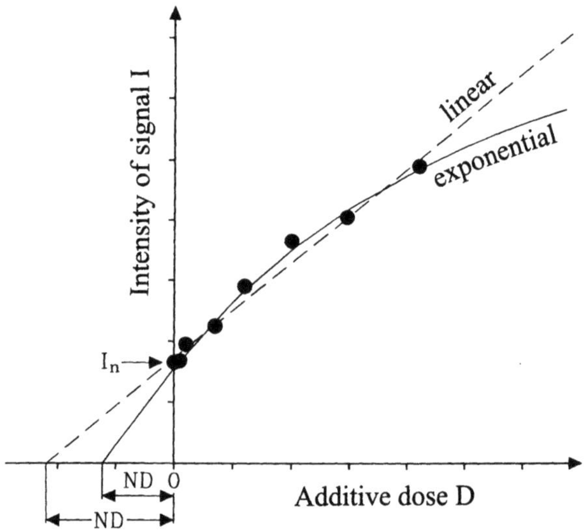

Fig. 81. Ambiguity of the ND determination for the additive technique. The regression requires knowledge of the mathematical function of the signal growth. In the example presented here the data set is too small to allow a discrimination between linear and exponential saturation growth. Both functions fit through the data points, however they result in very different ND extrapolations. (After Barabas 1989)

the corresponding ND extrapolations may differ considerably from one another (Fig. 81). To reach a decision on the most suitable function it is helpful to construct the growth curve over a dose range that is as extended as possible.

Concerning regenerated growth curves, the ND is evaluated through that artificial dose which exactly recreates the intensity of the natural signal (Fig. 80). The obvious possibility of evaluating the ND in this way, especially if there is a non-linear signal growth, is frequently not applicable due to sensitivity changes caused by the resetting process. On samples which have been artificially heated or exposed to light, sensitivity increase or – more rarely – decrease are observed, which are explainable in terms of solid state physics by redistribution of charges. In the case of a sensitivity increase the regeneration dose, recreating the natural signal, underestimates the ND value.

Since the sensitivity is different for α-radiation compared to β- or γ-radiation, the general age equation [Eq. (46)] needs to be modified to

$$t = ED_\beta/(a \times DR_\alpha + DR_{\beta\gamma c}) \tag{48}$$

where the equivalent dose ED_β is the artificial β-dose which yields a TL, OSL, or ESR signal of equal value to the natural signal. The evaluation of

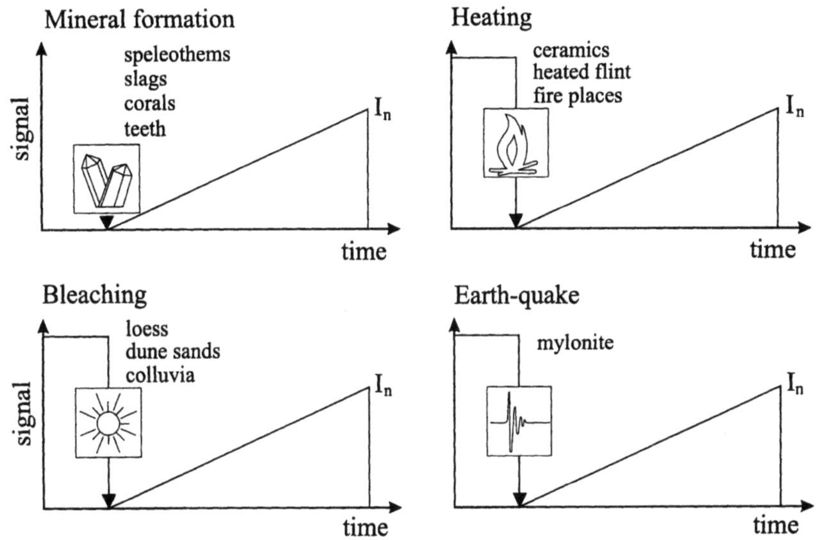

Fig. 82. Radiation dosimetry dating requires defined events during which the natural dosimeter was fully reset. Apart from mineral formation, this happens if minerals experience heating, light exposure or pressure. In this way different geological and archaeological events become datable

the dose rate DR requires a separate consideration of its α-contribution DR_α multiplied by the a-value and the remaining contribution $DR_{\beta\gamma c}$ consisting of β-, γ- and cosmic radiation.

Resetting. Usually the minerals to be dated are much older than the age range covered by these radiation dosimetry clocks. The age range comprises not more than the last 10^5 or at best 10^6 years. However, the radiation damage accumulated by these minerals may be extinguished by particular events. Partial or even complete resetting of the TL, OSL or ESR signals may be caused by heating, light exposure or pressure. Consequently, apart from recent mineral formation, such as the crystallization of cave sinter, the last resetting event – in the form of heat, light or pressure – can be dated (Fig. 82). Volcanic and archaeological events that are related to heating are datable. For the dating of sediments the possibility that the signals have been bleached during deposition is exploited. The pressure sensitivity of the ESR signal enables the age determination of tectonic faults.

Fading. Essential to all radiation dosimetric methods of dating is that the centers involved in the signal generation are stable over the complete age range in question. Like any other type of radiation damage the TL, OSL and ESR centers are subjected to fading whose kinetics is essentially thermally controlled. Assuming a first-order kinetics the thermal stability of a particular center at the ambient temperature T can be described by the mean lifetime τ_2, during which an initial concentration of centers decreases to the value $1/e$. With regard to the Arrhenius equation the following function is valid:

$$\tau_2 = 1/a_0 \times e^{E/(k \times T)} \tag{49}$$

where a_0 is the frequency factor [s^{-1}], E is the activation energy [eV], k is the Boltzmann constant and T the absolute temperature [K].

Apart from the saturation characteristic, the mean lifetime τ_2 controls the upper age-limit of the applicability of the clock. The value of τ_2 is determined by annealing experiments and subsequently extrapolated to the usually low ambient temperatures. Provided there is coexistence of radiation-induced build-up [Eq. (47)] and thermal-induced decay of centers, the signal I grows according to

$$I = S \times DR \times \tau \times (1 - e^{-t/\tau}) \tag{50}$$

where $\tau = \tau_1 \times \tau_2/(\tau_1 + \tau_2)$. Finally, when the rate of trap filling is balanced by the rate of escape, a thermal equilibrium is reached at the intensity $I_E = I_S \times \tau/\tau_1$ (Fig. 79). Consequently, due to the saturation and equilibrium phenomena, the measured age t_m underestimates the real age t according to

$$t_m = \tau \times (1 - e^{-t/\tau}). \tag{51}$$

Hence, with increasing age, the discrepancy between t_m and t continually increases, i.e., the clock is slow. The measured age approximates asymptotically to the limiting value τ (Fig. 83). If one tolerates 10% deviation of t_m

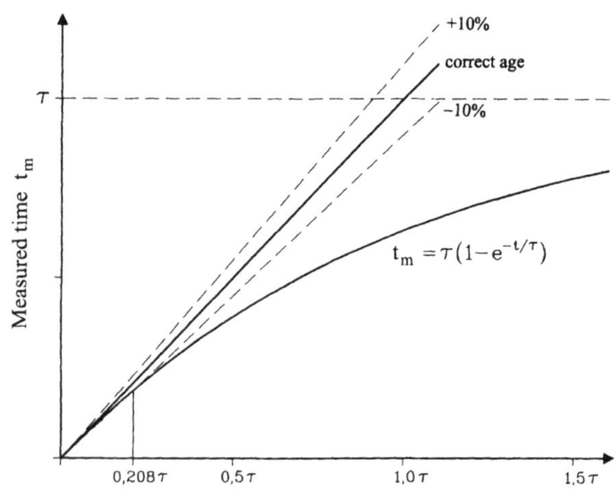

Fig. 83. The phenomena of saturation and thermal equilibrium cause an increasing difference between the determined apparent age t_m and the true age t, with t_m asymptotically approaching to the limiting value τ

from t, t_m has to remain $< \tau/5$, and in the case of 5% deviation even $< \tau/10$. As far as near-surface sediments with normal temperatures are concerned, most centers possess τ-values of a few Ma and more, so that the datable age range presumably comprises up to a few 100 ka. Efforts have been undertaken to correct for the fading-induced age underestimation (Mejdahl 1988).

Dose Rate. Once the natural dose ND has been evaluated, the natural dose rate NDR still needs to be determined for the age calculation with Eq. (46). The ionizing radiation level on the Earth's surface originates from the natural radioactivity and to a minor extent from the cosmic rays. Considerable difficulties may arise from spatial and temporal changes of the dose rate (Aitken 1985). These complications will be discussed subsequently. In any case, in addition to the micro-dosimetric investigations of the sample, on-the-spot measurements of the dose rate – precisely at the place where the sample is taken from – are required.

The essential radionuclides which contribute to the natural dose rate are ^{232}Th, ^{238}U, ^{235}U, their daughter products, as well as ^{40}K and ^{87}Rb. These nuclides emit α-, β- and γ-radiation each of which possesses a different penetration range. In silicates the penetration ranges comprise for α-, β- and γ-radiation several 10^{-2}, 10^{0} and 10^{2} mm, respectively (Fig. 84). Along their paths α- and β-particles as well as photons experience energy loss due to interactions with the lattice atoms, until they finally come to halt. The following two facts are crucial for the understanding of dosimetry. First of all, emission and energy absorption of the radiation do not occur at identical sites. Secondly, the energy of the particles diminishes along their path giving rise to a broad energy spectrum ranging from initial energy to zero.

If the radionuclides are uniformly distributed throughout the sample and the sample volume largely exceeds the penetration ranges of the radiation, the evaluation of the dose rate is straightforward. Inside the body a homogeneous radiation field exists, and the radiation energy emitted within a body is equal to the absorbed energy and hence the dose rate is quantifiable. In dosimetry the term quasi-infinite matrix (Fig. 85) is used to describe this situation.

Ideal conditions like this with respect to homogeneity and sample size are rarely found in the field of dating. Inhomogeneous distributions of radionuclides, which occur on very different scales, are rather the rule than the exception in nature. In zoned crystals strong inhomogeneities over short distances of several 10^{-3} mm may be present. Different minerals usually vary considerably in their radioactivity. A difference in the radioactivity usually exists between the sample and its embedding sediment matrix. Not only the sample, but also the enclosing sediment can be inhomogeneous. Due to geometric effects there exist dose rate gradients in

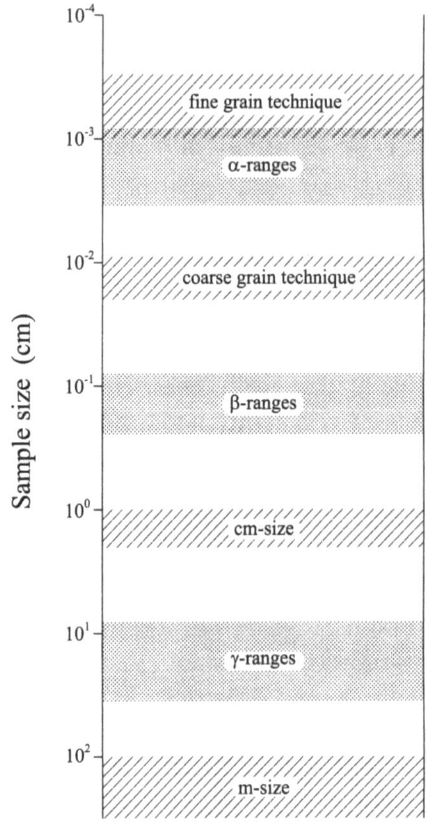

Fig. 84. Relationship between the α, β and γ ranges and the grain size of minerals to be dated. A choice of grain sizes falling in between the particular ranges is favorable for evaluation of the dosimetric contribution from the different types of radiation

the proximity of the interface between adjacent materials of different radioactivity (Fig. 86). The gradients are of the order of the penetration ranges and experience further complication by the energy spectrum of the radiation.

In dosimetry the term *homogeneity* needs to be put more precisely. An inhomogeneity, which is smaller than the penetration range of a particular radiation, does not express itself as an inhomogeneity in terms of the dose rate. However, if the inhomogeneity size is of the same order as the range, it will give rise to a significant spatial variation of the dose rate (Fig. 85). In this context homogeneity always needs to be related to the range of the radiation. Due to highly differing α-, β- and γ-penetration ranges, a sample which bears radioactive inhomogeneities of the order of 30 µm is inhomogeneous in terms of α-radiation (Fig. 84), but may be homogeneous in terms of β- and γ-radiation. A sample, e.g., a mammoth tooth, with variations in its uranium content on a microscopic scale, is inhomogeneous in terms of α- and β-radiation, but may be homogeneous in terms of γ-ra-

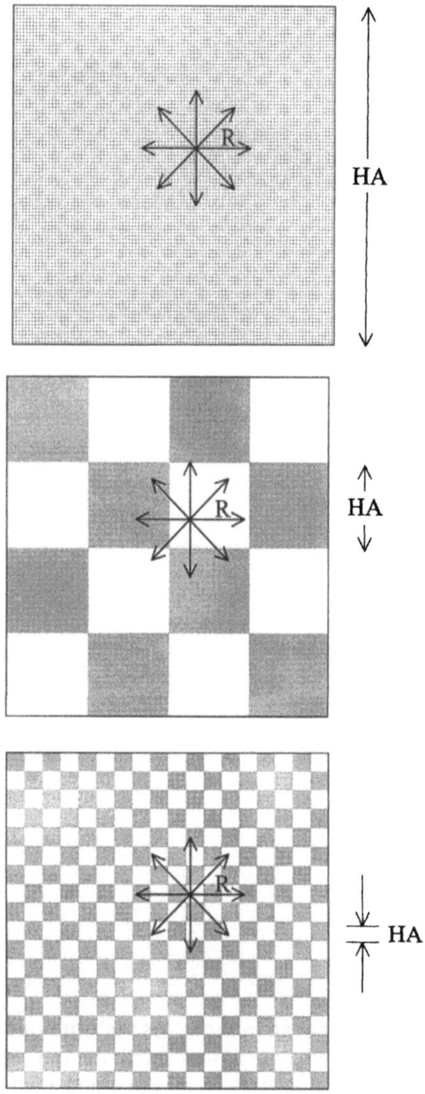

Fig. 85. Relationship between the size of the area HA with homogeneous dose rate and the radiation range R. To enable dating a uniform radiation field is required. If HA$\gg$R (*above*), a quasi-infinite, homogeneous radiation field prevails within HA. If HA $\approx$ R (*middle*), an inhomogeneous radiation field is generated as a result of geometry and attenuation effects. If HA $\ll$ R (*below*), a homogeneous radiation field exists across the areas

diation. The evaluation of the dose rate for a given sample of a particular size needs to take separately into account the internal component and the external one derived from the environment. Frequently, the sample size, e.g., a mollusk with a size of a few centimeters, is far below the γ-penetration range but still significantly above the β-range. The γ-component of the dose rate supplied to such a sample stems almost exclusively from the environment, whereas the α- and β-components are internal. For an even smaller sample, e.g., an individual lamella of tooth enamel with the size of

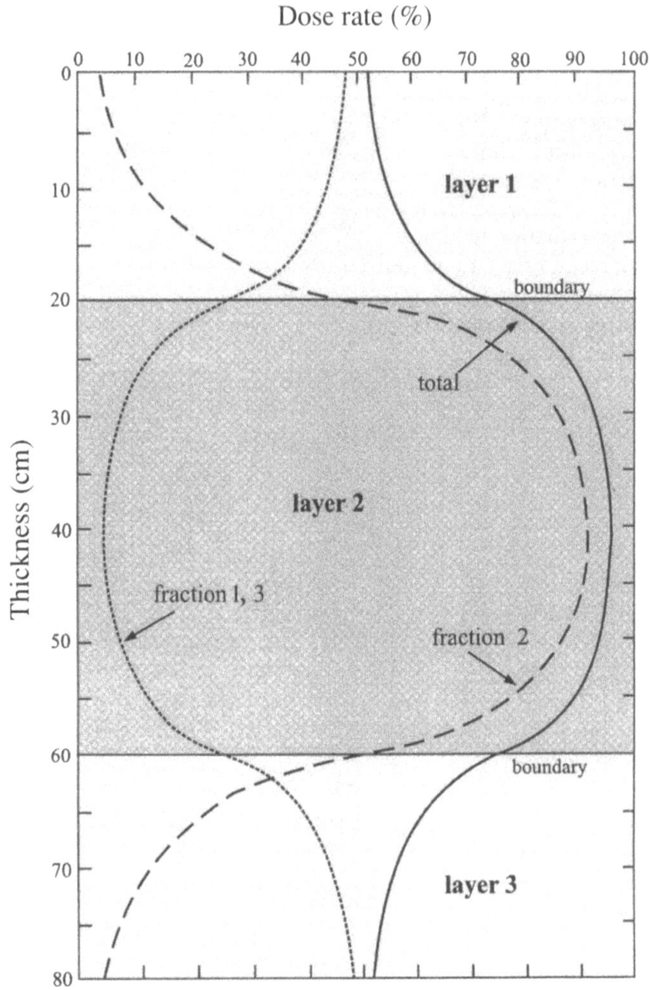

Fig. 86. Change in the γ-dose rate at boundaries of layers (layers of 2.0 g cm^{-3} density and 0.25 water content) with the radioactivity in *layer 2* being twice as much as in *layers 1* and *3*; 100% refers to a quasi-infinite thick layer 2. Dose-rate gradients express themselves in the order of the γ range. (After Aitken 1985)

a few 100 μm, only the α-component is internal whereas the β- and γ-components are external.

The age determination requires materials of uniform and definable dose rate. Complications arise if the scales of the radioactive inhomogeneities and the ranges of the radiation fall together. If the scale of the radioactive inhomogeneity is clearly larger or smaller than the penetration range of the radiation, the dose rate is uniform and quantifiable. Hence it is preferable to choose such grain or sample sizes which fall just in between the ranges in question. The fine-grain fraction of 2–10 µm stands out for a homogeneous dose rate in terms of all three radiation types. Coarse-grain fractions around 100–200 µm are large compared to the α-range, but small compared to the β- and γ-ranges. Another preferable size range comprises several cm, that is large compared to the α- and β-ranges, but small compared to the γ-range. The sample preparation should be designed to produce such advantageous subsamples of definable and homogeneous dose rate. If this can not be realized, appropriate correction has to be made.

Furthermore, the evaluation of the dose rate is complicated by the moisture content residing in the pore volume of the sample. Sediments may contain up to 50 wt% water. In comparison to silicates, water has significantly higher mass absorption coefficients for α-, β- and γ-rays, but negligible radioactivity, hence it attenuates the dose rate more or less, an effect which requires correction. For this purpose the moisture content of the sample and its environment has to be determined (Aitken 1985).

The intensity of the cosmic rays on-site has to be known. Although their contribution to the total dose rate comprises in most cases only a few percent the exact value may vary considerably. It increases with latitude and, more importantly, with altitude. Also, the depth of the sample location below the surface is an essential parameter (Fig. 87).

The age determination according to Eq. (46) implies constancy of the environmental and internal sample components of the dose rate ever since

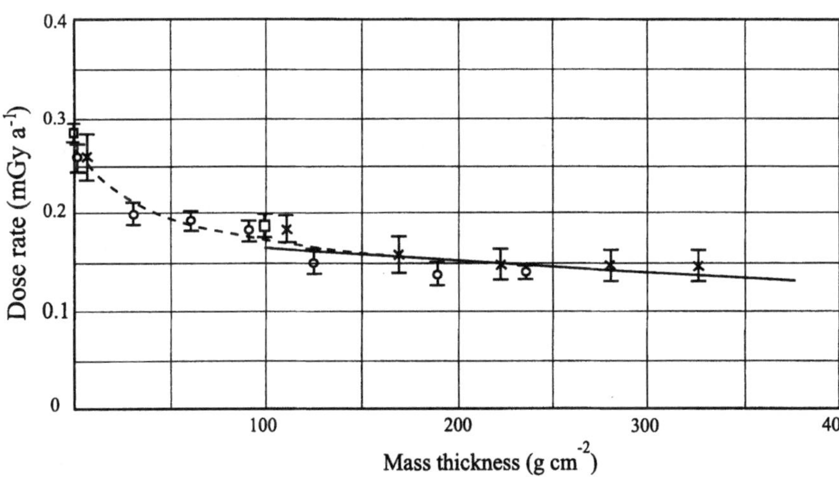

Fig. 87. Cosmic dose rate as function of depth beneath the rock surface expressed as mass thickness. (Reprinted from Prescott and Hutton 1988; copyright 1988, with kind permission from Elsevier Science Ltd., The Boulevard, Langford Lane, Kidlington OX5 IGB, UK)

time zero, otherwise the annual dose measured today is not representative of the past, leading inevitably to age errors. If there has been a change in the dose rate, not only is its extent relevant, but also the time of the change matters. The earlier the change in the dose rate has taken place the smaller is the resulting systematic error of the calculated age. Temporal variations in the dose rate may be induced by changes in moisture, decrease in porosity through compaction, variability of the content of radionuclides, disequilibrium of the decay chains and change in the thickness of the superimposed layers. To what extent these factors have affected the paleodose is particular difficult, if not impossible, to evaluate since the processes relevant to the dose rate may have taken place a long time ago and are hardly or not at all clear. One does one's utmost to recognize these variations and to consider them in the age calculation. Frequently there is nothing else to be done but estimate the variations and make allowance for them mathematically in the age error.

The determination of moisture of the sample is straightforward at the time of its collection. However, it is subjected to seasonal and secular changes. The latter ones are usually climatically assessable, e.g., if a water content of 80% in a sample's pore volume has been determined, the range of variation may be estimated to ± 20% and taken into account in the total error. Sometimes changes of moisture content on a longer term are deducible from hydrogeological, pedological and paleoclimatic reflections. Like any moisture decrease, compaction of the pore space also leads to increasing the dose rates. Compaction is known from loess, for instance. Effects of compaction are sedimentologically discernible; however, the time at which it occurred remains open. Paleoclimatically triggered calcareous precipitation in the pore space of soils and sediments may increase the dose rate (Singhvi et al. 1996).

Fossil teeth, bones, shells and corals possess considerably higher uranium contents than in vivo. After the organisms' death and burial the mineralized tissues gradually take up and incorporate uranium from the groundwater, increasing the uranium content of these tissues many times. Enrichment factors of 1000 and more are not unusual, resulting in a strong increase in the dose rate. The age determination, however, requires knowledge of the temporal function of the increase in the dose rate. Since this time dependence is not precisely known, models have to be employed. Early uptake models, i.e., those that assume that the uptake occurred soon after death, are usually opposed to models assuming linear uranium uptake. In rare cases the assumption of recent uranium uptake is also made.

Other causes of temporal shifts of the dose rate arise from the radioactive decay series starting from uranium and thorium. After a few 10^5 years at the latest – depending on the half-life of the particular daughter nuclide

– the state of radioactive equilibrium is attained, with fixed relative proportions of radioactive nuclides. The degree to which the equilibrium state is already re-established, is in its turn the basis for the uranium series dating methods (Chap. 4). During the formation of Quaternary calcareous rocks, e.g., cave sinters, uranium is incorporated in preference to its daughter products. In the course of time the radioactive daughter products gradually build up until the attainment of radioactive equilibrium. The associated increase in dose rate can be mathematically taken into consideration. Several mobile daughters exist in the radioactive decay chain starting with ^{238}U. Above all, the noble gas isotope ^{222}Rn is noteworthy, since its half-life of 3.8 days is long compared to the diffusion time. If there is radon escape from the sample the following members of the chain will not contribute as much to the dose rate as they were doing in the equilibrium state. The radon loss may vary with the soil moisture. Radium, a geochemically mobile element, reacts in a sensitive way to shifts in the pH value of the groundwater. Changes in the dose rate due to ^{226}Ra and ^{222}Rn mobility can be verified by means of α- or γ-spectrometry.

Dose Rate Evaluation. A whole series of techniques is available for the evaluation of the dose rate. The suitability of a particular technique depends on what shall be measured, the internal contribution to the dose rate, which mostly derives from the α- and β-radiation, or the external contribution of the γ- and cosmic rays. Inaccurate dose rate values import large systematic errors into the age determination, hence it is worth the effort to combine several techniques.

Provided the *element concentrations* of uranium, thorium, potassium and rubidium of the sample and its environment are known, the dose rate can be calculated, on the assumption of radioactive equilibrium in terms of the uranium and thorium decay chains. For this, conversion factors are available (Table 7). They provide the dose rate for a quasi infinite matrix per unit of concentration.

Table 7. Dose rate for radioactive elements (Nambi and Aitken 1986)

Element	Concentration	Dose rate (mGy a^{-1})		
		Alpha	Beta	Gamma
Uranium	1 µg g^{-1}	2.781	0.147	0.1136
Thorium	1 µg g^{-1}	0.739	0.0286	0.0521
Potassium	1 %		0.8140	0.243
Rubidium	100 µg g^{-1}		0.0486	

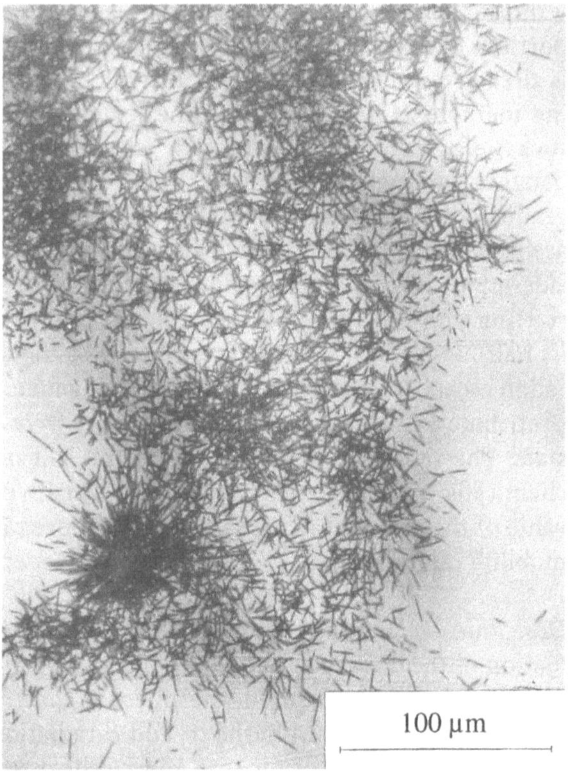

Fig. 88. Microscopic uranium distribution in ceramics revealed by induced fission tracks. The *cluster* of fission tracks is caused by a uranium-rich zircon inclusion within uranium-poor matrix

Analytical methods employed are mostly neutron activation or γ-spectrometry for U, Th, K, and atomic absorption for K and Rb. Fission track analysis for U and Th enables the mapping (Fig. 88) of the microscopic distribution of these elements at the same time (Wagner 1980a). All these methods are relatively laborious. The dose rate values obtained have to be subjected to various attenuation corrections. The contribution by the cosmic rays is estimated according to the burial depth, altitude and latitude of the sampling site (Prescott and Hutton 1994); it usually amounts to values around 0.20 mGy a^{-1} and needs to be added to the other components.

By means of *alpha counting* on sample material < 50 µm the natural α-activity is counted by ZnS scintillation. This simple method is well suited to the determination of the α-dose rate, since it takes into account an existing disequilibrium to some extent. As far as the β- and γ-contributions are concerned, this method is of limited use, since knowledge of the U/Th ratios is required, the assessment of which has been attempted by pair counting. Alpha counting can be used to prove radon escape

(Aitken 1978). *Alpha spectrometry*, which is experimentally laborious, can be applied for the verification of the radioactive equilibrium (Mangini et al. 1983).

The method of *beta counting* enables the direct determination of the sample's natural β-activity, from which the β-dose rate is deducible. Radioactive equilibrium is fairly assessed. This method is not suitable for the α- and γ-contributions. From the view point of sample preparation the method is straightforward, however, Geiger–Müller counting instrumentation is required (Bötter-Jensen and Mejdahl 1988).

The energy spectrum of the natural gamma radiation of U, Th and K is assessed by means of *gamma spectrometry*. High resolution γ-spectrometry using Ge detectors is carried out on a ca. 30-g sample in the laboratory. Apart from the elemental quantification, this technique enables the checking of the radioactive equilibrium (Murray and Aitken 1988). The NaI-detectors are less powerful in resolution, but more sensitive, and are excellently suited for the on-site determination of the in situ environmental γ-dose rate, where ZnS-scintillation counters are also employed. A measuring duration less than 1 h is usually sufficient to obtain reliable results, including the contribution of the cosmic radiation and the correction for water content. The measurements demand expensive instrumentation, however little experimental expenditure.

7.1 Thermoluminescence

Thermoluminescence (TL) is the emission of "cold" light in excess to the incandescent glow that occurs when a non-conducting solid is heated. It is caused by radiation damage having been accumulated in the crystal lattice. Its intensity depends on the radiation dose and thus provides a tool for age determination. Thermoluminescence was already observed by Robert Boyle more than 300 years ago, and about 100 years ago Wiedemann and Schmidt (1895) discovered that it is induced by ionizing radiation. The idea of using minerals as natural TL dosimeters to enable age determination finally stems from Daniels et al. (1953).

Of the three radiation dosimetry methods of dating, TL dating certainly represents the best known and most intensively developed method. The first archaeological materials on which the new method was tested were ceramics and bricks (Grögler et al. 1958). Since then this method has become firmly established in the fields of archaeology and geology. During the 1960s and 1970s, its dominant application was the dating of ceramics. As soon as it became evident that this technique was also suited to the age determination of sediments, the TL method experienced in the 1980s a strong stimulus. The state of research in the middle of the 1980s was comprehensively described by Aitken (1985).

TL dating covers a wide age range of a few 10^2 a up to ca. 1 Ma, thus ranging far beyond the limits of the ^{14}C method. It permits the assessment of rather different events. As far as cave sinter, metallurgical slag and volcanic materials are concerned it is their crystallization or solidification, i.e., the event of their formation, that is dated. In the case of heated and fired objects, such as ceramics, bricks, fire places, stone implements and baked contacts, the last heating event is dated. The optical bleaching of the TL signal enables the dating of aeolian and fluvial sediments which have been exposed to the daylight during their transport and deposition.

Despite the fact that the TL method has already been largely developed and successfully applied many times, it is not suited as a matter of routine. There is a continuing need for basic research, especially with regard to high ages and improved accuracy. The age errors mostly comprise about 10%, but may be considerably smaller in particular cases. In comparison to the precision of ^{14}C ages this uncertainty appears rather large. However, beyond 10 ka, an age range for which generally accepted ^{14}C-correction curves are not yet available, the TL method is clearly able to compete with ^{14}C in dating accuracy. For the last millennium the TL ages are even superior to the ^{14}C ages in most cases, due to the equivocality of the latter.

7.1.1 Methodological Basis

TL Phenomenon. Thermoluminescence requires the existence of lattice defects. These defects serve as traps for positive and negative charges, liberated at lattice atoms by ionizing radiation. In this way during irradiation the solid can take up and store radiation energy in addition to its thermal energy. This step at which the solid acquires the latent TL is termed the *excitation phase*. Through exposure to heat or light, the non-thermal portion of the energy contained in the solid can be stimulated and released in the form of luminescence (*stimulation phase*). Depending on the type of stimulation thermally stimulated luminescence, customarily termed thermoluminescence, is distinguished from optically stimulated luminescence.

The individual steps responsible for the TL phenomenon can be illustrated by the energy-level model (Fig. 89). In this model an electrically insulating crystal is characterized by an occupied valence band and an empty conduction band with an energetically forbidden zone in between. This zone is inhabited by defects with charge deficits (Fig. 78). The ionizing radiation removes electrons from their parent atoms in the valence band and lifts them into the conduction band. From the conduction band, where

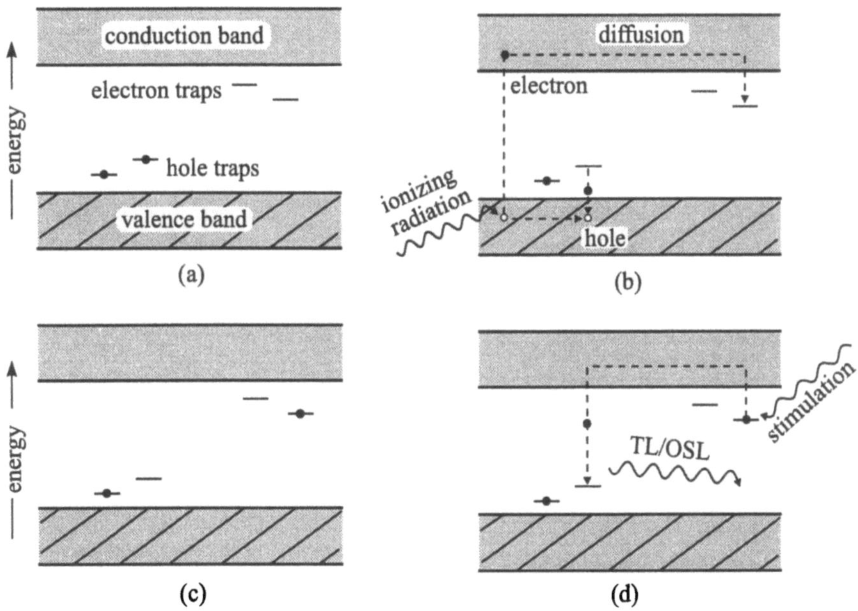

Fig. 89 a–d. Energy band model of the luminescence phenomenon. **a** Ground state with empty charge traps; **b** ionizing radiation removes an electron from its parent atom and lifts it to the conduction band from where the electron is trapped; the remaining positive charge (electron-hole) is trapped at a hole trap; **c** metastable state with occupied charge traps; **d** the electron is liberated from its trap by thermal or optical stimulation and recombines with a hole center (luminescence center) under emission of luminescence (*TL/OSL*)

they can freely diffuse, most electrons return quickly to the ground state, but some of them are trapped at defects with negative charge deficits (electron traps). The remaining holes are trapped at lattice defects with positive charge deficits (hole traps). The trapped electrons stay in metastable state until they are evicted by thermal stimulation. The two luminescence phenomena TL and OSL are distinguished by the stimulation of the trapped electrons. Thermal stimulation is induced by heating whereas optical stimulation is caused by exposure to light. Via the conduction band the electrons recombine with the hole centers (luminescence centers) under emission of luminescence (Fig. 90). In addition to luminescent, radiative recombination non-luminescent recombination also occurs. The depths of the electron traps determine the amount of activation energy necessary for releasing the electrons from the traps. Traps of different depths are sequentially emptied during progressive heating of the solid, the shallow ones first and the deeper ones later. Recording the TL intensity as a function of the heating temperature yields a glow curve (Fig. 91). Thus the glow

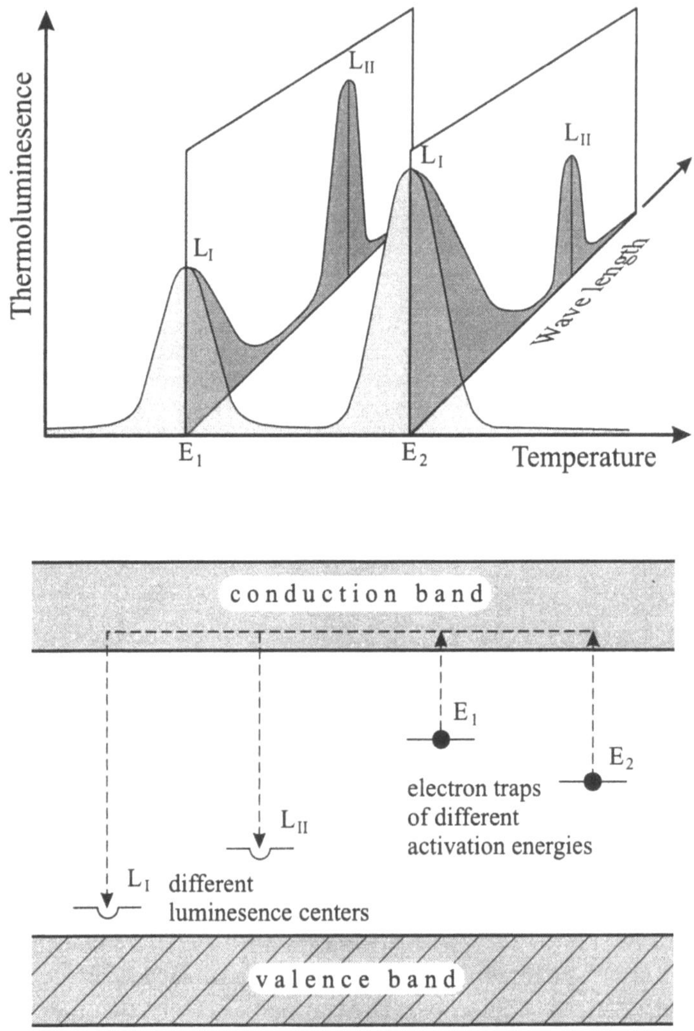

Fig. 90. Stability and color of the TL signal. The stability depends on the activation energy E of the electron traps. A shallow trap (E_1) is emptied at lower glow curve temperature than a deeper trap (E_2). The wavelength, i.e., the color of the luminescence emission is determined by the type of luminescence center (L_I and L_{II})

curve temperature reflects the thermal stability of the electron traps involved. The color of the emitted TL is governed by the type of luminescence center at which the recombination takes place (Fig. 90). The TL spectrum thus provides information on these centers. To obtain full information on the centers involved in the luminescence emission, spectral recording is desirable (Fig. 92), for which filters, monochromators and CCD-cameras

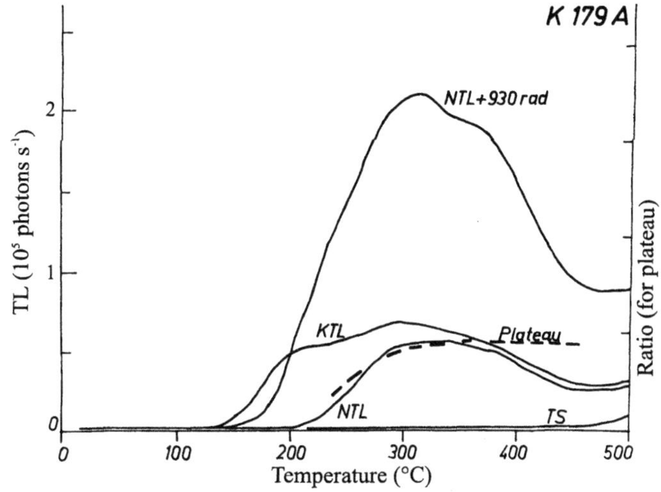

Fig. 91. TL glow curve of the fine-grained fraction of a medieval ceramic object. The TL intensity is plotted as a function of the heating temperature (*NTL*: TL after natural dose, *KTL*: TL after resetting NTL and artificial dose, *NTL+930 rad*: TL after 9.3 Gy β-irradiation additive to the natural dose; the *plateau curve* – serving as a stability test – represents the ratio NTL+930 rad–NTL)/NTL, *TS*: background

Fig. 92. Three-dimensional TL spectrum of microcline from Pitcain, New York, which was β-irradiated with 2200 Gy, stored for 4 months at 23 °C and subsequently recorded by a CCD-camera. (From Krbetschek et al. 1996)

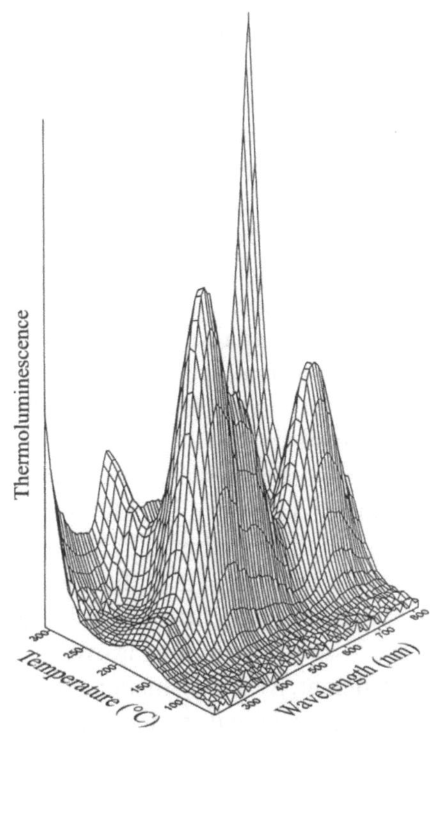

are utilized. The intensity of the TL signal does not only depend on the concentration of released electrons, but also on the probability with which these electrons find a luminescence center. Therefore, the TL sensitivity is essentially determined by the concentration of the luminescence centers, and any changes affecting them – for instance by radiation, by heat or light exposure – become apparent as changes in TL sensitivity.

Thermal and Optical Stability. If a solid with occupied electron traps and the potential to produce thermoluminescence is exposed to elevated temperatures or to light, more or fewer electrons can escape from their traps according to the degree of activation and trap depth. The latent TL signal fades partially or completely. Principally, this effect is desired since it leads to the resetting of the TL system and hence forms the prerequisite for the dating events of heating or bleaching. On the other hand, it may significantly reduce the TL signal resulting in too young TL ages.

The TL signals are erased by sufficiently long and/or intensive heating. The *annealing* of the TL signal obeys the laws of kinetics as formulated in the Arrhenius equation [Eq. (49)]. The TL signal appearing at a given glow curve temperature can be characterized with respect to its thermal stability behavior by a specific activation energy and a frequency factor, both being deducible from experiments. From these parameters the corresponding mean lifetime is obtained for a given temperature. One hour of annealing at 500 °C is sufficient to achieve zeroing of the TL system. At lower heating temperatures, correspondingly lower regions of the glow curve are affected by annealing. At frequently met ambient surface temperatures of around 20 °C, but geologically long residence times, the TL signals appearing up to ~280 °C glow curve temperature are usually affected by fading. Below 200 °C glow curve temperature the naturally induced, latent TL has already completely decayed and hence no signal appears at all. Because at glow curve temperatures above 280 °C, it is not known a priori whether a measured TL signal still contains thermally unstable components, stability tests and thermal pretreatment are required (Sect. 7.1.2). The long-term fading of TL signals is the chief cause for the upper limitation of the age range datable by TL.

Also exposure to light, *bleaching*, gradually reduces the latent TL signal. In contrast to annealing, bleaching does not fully erase the TL signal, even when the duration of light exposure is extended, and an unbleachable residual is left. The rate of the TL reduction depends on the light intensity. In sunlight the bleaching of the TL signal, except the residual, is completed after 10–20 h. When there is overcast weather, the bleaching by daylight is less efficient. An even smaller bleaching effect is observed below a water column, especially if the light intensity is additionally reduced by suspended dimming particles (Ditlefsen 1992; Rendell et al. 1994). Under such conditions only partial bleaching is to be expected. Distinct TL signals of the same mineral possess different light sensitivity. For instance, the TL signal appearing at 325 °C glow curve temperature in quartz is particularly sensitive to bleaching. There are also differences from mineral to mineral, e.g., the TL signal in feldspars is more liable to bleaching compared to quartz. Furthermore, the spectral composition of the light influences the bleaching behavior. Particularly efficient are the short-wave components of the visible light and the ultraviolet component. As a whole, the bleaching behavior of the TL signals is extraordinarily complex and not yet sufficiently explored. The existence of the unbleachable residual suggests that only a fraction of the TL centers is light sensitive. The physical understanding of the bleaching phenomenon is complicated by the observation that the energies required for the thermal and optical stimulation of the same TL signals generally do not correspond to each other. McKeever (1991) explains the comparatively high optical activation energies in the case of quartz by a model, in which during bleaching electrons are released from energetically deep traps which are not thermally affected.

A phenomenon related to bleaching is *phototransfer thermoluminescence* (PTTL). Under PTTL the increase in certain TL signals after light exposure is understood. A possible explanation is that not all electrons that are released from deep traps by optical stimulation and lifted into the conduction band recombine. On the contrary, some of them are retrapped, resulting in the occurrence of PTTL during subsequent thermal stimulation. Attempts have been made to exploit this phenomenon for the dating of ceramics (Bowman 1979), loess (Wallner et al. 1990) and fluvial sands (Murray 1996).

The phenomenon of *anomalous fading* represents a disturbing feature for the purpose of dating (Wintle 1973). It is a decrease in the TL signal not explainable by annealing according to the laws of kinetics. It has to be explained by direct transitions between adjacent centers. Anomalous fading is observed – although not always – on feldspars, zircon and apatite, whereby freshly irradiated samples show a clear reduction of thermally stable TL during the first weeks after irradiation although they have been kept in darkness and at room temperature. Since anomalous fading can strongly diminish the TL age, tests for examining this behavior are necessary.

Natural Dose. The natural dose ND received by a sample since its formation or last zeroing is deduced from the natural TL signal and the dose-dependent TL growth. For this, both, the additive and the regenerative techniques are applied (Fig. 80). In the lower dose range (< 50 Gy) the TL signal mostly shows quasi-linear increase with dose. Straight lines of additive TL growth are optimally suited for the extrapolation of the natural dose of young samples (< 10 ka). For regenerative TL growth at doses of a few Gy one frequently observes supralinear steepening before the growth curve merges into linear ascent. This deviation from linearity is added to ND′, derived additively, as the *intercept* correction D_0 in order to obtain ND. At higher doses (> 50 Gy) complications occur such as saturation and thermal equilibrium, which cause sublinear flattening of the TL growth curve. But renewed steepening is occasionally observed also (Packman and Grün 1992). Only if the experimental data set can be fitted mathematically by a physically based function, can the equivalent dose be reliably extrapolated. If this approach is not successful, the TL system is artificially reset by heating or bleaching. By irradiating with various doses one assesses that dose (*regeneration dose* ND_r) which exactly regenerates the natural TL signal (Fig. 80). This technique, however, requires equal sensitivities with respect to the TL growth of the regenerated and natural TL.

The phenomenon of changing TL sensitivity can actually be utilized for dating, known as the *pre-dose technique*. The pre-dose effect describes an increase in sensitivity of the 110 °C TL peak in quartz that is observed after radiation exposure followed by rapid heating up to 500 – 600 °C. The dose-dependence of the sensitivity change is the basis for the age determination. The potential and limitation of the pre-dose effect for dating purpose has been recently summarized by Bailiff (1994). Essentially, it is useful for dating ceramics of the last two millennia.

If sediments are to be dated by TL, attention has to be paid to the fact that the natural signal I_n consists of two components. One is the initial signal I_0, which was already present at the time of deposition, and the other is the time-dependent signal I_t, which has been induced since the moment of deposition. Only $I_t = I_n - I_0$ is of relevance to the age determination and hence the knowledge of I_0 is required (Fig. 93). Depending on the deposition environment, I_0 represents a partial degree of bleaching or the unbleachable residual (total bleaching). Whereas total bleaching can be assumed for aeolian sediments, this is, of course, not necessarily the case for fluvial, limnic and glacial sediments. To determine the initial signal I_0, the I_n signal is experimentally bleached by sun simulators or natural sunlight. Depending on the sediment type, partial bleaching and total bleaching techniques are distinguished. It is worth emphasizing that the uncertainties of the I_0 determination in the case of partially bleached sediments represent a significant source of error in the TL dating. For this reason fully bleached aeolian sediments such as loess and dune sands are preferable.

Grain Size Fractions. Usually fine-grain (2 – 10 µm) and coarse-grain fractions (100–200 µm) are used for TL dating (Fig. 84). Fine-grain fractions,

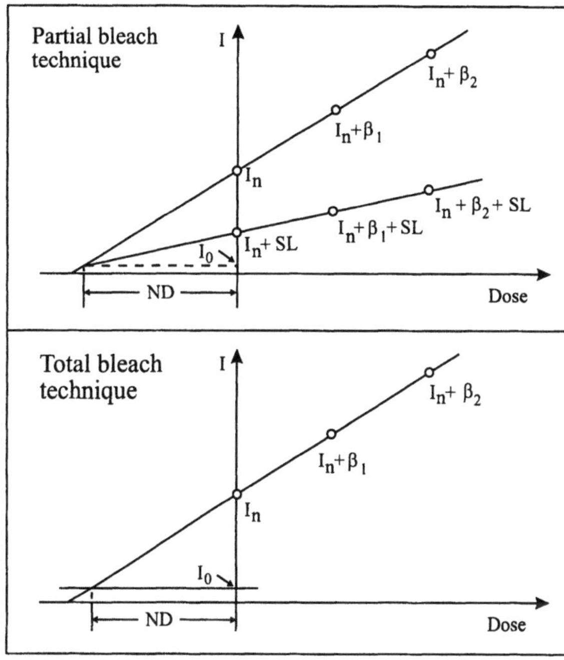

Fig. 93. Dating of sediments requires a knowledge of the initial component I_0, which has to be subtracted from the natural TL signal I_n. According to the sediment type, I_0 may represent partial or complete natural bleaching. To simulate the natural conditions partial and complete bleaching techniques are employed in the laboratory, usually by means of sun lamps (*SL*). Laboratory overbleaching bears the danger that ND and thus the TL age is overestimated

which are commonly of polymineralic composition, have the disadvantage of non-applicability of a mineral separation. It is possible, however, to obtain monomineralic coarse-grain fractions by means of mineral separation. In the case of quartz and plagioclase feldspars, which contain negligible concentrations of radionuclides and hence receive only an external dose rate, ~20 µm of the outer surface rind is removed by etching with hydrofluoric acid. In this way the external α-component, which is geometrically difficult to assess, is eliminated and thus the remaining core contains only the external β- and γ-fractions of the dose rate. The situation is quite different with zircon grains, which emit internal β- and γ-rays essentially to the environment and only retain their α-component. This dosimetric advantage – in addition to the high uranium concentrations – enables the dating of zircons by TL autoregeneration, where a TL signal regenerates itself within months after the NTL measurement (Templer 1993). Potassium feldspars, which contain a significant internal β-contribution, ideally should be larger than the β-ranges.

Age Calculation. The analysis of the TL glow curves for age determination requires certain criteria for the choice of suitable TL signals:

1. The radiation-induced signal has to be thermally or optically reset by the event which shall be dated.

2. It must have been stable during the time span in question.
3. Its growth characteristics have to follow a mathematical function.

On the basis of these criteria one's choice of suitable TL signal can be taken if a glow curve shows several maxima. An essential advantage may derive from a restriction to defined spectral regions of the TL emission since the above mentioned criteria might be fulfilled only for narrow sections of the spectrum. The age is calculated according to Eq. (48).

7.1.2 Practical Aspects

Different grain size and mineral fractions of the same sample, e.g., from a sediment or ceramic sherd, can be dated by TL. Since the ages of the fractions are to some extent independent from each other with respect to the TL characteristics and the dosimetry, their correspondence represents an important indication of the reliability of the age determination. Although this procedure requires a higher experimental expenditure, it is advisable to employ a sampling strategy which leaves this possibility of multiple age determination open. For a single age determination by TL, ~100 mg of separated sample material is required. This corresponds to an initial sample weight comprising ~10 g to 1 kg. Apart from the actual TL measurement material for the determination of the dose rate also has to be considered. The original material is wrapped in a watertight fashion, so that the original moisture can be measured afterwards in the laboratory. The samples have to be protected against heat and light, especially in the case of sediments of which the last light exposure is to be dated. In order to eliminate as far as possible any exposure to light, the sediments must be taken under shady conditions by means of steel cylinders or other types of opaque containers (Fig. 94).

In the laboratory the samples are separated under strictly subdued light into fractions of suitable grain sizes and minerals according to standardized methods. Commonly the coarse-grained fractions consist of separates of either quartz, various feldspars, zircons or the glass phase. Strain during crushing of the samples occasionally causes triboluminescence – a type of luminescence unwanted for dating purposes – which can be removed by washing the sample grains in diluted acids. As the TL signal is erased during each TL measurement, TL dating requires a series of subsamples. Usually around 50 subsamples of 1–4 mg each are prepared. The individual subsamples are put on aluminum or steel discs. The sample processing is accordingly laborious and time consuming.

The measuring instrument consists essentially of a plate on which the sample is heated and a photomultiplier, over which the electronically in-

Fig. 94. Sampling with the steel cylinder hammered into the loess profile at Bruchsal-Aue, Germany. This coring technique protects the sample from light exposure

tensified light emission is recorded as a function of the heating temperature. After each measurement the heating is repeated to record and subtract the background (Fig. 91). It is crucial to conduct the TL measurement under purest nitrogen or argon atmosphere to prevent unwanted chemiluminescence which is energetically fed from exothermic chemical reactions. Optical filters which can be inserted between sample and photomultiplier permit the recording of certain spectral regions. Computer-aided modern TL instruments permit the fully automated operation and the analysis of a larger series of subsamples. The fine-grain technique requires the determination of the a-value and thus the use of an α-source besides a β- or γ-source.

In order to test if the signal of natural thermoluminescence has been stable over the age range of interest, the *plateau test* is applied (Fig. 91). In the plateau test the equivalent dose or the TL age is calculated in dependence of the glow curve temperature. The begin of the plateau value is indicative of the thermal stability of the analyzed TL signal. Thermally unstable components can be thermally "washed" through preheating to obtain a better developed plateau.

Various experimental aspects, such as type of mineral, grain size, preheating, heating rate, spectral range, growth function, additive or regenerative technique and bleaching, are involved in deriving a TL age. Their choice is largely empirical and has not yet been optimized or standardized. Various combinations of these parameters furnish manifold variants of the TL technique of dating with the consequence that the dating of a particular sample by several laboratories will not necessarily yield directly comparable TL ages. Therefore it is necessary to present the experimental procedure together with the age-data. The results of the measurements can not be judged without this important information. This certainly applies also to the techniques of dose rate determination. Only then can it be evaluated whether the quoted error is realistic.

7.1.3 Application

Ceramics and Burned Clay. Burned clay ware has been the preferred object of TL dating for a long time. Important concepts of the TL method of dating were established on ceramic sherds, bricks and kilns during the 1960s and 1970s. Of special interest to the archaeologist is the fact that by TL the firing event, i.e., the ceramic production, is directly dated; this is because the ceramic typology is an important pillar of the chronological framework. Consequently, with TL one dates the archaeological "index fossil" itself and thus circumvents possible correlation errors, which can easily arise if associated materials are dated. Despite this advantage the TL method has lost ground in the field of ceramic dating during recent years. This is due to the relative large error of 6–10%. Most questions concerning the chronological position of Neolithic and Bronze Age European cultures can be more accurately answered by ^{14}C dating and hence the rougher time resolution of the TL age is frequently not sufficient. However, there are problems with ceramics for the solution of which the application of TL is useful and advantageous. This is especially valid for archaeometallurgic kiln remains, which otherwise are hardly datable and which may be precisely TL dated to half or even quarter of a century.

For the dating of ceramics it is advisable to take several sherds of palm size from a single find context taking the variability of ware types into consideration, because their TL characteristics can be rather different. Adhering soil is desirable since it is useful for the determination of the environmental dose rate. If possible in situ measurements of the environmental dose rate by portable detectors are also conducted. If the ceramic ware contains quartz temper the coarse-grained fraction is separated and dated in addition to the fine-grained one, thus the resulting age can be determined as a combination of both TL ages. The ages of several sherds are combined into a context age.

TL results obtained from Bandkeramik sherds of Lamersdorf, northwest Germany, will be mentioned as an example. All of them belong to the younger phase of the Early Neolithic Bandkeramik. The nine single dates vary between 6400 and 7840 a, but do not significantly differ within their error limits. The mean context age is 5050 BC (± 130 a, ± 600 a, HdTL 30), where the first error represents the statistical 1 $\bar{\sigma}$-standard error (precision) and the larger second one represents the analytical total error (accuracy). The latter is relevant for the comparison with independently obtained ages. The Bandkeramik period of the Lower Rhine and Maas region is placed, according to calibrated ^{14}C ages, in the time between 5300 and 4950 BC, which is in good agreement with the TL date (Wagner and Lorenz 1997). Comparable TL age determinations were conducted on Bandkeramik pottery sherds from Lichtenwörth and Hainburg, Lower Austria (Fig. 95) as well as from Ulm-Eggingen, South Germany. The TL dates of all

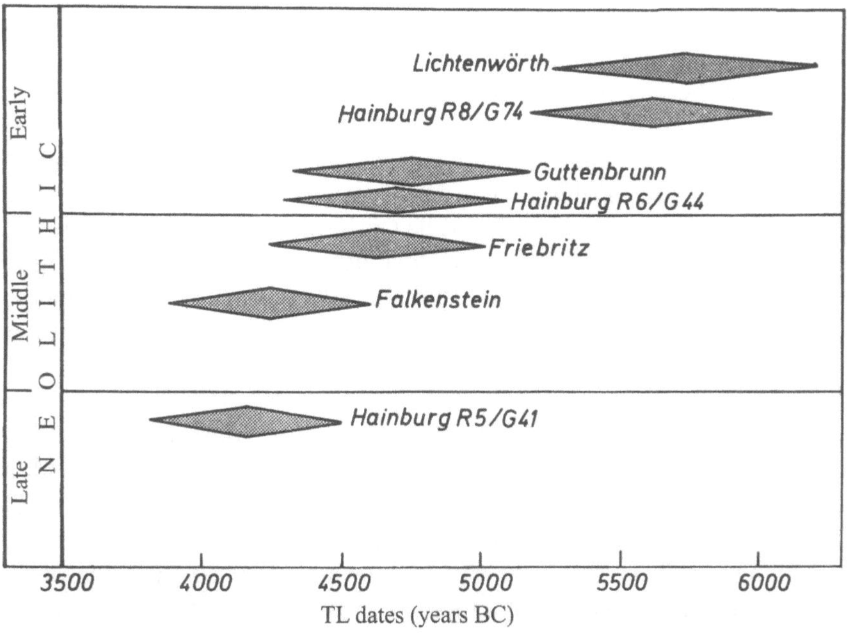

Fig. 95. Thermoluminescence dates of ceramic sherds from Neolithic sites in Lower Austria. The context ages are mean values calculated from several individual age determinations of the sherds. The lengths of the *diamonds* indicate the total 1$\bar{\sigma}$-error. (After Pernicka and Wagner 1983/84)

these sites show that the climax of the Middle European Bandkeramik culture falls into the 2nd half of the 6th millennium BC (Wagner et al. 1993).

Another example are the ceramics of the horizon of the Late Uruk to Early Bronze Age of Hassek Höyük at the Upper Euphrates in East Anatolia (Wagner et al. 1992). These measurements were carried out to test by an independent method the reliability of ^{14}C dates obtained from charcoal and cereal grains (Willkomm 1992). For TL dating the sherds were separated into the fine-grained and the quartz-inclusion fractions. The combined TL ages of both fractions fall into the 4th–3rd millennia BC for the various settlement phases and are in agreement with the ^{14}C dates.

TL has often been successfully applied to medieval kiln structures. For this age range the accuracy of ^{14}C dating is frequently not sufficient due to the complications arising from the old-wood effect and the ambiguity of the calibration curve. An example of this kind is the TL age determination of a smelting furnace in the vicinity of Metzingen, South Germany, with evidence of early iron casting technology. The fine-grained fraction and coarse-grained quartz were investigated from a sample taken from the burned furnace lining. The TL date 996 AD $\pm$ 105 a places the iron smelting into the High Medieval Period (Wagner and Wagner 1994).

Another application of TL is authenticity testing of ceramic objects of importance to the art historian. Here, it is not the best possible age estimate which is aimed at, but discrimination between the high age of a genuine object and the recent age of an imitation or forgery. From the object in question 200 mg of drilling powder is collected under subdued red light. In most cases the natural dose is sufficient as criterion of authenticity, and occasionally the internal dose rate needs to be evaluated. An example of a TL-proved imitation is the alleged Middle Corinthian amphoriskos of the 6th century BC displayed in Fig. 96. It does not show any natural TL signal which unveils its production within the last century (Wagner 1980b). For the purpose of deceit this piece has been skillfully provided with an artificial sinter.

Burned Flint and Stones. Due to its favorable TL and dosimetry characteristics, flint is well suited especially for the time span comprising the Middle and Lower Paleolithic, which is not assessable by means of ^{14}C dating. TL dating requires the zeroing of the TL clock of the artifacts by heating during Prehistory. This is, indeed, the case for a considerable portion of the flint artifacts which exhibit macroscopic evidence of heating (Fig. 8).

For the TL measurement, the flints are processed to furnish fine-grained or coarse-grained fractions, where the carbonate admixtures have to be removed by diluted hydrochloric acid. The characteristics of saturation, thermal stability and growth curve of the TL signal should enable the age

Fig. 96. Thermoluminescence authenticity test. The allegedly genuine Corinthian amphoriskos (16 cm high, sintered) was recognized as a modern forgery by TL measurements on drilled powder

determination up to 1 Ma in dependence on the dose rate. The zeroing of the signal requires at least a temperature of 450 °C. The question whether such temperatures were obtained during the heating event, can be evaluated by means of TL tests (e.g., the plateau test). Usually the radioactive elements are homogeneously distributed within the flint, and their contents frequently comprise about 1 $\mu g\ g^{-1}$ U, < 0.5 $\mu g\ g^{-1}$ Th and < 0.1% K, resulting in an internal dose rate of around 0.5 $mGy\ a^{-1}$.

It is advisable to collect several flint samples from each layer to be dated. Among the oldest flint artifacts, which have been TL dated so far, are those from the site Maastricht-Belvedere, Netherlands, belonging to the Acheulian cultural complex. By means of burned flint artifacts, Huxtable and Aitken (1985) determined a mean TL age of 270 ± 22 ka for the interglacial layer IV of this site. For the Lower Paleolithic site Schöningen 13/I, North Germany, Richter (1997) reports even a TL age of 453±39 ka for a heated flint tool. Most TL data of flint are available for the Mousterian culture, which is commonly associated with Neanderthal Man. For some of these Mousterian sites high TL ages of up to 200 ka were determined. Obviously, this culture is not restricted to the last glaciation but dates back to the penultimate glaciation (Valladas 1992). At the site of Tabun, Israel, TL ages of flint indicate that the technological transition from the Acheulian to the Mousterian already occurred 270 to 250 ka ago (Mercier et al. 1995). The common idea, won in Europe, according to which Modern Man follows Neanderthal Man needs to be revised on the basis of the sites of Qafzeh, Kebara and Skhul, Israel, where TL dating played a decisive role (Valladas et al. 1988; Mercier et al. 1993). At Kebara a skeleton of Neanderthal Man and at Qafzeh and Skhul fossil remains of Proto-Cro-Magnon, a predecessor of Modern Man, have been unearthed. TL dating on 20 flint fragments from a hominid-containing layer in Quafzeh yielded 92 ± 5 ka, whereas 30 flint artifacts from several layers in Kebara provided TL ages ranging from 50 to 70 ka; the Neanderthal skeleton comes from a layer dated to 60 ka (Fig. 97). At Skhul a mean TL age of 119 ± 18 ka was determined for six burnt flints from level B from which remains of a Proto-Cro-Magnon were recovered. The early age of anatomically modern Man, which is supported by ESR (Sect. 7.3.3) and uranium series dating (Sect. 4.3), bears consequences for hominid evolution far beyond the Levant. Consequently, in opposition to so far accepted assumptions, the early forms of modern Man existed significantly earlier and synchronously with Neanderthal Man (Bar-Yosef and Vandermeersch 1993). Therefore Neanderthal Man cannot simply be the evolutionary precursor of modern *Homo sapiens,* as it appears from European find contexts. On the contrary, the Levantine evidence suggests a development of Proto-Cro-Magnon Man independently from Neanderthal Man. An example of the lower temporal range of TL application are six flints from the Chalcolithic layers of Hassek Höyük in East Anatolia yielding ages of 5400 ± 300 a (Göksu et al. 1992).

In addition to flint other types of rock have also been – intentionally or not – exposed to heat. One of the first TL applications to such kinds of materials is the dating of stones used as pot-boilers for cooking. The stones were first fired and then put into cooking vessels to bring the water to boil. Huxtable et al. (1976) dated fired sandstone from the "burnt mounts" of Orkney. Most of the stones did not contain sufficient quartz grains for quartz-inclusion dating so that the majority of the samples were dated with the fine-grain technique. According to the TL dates these boiling sites belong to the 1st half of the 1st millennium BC.

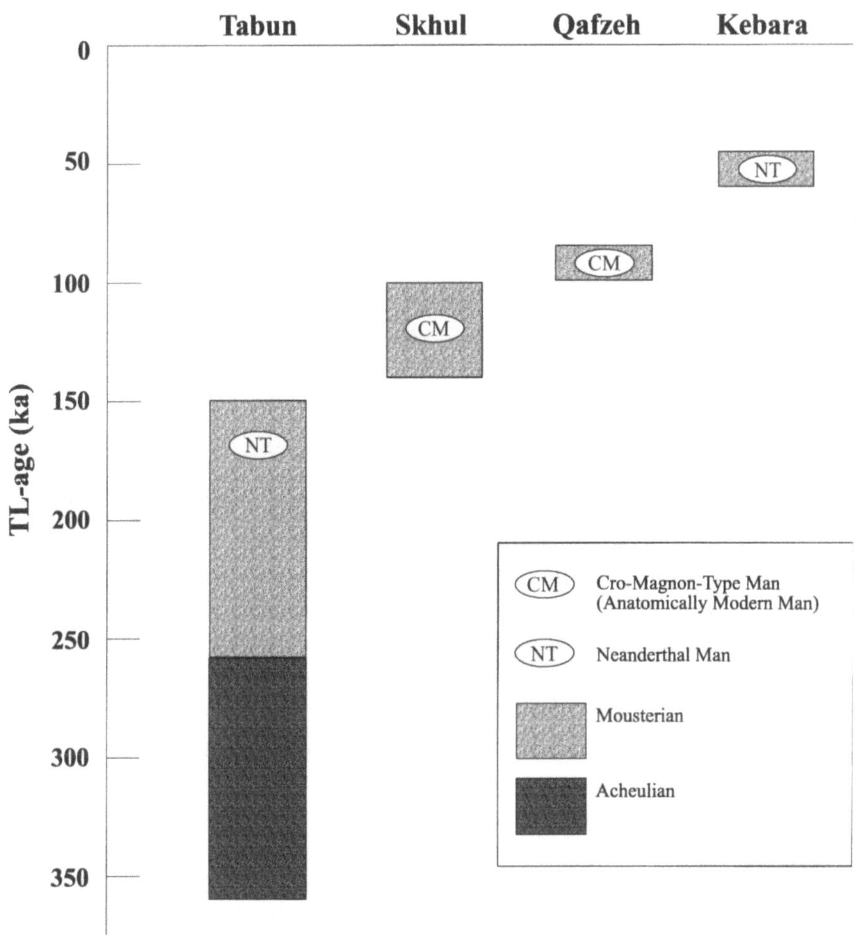

Fig. 97. Thermoluminescence ages of burned flint artifacts from Levant Paleolithic sites, Israel. (After Valladas et al. 1988; Mercier et al. 1993, 1995)

At the Paleo-Indian campsite Caverna dal Pedra Pintada in the Brazilian Amazon, ten burned quartz and chalcedony flakes were found, among other materials. Their TL dating yielded ages between 9.5 and 16 ka, which agree with ^{14}C dates of 10.5–14.2 ka for that site (Roosevelt et al. 1996). This discovery of Paleo-Indians along the Amazon during the late Pleistocene indicates that the Paleo-Indian radiation was more complex than current theories provide for.

Vitrified Forts. Prehistoric walls consisting of vitrified quartz- and feldspar-bearing rocks are of widespread occurrence in western Europe. Presumably, they represent fortification walls constructed of rock and timber which were welded by burning (> 1000 °C). The TL ages obtained from feldspars of six Scottish forts range from the late 3rd millennium BC to the late 1st millennium AD, which contradicts an association to a particular culture (Sanderson et al. 1988).

Artificial Glass. There have been few attempts to date man-made glass by TL. Complications are caused principally by its non-crystalline state. The TL of glass seems to be reproducible only if the glass is not heated above its glass transition temperature because at that temperature the structure of the sample begins to change. A clear connection between the TL sensitivity and the degree of crystallinity was observed (Müller and Schvoerer 1993). Therefore, TL is not a general dating method for man-made glass, but may be applicable in particular cases.

Slags. Attempts have been made to date archaeometallurgic slag by thermoluminescence and thus to directly date the smelting process. Principally suitable for this should be the commonly present glass phase in slags as well as the crystalline phases pyroxene and fayalite. Unfortunately, the TL characteristics of the glass phase proved to be insufficient for dating purpose presumably due to the high metal content of the glass. The crystalline silicate phases of slag are particularly complex in their chemical composition and structure. The intimate intergrowth patterns of these phases are especially problematic with respect to dose rate determination, so that rather large age errors of around 20 % result (Elitzsch et al. 1983; Lorenz 1988). For TL dating of archaeometallurgic sites it is recommended to use heated quartz charge which is enclosed in the slag or fragments of loam adherent to the furnace wall, instead of slag material itself.

Volcanites. Associated with volcanic activities are both, the newly formed rocks such as lava and ashes as well as the heated pre-existing rocks. Both types of heated material can be TL dated provided they bear quartz, feldspar or glass. The direct TL dating of lava streams causes difficulties, however, owing to their rarity of quartz or to the tendency to anomalous fading of their feldspars. Systematic efforts to TL date basaltic rocks were conducted by May (1979). Plagioclase separates of the alkali basaltic series on Hawaii covering an age range of 4.5 ka – 3.3 Ma, showed an exponential growth of the NTL signal, normalized to TL sensitivity and dose rate, of up to 200 ka (Fig. 98). Tholeiitic basalts, however, exhibited unfavorable TL behavior. Guerin and Valladas (1980) succeeded in dating plagioclase of various lava streams of the Chaine des Puys, France (Fig. 16) by using only the UV-component of the TL emission in the high temperature range between 500 and 700 °C. Ages between 9 and 216 ka were determined with an accur-

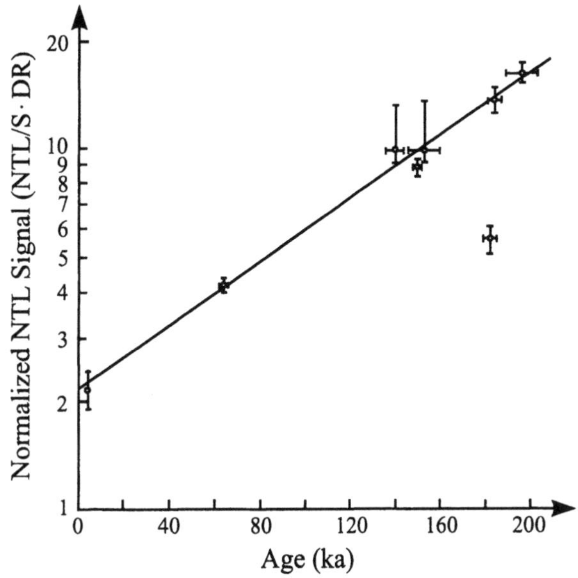

Fig. 98. Relationship between NTL signal (normalized to TL sensitivity S and dose rate DR) and geological age of alkali basalt from Hawaii. The NTL signal grows exponentially with age, which expresses itself as a straight line in the half-logarithmic diagram. (After May 1979)

acy of 10 %, which is in agreement with independent data. If available, TL dating may make use of heated quartz derived from quartz-bearing xenoliths enclosed in the lava or fritted contact rocks. Such quartz minerals provide good dating results especially if the red spectral region of the emitted TL is analyzed. This was demonstrated for various volcanic eruptions in the Chaine des Puys comprising ages up to 500 ka, among others for the Tartaret basalt flow with a TL age of 13.7 ± 0.8 ka (Pilleyre et al. 1992) and the Sancy pumice with a TL age of 544 ± 42 ka in accordance with the K–Ar age (Miallier et al. 1994). Suitable as well are baked argillaceous shales in volcanic slag. This was shown by Zöller (1989) who determined a TL age of 43 ± 3 ka for the Middle Würmian eruption of the southern crater of the Mosenberg volcano in the Eifel, Germany.

The fine-grained glass phase of volcanic ash bears a considerable potential for TL dating. Due to their fine grain size, distal ash deposits are workable as well, providing an excellent opportunity for tephrostratigraphic correlation over a large distance. Employing this technique, an age range from several up to 400 ka is covered, as was shown by Berger (1991) for various North American tephra deposits (Fig. 99). Presumably, ashes only a few 100 years old may be dated in this way as well. The extraction of the suitable fraction, consisting of glass fragments of 4–11 µm size, requires a separation by a heavy liquid (2.45–2.50 g cm^{-3}). Furthermore, the preheating procedure and the TL growth behavior demand special attention.

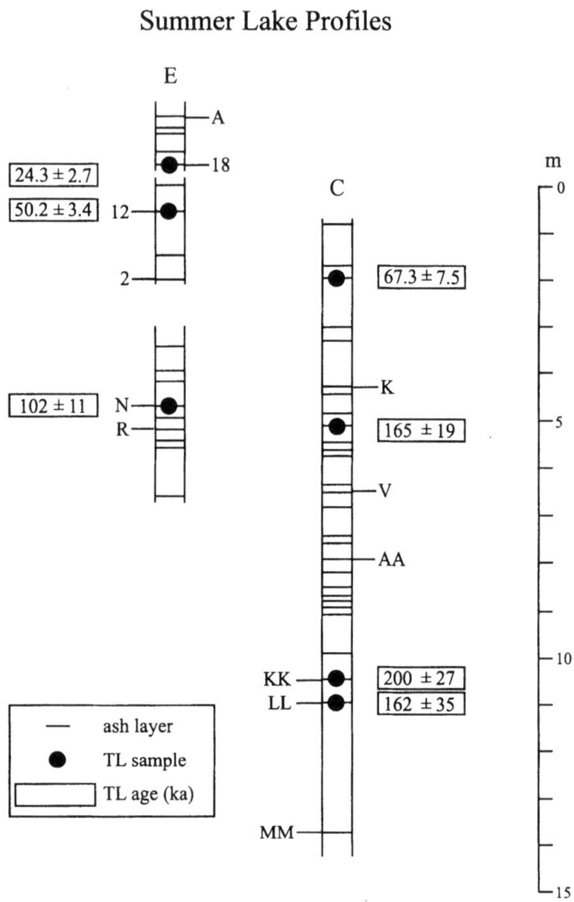

Fig. 99. Thermoluminescence ages (in ka) of the glass component in volcanic ashes from two profiles at Summer Lake, Oregon. The tephra layers are intercalated in clays, silts and sands. The TL data indicate strong temporal changes of the sedimentation rate. (After Berger 1991)

Impactites. Big meteorite impacts generate high shock-wave pressures and temperatures, which may extinguish the TL signals of affected rocks so that the impact event may be datable by means of the subsequently formed TL signal. This potential of the TL method was tried out for the Arizona Meteor Crater. Sutton (1984) determined that a shock-wave intensity of 10 GPa and 680 °C was sufficient to reset the TL signal for quartz of the Coconino sandstone. As impact age for the Arizona crater 49 ± 3 ka was determined by TL on quartz of sufficiently shocked target rocks (Sutton 1985).

Pseudotachylite and Fault Breccia. The high temperatures associated with the formation of friction melts during intense fault movements are sufficient to extinguish the TL signal in the quartz and feldspar relics. The glass phase of the pseudotachylite also offers a potential for TL dating. From a pseudotachylite of Langtang, Nepal, the fine-grained fraction of the glassy matrix (containing 10–15% quartz fraction < 1 mm) was prepared for TL measurement. Linear growth was observed. After evaluating the natural dose (538 Gy) and the dose rate (7.557 mGy a^{-1}) the TL age of the pseudotachylite was calculated as 71.2 ± 5.6 ka. In order to reconstruct the neotectonic evolution of the frontal borderzone of the Himalaya, Singhvi et al. (1994) attempted to date fault gouge by TL. Mineralogical data on illite and chlorite were used as indicator for sufficient heating within the fault planes. The coarse- and fine-grained fractions of a gouge sample from the Nainital fault, India, revealed ~40 ka.

Fulgurite. If lightning strikes a sandy surface the heat effect is sufficient to cause fritting and melting of the quartz grains. Therefore the TL signal of the heated quartz can be expected to be reset. Any TL applications on fulgurites have not yet been tried out.

Loess. For TL dating of loess the polymineralic fine-grained fraction consisting predominantly of quartz and feldspar is used; for sandy loess the coarse-grained fractions of quartz or feldspar are also used. In central Europe – as well as in other periglacial areas – numerous loess profiles have been investigated (e.g., Zöller et al. 1988). In this connection it became obvious that reliable TL ages are obtainable for the last Glacial–Interglacial cycle, i.e., the last 120 ka. The TL ages of older loess show tendencies toward greater or smaller underestimation. On the other hand, Berger et al. (1992) obtained concordant TL ages up to 800 ka for tephrochronologically classified loesses of Alaska and New Zealand. The underestimation of the TL ages is presumably caused by the loss of the TL signal of a particular component of the polymineralic fine-grained fraction. Therefore loess of one area may tend to age underestimation whereas loess of another area does not. An example of TL dating is the loess/paleosoil sequence within the Tönchesberg caldera in the Quaternary Eifel volcanic field, Germany (Zöller et al. 1991). Because this caldera was formed 200 ka ago during a volcanic eruption, it acted as a sediment trap for Rissian and Würmian loesses, which are separated by an Eemian soil. Furthermore, several humus zones are intercalated in lower Würmian loess, suggestive of boreal climatic conditions. Four find horizons show that Neanderthal Man repeatedly visited this caldera as a hunting site. The TL dates (Fig. 100) fit well into the chronological framework established by independent chronostratigraphic results.

At the Lower Paleolithic site of Diring Yuriakh, central Siberia, TL dating of loess revealed that the cultural horizon with unifacial choppers is earlier than 260 ka, and thus an order of magnitude older than other documented Siberian Paleolithic sites (Waters et al. 1997). Loess and cover sand overlie the find horizon. The fine-grained feldspar-dominated fraction was used

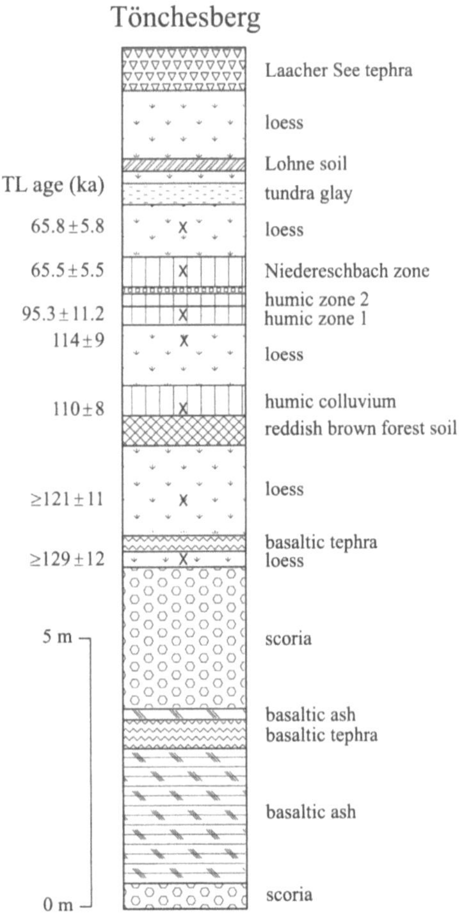

Fig. 100. Thermoluminescence ages of loess from Tönchesberg in the Eifel area, Germany. The caldera of the inactive volcano was gradually filled by loess during the last two Glacials and repeatedly occupied by Neanderthal Man as a hunting site. (After Zöller et al. 1991)

for TL analysis. No indications of saturation or instability of the TL signal were observed. The relatively low dose rate of 4 mGy a^{-1} contributed to rendering the TL ages reliable beyond 100 ka. The result is of great significance for understanding the timing of human expansion to the north and the early adaptation to a cold climate.

Dune Sand. This aeolian sediment has the advantage over loess that it permits the separation of a monomineralic coarse-grained fraction for TL measurements. Concentrates of quartz or potassium-feldspar are used. The feldspar excels in having a high TL sensitivity, which is important for Holocene sands. The thinness of the vegetation blanket at the transition late Glacial–early Holocene favored the formation of dunes in those areas with sufficient sand supply. For the dunes in Sandhausen, southwestern Germany, the quartz and the feldspar fractions yielded TL ages of 12.2 ± 1.9

to 0.9 ± 0.4 ka and 11.9 ± 1.8 to 0.9 ± 0.4 ka, respectively (Baray and Zöller 1993). In each case, the TL analysis provides the age of the last sand transport. The wandering of the dunes was revived through anthropogenic influences, such as wood clearing. TL ages around 900 a substantiate medieval activities affecting the Sandhausen dune. The temporal development of desertification provides important paleoclimatic data as has been proved by Chawla et al. (1992) for the Northwest Indian Thar Desert. Aeolianites and beach sands offer interesting TL applications – especially with respect to the variation of the sea level over time, which can be recorded up to an age range of 800 ka (Huntley et al. 1993a). The TL ages of a series of South Australian beach dunes correlate not only with other age data but also with interglacial sea level maxima which are responsible for the formation of the dunes (Fig. 101).

A TL study of considerable significance represents the dating of the sandy footslope deposits at the sites Malakunanja and Malangangerr at the

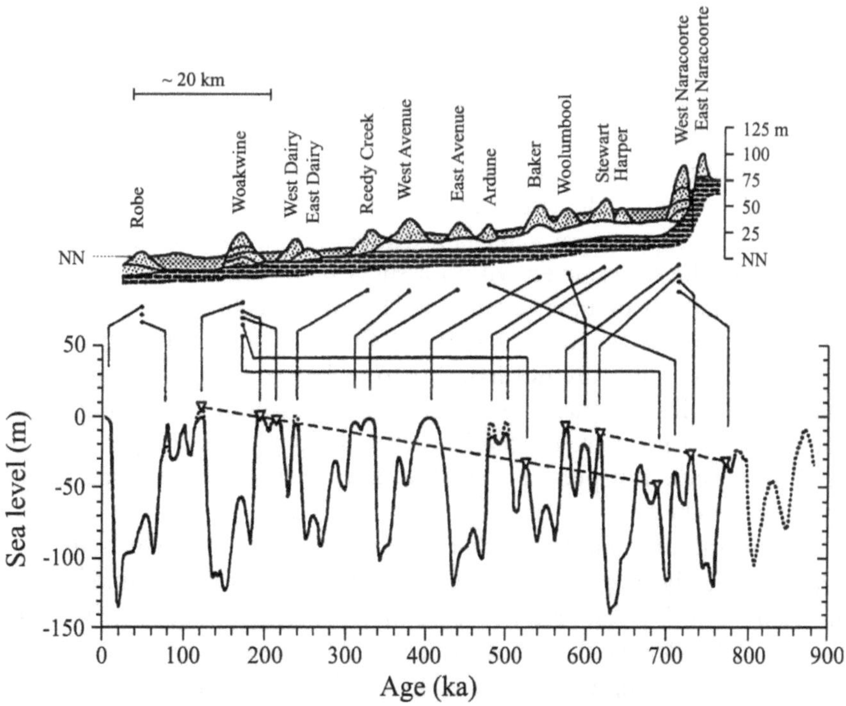

Fig. 101. Section through coastal dunes at Robe, South Australia. The TL ages correlate with the sea level maxima (deduced from the marine $\delta^{18}O$ curve). The arrangement of the dunes is the result of sea level fluctuation and tectonic uplift. (Reprinted from Huntley et al. 1993; copyright 1993, with kind permission from Elsevier Science Ltd., The Boulevard, Langford Lane, Kidlington OX5 IGB, UK)

Arnhem Land escarpment, northern Australia (Roberts et al. 1990). The deposits contain traces of an early human occupation with numerous stone tools. The TL ages of the coarse-grained quartz fraction of the sand substantiates the hitherto earliest known presence of humans on the fifth continent 50–60 ka ago. These high ages have in the meanwhile been confirmed by OSL dates from another site in the same region (Roberts et al. 1994). Lately even higher TL ages were reported from a rock shelter site in the Jinmium area further west (Fullagar et al. 1996). This former occupation site is also characterized by stone artifact finds and rock paintings. The TL samples were collected from auger holes in the sand sheet and from walls in excavated trenches. From the sediments, probably of "colluvial" genesis, the 90–125 μm quartz grain fraction was separated. The TL ages obtained range from 2.3 ± 0.7 to 116 ± 12 and 176 ± 16 ka, and are claimed by the authors to represent the time of deposition. If true, ancient Man's presence in northwestern Australia before 116 ka ago would have far reaching implications for the early human migration pattern. However, due to the danger of incomplete resetting of the TL signal in the analyzed quartz grains, these data definitely need confirmation by OSL dating.

Sand (*Aquatic*). The water column partially absorbs light, particularly in the short-wave spectral region. In addition, suspended particles reduce the light intensity. Hence mineral grains which have been transported and deposited in water are less profoundly bleached compared to aeolian ones. Since the light exposure of water-lain sediments may vary considerably depending on the facies, even within a short deposition distance a rather variable suitability of aquatic sand for TL dating has to be expected. More systematic studies on the suitability of fluvial, glaciofluvial, limnic and coastal marine sands are needed. Due to its better TL bleachability, feldspar is generally preferable to quartz. The light conditions during transport and deposition are crucial in each individual case (Berger 1984). If sands have been transported in shallow, clear rivers even the quartz grains are fully bleached.

For instance, the *fluvial* sands of Steinheim, Germany, which are biostratigraphically assigned to the Eemian stage, and overlie the find layer of the Steinheim Man, an archaic *Homo sapiens*, yielded a quartz TL age of 109 ± 16 ka (Zöller 1994). During the late Middle to early Upper Pleistocene the archaic *Homo sapiens* is replaced in Africa by the anatomically modern Man *Homo sapiens sapiens*. Alluvial sands from Middle Stone Age sites in the Upper Semliki Valley, Zaire, with early remains of anatomically modern Man were dated by TL. The quartz fraction from sands, above the find horizon, gave an age of 82 ± 8 ka which is supported by uranium series and ESR dating and represents a *terminus ante quem* for the find horizon

(Brooks et al. 1995). For dating river sands in the age range of 100 to 2000 years the phototransfer-thermoluminescence (PTTL) method, in combination with OSL, was successfully employed (Murray 1996). *Marine* beach sands are well exposed to light by repeated displacement in the shallow coastal regions as was shown through TL dating on quartz and feldspar separates from middle Pleistocene sands of the Channel coast in northern France (Balescu et al. 1992). It was established by TL dating of the quartz fraction that the middle and upper Pleistocene littoral deposits from the NW coast of Sicily, Italy, represent the highest sea levels of the last three Interglacials (Mauz et al. 1997). Not much is known about the suitability of *glaciofluvial* sediments for TL dating. The degree of TL bleaching strongly depends on the type and distance of transport (Forman 1988; Gemell 1988). The coarse-grained feldspar fraction of the Kames deposits of the last Glacial in the vicinity of St. Petersburg, Russia, displays a degree of partial bleaching equivalent to an exposure of 10 min on a sunny day (Hütt and Jungner 1992).

Carbonates. Inherent in the TL method is an interesting potential with respect to the dating of calcitic sinter, as was shown by Debenham and Aitken (1984). The 280 °C TL signal in the blue spectral region possesses good TL characteristics. An age range of up to 1 Ma should be covered by means of this signal. A rather extensive series of stalagmite samples which were taken from various strata of the cave deposits in Caune de l'Arago, Tautavel in France, an important Lower Paleolithic site with hominid finds, yielded TL ages between 86 ± 30 and 298 ± 33 ka. These data are in agreement with the stratigraphic concept and the U–Th ages (Fig. 102). TL age underestimation was found for samples whose dose rate is dominated by the internal α- and β-components. Singhvi et al. (1996) proposed a luminescence method for dating "dirty" pedogenic carbonates. The method uses changes in the dose rate to quartz grains when the sediment becomes carbonated. This time of carbonate formation is assessed from the difference of the natural doses as well as dose rates in carbonated and uncarbonated quartz separates. Applied to pedogenic carbonates of the Thar Desert, India, the method revealed carbonate formations 17, 5 and 2 ka ago.

Because the samples of mollusk shells suffer structural alteration at usual glow-curve temperatures only sporadic TL studies – in contrast to ESR studies – exist for this material. However the calcite shells of the Pectinidae family are affected to a smaller degree by alteration, as the investigation by Ninagawa et al. (1992) has demonstrated. A suitable TL signal for dating appears at the glow curve temperature of 240 °C in the red spectral region. A correspondence between the TL ages of several Japanese species of the

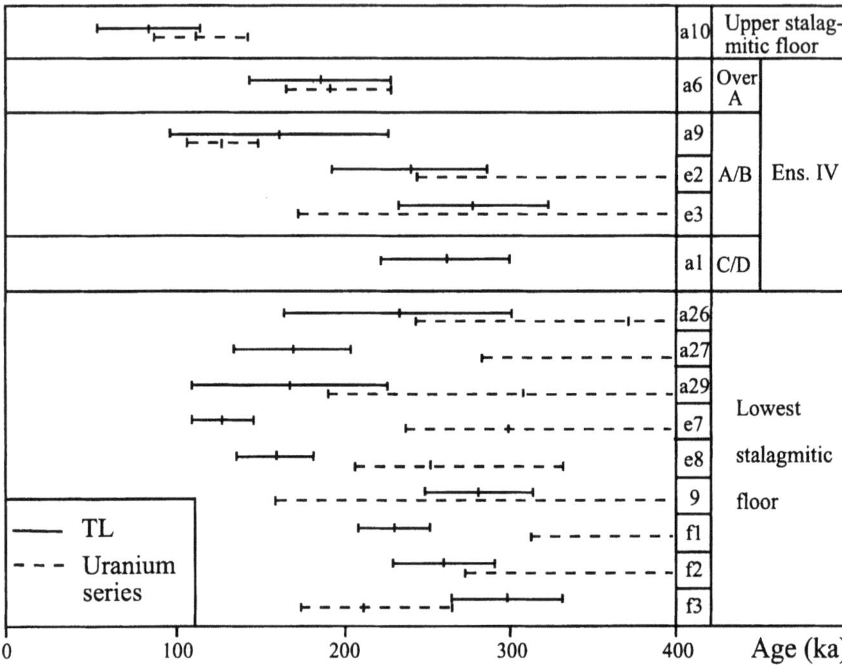

Fig. 102. Thermoluminescence and uranium series ages of stalagmites from the Caune de l'Arago cave, Tautavel, southern France. The *bars* represent the 1$\bar{\sigma}$-error. (After Debenham and Aitken 1984)

Pectinidae and independent age data were observed up to 600 ka; beyond this range the state of thermal equilibrium is reached. TL dating was successfully applied also to calcitic shells of the Ostreidae family (Ninagawa et al. 1994). Due to the higher TL sensitivity, TL dating may be advantageous over ESR dating of Pectinidae within the lower age range (< 100 ka). However, difficulties arising from temporal variations of the dose rate affect both methods of dating equally.

Colluvial and Alluvial Silts. Silty, polymineralic sediments possess different degrees of bleaching depending on the depositional conditions (Berger and Easterbrook 1993). Investigating a colluvium with a loess-derived component in NE England, Wintle and Catt (1985) obtained a TL age of 17.5 ± 1.6 ka, which is in agreement with the ^{14}C date and the stratigraphic age. Forman (1989) reported TL ages on fine-grained distal and coarse-grained proximal Holocene colluvia from Utah. It became clear that due to the longer transport distance only the distal colluvia were sufficiently bleached and suitable for TL dating. Lang et al. (1992) systematically studied complete profiles of Holocene alluvia from the Elsenz river, Germany. Applying the partial bleach method, ages of 5.9 – 1 ka were obtained. Although the TL data substantiate the Holocene redeposition of the material, the TL age of the upper alluvial sediments proved to be too old in comparison to their stratigraphic position. Obviously, the younger alluvial sediments were not sufficiently bleached during their final deposition, a fact which is presumably related to an increased load of fine-grained material in the receiving stream caused by an increased soil erosion in the agricultural intensively used Kraichgau region.

Archaeological Sediments. Inorganic archaeological deposits are continuously turbated at least in their uppermost layer by human activities, e.g., trampling. For this reason a part of the quartz and feldspar grains was exposed to light and their TL signal became bleached. Chawla and Singhvi (1989) tried to explore this dating potential on four archaeological sites in India. The sampled layers belong to the first two millennia BC according to ceramic finds and ^{14}C data. For the TL measurement the fine-grained fraction was separated. Since it was assumed that during their formation the archaeological sediments had been bleached to a certain degree, but not necessarily fully, the ND was determined by means of the partial bleach technique. For all of the nine sediments that were analyzed defined bleaching events were determined. The good correlation (Fig. 103) between the

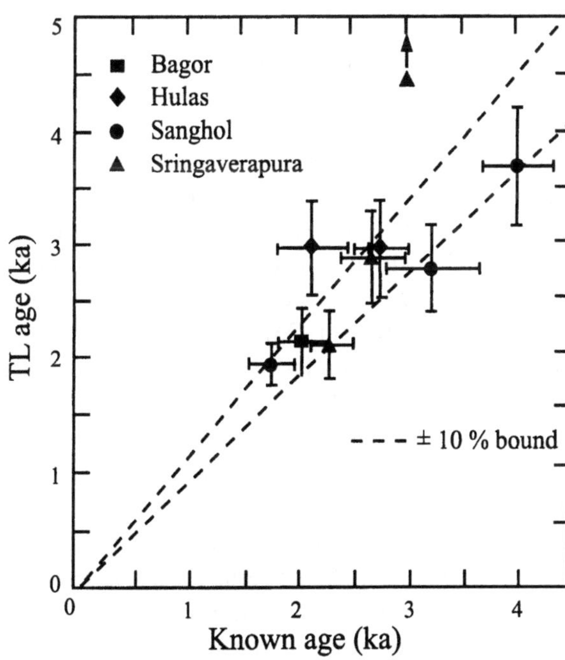

Fig. 103. Comparison between thermoluminescence and independent ages of archaeosediments. The samples originate from nine layers of four prehistoric sites in India. The single *outlier* stems from a flood sediment which was not sufficiently exposed to light. (After Chawla and Singvhi 1989)

TL ages and independent ages demonstrates that these events correspond to the formation of the strata. Despite the error limits of the TL ages, which are too large for archaeological demands, this investigation shows the suitability in principle of the TL techniques for the dating of archaeosediments, and that it is worth making more efforts in this direction.

Phytoliths. These tiny (10 – 200 μm) opaline bodies, secreted in plants, commonly occur in archaeosediments as the result of decay or burning of plants. Attempts have been made to date them by TL (Rowlett and Pearsall 1993). Difficulties may arise from changes in the crystalline structure of silica above 286 °C. Since the TL measurement involves heating to at least 400 °C, changes which disturb the TL sensitivity characteristics can be expected. The phytoliths were separated from archaeosediments collected at two pre-Columbian sites in the Chandi valley, Ecuador. Generally, the obtained TL dates between 800 and 3020 BC agree well with the ^{14}C-ages. The method turned out to be quite feasible, especially if the phytoliths have been heated to > 300 °C in antiquity.

7.2 Optically Stimulated Luminescence

The phenomenon of optically stimulated luminescence (OSL) is closely related to that of thermoluminescence, and is also induced by radiation. However, unlike TL, the electrons are not thermally but optically evicted from their traps (Fig. 89). If the liberated electrons recombine at luminescence centers, light is emitted as OSL. In addition to OSL the terms *photoluminescence* or more precisely – because of the induction by ionizing radiation – *radiophotoluminescence* are also used. Since the term photoluminescence is used phenomenologically otherwise, it is preferable to avoid confusion and to talk about *optically stimulated luminescence* (*OSL*), an internationally widely accepted term. Occasionally also *optical dating* is used. According to the spectral region of the stimulation light applied, one distinguishes different OSL techniques, namely IRSL with infrared and GRSL with green stimulation. Analogously to TL, the OSL signal is used to evaluate the natural dose which a mineral has received since its formation or last resetting. This enables the determination of the age, provided the dose rate is known. The evaluation of the dose rate is identical to the TL method.

Of the three radiation dosimetry methods of dating, the OSL method was developed last. The impetus to exploit the long-known OSL phenomenon for dating goes back to Huntley et al. (1985). Since that time enormous efforts have been undertaken to work out OSL dating methodologically because it possesses physical advantages over TL with respect to the dating of bleached material, mainly sediments. The OSL method is based on the easily bleachable electron traps which are already emptied after minutes of exposure to daylight. In comparison, the TL yielding traps need about one day of sunlight exposure (Fig. 104). Thus events involving weak light exposure come into the range of datability. For bleached material the OSL method is the physically more direct dating method since the same traps that were affected by natural bleaching are analyzed. A further crucial advantage is the almost complete optical resetting of the OSL signal, i.e., in contrast to TL no unbleachable residual signal of significant intensity is left. This especially enables the dating of very recent events (> 100 a). The upper limit of OSL applicability should be similar to TL, which still has to be explored. With respect to dating accuracy, the OSL method promises higher precision than the TL method for special techniques of measurement. The OSL signal can be measured during short periods of stimulation

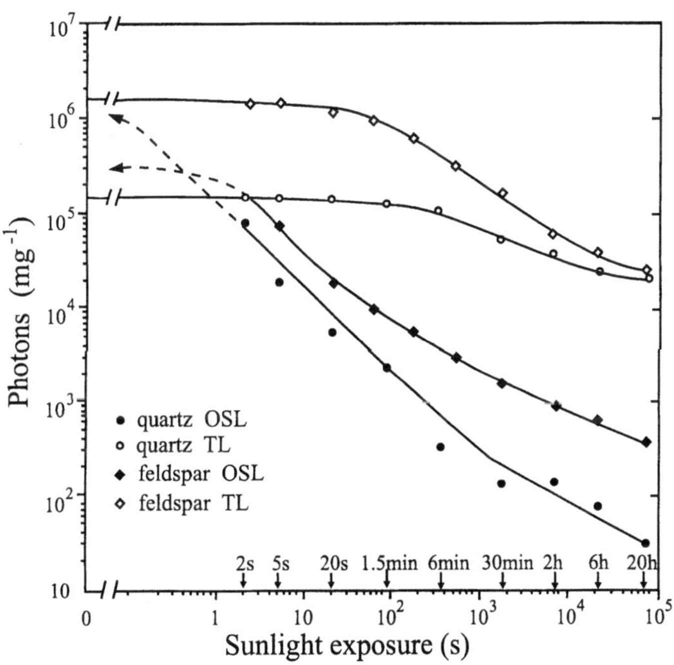

Fig. 104. The optical sensitivity of the OSL signal of quartz and feldspar is extraordinarily high in comparison to the TL signal (quartz 320–330 °C, feldspar 310–320 °C glow curve temperature). After 10 s and 9 min, respectively, of sunlight exposure the OSL signals of quartz and feldspar are bleached to 1% of their initial values. (Reprinted from Godfrey-Smith et al. 1988; copyright 1988, with kind permission from Elsevier Science Ltd., The Boulevard, Langford Lane, Kidlington OX5 IGB, UK)

without a significant diminution of the population of OSL yielding electron traps. This enables the dating of single aliquots and grains.

Good bleaching behavior was observed for quartz, feldspars and zircon. The most important dating objects are sediments containing these minerals, which have been exposed to light during their transport. Bleached surfaces of rocks, such as stone implements and building blocks, should in principle become datable by OSL. Owing to the fact that OSL does not require any strong heating procedures – as opposed to TL – organic materials such as bones, teeth or calcareous mollusk shells might also become datable through this method. To what extent OSL excels TL if applied to the dating of burned objects, such as ceramics, still needs to be fully explored. OSL requires smaller samples and provides a better precision of measurement. As far as the sampling procedure is concerned the same rules as for TL dating apply, but because the OSL phenomenon is much more light-sensitive, any light exposure has to be strictly avoided.

7.2.1 Methodological Basis

OSL Phenomenon. The close relationship between both luminescence phenomena permits a reference to the explanations given in the section on TL (Sect. 7.1.1) in order to treat the OSL method. The OSL method requires the presence of atomic defects in the crystal lattice, which form traps for the electrons which have been liberated by ionizing radiation. For illumination of the particular energy levels involved, the band model is referred to (Fig. 89). The charges are entrapped by the hole and electron traps. The mean lifetime of these metastable states is determined by the energetic depth of the electron traps. During optical stimulation the trapped electrons are activated so that they leave the traps and are shifted to the conduction band from where they recombine with luminescence centers under light emission (OSL). For stimulation the wavelengths of green light, from argon laser (514.5 nm), green diodes or halogen and xenon lamps equipped with appropriate green filters, are used for quartz and zircon, but infrared light of IR-diodes (880 ± 80 nm) for feldspar. To avoid scattering light the OSL is measured at significantly shorter wavelengths than those of the stimulating light (Fig. 105).

If the OSL intensity is directly measured as the stimulation proceeds, a decreasing curve, the so-called shine-down curve, is observed with increasing exposure time (Fig. 106). Stimulation light intensities of ca. 10 mW cm^{-2} cause zeroing of the OSL signal in quartz and feldspars within a few minutes. A minor residual signal may originate from the dark current of the photomultiplier, scattered light and unbleachable OSL components.

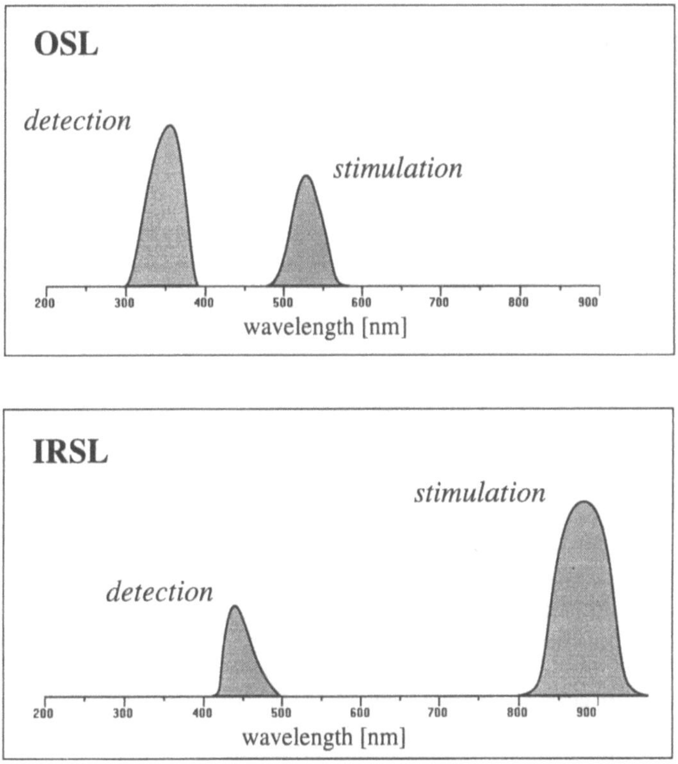

Fig. 105. Optical stimulation of luminescence by different wavelengths: (*top*) by green light; (*bottom*) by infrared. The OSL emission (GRSL and IRSL, respectively) is detected with inserted suitable filters at significantly shorter wavelengths as the stimulation light. (Courtesy of A. Lang)

The shine-down curve does not follow a single exponential decay function. The non-exponential character of the curve may be due to contribution of traps which are harder to stimulate and to electron retrapping.

Bleaching experiments associated with the TL and OSL measurements, which have been conducted so far, seem to prove the non-equivalence of thermal and optical stimulation energies of the electron traps of quartz and feldspar, in contrast to what would be expected on the basis of the energy-level model (McKeever 1991; Duller and Bötter-Jensen 1992). In this connection allowance has already been made for the different di-electric constants for thermal and optical processes. In the case of quartz the required optical energies (green stimulation light around 500–550 nm) are too high in comparison to the thermal activation energy of the bleached traps. Only for the very light-sensitive TL signal at 325 °C, which already begins to bleach under red light at 700 nm, is it justified to talk of optical

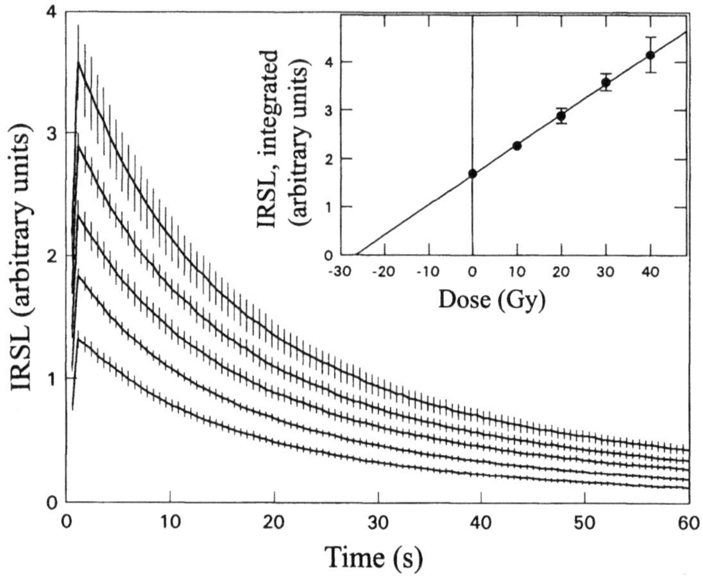

Fig. 106. IRSL shine-down curves (with 7 mW cm^{-2}) of a potassium-feldspar fraction (125–200 μm) from an early Holocene dune sand at Walldorf, Germany. Curves for the natural dose (*lowest curve*) and various additive doses are presented. The additive growth curve (integrated over 5–35 s stimulation time (cf. Fig. 107) is displayed in the *insert*. Its extrapolation to the dose axis gives the natural dose. (Reprinted from Wiggenhorn 1994; copyright 1994, with kind permission from Elsevier Science Ltd., The Boulevard, Langford Lane, Kidlington OX5 IGB, UK)

and thermal energy equivalence. On the other hand, in the case of feldspars, which are already stimulated in the infrared region of the spectral region around 850 nm (IRSL), the optical stimulation energy is too small compared to the thermal activation energy of the bleached traps. Hence, from the energetic point of view, a thermal support of the IR stimulation has to be assumed. Thus the IR stimulation is not merely a simple electron transfer into the conduction band, and the term thermo-optical process has been coined (Hütt et al. 1988). A coupled phonon–photon process is also supported by the observation of a strong increase of the OSL signal with increasing temperature (Duller and Wintle 1991). To sum up, it can be said that both luminescence phenomena TL and OSL cannot easily be energetically correlated and are partly fed from different electron traps.

A restoration of the OSL signal, which was first observed by Urbach (1930), is called *recuperation* (Aitken and Smith 1988). If after shine-down to, e.g., 1% of the initial signal an OSL measurement is repeated, a new OSL signal which may be a multiple of the final signal of the first measurement is observed. This effect occurs, in particular, for quartz and causes a nonradiation-induced OSL component, a fact which complicates the dating of young samples. The recuperation phenomenon is explained by optical transfer of a part of the electrons from the bleachable 325 °C trap to a shallower unbleachable trap during the shine-down. After completing the optical stimulation a portion of the electrons is again thermally retransferred from the latter trap to the 325 °C trap by the ambient temperature. There are also other forms of OSL recuperation (Aitken 1992).

The OSL emission spectra of various feldspar and quartz samples have been investigated. The OSL and TL emission spectra essentially look similar when recorded from the same sample, as was shown by Wiggenhorn and Rieser (1996) for potassium-rich feldspar with dominant wavelengths around 400–420 nm (blue) and 540–600 nm (yellow). This suggests the involvement of identical luminescence centers. By means of suitable filters the IRSL emissions of orthoclase (blue/ultraviolet) and plagioclase (green) of a polymineralic sample may be separated, for example.

Thermal and Optical Stability. The OSL shine-down curve does not contain any information on the thermal stability of the analyzed electron traps. This is a disadvantage compared with the TL glow curve which reflects the stability spectrum of the emptied traps as a function of increasing heating temperature. The initial hope to control the thermal stability by the wavelength of the stimulating light, has not been fulfilled due to the complex transfer mechanisms. For dating purpose, the fact must be accounted for that a fraction of the thermally unstable components of the natural OSL signals has been subjected to decay over time, whereas the freshly induced OSL signals have not suffered any loss by thermally unstable components. This is achieved by the application of preheating techniques in order to wash out the unstable components. Quartz is usually heated for 16 h at 160 °C. The heating of the artificially irradiated aliquots serves at the same time to transfer electrons thermally from shallow traps to the 325 °C-trap, what has already happened in the naturally irradiated aliquots during the millennia at ambient temperatures (Rhodes 1988). For the thermally stable OSL signal of quartz the mean lifetime is estimated as 20 Ma. Variously strong preheating is applied to feldspars, e.g., 1 h at 160 °C, followed by a storage period over 3 days at 100 °C to allow for anomalous fading. In addition, stepwise preheating procedures, until the attainment of a plateau-value of the natural dose, have been proposed for feldspars and polymineralic samples (Huntley et al. 1985; Godfrey-Smith 1994). The mean lifetime of the OSL signal is presumably smaller for feldspars compared to quartz, so that a thermal equilibrium might be attained within the range of several 100 ka.

The optical sensitivity of the OSL method is extraordinarily high. Under sunlight the OSL signal of quartz is bleached after 10 s, that of feldspar after

9 min, in each case to 1% of its initial value (Fig. 104). Under overcast sky or under water the bleaching times are longer according to the lower light intensity, as experimental investigations on potassium feldspars show (Ditlefsen 1992). The IRSL signal from a fine-grained loess was observed to decay completely within 30 min when exposed to light on a cloudy, foggy winter day (Lang and Wagner 1996). In order to ascertain that the OSL signal has in fact been completely extinguished during the deposition, and has not accidentally been diminished during the sampling procedure, the analysis of the shine-down curve is carried out in time intervals, and for each single interval the natural dose is calculated (Fig. 107). Here the phenomenon of decreasing bleachability of the traps with increasing shine-down time is exploited. The existence of a plateau is evaluated as an indicator for the absence of both disturbing factors (Huntley et al. 1985).

Data Evaluation. Additive and/or regenerative OSL growth curves are established (Fig. 80). For complete shine-down the OSL signal can be evaluated either by integration of the total area or of single areas of short succeeding time intervals of the shine-down curve. The differential analysis (Fig. 107) provides information on the bleaching behavior of the sample by means of the ND-plateau test. Both techniques require a large series of subsamples. In the case of a short, incomplete short-shine, e.g., 0.1 s, only an

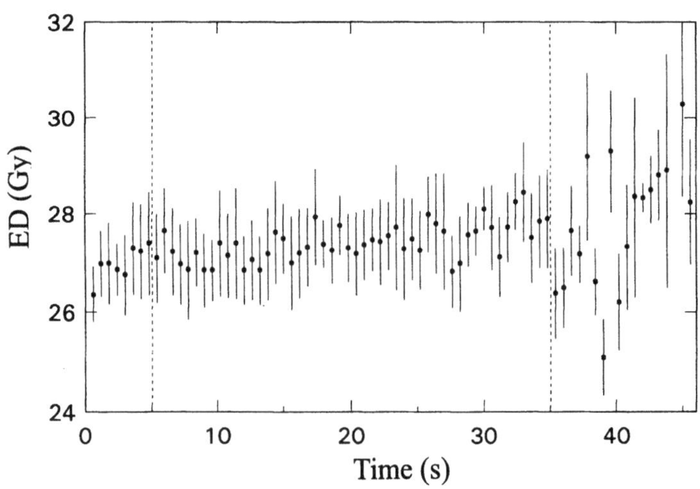

Fig. 107. Dependence of the equivalent dose ED (=ND) on the IR-stimulation time for the data displayed in Fig. 106. The existence of a plateau is indicative of complete resetting of the IRSL signal during sedimentation. (Reprinted from Wiggenhorn 1994; copyright 1994, with kind permission from Elsevier Science Ltd., The Boulevard, Langford Lane, Kidlington OX5 IGB, UK)

insignificant portion (<1%) of the OSL sensitive traps is emptied. Subsequently, the same subsample may be successively irradiated to establish the growth curve (Duller 1995). This enables the OSL dating of individual aliquots or even single grains (Murray et al. 1997). A grain-discrete age determination does not only increase the precision of the dating of sediments but also offers manifold possibilities of application. The techniques of evaluation of the dose rates are identical for TL and OSL dating. From this point of view it is suggestive to carry out both luminescence techniques on the same sample material in parallel; this can be done without too much additional effort. In the case of single grain dating by OSL, e.g., on potassium-feldspars, the grain-intrinsic dose rate (by potassium analysis) needs to be known separately for each single grain. The OSL age is calculated, as in other dosimetry methods, from the natural dose and the dose rate according to Eq. (48).

7.2.2 Practical Aspects

Like TL, the OSL method of dating can be employed on bleached quartz- and feldspar-bearing samples. In contrast to TL, less intensively bleached sediments can also be used, which makes the OSL method especially attractive for the dating of water-lain sediments. The sampling procedure is similar for both methods. Hence the relevant information (Sect. 7.1.2) on the sample size, grain size, soil moisture, dose rate and other factors is valid for OSL too. Since the OSL signal is extraordinarily light-sensitive, special attention has to be paid to avoid any light exposure. Profiles should be screened from sunlight during the cleaning and sampling procedures. In order to largely minimize their exposure to daylight the specimens are optimally collected with steel cylinders driven into the cleaned section face (Fig. 94). After removal the cylinders are quickly capped with tight steel lids.

The sample preparation in the laboratory has to be conducted under strictly subdued light of specific wavelengths (darkroom). As for TL, the sample material is separated into the polymineralic fine-grain fraction and the coarse-grain fraction consisting of quartz or feldspar. Usually ~50 aliquots of 1–2 mg are prepared and applied to steel or aluminum discs for measurement. On the one hand, the fine-grain discs of polymineralic sediments are advantageous owing to their better reproducibility, on the other hand, however, they prevent the separation of the sample into different mineral phases. Nevertheless, the OSL dating of such sediments without preceding mineral separation can be realized by the selective IR-stimulation of the feldspar fraction, since quartz, calcite and zircon do not show any IRSL, and a possible IRSL contribution by amphibole, pyroxene and tourmaline is presumably insignificant (Spooner and Questiaux 1989). For the coarse-grain fraction, quartz is separated by heavy liquids and cleaned by hydrochloric and hydrofluoric acids. The purity of the quartz fraction, that means the absence of feldspar impurities, has to be checked through the absent IRSL signal. Coarse-grain fractions of feldspars are obtained by heavy liquid separation. Feldspar concentrates may be contaminated to a certain degree with quartz without causing any significant disturbances, since often the OSL signal of feldspar is 100 times more intensive than that of quartz. Of all feldspars, potassium-feldspar is preferred due to its internal dose rate contribution. It is noteworthy that the composition of feldspars can vary considerably, which in turn might result in variable OSL characteristics. Suitable filters enable the separation of the OSL emissions of different feldspars. In contrast to quartz, the OSL signal of feldspar is liable to anomalous fading. The possibility of applying OSL dating to single grains is promising, as Duller (1991) has shown for sanidine grains of 200 µm size.

The instruments designed for OSL differ from each other in the light sources for stimulation. Sources may be argon ion lasers (Huntley et al. 1985), xenon arc lamps (Hütt et al. 1988; Spooner and Questiaux 1989), quartz halogen lamps (Bötter-Jensen and Duller 1992), infrared diodes (Poolton and Bailiff 1989; Spooner et al. 1990) and green diodes (Galloway 1997). The xenon arc lamps (1 kW) and quartz halogen lamps (75 W) yield a wide spectrum from which the desired stimulation wavelength has to be filtered out. Quartz requires green stimulation (GRSL). Feldspars are commonly subjected to infrared stimulation (IRSL), but may be stimulated by green light as well (Huntley et al. 1991). Most instruments are designed for permitting TL as well as OSL measurements. For infrared stimulation a TL instrument can be used if a ring with infrared diodes is inserted above the heater plate (Fig. 108). An electronic shutter permits short stimulation times of 0.1 s. In addition this apparatus can be equipped with a green light source in order to be used for combined TL, IRSL and GRSL measurements (Bötter-Jensen and Duller 1992). Provided that there is a large difference between the wavelengths of the stimulation light and the OSL emission, adequate color filters will allow only minor amounts of scattered light to reach the photomultiplier. Laser stimulation is more laborious and is rarely applied anymore. An argon ion laser beam is directed onto the sample and the OSL is measured in the blue-violet spectral region (around 380 nm). The separation of stimulating and emitted light is more difficult to realize for the green stimulation than for the IRSL, due to the rather small differences in wavelengths. The intensity of the scattered light is reduced by strongly absorbing color filters, which are inserted between sample and photomultiplier. Methodological improvements are expected

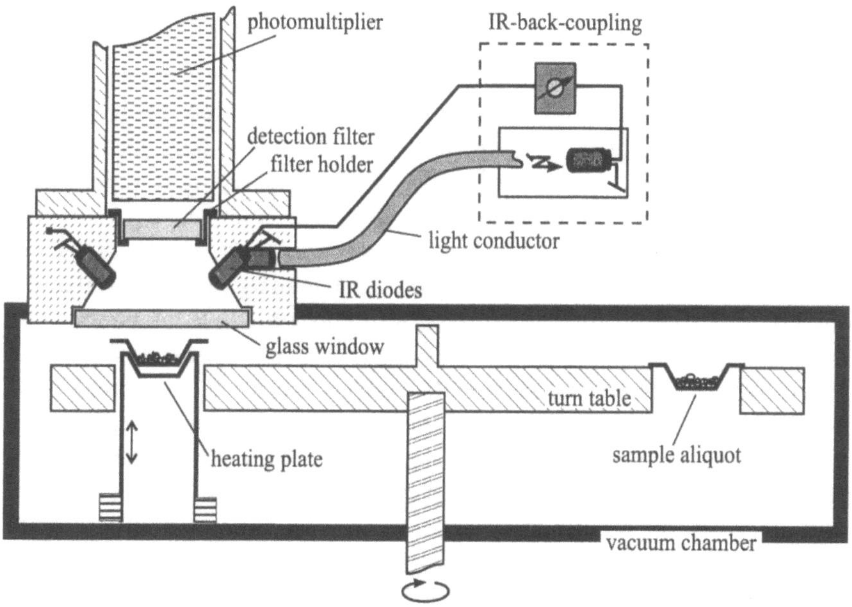

Fig. 108. Measuring instrumentation for infrared stimulated luminescence. (After Lang 1996)

from the wavelength-selective recording of the OSL emission spectra by means of monochromators and CCD cameras (Rieser et al. 1994). Since the optical stimulation hardly affects the TL signal of feldspars and can be conducted in the same measuring instrument used for TL, both OSL and TL dating can be carried out on the same sample without special experimental expenditure.

Thermally unstable OSL components are eliminated by preheating of the samples. Suitable annealing conditions are 16 h for quartz and 1 h for feldspar at 160 °C, but also 5 min at 220 °C is frequently applied. The OSL signal is either recorded as short-shine or as shine-down curve. Short stimulation times between 0.1 and 0.5 s cause only an insignificant portion of the OSL-supplying electron traps to be emptied. Hence, the OSL growth curve can be established on the same grain (single-grain technique) or subsample (single-aliquot technique). In this way complications which are linked to grain- or aliquot-dependent variations in the OSL characteristics and the dose rate can be eliminated, and the precision of the age determination is increased. Long stimulation allows the calculation of the natural dose for each interval of exposure. The existence of a plateau value of the natural dose is taken as the criterion for complete bleaching during deposition (Fig. 107). The age errors range around 10 %, as for TL dating.

7.2.3 Application

Dunes. The OSL signals of quartz and feldspar grains are very light-sensitive. It takes less than 1 h of exposure to sunlight for them to be totally extinguished. Even under conditions of subdued daylight redeposited aeolian sands may have experienced complete resetting of their OSL clocks. The quartz of a recent sand dune, situated in the northern part of Mali, gave an OSL age of 46 ± 30 a (Smith et al. 1990 b). The potential of the OSL method to date subrecent dune movements which are too young to be assessable by TL was demonstrated for the wandering dune of Pyla, France. Quartz from surface-near sediments yielded 350 ± 110 a, whereas sediments which stem from the basis of the dune were determined to be Early Würmian with 95 ± 14 ka (Smith et al. 1990 b). For the Late Paleolithic site Hengistbury Head in Dorset, Rhodes (1988) found the OSL ages of different sand deposits to be in good agreement with independent age data (Fig. 109). By OSL dating of quartz from Kalahari sand dunes, Stokes et al. (1997) identified several significant arid events since the last interglacial period, with dune-building phases 115–95, 46–41, 26–20 and 16–9 ka ago. Feldspars derived from aeolian sands were studied with green and infrared stimulation (Wintle 1993). Duller (1992) carried out comparative IRSL and TL measurements on potassium-feldspar from raised beach terraces in New Zealand and found concurrent natural doses. IRSL dating of feldspar inclusions in quartz from raised Australian coastal dunes yielded ages of up to 390 ka, which are consistent with TL ages (Sect. 7.1.3) and are in agreement with sea level highs (Huntley et al. 1993b). Of great potential is the possibility of applying IRSL dating to single aliquots, as was shown by Duller (1991) for sanidine grains of 200 μm in size, which were fractionated from Holocene dune sands. In principle this technique is applicable to individual grains. Age data ranging between 40 ± 17 and 4110 ± 330 a for Holocene dunes in the Mojave Desert, California, were obtained by the single aliquot technique on potassium-feldspars (Edwards 1993).

Loess. Since the TL signal of loess is nearly always completely bleached, full bleaching of the OSL signal is even more likely. In order to test the OSL datability of loess, Spooner and Questiaux (1989) chose the loess profile of Achenheim, eastern France. The stratigraphy of these loesses with intercalated Paleolithic find horizons has been investigated in detail and numerous TL data are available (Zöller et al. 1988; Packman and Grün 1992). For the OSL measurements the fine-grain fraction was separated. Green as well as infrared stimulation was carried out. Old loess (> 120 ka)

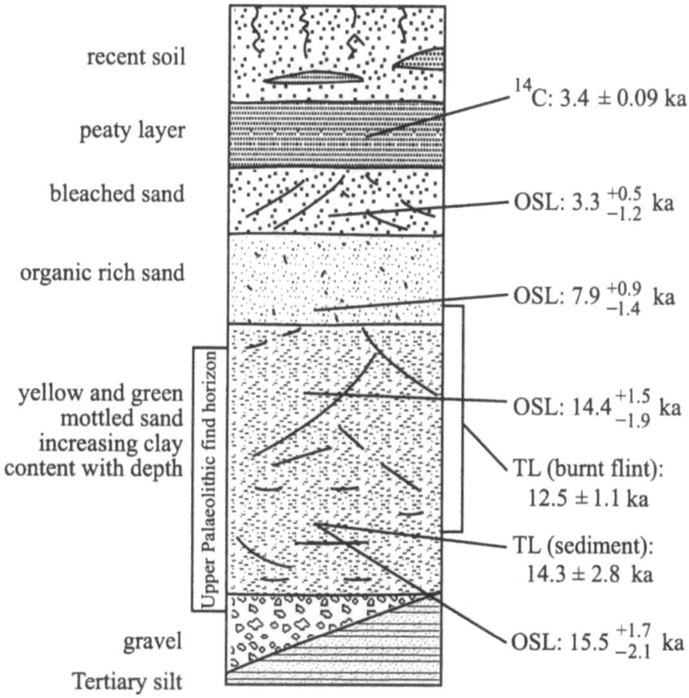

Fig. 109. OSL age determination on quartz from late Glacial and Holocene sands of the Upper Paleolithic site Hengistbury Head, UK. The OSL ages are in agreement with independent TL and ^{14}C data. (Reprinted from Rhodes 1988; copyright 1988, with kind permission from Elsevier Science Ltd., The Boulevard, Langford Lane, Kidlington OX5 IGB, UK)

yielded too young OSL age results irrespective of the stimulation and emission wavelengths. The age underestimation is obviously due to long-time loss of the OSL signal in the polymineralic fraction. Upper Würmian loess from Achenheim was dated by Questiaux (1991) using the polymineralic fine- and medium-grain fractions for green and infrared stimulation. The medium-grain fraction (43–54 μm) which contains few clay minerals is representative of loess. All OSL ages (15.6–21.5 ka) are consistent in themselves as well as geologically. On the basis of their investigations of numerous loess samples, Li and Wintle (1992) suggested using the coarse fraction of potassium-feldspar instead of the polymineralic fine-grain fraction for IRSL dating of older loess, because of its higher IRSL stability.

Sand (Aquatic). The fact that the OSL signals of quartz and feldspar are easily bleachable is a promising prerequisite for the determination of the deposition ages of fluvial, glaciofluvial and littoral sands. The light conditions vary much from one deposition milieu to the other. The intensity and the spectrum of the light changes with water depth and turbidity. The resetting of the TL and OSL signals during the transport is not necessarily always complete. In comparison to aeolian sand, water-lain sand requires a correspondingly longer exposure time. One has to proceed from the assumption that there are fluvial sands of rather different suitability for both luminescence methods, whereby the OSL signals are optically more sensitive than the TL signals (Fig. 104). A long period of transport in clear, shallow water is optimal, as e.g., in the littoral zone along shores and river deltas (Godfrey-Smith et al. 1988). GRSL dating on coarse-grain fractions of quartz from various fluvial sands, however, are encouraging (Smith et al. 1990b; Perkin and Rhodes 1994). Application of this technique to a subrecent sequence of flood deposits in southeastern Australia allowed Murray (1996) to obtain quartz GRSL – in combination with PTTL – ages as low as 100 ± 13 a.

With respect to fluvioglacial deposits, IRSL dating was performed on 100–200 µm large feldspars from Kames deposits in Karelia by Hütt and Jungner (1992). A satisfactory result in agreement with the late Glacial formation of the sediments was obtained for only one sample. Obviously, the light intensity in this depositional milieu was not sufficient to completely zero the IRSL signal. On the other hand, Saalian and Weichselian os deposits in west Finland yielded a stratigraphically relevant IRSL date (Hütt et al. 1993). Single aliquot analysis may be necessary because of the possibility that fluvioglacial sediments may contain a mixture of bleached and unbleached grains (Duller 1994).

IRSL dating of fluvio-lacustrine sediments from the Spiti valley, Himachal Pradesh, India, revealed that the seismic activity in that region dates back to the late Pleistocene (Banerjee et al. 1997). The deformation structures in these sediments (*seismites*) are thought to be caused by seismic events when the freshly deposited sediments were still soft.

Colluvial and Alluvial Silts. Silty deposits may form under very different conditions of intensity of light exposure, and it needs a larger data base before one can say which deposits are suited and which are not. However, the already existing dates for silty sediments show that OSL dating is indeed very useful, even for young deposits in the archaeological age range (Aitken and Xie 1992). IRSL dating is well suited for loessic colluvial sediments using the fine-grain fraction, as was demonstrated for various sites in southwestern Germany by Lang (1996) and Lang and Wagner (1997). The 410-nm emission of the IRLS signal shows good properties for dating purpose. The ages obtained enable the reconstruction of the geomorphic development from the late Glacial loess accumulation up to anthropogenic changes during the last few centuries (cf. archaeological sediments). Working on a number of British geological and archaeological sites, ranging in age from Middle Pleistocene to the medieval and comprising different types of sediments, Rees-Jones and Tite (1997) demonstrated the potential of OSL dating. Polymineralic as well as quartz fine-grain fractions of the sediment were dated by IRSL and GRSL, respectively. A major problem arose from insufficient bleaching caused by rapid deposition, in particular for alluvia, or by mixtures of components bleached at different times, as found in collapsed ditch fills. The presence of recuperation occurring after preheating was allowed for by applying an intercept procedure. Nevertheless, the dates show major episodes of alluviation in the Nene and Thames valleys caused by farming activity during the Neolithic, Bronze Age, Roman and Saxon periods.

Archaeological Sediments. Compared to TL, the OSL signal is, due to its higher light sensitivity, of particular interest when dating archaeological sediments which may have experienced only a little light during deposition. Among such poorly bleached deposits are archaeological colluvia, which are hardly datable by TL. An investigation of colluvial loess derivatives of a Neolithic settlement near Bruchsal, southwestern Germany, demonstrates the possibilities inherent in the OSL technique (Lang and Wagner 1996). This archaeological site is situated on the top of a hill. It consists of an early Neolithic (Bandkeramik period) ditch and a late Neolithic (Michelsberg period) double ring of ditches, which had been dug into the outcropping late Glacial loess. After the settlements had been abandoned the ditches filled up with colluvial sediments and were subsequently covered by additional colluvial layers (Fig. 110). Because of the elevated geomorphic situation, rather short transport distances (~100 m) have to be assumed, with correspondingly weak light exposure for the redeposition process. Samples for IRSL dating were taken from trench fillings and overlying colluvia. Different laboratory procedures were tried in order to overcome age underestimation. Preheating at 220 °C for 5 min and narrow wavelength IRSL detection seem to be best for reliable IRSL dating of these sediments. The results demonstrated that the light conditions during redeposition had been sufficient to reset the IRSL signal. The IRSL data clearly imply that the events of colluvial deposition are related to soil erosion triggered by human activities such as wood clearing and agriculture.

An example of an archaeological OSL study that has far-reaching implications for the original colonization of Australia is the dating of rock shelter sediments in northern Australia (Roberts et al. 1994). Flaked stone artifacts and ground pigments are associated with the deposits at the

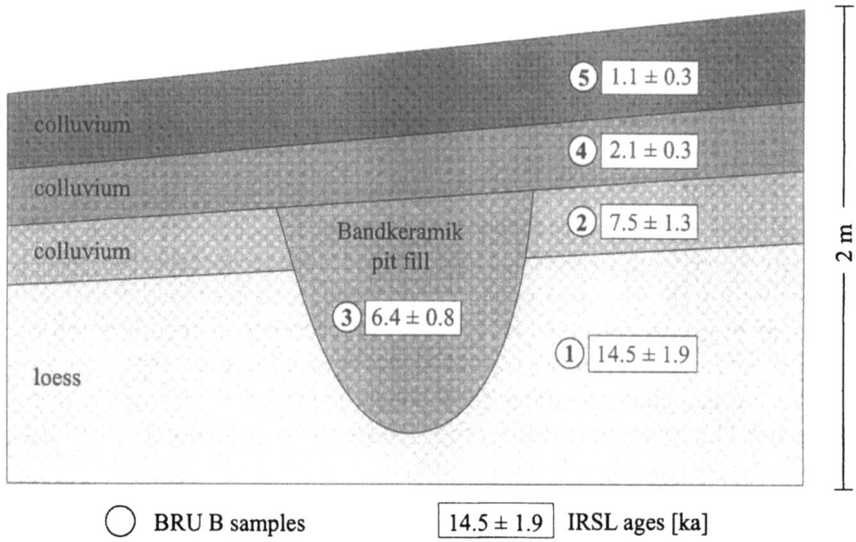

Fig. 110. Schematic cross section of the early Neolithic site Bruchsal-Aue, Germany, with IRSL ages of late Glacial loess, trench fill and various colluvia. (After Lang and Wagner 1996)

Nauwalabila site in Deaf Adder Gorge. The quartz-bearing sediments were analyzed by the GRSL technique. The lowest human occupation levels are bracketed by dates of 53.4 ± 5.4 ka and 60.3 ± 6.7 ka. For the upper levels, to which ^{14}C could also be applied, good agreement of both dating methods was obtained. The OSL ages independently confirm the evidence for an early human presence in Australia, as had been already revealed by TL dating of a nearby site further north (Sect. 7.1.3).

Wasp Nests. Nests being built by mud-wasps may contain – apart from ecologically relevant constituents such as pollen and phytoliths – quartz grains that had been exposed to light during construction, but protected from light since then. Roberts et al. (1997) applied OSL dating to various petrified wasp nests from rock shelters in northern Australia. The nests overlie, and occasionally underlie, prehistoric rock paintings. OSL ages ranging from 23.8 ± 2.4 ka to 100 ± 10 a were found.

Tephra. Fine-grain glass shards of volcanic ash deposits from various sites were examined for their suitability for OSL dating (Berger and Huntley 1994). Wavelengths of 514 nm (green), 633 nm (red) and ~880 nm (infrared) were selected for stimulation. Only weak OSL signals were observed, but in one case, the Mazama ash, an encouraging high IRSL signal, corresponding to a geologically reasonable natural dose, was obtained.

Ceramics. The OSL method of dating has also been applied to quartz of ceramic sherds. Here, otherwise as in sediments, the resetting of the OSL signal is a result of the firing process during the ceramic production. On ceramic of the Viking Period from Nicolaigade, Denmark, both TL and OSL methods resulted in consistent ages, and were also in agreement with the archaeological data (Mejdahl 1992). If compared with each other, the OSL method exceeds the TL method in this case, since it requires significantly less initial sample material (10 – 20 instead of 200 g for TL). Associated with this is not only a smaller experimental expenditure, but also a significantly higher precision of measurement (1 – 2% for OSL instead of 5 – 7% for TL). When using OSL instead of TL, single aliquot techniques become applicable which further reduce the required amount of sample material. By applying the *single aliquot regeneration added dose* (SARA) technique, encouraging results with GRSL on quartz from archaeological ceramics were obtained by Mejdahl and Bötter-Jensen (1997). An attempt to date heated colluvium, probably the remains of a camp fire, by IRSL turned out to be successful (Lang and Wagner 1996). At the already mentioned Bruchsal–Aue site, the fine-grain fraction of both, the burned and the unburned colluvium placed the fire in the Roman period.

7.3 Electron Spin Resonance

Electron spin resonance (ESR) dating is based on the accumulation of radiation-induced paramagnetic centers in minerals. It has close links with the luminescence methods of dating, especially in regard to dosimetry. Although the first attempts at exploiting the ESR phenomenon for dating go back to the 1960s (Zeller et al. 1967), ESR as a dating method did not begin to flourish before the 1980s, as summarized by Grün (1989), and is still undergoing rapid development.

The age range covered by ESR dating, although shifted towards higher ages, is comparable to luminescence dating. In special cases a lower age limit of a few hundred years can be attained. On the other hand, the ESR method permits age determination up to several 10^5 and even 10^6 a, which are not accessible by the luminescence methods. Hence ESR dating fully covers the Quaternary. It becomes increasingly meaningful for continental Quaternary deposits, of which calcareous cave sinters and travertine deposits are especially worth mentioning. Also mollusk shells, teeth and – to a lesser degree – bones are suitable for ESR dating. With respect to the marine realm, corals and foraminifers can be dated by ESR. Siliceous materials on which ESR measurements have been carried out successfully comprise quartz and flint.

ESR ages determine either the formation of calcareous sediments and organogenic carbonates or the secondary resetting of an already existing system as a result of heating, exposure to light or shearing pressure. Volcanic and archaeological events associated with heating processes are datable, as well as in favorable cases sedimentation processes, in the course of which bleaching of the ESR signal by sunlight has taken place. The pressure sensitivity of ESR permits the dating of neotectonic faults.

Errors of 10–20% are commonly stated to characterize the accuracy of ESR data. Large errors result mainly from the evaluation of the natural dose, due to uncertainties in the exact mathematical function describing the growth of the ESR signal with radiation dose. The intensity of the dose rate, both internal and external to the sample, can vary considerably over space and time. Such systematic errors are hardly assessable, and they are often not taken into account when estimating errors for ESR ages.

7.3.1 Methodological Basis

ESR Phenomenon. Under the term electron spin the particular characteristic of an electron is understood to be able to occupy two discrete quantum mechanical states. The spin may be conceptually regarded as an angular momentum, which results from the self-rotation of the electron. Its spin causes an individual electron to possess a magnetic moment and hence to be an elementary magnet. The electrons of an ordinary lattice are paired so that their magnetic moments compensate each other. However, a paramagnetic center possesses an unpaired electron. Radiation-induced free charges that are trapped at lattice defects constitute such paramagnetic centers. A particular crystal contains different centers, which can be energetically characterized by the g-value. This dimensionless factor is defined as the ratio between magnetic moment and angular momentum. For electron-hole centers it varies around the g-value of the free electron ($g = 2.00232$). Being exposed to an external magnetic field the electron as an elementary magnet adjusts itself in a way that its spin axis carries out a precession movement around the direction of the magnetic field (Fig. 111). The precession frequency is called the Lamor frequency and is proportional to the center-specific g-value and the strength of the applied field. If during the precession a microwave perpendicular to the magnetic field with a frequency equal to the Lamor frequency is applied, resonance of the electron spin results. During resonance the spinning direction reverses

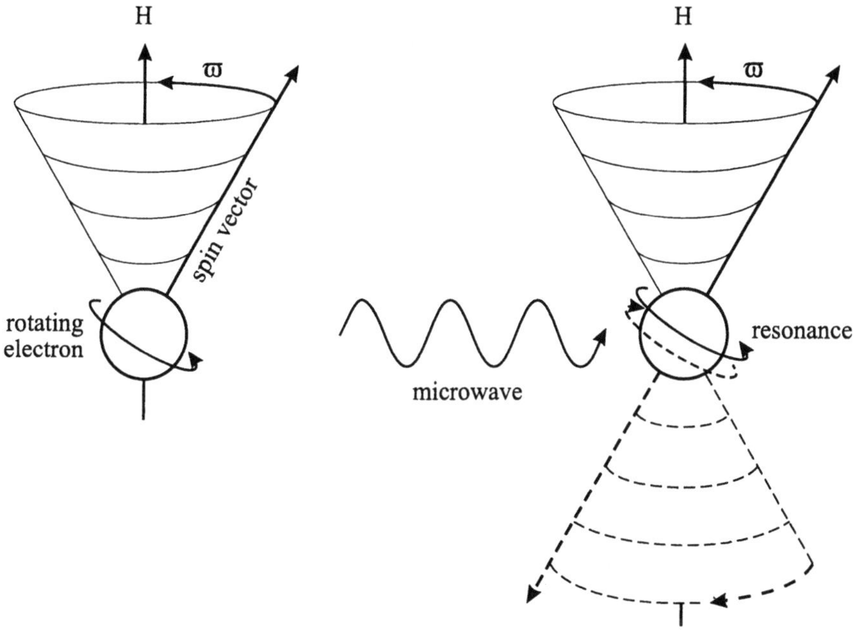

Fig. 111. Schematic of the ESR phenomenon. A paramagnetic center possesses an unpaired electron. If an external magnetic field is applied, the electron behaves as an elementary magnet, whose spin axis carries out a precession movement around the magnetic field direction H (*left*). If during the precession a microwave of defined frequency is applied perpendicularly to the magnetic field, resonance of the electron spin results, whereby its spinning direction reverses quantum-mechanically with the wave frequency (spin-flip) (*right*)

(spin-flip). The energy necessary for this spin-flip is met by the microwave, hence a portion of its intensity becomes absorbed. The resonance frequency identifies the type of center (g-value) and the degree of absorption is a measure for the concentration of this type of center. For practical reasons, not the microwave frequency is varied but the strength of the magnetic field. By continuously increasing the field strength, centers of increasingly smaller g-value are analyzed. The resulting ESR spectrum displays the specific microwave absorption for the centers of different g-values (Fig 112). For technical reasons to do with measurement, the ESR spectra of the microwave absorption are not directly recorded, but instead their first derivation as a function of the field strength is plotted (Fig. 113).

The concentration of the radiation-induced ESR centers is a measure of the radiation dose absorbed. Besides these ESR centers relevant to dating there may be interference with ESR signals shown by free radicals of or-

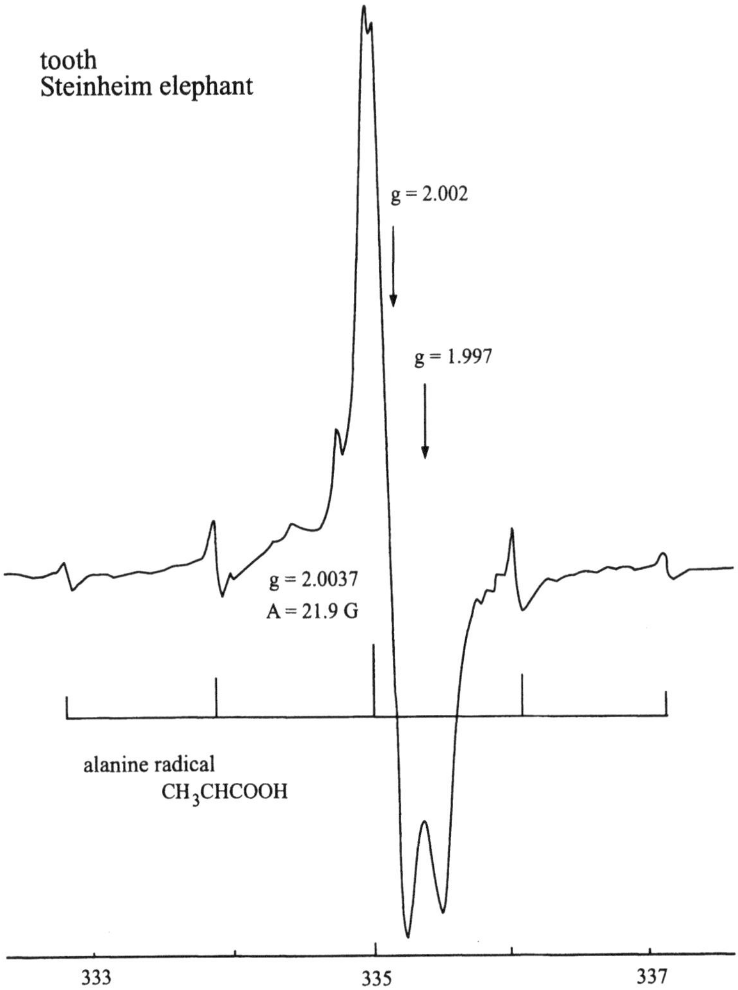

Fig. 112. ESR spectrum of tooth dentine of a fossil elephant from Steinheim/Murr, Germany. The individual ESR centers with their characteristic g-values are subjected to resonance by the variation of the magnetic field intensity H. (From Ikea 1981)

ganic molecules as well as the transition elements manganese and copper, resulting in rather complex ESR spectra (Robins 1991). Concerning quartz and feldspars, most of the atomic defects at which radiation-induced ESR centers form are known. These are the impurity atoms aluminum, germanium and titanium substituting for silicon. These lattice defects are already primarily manifested in the mineral. Additionally, in the course of time secondary defects are generated by collision of α-recoils with the oxygen

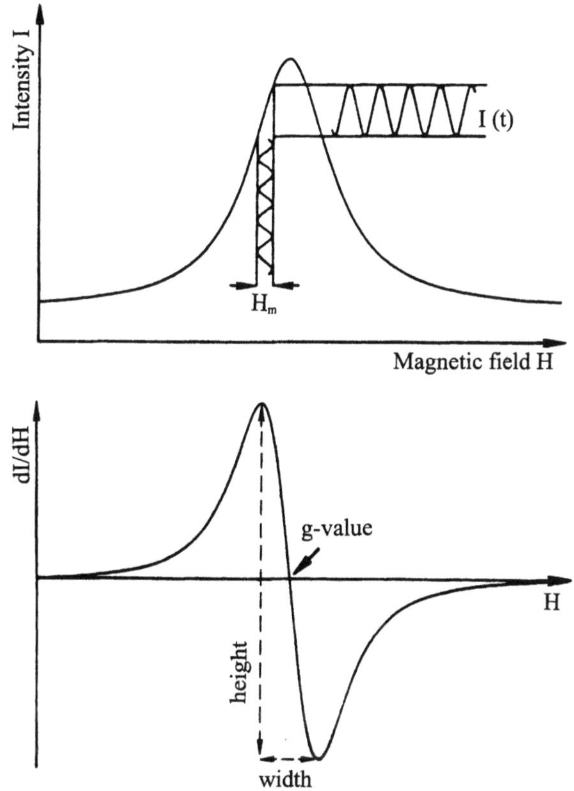

Fig. 113. The ESR spectrum (*bottom*) is the first derivation of the microwave absorption signal vs. the external magnetic field intensity H through an additional, oscillating magnetic field H_m (*top*). The *arrow* marks the position determining the g-value. (After Grün 1989)

atoms of the silicates. This process produces Schottky–Frenkel pairs, which consist of an oxygen vacancy (E′-center) and an interstitial oxygen atom (peroxy center; Fig. 114).

Natural Dose. The evaluation of the natural dose ND is done in the following way. On top of the natural dose, which a sample has received ever since its formation or its last resetting, additive doses are applied to the sample. The respective ESR signals are measured and a growth curve is constructed (Fig. 80). As opposed to TL, the ESR has the advantage that the concentration of the investigated centers is not disturbed, thus permitting the growth curve to be built up on the same aliquot. Most samples show exponential saturation with complex mathematical functions (Grün 1991). In any case one will try to exploit only those radiation-induced ESR centers of a sample with optimal growth characteristics. The problem inherent in the dose-dependent ESR generation behavior requires systematic investigations on the particular sample material and represent a major source of error for the age determination.

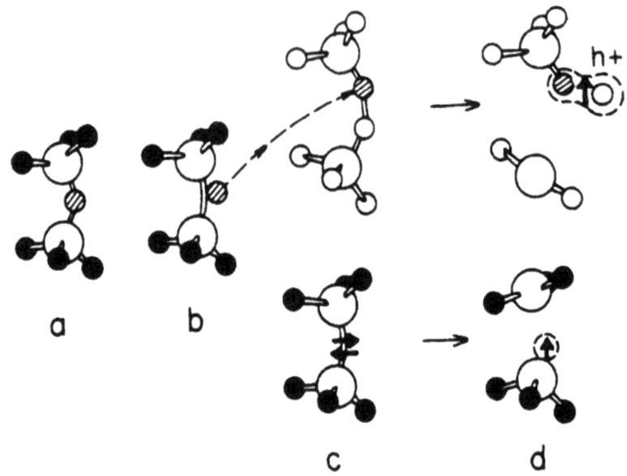

Fig. 114a–d. Model of the formation of radiation-induced defects and their transformation to ESR centers in quartz: *small circles* oxygen atoms, *big circles* silicon atoms. **a** Undisturbed lattice; **b** collision of an α-recoiling nucleus with an oxygen atom produces a Frenkel defect, consisting of an oxygen vacancy, charged-satisfied by trapping two electrons (**c** *below*) and an oxygen atom at an interstitial lattice position (**c** *above*). Through electron–hole trapping, defects are transformed to ESR centers: E′ center (**d** *below*) and peroxy center (**d** *above*). (Reprinted from Rink and Odom 1991; copyright 1991, with kind permission from Elsevier Science Ltd., The Boulevard, Langford Lane, Kidlington OX5 IGB, UK)

Stability. Dating, on the one hand, requires sufficiently stable ESR centers over the period to be dated. On the other hand, the centers have to be thermally sensitive enough so that the ESR clock can be zeroed by heating events. At ambient temperatures in sediments, as they reign close to the surface, most ESR centers possess mean lifetimes of 10^6 a and more, thus enabling the dating up to several 10^5 a. In the case of quartz a short heating up to at least 400 °C is sufficient for the complete annealing of the ESR signals derived from E′, Al and Ti centers. The absence of the Ge signal in a 500-ka-old sample suggests a maximal lifetime of < 100 ka which is less than the stability derived from annealing experiments. Apparently, other processes, not yet understood, extinguish the signal (Walther and Zilles 1994). Since lattice defects are thermally considerably more stable than centers, the radiation-induced creation of Frenkel defects in quartz seems to be suited for ESR dating far beyond 10^6 a (Rink and Odom 1991).

ESR centers in quartz are more or less light-sensitive. Experiments revealed that the Ge center bleaches rapidly and thus is promising for dating sediments which had been exposed to daylight over several weeks (Walther and Zilles 1994). Pressure may also affect the ESR signals (Fukuchi 1989, 1992), and the most sensitive one seems to be that of the peroxy center in quartz (Buhay et al. 1988), but the mechanisms involved are poorly understood. It is still unclear whether mechanical shearing or friction heating resets the signals. Notwithstanding that, this phenomenon enables fault dating by ESR.

Dose Rate. Spatial and temporal dose-rate variations are frequently encountered in ESR dating. They may introduce considerable systematic uncertainties to the age evaluation. This applies in particular to carbonates. In bones and teeth strong variations in the uranium concentration can occur on a scale similar to the penetration depth of α-radiation, causing strong gradients in the dose rate. Analogous difficulties may arise on a larger scale for β- and γ-radiation. The a-value has to be determined, making allowance for the weaker efficiency of α-radiation compared to β- or γ-radiation. Considerable differences in the radioactivity content may exist between sample and environment, a fact which makes a careful separation of the internal and external contribution to the dose rate necessary. The influence of the water content in the sample and environment has to be considered too. Specifically in the case of carbonates with low internal dose rate, the estimation of the cosmic dose rate contribution deserves attention.

A most serious impediment to the ESR dating of calcium phosphates and carbonates is the temporal variation of the dose rate. With respect to bones and teeth being exposed to groundwater, significant variations do occur due to uranium uptake. To allow for the time dependence of the associated dose rate increase, distinct models of uranium uptake are assumed, where the early uptake model is more favored than the linear one. Calcareous deposits frequently provide difficulties caused by an increasing dose rate as a result of the build-up of radioactive equilibrium, for instance of ^{230}Th (Sect. 4.1.1). In addition, increased radioactivity is observed on the surface of calcareous cave sinters. This increase is explained by the precipitation of daughter products of the ^{222}Rn present in the cave air.

Age Calculation. Only those ESR signals which fulfill particular criteria are suitable for the purpose of dating. They must

1. be radiation-induced, where their growth function with increasing dose should be mathematically as simple as possible;
2. show long-term thermal stability;
3. be sufficiently light-, heat- or pressure-sensitive depending on the event to be dated.
4. Their natural ESR signal should be significantly below the saturation level.

From the determination of both, the natural dose and the dose rate, the ESR age is calculated according to Eq. (48).

7.3.2
Practical Aspects

Sample sizes of a few grams are sufficient for ESR dating, provided the sample material is homogeneous. In most cases one will try to collect larger sample sizes to separate the suitable material in the laboratory and to be in a position to conduct micro-dosimetric measurements. As in the case of the luminescence dating methods, the sample locality should to be positioned at least 30 cm – and this has to apply over the whole natural burial period – beneath the terrestrial surface to enable quantification of the dose rate contribution from enclosing rocks and cosmic radiation. Due to the contribution of the external dose, representative samples of adjacent sediments (10 g) within 0.5 m distance have to be collected. An in situ γ-dose rate measurement should be carried out, principally at the sample locality. Since the water content is determined in the laboratory, the sample containing its moisture has to be kept in watertight bags. It is suggested that the sample should be largely screened from light, in particular if an event of light exposure is to be dated. Certainly the sample has to be protected from immoderate heat.

During the sample processing in the laboratory, one tries to gain homogeneous material of defined dosimetry and mineralogy. Quartz and feldspar are concentrated by heavy liquids and a magnetic separator. The ESR measurement is usually carried out on aliquots of 100-mg sample material of the grain-size fraction around 100 – 400 µm. The sample processing has to make allowance for the fact that crushing and grinding might have an imprint on the ESR spectrum. These unwanted effects can be suppressed by etching the surface in diluted acids. If quartz is processed for the dating of sediments, the sample preparation has to be conducted under subdued light in the darkroom. By applying special preheating techniques (e.g., 2 h at 100 °C) one tries to eliminate thermally unstable signals from the ESR spectrum.

The measurements are conducted in an ESR spectrometer consisting of a stabilized magnet which produces a homogeneous field at the sample's site, a microwave bridge and an electric signal analyzer. The sample, weighing a few 100 mg, is exposed to the magnetic field. Its various ESR centers are stimulated to resonance, in a way that at fixed microwave frequency the magnetic field strength is gradually changed. The majority of ESR spectrometers work at a microwave frequency of ~9 GHz and magnetic field strengths around 3500 Gauss (= 0.35 Tesla). The microwave absorption is measured during the shift of the magnetic field. In order to increase the resolution of the absorption signal, not the absorption curve itself but its first derivation is recorded. This is achieved experimentally by modulation of the magnetic field through an additional oscillating magnetic field (Fig. 113). Subsequently, the signal is electronically amplified and plotted as a function of the field intensity which gives the ESR spectrum. Instead of the magnetic field intensity *H* many ESR spectra only contain the *g-value* which implicitly reflects the magnetic field intensity. In the ESR spectrum the g-value characterizes the type of the center, whereas the height of the signal represents a measure for concentration of that center. Since ESR measurements do not influence the center concentration, the signal growth can be observed on the same aliquot after additive artificial irradiation. Suitable centers are chosen and their signal height vs. the radiation dose is plotted. From the growth curve the value of the natural dose is determined by back-extrapolation (Fig. 80).

With respect to the numeric presentation of the age data, certain minimal requirements should be observed to enable the estimation of their analytical quality. Grün (1991) suggests that – in addition to the age value and its error – the experimental conditions of the ESR analysis, the artificial irradiation dose, the mathematical growth function, the dose rate analysis and the error calculation should be described. Frequently the stated errors do not make allowance for systematic sources of error. The missing standardization of the data format makes it difficult for the non-specialist to come to a critical judgment on the age data. The accuracy of an ESR age is usually around 15 %.

7.3.3 Application

Calcareous Cave Deposits. Calcareous precipitation found in caves usually consists of calcite and more rarely aragonite. ESR permits the dating of their formation. Samples of calcitic sinters show complex ESR spectra which are partly attributed to organic radicals. The majority of ESR lines is excluded from dating purposes due to insufficient or absent radiation sensitivity; others are thermally unstable. The signal at $g = 2.0007$ is frequently used for dating with a mean lifetime of several Ma at an ambient temperature of 10 °C (Hennig and Grün 1983). Of aragonitic sinters only the line $g = 2.0020$ has been exploited for dating (Grün 1986). Frequently cave sinters are not datable due to insufficient ESR characteristics. Ages younger than 10 ka are rarely assessable owing to low dose sensitivity.

The sampling procedure has to make allowance for the slow growth of cave sinters of around several mm ka^{-1}, hence the major extension of the specimen must be parallel with the growth lamellae. Calcitic cave sinters contain a little less than 1 $\mu g\ g^{-1}$ uranium; sinters of aragonitic composition usually have a little more. What has to be taken into account as well is the precipitation effect of the radon daughter products from the cave air. Consequently the samples must be taken from places possessing a minimum distance of 2 mm from stable surface layers which can be identified by increased clay contents and dark coloration. Pure, contamination-free layers are suitable. Thorough consideration needs to be spent on questions dealing with the local dose rate and its temporal increase due to radioactive re-equilibration in the uranium series. Complications of a different kind arise from detrital calcareous inclusions. Recrystallization processes, becoming increasingly pronounced with age, are expected to complicate ESR dating beyond 0.5 Ma.

Important Paleolithic cave sites have been dated by ESR, including Caune de l'Arago, Tautavel in southern France (Sect. 7.1.3). The find layer containing cranium fragments is intercalated in calcareous sinters. ESR dating of calcareous deposits under- and overlying the hominid-bearing bed enabled its temporal bracketing between 242–313 ka as the upper and 147 ka as the lower age. This result is in good agreement with uranium series dates of 315–220 ka (Hennig and Grün 1983). Other ESR applications concern, for instance, the Grotte du Lazaret and Grotte du Vallonet as well as Abri Pie Lombard (Yokoyama et al. 1983). In the age controversy with respect to the skull unearthed in the Petralona cave, northern Greece, (Sect. 4.3) the ESR dating of remnants of calcitic sinter, precipitated on the cranium, represents an essential contribution. The ESR age of 198 ± 40 ka (Xirotiris et al. 1982) has to be regarded as a minimum age of the skull. This date supports the paleoanthropologic classification of the Petralona skull as an Early Neanderthal Man and contradicts the high age estimation of up to 700 ka on the basis of associated finds of uncertain stratigraphic position.

Travertine. The formation of travertine deposits can be dated by means of ESR. However, not every travertine is suitable for ESR measurements. Compared to cave sinters, travertine forms relatively fast. The involvement of bacteria and mosses in the precipitation process is manifested in a relatively high organic component, which in turn finds expression in complex ESR spectra. Favorable, according to Grün et al. (1988), are highly crystalline calcareous crusts. For the dating of travertine, in the same way as for cave sinters, the ESR signal $g = 2.0007$ is used. The uranium contents frequently comprise around 0.1 $\mu g\, g^{-1}$, but may be significantly higher in clay-rich layers. The in situ determination of the dose rate is always necessary due to its spatial variation. Also, its temporal variation due to radioactive disequilibration has to be taken into consideration. The datable age range comprises – similarly to cave sinter – 10 ka to 1 Ma.

ESR dating of travertine was conducted, e.g., for the profiles Weimar-Ehringsdorf and Bilzingsleben, central Germany, with important hominid fossils (Schwarcz et al. 1988). The ESR data show that the travertine complex of Weimar-Ehringsdorf has been deposited in the course of two major phases, namely the Lower Travertine with remains of a Pre-Neanderthal Man during the penultimate Interglacial ~200 ka ago and the Upper Travertine during the last Interglacial 120 ka ago (Fig. 32). Considerably higher ESR ages of ~400 ka were obtained for the travertine deposits of Bilzingsleben (Fig. 33). Consequently, at least one if not two Glacial periods have to be intercalated between the two hominid-bearing travertine deposits at Bilzingsleben and Weimar-Ehringsdorf (both interglacial). The ESR ages are supported by the results of uranium series dating conducted on these travertine deposits (Sect. 4.3).

Mollusk Shells. Their biostratigraphic implications make mollusk shells a most interesting dating object. They are found both in marine and continental deposits. The individual genera are not equally suited for ESR dating, hence certain genera, as e.g., *Glycimeris* and *Cerastoderma,* are preferred. Because the shells are composed of aragonite, they show complex ESR spectra consisting of different signals. The choice of suitable signals is especially complicated by the shape of the growth curve and the low thermal stability of ESR signals. Usually the ESR signals at $g = 2.0006$ and 2.0014 are exploited, but the latter one is problematic due to its composite character (Barabas 1989). In aragonite shells the ESR system is disturbed by the recrystallization of aragonite to calcite. This risk has to be expected especially in older samples. Recrystallization can be recognized by means of X-ray diffraction.

The major portion of the dose rate originates from the enclosing sediment. The shells themselves contain almost no thorium and potassium, hence only uranium needs to be considered. To eliminate the external α- and β-dose rate gradients, the outer 20–50 µm and 2 mm, respectively, are cut off according to the shell thickness. The determination of the environmental dose rate requires in situ measurements. With respect to mollusk shells there is the risk of uranium uptake. Thin-shelled mollusks are particularly affected by this. If the uranium content exceeds 1 µg g^{-1}, the temporal increase in the dose rate should be taken into account by both models of early and linear uptake.

The most conspicuous uncertainty in ESR dating of shells arises from problems associated with the signal definition and the regression function, as is demonstrated by the controversy on the temporal placement of the Holsteinian interglacial on the basis of ESR determinations on shells. Whereas Linke et al. (1985) found ESR ages around 200 ka, Sarnthein et al. (1986) obtained ESR ages between 350 and 370 ka, with far-reaching consequences for the correlation of the Holstein interglacial and $\delta^{18}O$-chronology (Barabas et al. 1988; Schwarcz and Grün 1988). The age range covered by ESR on mollusks reaches from a few ka to several 100 ka. Towards older ages it is increasingly limited by thermal instability and aragonite recrystallization.

Teeth and Bones. The dating of tooth enamel represents one of the most remarkable advances of the ESR method (Grün and Stringer 1991). Interesting applications are opened thereby, especially in the Middle and Lower Paleolithic as well as in Quaternary geology beyond the reach of ^{14}C dating. The suitability for ESR dating depends on the mineral hydroxyapatite. Tooth enamel consists of nearly pure hydroxyapatite. Cement and dentine of teeth contain higher contents of organic tissue and consequently are not used in ESR dating. The ESR spectrum of tooth enamel has at g = 2.0018 (Fig. 112) a suitable signal of good sensitivity and high thermal stability. These characteristics enable its application in principle to a wide age range of several 10^2 up to 10^6 a, i.e., the whole Quaternary.

The ESR dating of tooth enamel is greatly complicated by the dose rate which is subjected to strong temporal and spatial variations. Whereas the uranium content of recent teeth is $<1\ \mu g\ g^{-1}$, fossil ones may contain up to 0.15%. Dentine and cement of fossil teeth contain ten times more uranium than the enamel. The time-dependent increase of the dose rate is modeled by the functions of early and linear uranium uptake. The model of early uptake results in a lower ESR age compared to the linear model because of a higher average dose rate. Both model ages may differ considerably from each other – this is especially the case for high uranium contents (Fig. 33). Samples low in uranium are preferable. Because of the fact that the diffusion front gradually advances into the tooth during uranium uptake it is suggested to sample either individual lamellae embedded in the sediment or the shielded inner portions of big teeth, such as those of the mammoth. This is favorable as well for the quantitative consideration of spatial dose rate differences from the tooth towards the sediment and from the enamel towards the cement, but requires a specific sample preparation.

Two fragments of tooth enamel of *Palaeoloxodon antiquus* (fossil elephant) from the immediate neighborhood of the cranium fragment at the Grotte Caune de l'Arago, mentioned earlier in the section on calcareous cave deposits, yielded ESR ages of 210–240 ka (Hennig and Grün 1983). Schwarcz et al. (1988) determined the ESR ages on rhinoceros molars collected from the travertine sand which bears hominids and artifacts at Bilzingsleben (Fig. 33). The ages based on the assumption of a linear uranium uptake are on the average 30% older compared to the early uptake model ages and are in better agreement with the already mentioned ESR and uranium series ages of the travertine. They support a temporal placement of around 400 ka for this important Lower Paleolithic site.

The Middle Paleolithic sites Kebara, Skhul, Qafzeh and Tabun in Israel play a fundamental role with respect to the origin of anatomically modern Man (Proto-Cro-Magnon) and his relations to Neanderthal Man. The enamel of mammal teeth from these sites was subjected to ESR dating (Schwarcz et al. 1989; Stringer et al. 1989; Grün and Stringer 1991). Two results are worth mentioning in particular. Firstly, the ESR ages support the early dating (Sect. 7.1.3) of the anatomically modern Man to 80–120 ka, with the consequence that in the Levant *Homo sapiens sapiens* and *Homo sapiens neanderthalensis* co-existed for a long period of time. Secondly, the predominant portion of those ESR ages calculated on the basis of the early uranium uptake model are concordant with the more precise mass spectrometric $^{230}Th/^{234}U$-ages (Sect. 4.3). On the other hand, the ESR ages calculated on the assumption of linear uptake are almost exclusively too old.

At the Chinese Longgupo Cave, Sichuan, where very archaic hominid dental fragments and primitive stone tools were found, ESR dating contributed to establishing that the finds belong to the earliest Pleistocene, implying the very early arrival of the genus *Homo* in Asia. The linear uptake ESR age gave 1.02 ± 0.12 Ma for a cervid premolar fragment from a layer above the find horizon (Wanpo et al. 1995). ESR dating, combined with $^{230}Th/^{234}U$ dating, of bovid teeth collected from hominid-bearing layers at the Ngangdong and Sambungmacan sites, central Java, revealed mean ages of 27 ± 2 to 53.3 ± 4 ka, while the scatter reflects the uncertainty about uranium uptake (Swisher et al. 1996). These ages are of particular interest since the associated hominids are believed to belong to an advanced representative of *Homo erectus*. These ages imply that *Homo erectus* may have been contemporaneous with *Homo sapiens sapiens* in southeast Asia.

Attempts to date bones by ESR have been made repeatedly despite the associated extreme difficulties (Grün and Stringer 1991). Compared to tooth enamel, bones are mineralized to a smaller extent and absorb more uranium. In addition they are more liable to alterations during their burial. The ESR spectra are extraordinarily complex and variable. This can be ascribed to superposition of different ESR signals which to a major part result from organic radicals. A stable dose-sensitive ESR signal at $g = 2.0000$ is known from hydroxyapatite. However, basic research which would enable the identification of sufficiently suitable ESR signals has not yet been carried out. Further complications have to be expected with respect to the dose rate, since bones are well known for uptake of uranium and strong enrichment in this element. Also recrystallization from hydroxy- to fluorapatite takes place. Nevertheless, there exists a still unexploited potential for application of the ESR method to bone. This is demonstrated by such successful ESR determinations as in the case of the site of Peking Man (*Homo erectus pekinensis*) at Zhoukoudian. Horse and stag bones were dated to 260–570 ka from different layers of the profile at Zhoukoudian (Sales et al. 1989).

Deep-Sea Sediments. ESR dating of deep-sea sediments is conducted on foraminifers. For the determination of ages of up to 800 ka the ESR signal at g = 2.0036 appears to possess more suitable characteristics than the one at g = 2.0006 (Mudelsee et al. 1992). The ESR growth curves follow the function of exponential saturation. Figure 115, for instance, displays an example of ESR dating conducted on a deep-sea core. Concerning foraminifers, the signal at g = 2.0006 yields too low ages beyond 100 ka due to thermal instability. Above all, the accurate evaluation of the dose rate poses difficulties due to initial ^{230}Th, radioactive disequilibrium and decrease of pore water in the gradually compacting sediment.

Corals. For the dating of corals ESR is frequently applied, often together with uranium series techniques. The typical ESR spectrum of aragonite shows three lines of which the one at g = 2.0007 is preferred for dating. It is said to be thermally stable over the period 500 ka and longer at tropical ambient temperatures of 26 °C. Difficulties for the extrapolation of

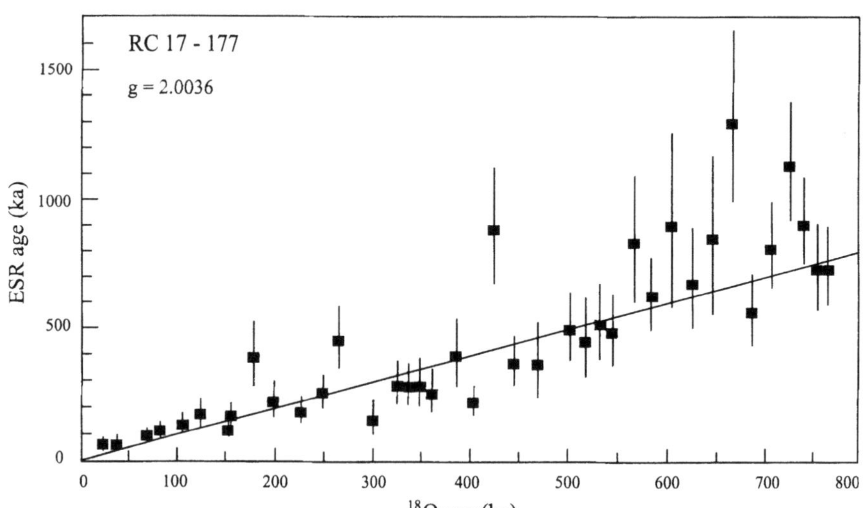

Fig. 115. Comparison of ESR ages of foraminifers (ESR signal g = 2.0036) with the $\delta^{18}O$ ages of a deep-sea core from Salomon Rise in the West Pacific Ocean. The *straight line* indicates equal ages. (Reprinted from Mudelsee et al. 1992; copyright 1992, with kind permission from Elsevier Science Ltd., The Boulevard, Langford Lane, Kidlington OX5 IGB, UK)

the natural dose from the additive growth curve arise out of deviations from a simple exponential function (Grün et al. 1992; Walther et al. 1992). Since the ESR spectrum is likely to be disturbed by the instability of aragonite and its transformation to calcite in the course of time, it is important to check for the presence of calcite by means of X-ray diffraction. Corals contain negligible quantities of thorium and potassium and hence the dose rate is predominantly controlled by the uranium. The uranium contents typically range around 3 µg g^{-1} and are evenly distributed over the coral colony. Therefore the calculation of the dose rate is straightforward. This advantage, together with the good outlining of the ESR signal, enables the dating of corals with an accuracy of 10%. In favorable cases the ESR ages of corals almost approach 1 Ma and thus range far beyond the age range of uranium series dating. Towards high ages the transformation of aragonite and thermal instability increasingly come into play and hence limit the applicability of ESR dating. An illustrative example of the potential of this method are the coral reef tracts in Barbados (Fig. 116). They are located up to 100 m above the present sea level. With increasing altitude the ESR ages increase up to 625 ka (Radtke et al. 1988). The recent uplift of Barbados can be reconstructed on the basis of these age–altitude-data.

Clastic Sediments. Light-sensitive ESR signals in quartz and feldspar grains, which are exposed to daylight during redeposition, may be bleached, a fact enabling the dating of the last exposure to light. In quartz such centers are the Ge center (g = 1.997) and to a smaller degree the Al center (Walther and Zilles 1994). After 7 h of sun irradiation the Ge center is fully

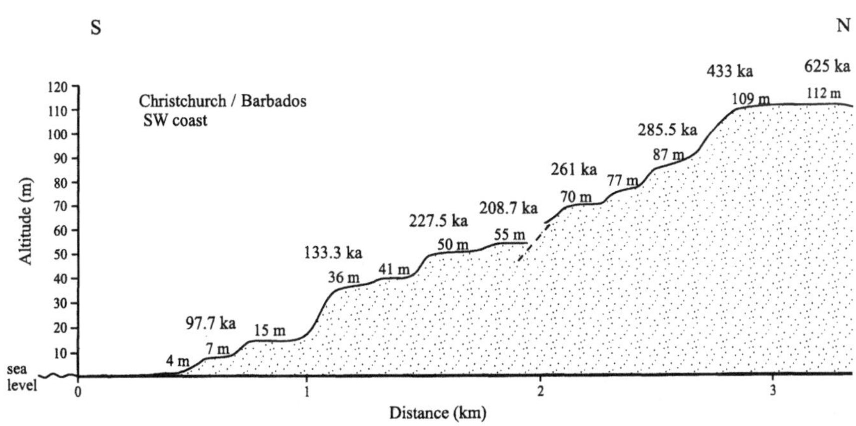

Fig. 116. ESR ages of raised reef tracts on Barbados. (After Radtke et al. 1988). The ESR ages increase up to 625 ka with elevation. The age–elevation correlation allows the reconstruction of the recent uplift of Barbados

bleached (Buhay et al. 1988). The thermal stability of this center is stated to comprise 30 Ma (Shimokawa and Imai 1987), which should permit the dating of samples of Lower Pleistocene age, as Huang et al. (1988) demonstrated for Chinese loess, which yielded ESR ages of 0.7–1.4 Ma. On the other hand, quartz of the Middle Pleistocene sands of the *Homo heidelbergensis* find horizon at Mauer did not yield any observable ESR signals of the Ge center (Walther and Zilles 1994), suggesting that the lifetime of the signal is too short. The Al signal of this sample still increased with additional dose indicating that it was not yet saturated. However, it is of limited use for dating purposes due to its hard and incomplete bleachability. If bleaching under water is assumed, a model age of 740 ± 500 ka for the Mauer sands is obtained (Walther 1996). Further basic research is required to judge the suitability of the Ti signal in quartz for dating purpose.

Flint. Microcrystalline quartz makes up the major constituent of flint. A considerable number of the flint artifacts handed down displays signs of heating. Since heating erases most of the ESR signals, heating processes may not only be proven by ESR but also be dated in this way. Most appropriate to dating are the E′ and the Al centers, which both possess high thermal stability at normal burial temperature. However, signals of organic radicals are superimposed on the Al signal. At heating temperatures above 425 °C the E′ signal is disturbed by a signal of a carbon radical, thus it is a heating temperature around 400 °C which optimally meets the requirements for dating. The elimination of the zone that has been exposed to light and to the external components of the α- and β-doses is achieved by removing at least several mm of the artifact's surface. Grinding in a mortar seems to have no influence on the ESR signal. For the determination of the dose rate, in situ γ-radiation measurements (external component) and U, Th and K analyses (internal component) are suggested. Porat and Schwarcz (1991) were using the E′ signal ($g = 2.0001$) of flint samples from the Mousterian site at Ibrahim, Lebanon, and determined ESR ages between 80 and 92 ka, which were in good agreement with independently determined ages. On the other hand, burned flint from the Lower Paleolithic site Yabroud, Syria, furnished too young ages which might be due to healing or dose rate effects. Compared to TL, the ESR dating of burned flint is less dose-sensitive, which is above all adverse to the younger age range. With respect to larger ages of ca. 600 ka the higher stability of the E′ signal might be favorable. But E′ center ESR ages of burned flint, in average 48.5 ± 5.1 ka, from levels VII–XII at Kebara, Israel, turned out to be systematically lower than the Al center ages, in average 64.6 ± 12 ka, and the TL ages (Sect. 7.1.3), whereby the discrepancy may be due to some loss of the E′ signal (Porat et al. 1994).

Mylonite. Associated with the shearing zones of tectonic faults the rock is immensely pressure-strained and intensely milled. The pressure sensitivity of the ESR signals of quartz is exploited for dating the last motion of the fault. Most sensitively responds the peroxy center (the so-called OHC center), followed by the Al , Ti and E′ centers. In addition to mechanical cataclasis, friction heat might contribute to the resetting of the ESR systems. Rock flour of grain sizes around 100 – 200 µm of mylonites meets the requirement for ESR dating of pressure strain. The heterogeneity within

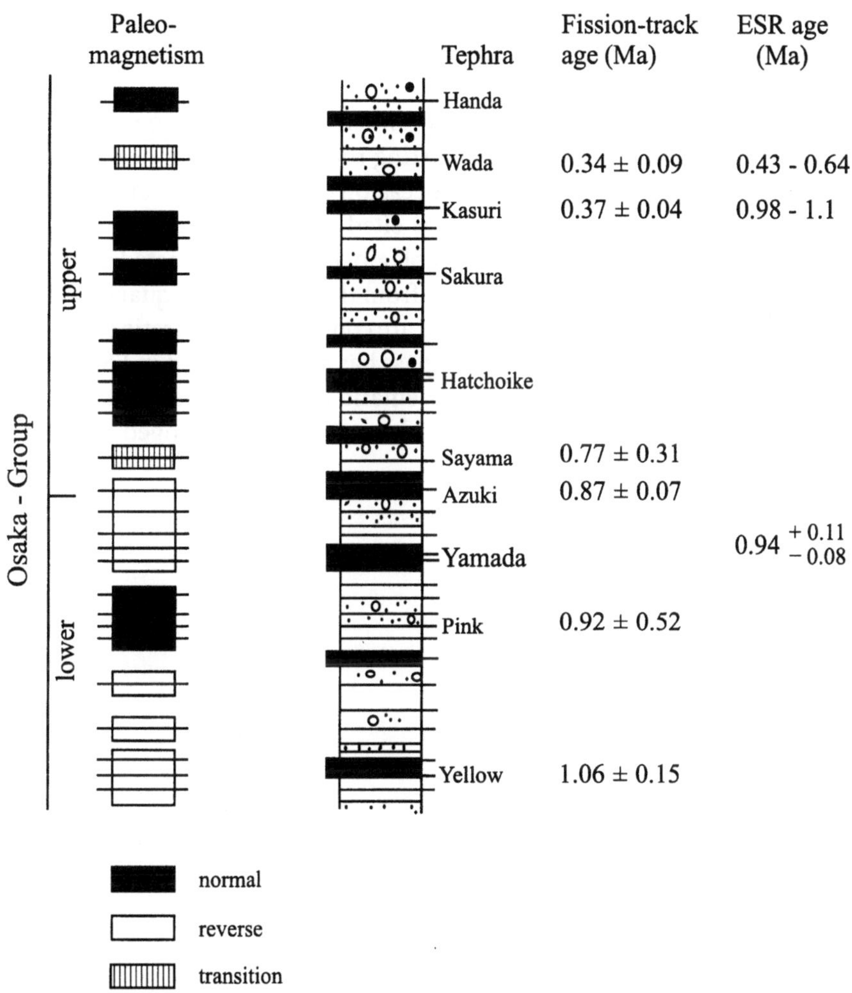

Fig. 117. ESR ages of volcanic ashes and clastic sediments from the Osaka succession, Japan. The ESR dates are consistent with stratigraphy and fission track ages. (Reprinted from Toyoda and Ikeya 1991; copyright 1991, with kind permission from Elsevier Science Ltd., The Boulevard, Langford Lane, Kidlington OX5 IGB, UK)

zones of faults requires in situ measurement of the dose rate. Quartz and feldspar concentrates of the Atotsugawa fault in Japan were used by Ikea et al. (1982) to date the last movement to 65 ka, where the ESR signal of the E′ center (g = 2.001) showing stability over several Ma was analyzed. When working on the San Gabriel fault zone, southern California, Lee and Schwarcz (1994) demonstrated that the zeroing of the ESR signal depends on the grain size of quartz, which suggests that shear stress alone is capable of zeroing the signals. They obtained ages of 1173 ± 130, 824 ± 70 and 357 ± 19 ka in accordance with the geologically determined sequence of faulting events. The potential of ESR for the dating of pseudotachylites associated with intense fault movements has not yet been investigated.

Volcanites. Volcanism can be dated by ESR either directly on newly formed rocks or indirectly on previously existing rocks which were heated in the course of the volcanic event. On this occasion quartz, feldspar and glass are used. The Japanese Osaka sequence is cited as an example of ESR dating (Toyoda and Ikeya 1991). Quartz and plagioclase were extracted from two tephra layers. Stable ESR centers stem from the E′ centers in quartz and the Ti centers in plagioclase. The ESR ages are consistent with stratigraphy and fission track ages (Fig. 117). Volcanic glass has also been used in ESR dating (Imai et al. 1985). However, more systematic investigations must be done before the potential of ESR for tephrochronology can be judged.

8 Chemical Reactions

This group of dating methods is based on time-dependent chemical reactions. Provided the rate of the chemical reaction is known then the duration of the reaction process, i.e., the age of the system, can be inferred from the progress of the reaction. The chemical reactions used for dating are essentially parts of chemical weathering processes, concerning chiefly diffusion, selective ion exchange and oxidation. One reaction which is largely independent of weathering is amino acid racemization. Chemical methods of dating can be assigned to two groups. One group (e.g., hydration) permits age determination of the surface exposure of materials consisting of silicates. By means of the other methods (e.g., racemization), the burial period of skeletal remains in the soil is measured.

If silicate-bearing materials, as for instance obsidian, are exposed to moisture in the soil or air their fresh surfaces are altered. Both, diffusion from and into the silicate results in the formation of weathering rinds of altered chemical composition. The primary petrologic texture remains more or less preserved. The thickness of the rind increases with the duration of the weathering process. In addition, encrustations composed of oxides and hydroxides may also be formed on the surface. Frequently the term weathering rind is employed without precise differentiation between selectively altered material and the newly formed crust material. The weathering rinds may be discernible either on a macroscopic (e.g., patina on silices) or only microscopic (e.g., hydration rind) scale, or they may be discerned by specific analytical techniques, such as nuclear magnetic resonance. The inprecise term *patina* is frequently used for surface films with clear color alteration as a result of aging. Age-dependent parameters may be the thickness or the composition of the rind. The period being dated is the extent of exposure time of a surface to the weathering agents.

Skeletal remains which are embedded in sediments are liable to get penetrated by the soil moisture – especially within the range of the groundwater – because of their porosity. The inorganic bone matrix is subjected to material exchange reactions. The organic component is affected by degradation processes. The amino acids, as degradation products of the proteins, are subject to racemization. The analytical determination of the time-dependent degree of enrichment or depletion of those elements and molecules involved in the chemical reaction permits the estimation of the time elapsed since burial.

A basic difficulty inherent in the dating application is the dependence of the reaction rate on multifarious factors. The progress of the reaction is influenced – apart from the compositional and structural constitution of the object to be dated – by environmental factors such as temperature, water content, pH value, redox conditions, nature and concentration of dissolved substances, permeability of the embedding milieu, and others. In addition to the fact that the individual influences are hardly quantifiable, their temporal variation further complicates the situation. Solely in exceptional cases can chemical clocks be "read" quantitatively. They are more appropriate for semiquantitative and relative age determination. In order to increase the accuracy of the chemical clocks, efforts are being made to achieve a better understanding of the kinetics of the particular processes. Some basic terminology with respect to the reaction kinetics will be briefly explained in the following.

Reaction Kinetics. A chemical reaction (A → B) from the educt A to the product B proceeds at a certain rate v. The reaction rate changes continuously in the course of the reaction. If the symbols $[A]$ and $[B]$ stand for the concentrations of the two reactants, the reaction rate is equivalent to the consumption of the educt $-d[A]$ or the formation of the product $+d[B]$ per time interval dt, i.e.,

$$v = -d[A]/dt = +d[B]/dt. \qquad (52)$$

The rate of a unidirectional reaction is proportional to the concentration of the educt where the proportionality factor a is called the rate coefficient of the reaction, i.e.,

$$v = -d[A]/dt = a \times [A]^n \qquad (53)$$

which possesses a specific value for a certain reaction at a given temperature. The exponent n of the reactant in Eq. (53) is indicative of the order of the reaction. Accordingly reactions of first order ($v = a \times [A]$) are distinguished from those of second order ($v = a \times [A]^2$) and generally of order n ($v = a \times [A]^n$). The order is dependent on the reaction type.

Examples of first-order reactions are racemization and radioactive decay. Rearranging of Eq. (53) yields for $n = 1$

$$[A] = [A]_0 \times e^{-at} \qquad (54)$$

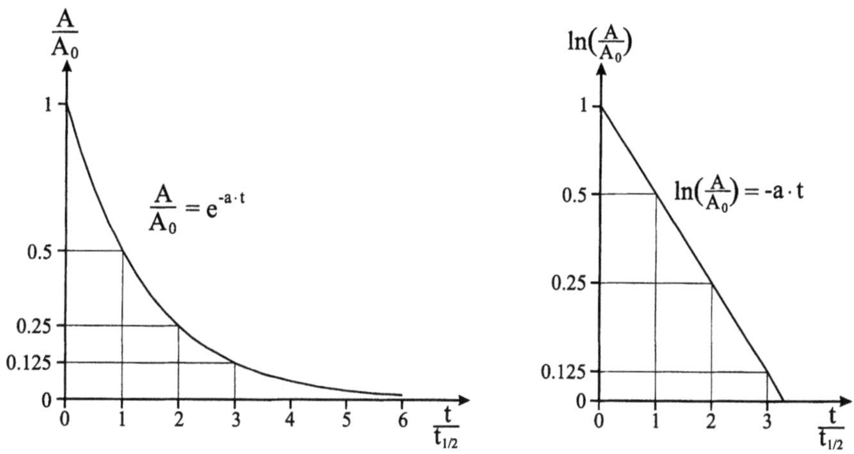

Fig. 118. In a kinetic first-order reaction the concentration of the educt $[A]$, normalized to its initial value $[A_0]$, declines exponentially to half as much after one half-life $t_{1/2}$ (*left*). The representation of this function in a semi-logarithmic diagram $\ln [A]/[A_0]$ vs. t results in a straight line. The numerical value of its slope is equal to the reaction rate coefficient a (*right*)

i.e., the concentration $[A]$ shows an exponential decrease from its initial value $[A]_0$ with time (Fig. 118). It is suggestive to use – instead of the linear representation – a half-logarithmic diagram where $\ln([A]/[A]_0)$ is plotted against the linear time scale. In this way the data fall upon a straight line and not, as they otherwise would do, on an exponential curve. The slope of the straight line represents the rate coefficient. In the case of a first-order reaction the rate coefficient can be expressed either as half-life, the time required for one-half of the initial concentration of the reactant to be consumed independently of its initial concentration, or as mean lifetime, the time after which the reactant has been consumed to $1/e$ of its initial concentration.

If the reaction does not only progress unidirectionally but also in the reverse direction, the rate of the total reaction equals the difference between forward and backward reaction v_1 and v_2, respectively, i.e.,

$$v = v_1 - v_2 \tag{55}$$

$$v = a_1 \times [A] - a_2 \times [B] . \tag{56}$$

In the course of time, reversible reactions attain a state in which forward and backward consumption balance each other. That is, when the system has reached a dynamic equilibrium

$$a_1 \times [A] = a_2 \times [B] . \tag{57}$$

In the state of equilibrium $[A]$ and $[B]$ are constant and $v = 0$; the reaction seems to have ceased. The concentration ratio $[B]/[A]$ is defined as

$$[B]/[A] = a_1/a_2 = K \tag{58}$$

where K is the temperature-dependent equilibrium constant. If forward and backward reactions possess the same rate coefficients, then $K = 1$, with equal abundance of both reactants in the state of equilibrium.

Diffusion. Concentration differences in solids cause equalizing mass transport, the so-called chemical diffusion. The diffusion processes are controlled by kinetic laws. The transport of matter is described by the two Fick's laws. The flux of ions or molecules dn/dt diffusing through a unit cross section is proportional to the diffusion coefficient D ($cm^2\ s^{-1}$) and the concentration gradient, which in non-stationary systems depends on the time t. Under certain boundary conditions the mean diffusion length x of the diffusing atoms can be described by a parabolic time-dependent law

$$x = \sqrt{2D \times t}\,. \tag{59}$$

The diffusion process in solids has to be imagined as changes of atomic positions, where the mobility of the endogenous and exogenous constituents of the solid is more expressed at the surface (surface diffusion), due to the lower binding energy as compared to the interior of the solid body (lattice diffusion). The least expenditure of energy is required for diffusion along grain boundaries (grain-boundary diffusion) which is schematically displayed in Fig. 119.

The hydration of obsidian, for instance, is a diffusion process. Environment-derived hydrogen ions diffuse into the rock surface. A surface-parallel diffusion front develops. The thickness M (= x) of the hydration rind grows with the age t according to

$$x = \sqrt{a_h \times t} \tag{60}$$

i.e., the growth function of the rind follows the square root of the time t (Fig. 120). If the square thickness M^2 is plotted against the linear time axis the hitherto parabolic growth curve is transformed into a straight line with the hydration rate a_h (= $2D$) as the slope.

The temperature sensitivity of both the rate and the diffusion coefficients is governed by the Arrhenius equation [cf. Eq. (49)]

$$a = a_0 \times e^{-E/kT}. \tag{61}$$

Accordingly the reaction rate increases exponentially with the absolute temperature T. The relation between T and a is illustrated by the Arrhenius diagram of the racemization in Fig. 121, where $\ln a$ is plotted against T^{-1}. The slope of the straight line directly yields E/k.

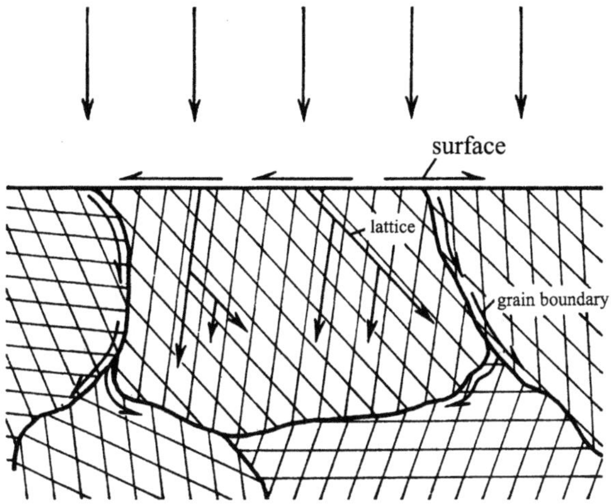

Fig. 119. Schematic of diffusion on the surface, through the lattice and along grain boundaries near the surface of a polycrystalline rock (after Meyer 1968). The extrinsic matter enters the rock due to concentration differences. The different mechanisms of diffusion have different rates, with the fastest along grain boundaries. Within a grain the rate depends on the type and orientation (anisotropy) of the crystal

Fig. 120. During hydration the rind thickness M grows with the square-root of time $\sqrt{t}$ (*above*). A diagram where t is plotted against M^2 results in a straight growth function with the slope being equal to the hydration rate a_h (*below*)

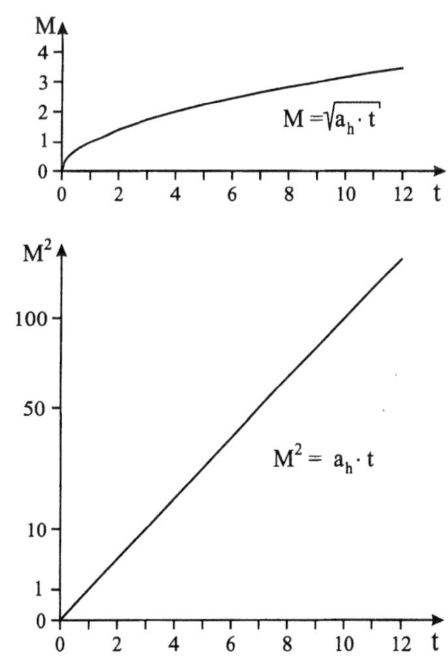

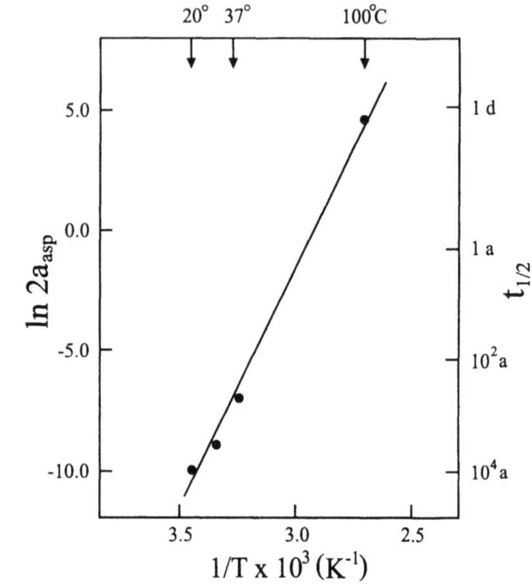

Fig. 121. Correlation between the reaction rate coefficient $2a_{asp}$ (*left axis*) or the half-life $t_{1/2}$ (*right axis*) of aspartic acid racemization and the temperature T (from Bada 1985). In an Arrhenius diagram, in which the reaction rate is logarithmically plotted vs. the reciprocal of the temperature, the data points can be fitted by a straight line. Its slope is proportional to the activation energy

8.1 Weathering Rinds

One of the classical geological approaches to the deduction of elapsed time is the evaluation of the state and rate of chemical weathering processes. Manifold weathering phenomena have been exploited for solving chronological questions in the Quaternary (Brooks 1985). For instance repeated attempts have been made to determine the duration of the soil formation from the soil thickness, or to delineate the age of rock surfaces from the thickness of weathering rinds.

When fresh surfaces of rocks, such as glacial debris or silex artifacts, are exposed to moisture from the soil or air, a macroscopically recognizable weathering rind frequently develops. With persistent weathering the front of the rind advances into the rock at the expense of the unweathered rock. The idea of using the rind thickness as an age indicator has been doubted for good reasons, since the rate of weathering processes depends on multifarious factors. In situations in which many of these factors are constant, e.g., petrographically identical objects, identical ambient temperature and soil chemistry, the weathering rind may reflect the age sequence of the surfaces. The range of application covers 1 to 500 ka. Without a more extended data base and a better theoretical understanding of the physical–chemical processes underlying the formation of weathering rinds, the practical application of this dating approach cannot be evaluated. In connection with weathering, *lichenometry* dating – although being beyond the scope of this book – is worth mentioning. Lichen sizes can be used to evaluate the ages of young (up to several 100 a) surfaces of rocks (Bull and Brandon 1998).

8.1.1 Methodological Basis

Most well-known are the weathering rinds on silices, also referred to as patina. They consist of a macroscopically recognizable surface layer at most 1 cm thick, which forms a marked color contrast to the unweathered matrix. The silex patina is white, however, staining is common. If the patina is smooth and glossy, it is called glossy patina. Analytical profiles laid through the patina and underlying matrix show significant compositional differences between patina and matrix (Fig. 122). The patination processes are complex and not fully understood. Most authors discuss the leaching of amorphous silicic acid out of the microcrystalline quartz skeleton as the dominating factor in the patination process. Presumably the diffusion of H^+-ions (hydration) into the surface in exchange for alkali ions (selective leaching) is also of importance. There may also be additive adsorption of iron and manganese oxides as well as hydroxides on the surface. In the course of their burial, silices gradually form patination rinds, and hence it is obvious to associate the color intensity or thickness of these rinds, which

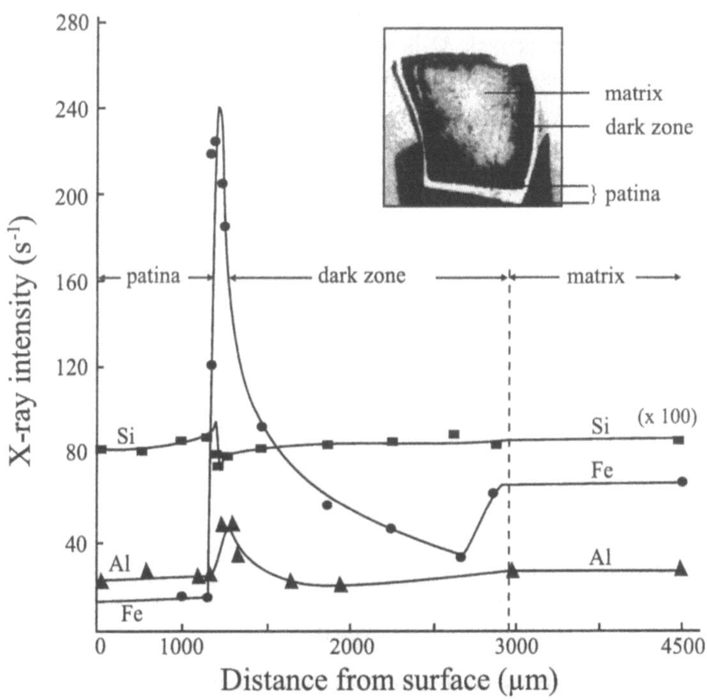

Fig. 122. Chemical profile through the weathering rind of a chert from Florida, established by electron microprobe analysis. (After Purdy and Clark 1987)

can reach several mm, with the duration of the weathering process. However, the thickness of the weathering rind is by no means exclusively controlled by the parameter of time. Other factors such as the mineralogy and structure of flint, the soil chemistry and the water content as well as the ambient temperature may exercise even stronger influences on the patination, compared to the mere aging effect. Rottländer (1977) concluded on the basis of inconsistent results from patination experiments, that the thickness of the silex patina does not represent a chronological criterion but reflects the dynamics of the embedding sediment. Purdy and Clark (1987) took the opposite view, according to which the patina formation is suitable for dating of the surface under certain burial conditions, and that in particular age determination should be possible through experimental determination of the kinetic parameters – in analogy to the artificial hydration of obsidian (Sect. 8.2.1).

The surfaces of fine-grained basalt fragments frequently bear weathering rinds. The rind thickness can comprise more than 1 cm and increases with the age of the blocks, as Cernohouz and Solc (1966) were able to observe on northern Bohemian basalt falling into the age range between 600 a and 870 ka. The fitting of the data is possible by the mathematical function $M = c_1 \times \log(1 + c_2 \times t)$, where $c_1 = 4.64 \pm 0.05$ and $c_2 = 0.010 \pm 0.001$, if the rind thickness M is substituted in mm and the age t in ka (Fig. 123). Numerous measured data on glacial blocks of andesite and ba-

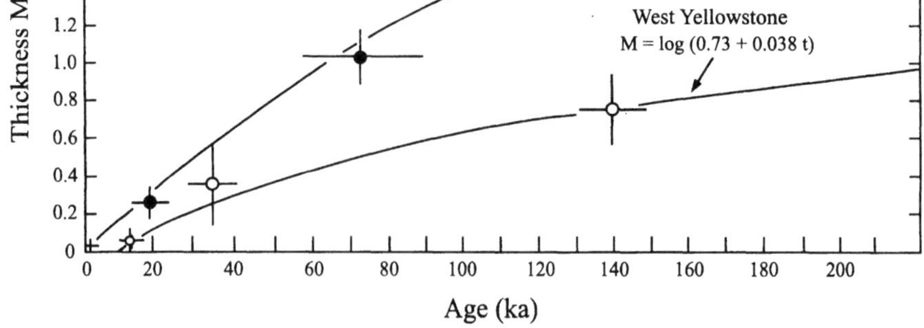

Fig. 123. Relationship between the thickness of weathering rind and age for glacial basalt blocks from Bohemia (after Cernohouz and Solc 1966) and the Yellowstone area (after Colman and Pierce 1981). The measured data can be empirically fitted by a logarithmic function

salt from the last two Glacial periods in western North America show similar curves although with different values for the two non-variables c_1 and c_2. Both studies revealed that the progression rate of the weathering front decreases with age. The curves show some similarity to those of the hydration of obsidian (Fig. 120), although the growth rates differ by two orders of magnitude and the growth functions are not identical. It has to be emphasized that the growth curves were empirically established. As long as the physical-chemical processes leading to the formation of the rinds have not been investigated in detail, the curves cannot be transferred to other conditions and can be exploited at the most for relative dating. Finally, the risk should be mentioned that with increasing age the weathering rind becomes increasingly more liable to chemical dissolution and mechanical abrasion. Colman and Pierce (1981) assumed for basalt that these phenomena prevail after 0.5 Ma and hence do limit the datable age range.

8.1.2 Practical Aspects

Only fine-grained rocks such as flint, basalt and andesite are suitable for dating. It is crucial that the sample material derives from controlled environmental conditions (moisture, temperature history, pH value and others). Since this method is supposed to date the age of the surface – and not that of the material – the surface to be dated should not display any features of dissolution or mechanical abrasion. Surface areas bearing encrustation of iron or manganese oxides – indicated by their dark color – also have to be avoided. The rocks are cut perpendicularly to the surface. The macroscopically recognizable weathering rind is measured by means of a stereo magnifier equipped with a micrometer ocular. For microscopic investigation thin sections are used. Because of the variable thickness of the rind, several measurements have to be conducted for each surface area. Comparison of the thickness of weathering rinds only makes sense if applied to petrographically uniform material. Hence the conduction of accompanying petrographic and chemical analyses is a prerequisite prior to any determination of the relative age sequence based upon the rind thickness.

8.1.3 Application

Silex Artifacts. There is abundant literature dealing with the potential of dating of silex artifacts, but hardly any age data have been presented so far, despite the apparent simplicity of the dating principle, a fact that clearly demonstrates the difficulties to grasp the complex environmental factors of the weathering process. As soon as hewed stone tools or flakes are buried the fresh rock surface starts to be attacked by chemical weathering. The thickness of the weathering rind should reflect the soil exposure time of the rock surface. Owing to environmental influences on the patination process, the technique is most appropriate for the same type of silex material stemming from the same find site. Purdy and Clark (1987) presented a patina age of 26 ka for a silex fragment from Florida, where the patina has been compared with that of an 8.5-ka-old arrowhead. On the basis of the patina color of silex artifacts from Speckberg, southern Germany, Rottländer (1989) differentiated between Middle and Upper Paleolithic origin. In this case the thickness of the patina does not correlate with the stratigraphic age.

Glacial Debris. Rock debris in the neighborhood of glaciers is mechanically strained. Expansion of freezing water in the pore spaces and cracks of a rock cause its splitting as well as abrasion and hence the formation of fresh rock surfaces. Subsequent to deposition and burial of the debris chemical weathering – favored by the mild interglacial climate – comes into play and results in the formation of weathering rinds. Owing to its longer burial period the debris of previous Glacial periods should show thicker weathering rinds compared to those of the debris of subsequent glaciation. Hence those weathering rinds seem to bear potentialities for relative dating. Colman and Pierce (1981) measured 7355 rinds on a large number of basalt blocks from the western USA, sampled from glacial deposits at depths of 20–50 cm. The thickness of the rinds varied between 0.1–4 mm and, apart from their age, were influenced by such factors as petrography (andesite or basalt, grain size), embedding material (moraine or sandars) and climate. Nevertheless, the rind thickness did correlate with the stratigraphic age of the glacial deposits and proved the burial time to be the most important parameter of the rind growth, thereby demonstrating the possibility of differentiating between the glacial debris of individual Glacial periods.

8.2
Hydration

If fresh glass surfaces are exposed to the humidity of the air or soil, water diffuses from the environment into the glass. A hydrated rind develops and grows in depth in the course of time. From the thickness of this microscopically visible rind the onset of the diffusion, i.e., the age of the surface, can be deduced. If the glass surface is the original one, the particular time of glass formation is determinable in this way. Most applications, however, deal with secondary events in which fresh surfaces have been exposed, as for instance fractured surfaces formed during the manufacture of the artifact.

The hydration method has gained in importance mainly for dating of obsidian. Obsidian rinds up to 50 μm thick have been observed, enabling dating within an age range of a few hundred years to 1 Ma, depending on the hydration rate. The hydration rate in its turn depends on the chemical composition of the glass and the ambient temperature. Obsidians of different provenance show different hydration features according to their chemical compositions. The same kind of obsidian undergoes faster hydration under warmer depositional environments than under colder ones. The effects of the glass composition and the ambient temperature have to be thoroughly evaluated. Thus in any case the hydration method of dating requires the differentiation of the raw material, which can frequently be achieved by chemical analyses and sometimes on the basis of visual criteria, too.

For obsidian objects of identical geological source and burial site the age sequence can be directly inferred from the thickness of the rind. On the other hand, the age determination explicitly requires knowledge of the hydration rate. The goal of deducing the hydration rate directly from the chemical composition of the glass for a given ambient temperature is not yet achieved. The hydration rate is deducible in two ways. One feasible way is the calibration by means of identical material of known age from the same environment. Another possible approach is the experimental determination of the hydration rate and subsequent extrapolation to the ambient temperature. It must always kept in mind that the latter approach requires independent knowledge of the ambient temperature which involves considerable uncertainties.

The hydration method was introduced by Friedman and Smith in 1960. With a growing data base it soon became obvious that the chemical composition of the glass bears a significantly stronger influence on the hydration rate than had been previously assumed, and that it can even surpass the effect of the ambient temperature during burial. In the meantime more than 10,000 sets of obsidian data have accumulated. However, a quantitative correlation between the chemical composition of the glass and the hydration features has not yet been unveiled, a non-surprising fact in view of the huge variety of glass compositions. In the same way, the physical and chemical aspects of the hydration processes have not been sufficiently investigated (Ericson et al. 1976). In most applications the thickness of the hydration rind is readily determined under the optical microscope, a straightforward procedure which enables the examination of extended series of sample sets. For some time more sophisticated nuclear-physical techniques have been employed yielding higher depth resolution.

Naturally, obsidian dating is restricted to volcanic areas – the geological provenance of this rock – and nearby areas to where raw obsidian had been brought through prehistoric trading. This essentially concerns the Mediterranean, the Near East, East Africa, Japan, New Zealand and America. Apart from chiefly archaeological questions aiming at the dating of obsidian artifacts, the hydration method is applied to the field of geochronology, too, with preponderance on volcanic eruptions and glaciations. This dating method has also been tried on artificial glass objects.

8.2.1 Methodological Basis

Glass is a solid, essentially amorphous, inorganic material which predominantly consists of silicon and aluminium as well as alkali metal and alkaline earth oxides. It is formed naturally during volcanic eruptions, more rarely as a result of meteorite impacts and when lightning strikes the Earth's surface. Artificial glass has been produced for four millennia.

Glass slowly absorbs water from its environment and may thereby attain water contents of up to several percent, in the case of obsidian 3–4%. This water absorption, the so-called hydration, takes place in the form of diffusion into the immediate subsurface zone of the glass. The hydrated rind is set off against the unaltered inner glass as a sharp diffusion front running parallel to the surface and being recognizable under the polarizing light microscope (Fig. 124). The progressing diffusion gradually affects deeper zones so that the hydration rind becomes continually thicker with elapsing time. The thickness M of the hydration rind increases with the square root of exposure time t (Fig. 120)

$$M^2 = a_h \times t, \tag{62}$$

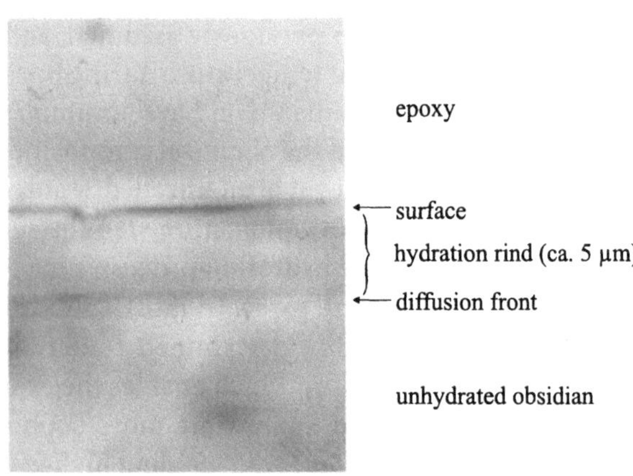

Fig. 124. Hydration rind of an obsidian artifact under the optical microscope. The artifact originates from the highland Maya culture and was produced presumably from raw material source El Chayal, Guatemala. (Courtesy of J.W. Michels)

where a_h represents the hydration rate (µm² ka⁻¹). The validity of this relation has been proved both by measurements on obsidian surfaces of known age and experimentally (Fig. 125). The deceleration of the growth rate with increasing rind thickness can be qualitatively understood by taking into account the increasing path length of the diffusive transport.

Owing to mechanical tensions, hydration rinds ~50 µm thick tend to chip off in flakes. Subsequent to the flaking a fresh surface is exposed from which hydration starts anew. Thus the application of hydration dating is limited to an age range equivalent to the maximum thickness of the rind. The hydration rind may be also affected by soil solutions and other influences causing weathering. On corroded surfaces rinds of different thickness may lie side by side.

The temperature dependence of the hydration rate is expressed by the Arrhenius equation [Eq. (61)]. This interdependence could be confirmed at elevated temperatures (Fig. 126). The appearance of T as exponent means that even minor temperature changes result in large changes of the hydration rate (ca. 10% per degree). Therefore detailed knowledge of the temperature history of a sample is an important factor in the hydration dating. The Arrhenius equation permits extrapolation from hydration ra-

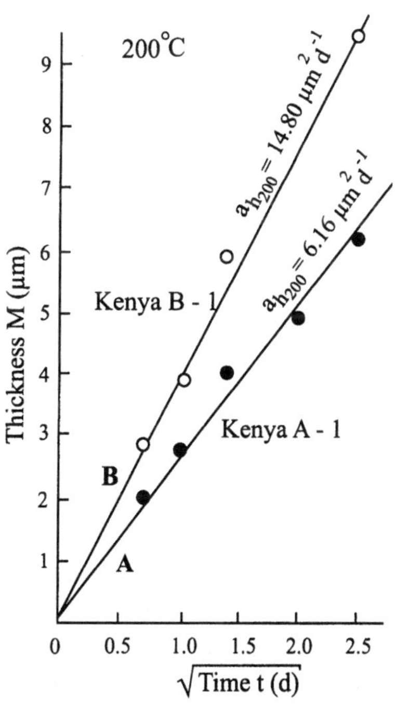

Fig. 125. Experimental determination of the hydration rate a_h of two chemically different obsidians from Kenya. Fresh obsidian surfaces were subjected to hydration in de-ionized water at 200 °C. (Reprinted with permission from Science; Michels et al. 1983a; copyright American Association for the Advancement of Science)

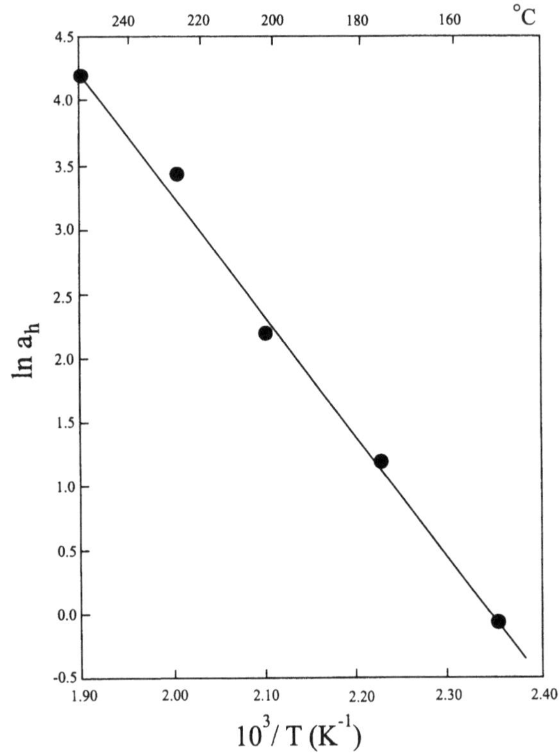

Fig. 126. Experimentally determined dependence of the hydration rate a_h on the absolute temperature T for obsidian from Cerro de las Navajas, central Mexico (after Michels et al. 1983b). In the Arrhenius diagram, which depicts $\ln(a_h)$ vs. T^{-1}, the data points fall on a straight line

tes which have been experimentally determined at higher temperatures, to normal ambient temperatures.

In depth profiles which have been measured by means of sputter-induced optical emission perpendicular to the obsidian surface, Tsong et al. (1978) have proved the hydration to be not merely water impregnation but to be connected with an exchange of material. Along the profiles in the hydrated rind – apart from increased water contents – depletion of the alkali metal concentrations has been found. Evidently the alkali ions are subjected to exchange by hydrogen according to the reaction

$$Na^+(\text{glass}) + 2H_2O \rightarrow H_3O^+(\text{glass}) + NaOH \qquad (63)$$

in the course of which the silicate framework essentially remains intact. Because of the complex substitution processes the chemical glass composition has a strong influence on the hydration behavior (Ericson et al. 1976). Hence, owing to the large geochemical variability of obsidian compositions it is not surprising that obsidians of different sources may differ from each other in respect to the hydration rate by up to a factor 25 ($<1-20\ \mu m^2\ ka^{-1}$). Friedman and Long (1976) found for a limited set of obsidian data a correlation between the hydration rate and the chemical in-

Fig. 127. Dependence of the hydration rate (at 10 °C ambient temperature) on the obsidian chemical composition expressed as "chemical index". (Reprinted with permission from Science; Friedman and Long 1976; copyright American Association for the Advancement of Science)

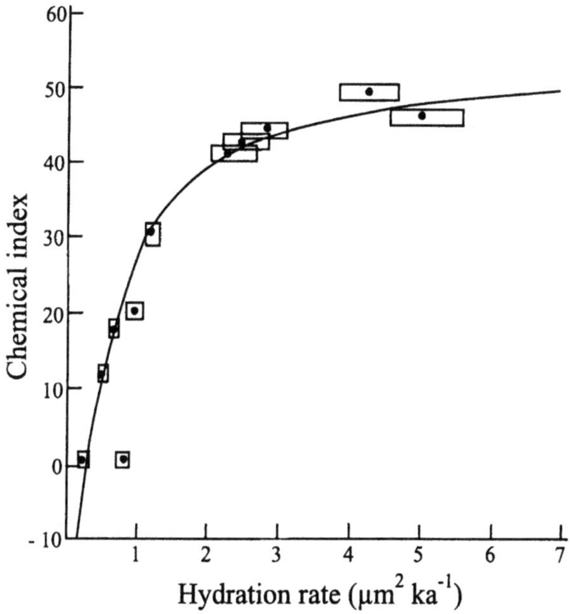

dex $SiO_2 - 45 (CaO + MgO) - 20 \times H_2O^+$, where the contents are inserted in [%] and H_2O^+ assigns the bound water present in the still non-hydrated obsidian material which is set free above 110 °C (Fig. 127). However, one still is far from being generally able to deduce quantitatively the hydration behavior from the chemical glass composition. The strong influence of the glass composition on the hydration rate entails that the glass chemistry has to be known. Commonly in use is X-ray fluorescence analysis. The refractive index or certain optically recognizable features, such as color, transparency and microlite content, are frequently helpful criteria in the differentiation of obsidian types.

Other environmental factors affecting the hydration rate, and worth mentioning in this context, are moisture and the soil chemistry. Although moisture in the air or soil influences the hydration rate, in most soil environments at depths below ca. 0.25 m the relative humidity is constant at 100 % so that its influence may be neglected (Friedman et al. 1994). Also, the soil chemistry may exert an influence on the hydration to a certain, presumably minor, degree.

Rewriting of Eq. (62) yields the age equations

$$t = M^2 / a_h \tag{64}$$

and

$$t = M^2 / (a_{ho} \times e^{-E/kT}), \tag{65}$$

respectively.

The chronometric dating requires – apart from the rind thickness M – the determination of the hydration rate a_h with two approaches existing for the latter:

1. The hydration rate is calibrated for each chemical group, i.e., separately for each source of raw material, and for the particular find site possessing its own temperature history, by means of the same sample materials for which independent age determination are available (above all ^{14}C ages but also K–Ar ages and historical data; Meighan 1976). Such a data set is only valid for the particular sources of raw obsidian and the particular burial site. In this approach the accuracy of hydration dating obviously depends on the reliability of the comparison ages.
2. Another approach is the conduction of the hydration-dating as an independent method by means of the experimental determination of the hydration rate according to Eq. (65). Apart from the rind thickness M the material-specific parameters a_{ho} and E have to be determined for each identified source of raw material (Michels et al. 1983b). In addition an assumption in respect of the ambient temperature is required. The hydration parameters a_{ho} and E are experimentally determined at elevated temperatures, at which the hydration progresses faster according to the reaction kinetics. The obsidian sample is split into several subsamples for obtaining freshly broken surfaces which are subjected to hydration for different time periods (up to several months) at different temperatures (up to 250 °C). For this, hydrothermal pressure vessels are used, in which the samples are tempered in water (Stevenson et al. 1989). Subsequently, the artificially produced hydration rinds are measured.

For the independent dating approach, the history of the ambient temperature for the episode of hydration needs to be known with 1 °C accuracy, if possible, owing to the sensitive influence of T on a_h in Eq. (65). The estimation of the ambient temperature is principally complicated by daily and annual variations. The hydration rate is an exponential function of T, hence the higher temperatures of the fluctuation range bear an overproportional influence on the progress of the hydration. Thus use of the mean air or soil temperature as ambient temperature is not appropriate. To avoid this difficulty the concept of the effective ambient temperature T_{eff} is used. It represents that temperature, which, if it were constant over the whole period of hydration, would yield the observed hydration. It always exceeds the average value of the fluctuation range.

If the hydration rate is not explicitly known a determination of the relative ages of those obsidian samples of equivalent chemical composition and the same find site can be conducted. A temporal placement reflecting only the relative age sequence of the obsidian artifacts is possible on the basis of increasing rind thickness. In the case of refitted tools with surfaces of different ages the sequence in which the individual faces were refitted can be delineated from the rind thickness.

Hydration results in an increase of the refractive index of the glass. In addition, a stress birefringence is developed in the hydration rind. Owing to these altered optical features the rind markedly contrasts to the non-hydrated glass under the optical – especially polarizing – microscope. Nuclear physical analytical techniques exploit the nuclear resonance of the hydrogen atoms founded on the reactions $^{1}H(^{15}N, \alpha\gamma)^{12}C$ or $^{1}H(^{19}F, \alpha\gamma)^{16}O$ (Lee et al. 1974; Lanford et al. 1976). The resonance occurs for the time being at a particular energy of the bombarding ^{15}N or ^{19}F ions. If a hydrated glass surface is bombarded with ions of the resonance energy, only the hydrogen atoms on the surface are affected by the resonance reaction since with increasing penetration depth the energy of the ions decreases and falls below the resonance energy. With continuous increase of the energy of the ion beam the resonance energy in successively deeper regions is reached. In this way it is possible to measure with a high resolution the hydrogen concentration along depth profiles perpendicular to the surface (Fig. 128). The rind thickness still assessable by the nuclear resonance technique ranges around 0.05 µm. This enables the determination of correspondingly young ages.

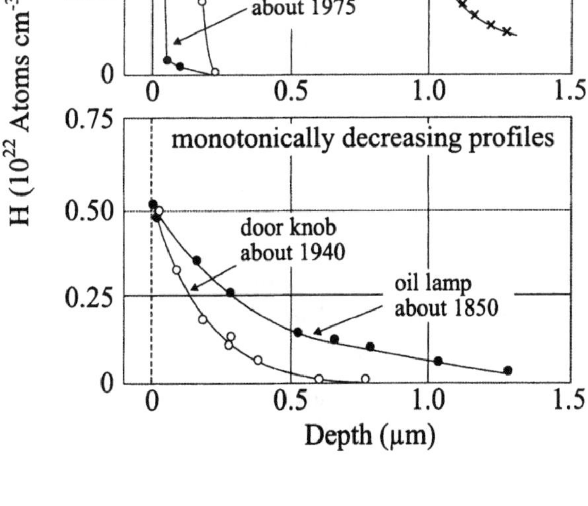

Fig. 128. Hydrogen concentration profiles through the hydrated surface of artificial glasses recorded by the nuclear resonance technique. The hydrogen concentration decreases with increasing depth. Glasses with stepwise decreasing depth profiles (*above*) have to be distinguished from those with monotonically decreasing ones (*below*). Older glasses exhibit thicker hydration rinds. (Reprinted with permission from Science; Lanford 1977; copyright American Association for the Advancement of Science)

8.2.2 Practical Aspects

If practicable, obsidian samples should be taken from depths > 30 cm below the surface to exclude the influences of the daily temperature course as much as possible. In particular, samples which have experienced any warming through sunlight should be excluded. Samples showing marks of subsequent heating equally have to be omitted. To avoid glass surfaces which have been corroded by the soil solution, the surfaces should be checked by scanning electron microscopy. Etching grooves and a rough morphology are characteristic of corroded surfaces.

For the measurement of the hydration rind a tiny fragment from the glass surface is needed, in order to prepare a cross section ~0.05 mm thick and perpendicular to the surface. The measurement is conducted under the polarizing microscope in oil immersion at 500- to 1000-fold magnification. Measurements are regularly conducted at different locations. Precision of ± 0.1 µm is obtainable by means of a micrometer ocular. If routinely done, the time required for sample preparation and measurement is 15 min. A precision of 0.02 µm is achieved with the considerably more elaborate nuclear-physical analysis. These techniques are not routinely applied in obsidian dating and are more suitable for particular questions such as the determination of extremely thin hydration rinds of very young glasses. In principle, they enable a non-destructive analysis, an advantage rendering possible the experimental investigation of the rind growth of an individual sample.

Any information attainable on the chemical composition and the source of the raw material is important to the dating laboratory. For chemical analysis (XRF, AAS, NAA) 100–350 mg sample material is taken from each glass object. In the laboratory the chemical composition of the object is compared to the database to check whether the hydration parameters are already known for a particular obsidian material. The conversion of the hydration rate from a given temperature T to T_{eff} follows

$$a_{\mathrm{h,eff}} = a_{\mathrm{h}} \times e^{(1/T - 1/T_{\mathrm{eff}}) \times E/k} \qquad (66)$$

Because of the sensitive temperature dependence of the hydration rate a precise knowledge of the spatial and temporal temperature variation on the sampling site is of extraordinary importance to the reliability of the ages in question. Several approaches exist for that purpose, but all have to account for the fact that the temporal temperature variations become stronger with decreasing depth (Stevenson et al. 1989):

1. The annual temperature course is recorded by temperature sensors in the layer from which the sample has been taken. The particular hydra-

tion rates are then calculated for sufficiently small time intervals, and an average value over a time period is determined (Friedman and Long 1976).

2. T_{eff} is evaluated by insertion of temperature-sensitive cells at the sampling site, and integrated – if possible – over a period of one year. In these cells temperature-dependent reactions analogous to the hydration take place. The cell constructed by Ambrose (1976, 1980) is based on the evaporation of water.
3. Alternatively, T_{eff} may be estimated from the mean annual temperature and the fluctuation range of the air temperature, according to the equation [Eq. (69)] given by Lee (1969). All these methods consider at best the annual temperature course, but not the long-term temperature fluctuations, which reached deeply into the subsurface during the Glacial period.

8.2.3 Application

Obsidian. Obsidians of the same lava flow commonly show uniform chemical composition resulting in identical hydration rates under the same temperature conditions. Obsidian materials from different sources are recognizable and identifiable on the basis of their chemical composition. Instead of chemically analyzing the obsidian materials, Friedman and Obradovich (1981) have tried to exploit the refractive index n for the determination of the geological source. In practice, usually just a few sources have to be taken into consideration so that macroscopic criteria may already be sufficient for the identification. The hydration method of dating offers the great advantage of permitting the investigation of an extensive series of samples without too much expense and waste of time. Almost all hydration measurements have been conducted on obsidian – both from geological deposits and archaeological find complexes. Extensive databases are available comprising thousands of individual measurements, e.g., for North and Meso America (Meighan and Scalise 1988).

The dating of volcanic eruptions is of interest to geology. Friedman and Obradovich (1981) determined ages between 1200 a and 1 Ma on primary glass surfaces from several obsidian flows in western North America; these were in agreement with the ^{14}C and K–Ar ages. Inherent difficulties are caused predominantly by the strong temperature fluctuations to which the surfaces are exposed. Dark obsidians may grow hot during exposure to the sun. Using obsidian and pumice samples, Friedman (1977) succeeded in dating several successive eruption phases of the Newberry Crater in Oregon. Individual samples could be classified into discrete groups according to their different rind thicknesses. The thickest surface rind dated the consolidation of the lava, and thinner ones along cracks dated the cracking, caused by tremors resulting from eruptions within the closer vicinity. This approach enabled the detailed chronological reconstruction of the eruption history of the Newberry Crater covering the last 6700 a.

If an obsidian flow is transversed by glaciers, tension cracks evolve within the obsidian along which hydration sets in (Friedman et al. 1973). In the Obsidian Cliff of Yellowstone Park rinds of two different thicknesses, 7.9 or 14.5 μm, occur as tapestries within the cracks (Fig. 129). However, on the obsidian surface the rind thickness comprises 16.3 μm. The latter is supposed to be related to formation of the obsidian 176 ka ago (K–Ar). The hydration rinds along the cracks represent records of glaciations ca. 40 and 140 ka ago. Obsidians that are constituents of till deposits show development of tension cracks during glacial transport. Soon after deposition of the moraine material, these cracks are affected by hydration. Friedman and Trembour (1978), working on obsidian till of the terminal moraines Pinedale and Bull Lake of the Yellowstone region, proved the occurrence of two Glacial episodes ca. 40 and 140 ka ago, which coincided temporally with the glaciations of the Obsidian Cliff mentioned above.

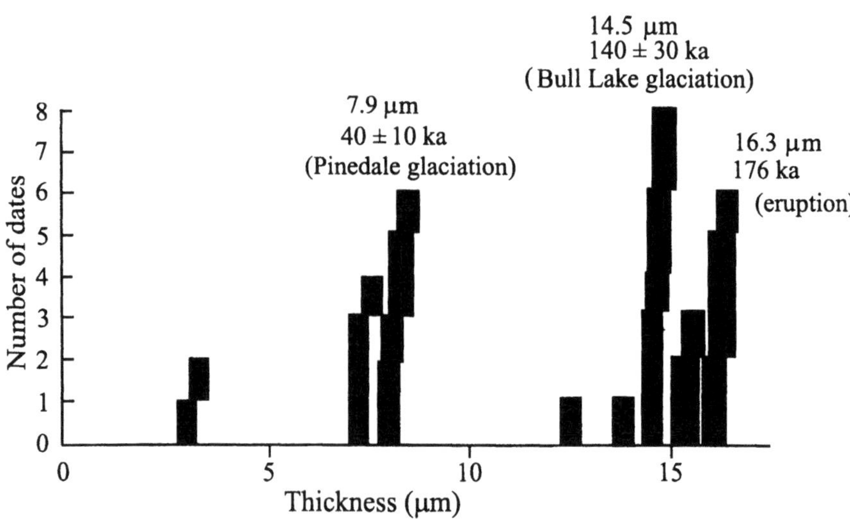

Fig. 129. Thickness of hydration rinds along cracks in obsidian from Obsidian Cliff, Yellowstone Park. The cracks with rind thickness of 14.5 and 7.9 μm were formed by the ice load 140 and 40 ka ago, respectively, and allow the dating of these two glaciations. The obsidian flow is 176 ka (K–Ar) old. (Reprinted with permission from Science; Friedman et al. 1973; copyright American Association for the Advancement of Science)

If the hydration rate is unknown, the thickness of the rinds can nevertheless be used to answer questions of relative chronology in archaeology. Given a stratigraphically disturbed accumulation of obsidian artifacts of identical chemical composition embedded in a sedimentary environment of uniform ambient temperature, the rind thickness yields key information on successive stages of tool production. Larger non-productive periods are also discernible (Michels and Tsong 1980). A project in East Africa shows the difficulties which may arise in chronometric obsidian dating. The East African rift system is characterized by young volcanism. Associated with it are many occurrences of obsidian which have been exploited for tool production since Acheulian times. On several sites at Prospect Farm in Kenya, Michels et al. (1983a) sampled 138 obsidian artifacts from Middle and Lower Paleolithic and Neolithic layers and additionally from adjacent geological sources of obsidian raw material. Two sources, A and B, were identified by AAS analyses. These analytical results allowed the artifacts to be grouped and correlated with these two sources. The correlation could also be carried out using visual features, such as transparency and color. The experimentally determined hydration rates for the two obsidian groups were 6.16 (type A) and 14.80 $\mu m^2\ d^{-1}$ (type B) at 200 °C (Fig. 125). The effective hydration temperature (= 18.62 °C) was calculated according to Eq. (69) for $T_a = 18.2$ °C and $R_T = 2.4$ °C (mean August temperature minus mean March temperature). The corresponding hydration rates are 2.60 and 16.14 $\mu m^2\ ka^{-1}$, respectively. In the scanning electron microscope obsidian type A showed numerous etching grooves indicative of surface weathering. For this reason, this type had to be partly omitted for dating purposes. In addition, one and the same surface showed different ages depending on the spot where the rind thickness was measured. These features were caused by corrosion grooves and local rind defoliation, leaving patches with thinner rinds alternating with areas still carrying the relics of the original surface. The hydration ages of the obsidian artifacts from Prospect Farm yielded several phases of settlement between 120 ± 1.7 ka and 2570 ± 100 a. The presented errors of 1–2% do not account for long-term temperature fluctuations.

Obsidian artifacts which were affected by fire during or subsequently to their production show fine temperature-induced cracks along which hydration sets in. Hence the rind thickness furnishes a tool for dating the heating event. Suzuki (1973) combined hydration and fission track dating to determine the hydration ages of heated obsidian artifacts of pre-ceramic archaeological sites in Japan. One obsidian tool belonging to the post-Jomon culture of Shirataki on Hokkaido (Fig. 130), for example, demonstrates refitting expressed by the occurrence of different rind thicknesses. Only the rind thickness of 1.9 µm is in agreement with the associated ob-

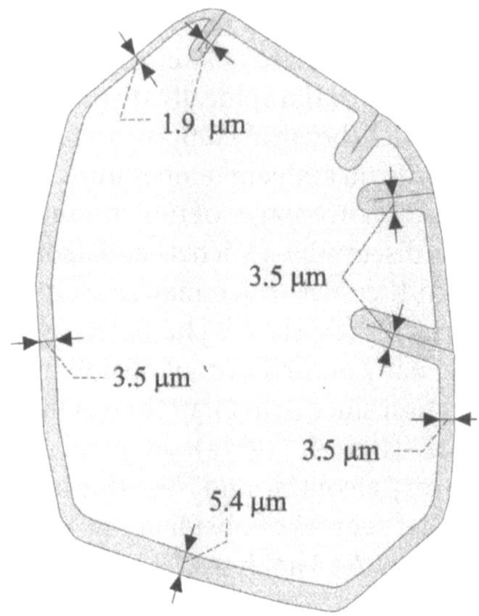

Fig. 130. Schematic of the hydration rinds of an obsidian artifact from Shirataki, Hokkaido, Japan (after Katsui and Kondo 1976). The difference in rind thickness indicates the age sequence of the surfaces and cracks. The stone implement was obviously twice refitted and reused. The youngest surface with 1.9 μm rind thickness corresponds to the associated obsidian artifacts from the post-Jomon find horizon; the thicker rinds reflect earlier cultural phases

sidian finds. The surfaces showing thicker rinds are indicative that this artifact experienced earlier production and reuse.

Hydration rinds were found also on basalt glass (Morgenstein and Rosendahl 1976). These authors reported hydration ages from Hawaii for both, basalt flows (19,086 ± 2180 to 9260 ± 1529 a) and flakes (down to ages as low as 144 ± 14 a), for which the hydration rate could be calibrated with historically known basalt eruptions.

Artificial Glass. The hydration process of man-made glasses apparently differs from that of obsidian, as can be deduced from the observation that they frequently do not exhibit sharp diffusion fronts, which otherwise would be clearly recognizable with the optical microscope. Furthermore, owing to their young ages, the hydration rinds of artificial glasses are frequently of smaller width than the wavelength of light, and are thus not visible by optical microscopy. According to Lanford (1977) the nuclear resonance technique enables the measurement of the hydrogen content along depth profiles, even of hydration rinds as little as 0.05 μm thick. In artificial glasses of known age two types of depth profiles were found, a step-like and a monotonically decreasing one (Fig. 128). Both glass types show larger rind thicknesses with increasing age. The correlation between age and rind thickness (Fig. 131), with hydration rates around 3.3 $\mu m^2\ ka^{-1}$, shows that age determination is possible even of glasses only a few decades old. Strongly variable glass chemistry is problematic and requires the eva-

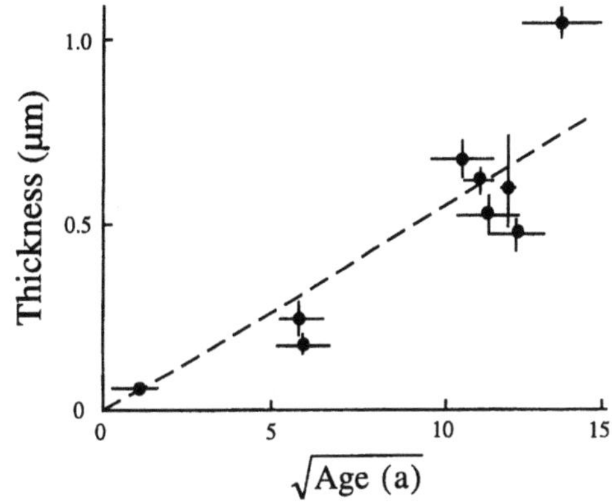

Fig. 131. Dependence of thickness of the hydration rind on the age t of artificial glasses. The slope yields $a_h = 3.3\ \mu m^2\ ka^{-1}$. (Reprinted with permission from Science; Lanford 1977; copyright American Association for the Advancement of Science)

luation of the hydration rate for each individual glass sample. Other complications arise from strong temperature fluctuations and the soil corrosion of glass finds. So far, precise age determinations of man-made glasses have not yet been accomplished by hydration. But it would be reasonable to use this method for authenticity testing of glass objects. This technique might also be applicable to ceramic glazes and glassy archaeometallurgic slag.

8.3 Glass Layer Counting

This method is founded on the weathering-induced formation of thin lamellae on the surface of artificial glasses during their soil burial. The lamellae are responsible for the colorful phenomenon of *iridescence*. Provided one single lamella forms per year, the age can be determined by counting the layers under the microscope. The age yields the period during which the glass has been exposed to weathering or more precisely the duration of soil burial. Experience hitherto, however, shows that owing to frequently occurring irregularities only few glasses are suitable for this approach and that the counting results do not always correlate with the age.

Although the weathering phenomenon of iridescence was already described in detail in the last century (Fowler 1880), only a few decades ago was the attempt made to exploit this phenomenon for age determination (Brill and Hood 1961), and since then this approach has been practiced only a few times. As long as the physical and chemical bases of this method have not been clarified, the method has no perspective of being successfully applied to the dating of glasses.

8.3.1 Methodological Basis

The surfaces of artificial glasses which experience any long-term soil burial frequently display rather different weathering features. Some glasses show a golden or multicolored iridescence, whereas others carry a weathering crust and yet others appear as if they had just been subjected to a fire polish (Bezborodov 1975). The intensity of weathering depends, apart from the age, on manifold factors such as chemical composition and production technique of the glasses, as well as extrinsic influences. Extrinsic factors may be humidity, soil chemistry, temperature and bacteriologic effects.

The iris consists of a porous, maximally a few millimeter thick layer of surface-parallel lamellae. Thin sections cut perpendicular to the glass surface display numerous superimposed glass layers of 0.5–20 μm thickness (Fig. 132). This layered structure, consisting of decomposed glass separated by reaction remnants of different refraction, causes the interference phenomenon of iridescence observable on old glass surfaces (Brewster 1863). Chemical analyses prove the weathering rind to be strongly hydrated (up to 23% water), as opposed to the glass matrix, and depleted in alkali and alkaline earth elements but enriched in iron, aluminum, calcium and titanium (Geilmann 1956; Newton 1971). The rhythmic composition associated with the layered structure can be well proved by electron microprobe analyses.

So far, the processes responsible for the formation of the lamella structure are still largely unclear. In particular, the question remains unanswered whether the growth of the lamellae proceeds from inside to outside or in the reverse direction. Here, one assumes a periodic leaching of the alkali metals out of the glass matrix starting from the surface or from centers, and the rhythmic character is attributed to changing physicochemical conditions. Geilmann (1956) describes two types of weathering in Roman and medieval glasses (Fig. 133). *Skin weathering* is characterized by the formation of surface-parallel lamellae causing the magnificent interference colors. In the case of *groove weathering*, deterioration starts from favorable points and results in hemispherical ring systems or cone-like grooves filled by convex glass layers. Geilmann (1956) suspects that the lamellae formation obeys a process bearing a certain analogy to the rhythmic precipitation of the Liesegang rings in gels. According to Brill and Hood (1961) the individual layers are formed by annual fluctuations concerning especially the temperature and moisture in the enclosing soil, a mechanism resulting in the growth of additional lamellae into the inner glass at the expense of the intact glass matrix. Accordingly, the lamellae glass

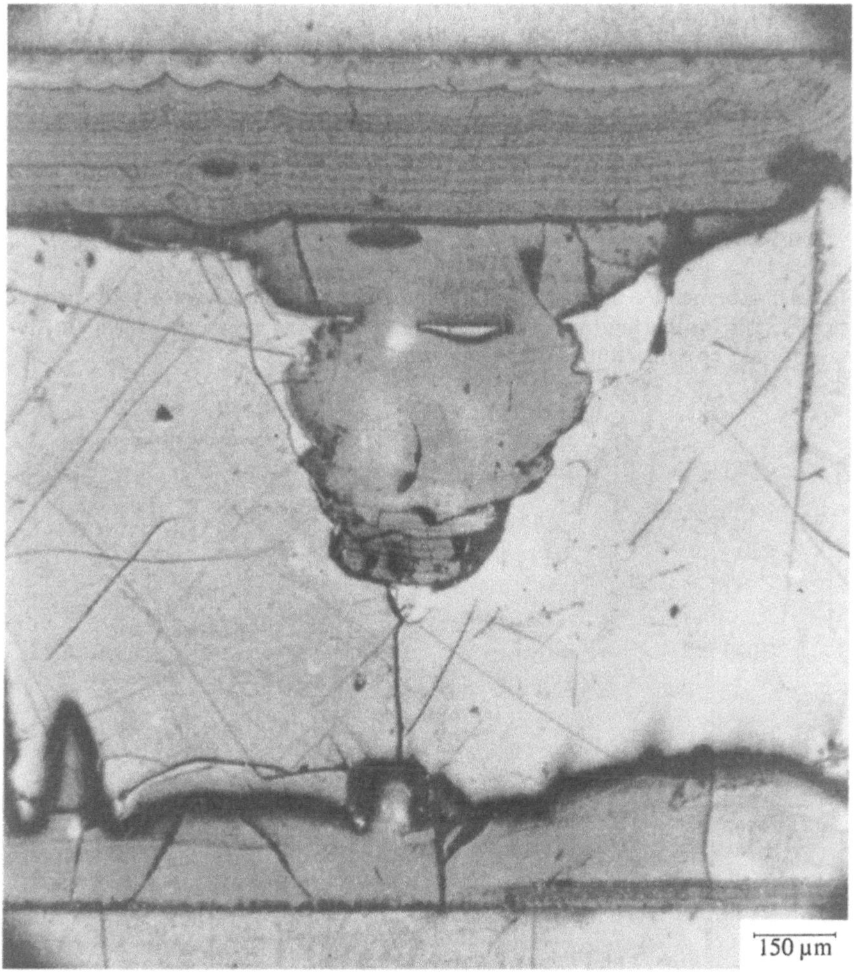

Fig. 132. Weathering-induced glass lamellae at the surface of a window glass, which was soil-buried over a period of 288 a. The micrograph was taken from a thin section perpendicular to the surface (from Newton 1971). On the surface 65 surface-parallel lamellae are displayed. The low-lying corrosion pocket contains further 220 lamellae. The bottom surface of the glass shows considerably fewer glass layers

structure reflects the seasonal change of active and non-active weathering episodes. As a proof of this hypothesis, these authors refer to the counting results of glass layers on glasses of known burial period. Various glass objects of colonial time show agreement between the number of lamellae and years of burial (Fig. 134). Correlations were proved even for some medieval and ancient glasses. The fact that among the glass objects mentioned some stem from shipwrecks is surprising because there should be no differences

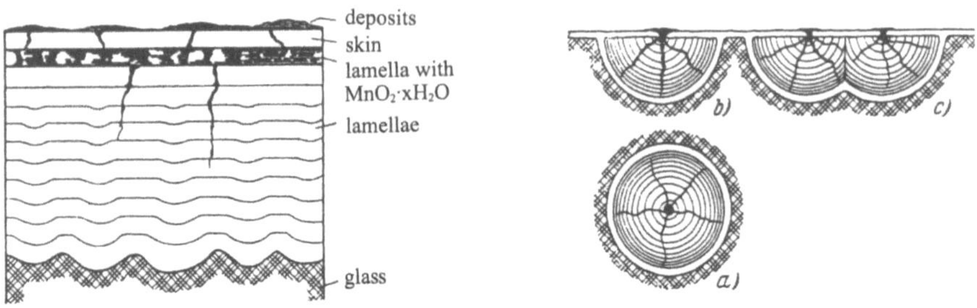

Fig. 133. Scheme of a microscopic section through the glass surface with surface-parallel skin weathering (*left*) and hemispherical groove weathering (*right*: *a* plan view; *b* and *c* cross section). (After Geilmann 1956)

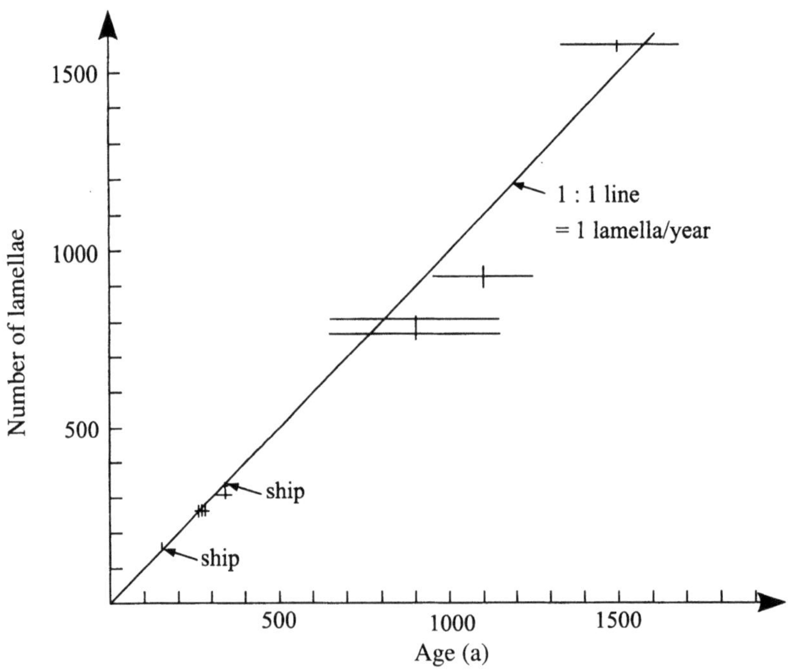

Fig. 134. Glass layer counting on glasses of known age. The diagram shows the number of lamellae vs. the burial time. The data display scattering around the 1:1 straight line with a lamellae formation rate of one per year. (Reprinted with permission from Nature; Brill and Hood 1961; copyright Macmillan Magazines Ltd.)

in seasonal fluctuations in moisture and temperature on the ocean floor. Complications also arose from the frequent absence or incompleteness of the weathering crust and the variation in the number of lamellae on one and the same sample. On the basis of his observations on window glasses which had been buried in the soil for 288 years (Fig. 132), Newton (1971) doubted that the formation of layers followed an annual rhythm, because even if layers exist, their number is almost always smaller than the duration of the burial period. Less or more than one lamella per year may be formed. In addition, the observation that the lamellae do not necessarily express a surface-parallel orientation contradicts the idea of an externally controlled periodic formation mechanism. Newton (1971) reported a laboratory experiment in which sodium-rich glass was exposed to CO_2-free water vapor in an autoclave for 4 h at 144 °C and 4 bar, resulting in the formation of ten lamellae. This result led him to the conclusion that the rhythmic lamella structure was not caused by a periodically changing physico-chemical process, but, on the contrary, by a continuously proceeding one. The formation rate of the lamellae is said to be slower than one lamella/year in most cases, and a formation rate of one lamella/year occurs by chance only under favorable temperature conditions.

8.3.2 Practical Aspects

Generally, only a restricted number of glass finds is suitable for this method. The glass should possess a sufficiently thick weathering crust. Only a small fragment comprising the complete crust down to the intact glass is needed for dating. Polished sections are prepared perpendicularly to the glass surface. The counting of the glass layers is conducted under the microscope using at least a 100-fold magnification. Because of the fact that the surface may carry different numbers of lamellae at different locations, several areas have to be investigated where those showing a surface-parallel lamella structure are favorable.

8.3.3 Application

Artificial Glass. Brill and Hood (1961) studied glass from three origins: seven glasses, manufactured during the colonial period and collected from shipwrecks and excavations in North America; three Islamic glasses from Nishapur, Iran; and one Byzantine glass lump from Sardes, Turkey. These authors counted the glass layers and found good agreement between their results and the known or at least estimated burial duration (Fig. 134). They evaluated the concordance as an argument for the formation of one weathering lamella/year. Although this original investigation was carried out more than 30 years ago, almost no further convincing data have been added. On the contrary, the formation rate of one lamella per year has been doubted by Newton's studies (1971) on English window glasses which had experienced a conflagration and had subsequently been buried for 288 years in the soil. Only a restricted area of one single glass fragment showed a lamellae number (285) in agreement with the age. This area consisted of 65 flat, surface-parallel layers situated directly beneath the surface and 220 domed layers in an underlying corrosion pocket (Fig. 132). Everywhere else the layer number was smaller. Many old glass surfaces do not show any layering at all. Other glasses, however, are completely weathered with no intact glass being left. According to Brill (1969), Roman and Byzantine glasses are usually too weathering-resistant and do not form suitable crusts. Presumably, the arid climatic conditions prevent ancient Egyptian glasses from forming suitable crusts. Finally, Mesopotamian glasses are mostly fully weathered, which makes them unsuitable for dating. Brill (1969) evaluated medieval and modern (until the beginning of the 18th century) glass as the most favorable for glass layer dating. Unfortunately, the suitability of this dating method for man-made glass cannot yet be judged because too few studies are available; in addition, these have led to contradictory results. The possibility of applying glass layers for authenticity testing of ancient glass objects seems to be realistic.

8.4 Fluorine Diffusion

This method makes use of the diffusion process of fluorine from the groundwater into the rock surface during the burial period. The fluorine diffusion front progresses into the rock matrix with increasing burial time. This correlation can be utilized for the determination of the burial period, i. e., the time over which the surface has been exposed to weathering. For this the thickness of the fluorinated rind needs to be measured. Principally, this technique can be used to date the manufacture of stone implements. However, the relatively few available case studies on silicate-bearing rocks (Taylor 1975) at best speak well for a method of relative dating. Also reasonable would be its application to the chronological positioning of moraine material.

8.4.1 Methodological Aspects

Fluorine exists in the groundwater dissolved as F^--ion rarely exceeding 1 μg g^{-1}, although in extraordinary cases reaching up to 60 μg g^{-1}. Generally, groundwaters in active volcanic regions are found to bear increased fluorine contents. Starting from the rock surface, fluorine diffuses towards the rock core forming a surface-parallel diffusion front, whose thickness gradually increases with burial time and can reach several µm. Such fluorinated rinds were observed by Taylor (1975) in various siliceous rocks, namely trachyte, phonolite, dacite, arkose, quartzite and hornfels. The fluorine contents of the crust range from 100–900 µg g^{-1}, which is many times the contents of the rock core. Several depth profiles are distinguishable on the basis of their shape: saturated, unsaturated and more rarely reversed ones. The profile types are controlled by the fluorine content of the groundwater and the presence of hydrated minerals on the rock surface.

Proving the fluorination of such rinds did not become possible until the nuclear resonance technique could be employed, allowing the recording of low fluorine contents with a precision of 8% and a depth resolution of 0.05 µm. The technique is based on the nuclear reaction $^{19}F(p, \alpha\gamma)^{16}O$, which is used in the reverse direction for hydrogen analysis in the hydration rinds of glasses (Lee et al. 1974). For the fluorine analysis the sample surface is bombarded with protons. The nuclear resonance sets in at 0.83 MeV. The γ-radiation generated by the excited ^{16}O-nucleus is proportional to the fluorine concentration – in fact, at the depth where the protons possess an energy just equal to the resonance energy. With gradual increase in the energy of the proton beam the resonance energies of successively deeper regions are reached. The variation of the proton energy enables high-resolution fluorine measurements along depth profiles at a right angle to the rock surface (Fig. 135). Because of interference effects at higher proton energies, the maximum rind thickness analyzable by this technique is restricted to 1.2 µm. This restriction limits the upper datable age range to 10 ka. Therefore an analytical technique for depths beyond 1.2 µm would be advantageous.

The essential prerequisite for dating is a regular increase in the rind thickness with the age of the rock surface. For a series of chipped stone tools from southern California of known ages ranging from 1 to 8 ka, Taylor (1975) presumed to recognize such a correlation quantitatively (Fig. 136). Apart from a broader data basis, knowledge of the mechanism of fluorine diffusion in polymineralic rocks would be essential. A favorable diffusion path probably follows grain boundaries. Furthermore, basic research that deals with the most crucial questions is still lacking; such questions include to what extent the diffusion process is controlled by the petrographic features of the object under investigation, by the fluorine content of the groundwater and by the ambient temperature. Consequently, the research on fluorine diffusion has not yet achieved a methodologically advanced

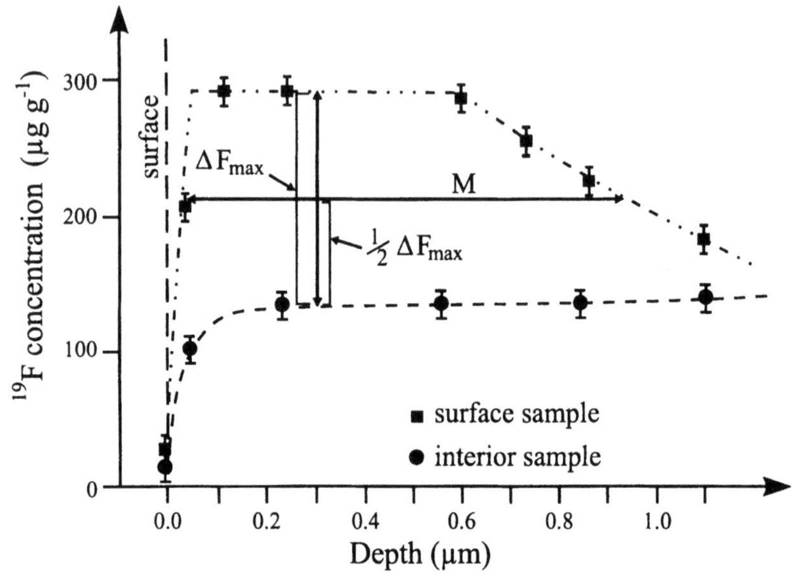

Fig. 135. Depth profile of the fluorine content perpendicular to the surface (*upper curve*) of a trachytic stone artifact from San Diego, California (after Taylor 1975). The fluorine was analyzed by the nuclear resonance reaction $^{19}F(p, \alpha\gamma)^{16}O$. For the measurement of the depth profile the energy of the proton beam was continuously increased. The depth of the diffusion front is defined as the thickness M at which the maximum difference ΔF_{max} of the fluorine content between the surface profile and a sample from the interior reaches half its value

enough state to be sufficient to enable quantitative analysis; at best, this method allows a rough relative dating. A wrong age estimation could arise from the loss of the μm-thin rind as a result of abrasion of the rock surface. However, the fluorinated rind was missing in only a single case among 40 studied rock surfaces, a fact which made Taylor (1975) judge this risk a minor one.

8.4.2 Practical Aspects

Siliceous rocks exposing chipped faces which have not suffered any abrasion since burial are suitable for fluorine diffusion dating. During burial, the faces must have been permanently exposed to soil moisture or the groundwater. Drill cores of ca. 12 mm in diameter and 10 mm length are taken from the surface. For comparative measurements a sample from the rock core is needed, too. Subsequently, the cores are chemically cleaned. These samples are bombarded perpendicular to their surfaces by protons

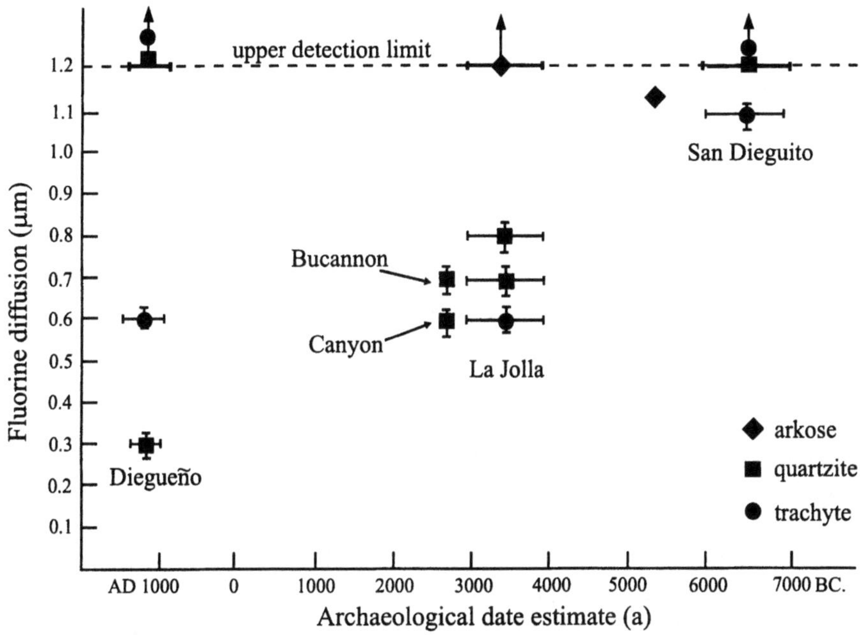

Fig. 136. Relationship between the thickness of the fluorine diffusion front and the archaeological age of chipped stone tools near San Diego, California (after Taylor 1975). The nuclear resonance technique employed to determine the fluorine content has an upper measuring limit at 1.2 µm

in the particle accelerator for fluorine analysis. For the measurement of the fluorine concentration along a depth, the energy of the proton beam is continuously increased from ca. 0.8 to almost 1 MeV. The proton beam affects a ca. 10 mm^2 large area on the rock surface.

8.4.3 Application

Stone Artifacts. Once a chipped stone tool or flake is embedded in the soil, fluorine commences to diffuse from the surroundings into the fresh surface. Hence the thickness of the fluorine diffusion rind should bear information on the exposure time of that face. This method seems to be applicable to surfaces of siliceous rocks, such as volcanic and plutonic rocks as well as arkose, as has been demonstrated by Taylor's investigations (1975) on chipped stone tools. Those artifacts stemmed from the vicinity of San Diego, USA. Rind thicknesses < 1.2 µm appear to show a correlation with their conceptional archaeological age, as deduced from ^{14}C dates, ranging from 1000 AD and 6000 BC. Neither in a qualitative nor quantitative respect does this insufficient data corpus permit drawing any conclusions on the practicability of fluorine diffusion for the age determination of chipped rock surfaces.

8.5 Calcium Diffusion

Calcium diffuses along the interface between mortar and bricks, from the calcium-bearing cement into the clay-bearing building material. The calcium diffusion front penetrates deeper into the bricks with increasing age. This process allows the deduction of the age of the masonry from the thickness of the diffusion rind. Although this technique is still in the early stages of development, it will be referred to within this context for completeness. This technique developed as a by-product of investigations of clay-bearing materials and their usefulness as sealing material for deposits of nuclear reactor waste (Waddell and Fountain 1984). So far, the few measurements conducted on fired and air-dried bricks, dating back to 4000 a, do not permit an evaluation of the suitability of this method for the dating of masonry. Among other reasonable potential applications, ceramic vessels decorated with calcium carbonate-bearing white pigments are worth considering.

8.5.1 Methodological Basis

Along the surfaces of contact between clayey materials and calcium-bearing cements, dissolved calcium from the cement diffuses into the clayey material as a result of the marked concentration gradient. Such interfaces are found in masonry constructed from air-dried or fired clay bricks which were erected using calcium-bearing binders. Binders coming into consideration are mortar, whitewash, cement and stucco containing $CaCO_3$ or $CaSO_4$ as major components. On clay bricks of different ages (70 – 4000 a) Waddell and Fountain (1984) found Ca-diffusion fronts 7 – 175 µm in thickness. The rind thickness is defined as the distance measured perpendicularly to the surface of contact along which a gradient in the Ca-content is observed. A continuous increase in the Ca-content of up to 40 % is found within the diffusion rind. The Ca-content is measured by electron microprobe analysis.

Processes involved in the calcium transport are: (1) decomposition of the calcium-rich constituent of the binder; (2) diffusion of Ca^{2+} ions from

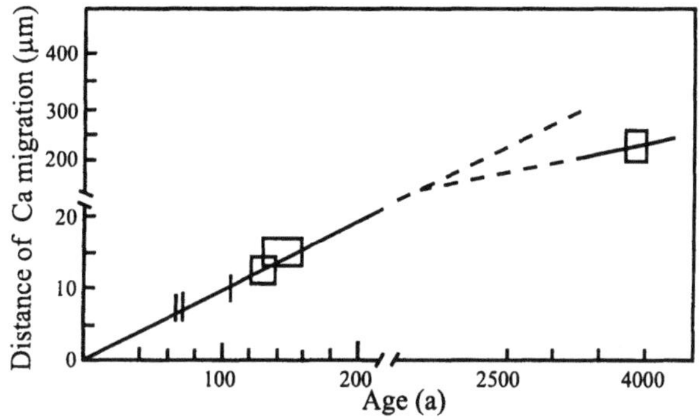

Fig. 137. Relationship between the thickness of the calcium diffusion front and the age of clay bricks, where the cement is still attached to the brick surfaces. (After Waddell and Fountain 1984)

the binder into the clay; and (3) absorption of Ca^{2+} ions in the clay in exchange with Na^+ and K^+ ions. The calcium diffusion is not a constant process. The diffusion rate is reduced by calcium absorption. In addition, impermeable inclusions in the clay, e.g., quartz grains, impede the diffusion, thus resulting in a curved diffusion path. Furthermore, the diffusion rate is delayed by increasing mineral precipitation in the pore spaces. These time-dependent kinetic complications lead to the more appropriate concept of an "apparent" diffusion coefficient rather than a "real" one. Apparent diffusion coefficients, evaluated from the calcium gradient in the diffusion rind, comprise $\sim 1\ \mu m^2 a^{-1}$ and do not significantly differ from one another for the investigated clay mineral.

The thickness of the calcium diffusion rind grows with the age of the bricks – at least in the case of the investigated samples (Fig. 137). In the age range < 200 a the slope is quasi-linear but the 4000-a-old sample falls beneath the extrapolated straight line. The available data are too poor to allow the mathematical description of the growth curve and an evaluation of its suitability for dating.

8.5.2 Practical Aspects

Both fired and air-dried bricks seem to be suitable for calcium diffusion dating – provided they have been permanently in contact with the cement. Sample processing and calcium analysis are relatively straightforward procedures. The interface between the clayey bricks and calcium-rich cement is surveyed for existing intact areas under the binocular. Suitable areas of sizes of 5 × 5 × 9 mm are sawed out. Subsequently to mounting in synthetic resin the sample is ground and polished perpendicular to the surface of contact. The calcium concentrations are measured along depth profiles by the electron microprobe technique. The diffusion rind is measured at several points to eliminate any influence of inhomogeneities. Provided the measurement of the diffusion rind is done with a precision of ± 2 μm, the obtainable accuracy of the age evaluation should be ± 25 a according to rather optimistic assumptions by Waddell and Fountain (1984). However, the existence of a uniform growth curve for different samples is more than questionable. Experience with other chemical dating methods makes it more likely that the diffusion rates vary from one sample to the other depending on the chemical and structural composition of the clayey material and the environmental conditions such as humidity, soil chemistry and temperature. Fundamental research still needs to be conducted prior to any dating applications.

8.5.3 Application

Bricks. Principally suitable for this method are mud bricks and fired bricks of which the surfaces have been permanently covered by a calcium-bearing binder – cement, plaster or stucco. Not the material itself is dated but the contact between brick and binder, i. e., the erection of the masonry. So far, only a few measurements exist on samples of known age: one unburned mud brick with thin lime coating from Tel Ifshar, Israel, and several fired clay bricks with mortar from North America (Fig. 137) (Waddell and Fountain 1984). Accordingly, the calcium diffusion could become a sensitive dating method for brick walls and ruins. However, the practicability of the method has not yet been sufficiently tested.

8.6 Cation Ratio

Characteristic weathering rinds are formed on rock surfaces which are exposed to arid or semi-arid climatic conditions. This so-called desert varnish, consisting of paper-thin, black–brown crusts, experiences with time chemical alterations which above all find their expression in the relative cation concentrations. This phenomenon allows the deduction of the age of the rock surface from the ratio of specific cations – in particular (K + Ca)/Ti – and thus bears key information on the denudation history and the morphogenetic landscape development. Strictly speaking, the cation method furnishes merely a lower limit (*terminus ante quem*) for the age of the surface, since it can never be fully excluded that the formation of the still existing desert varnish has started at a later state.

The cation method looks back to a short developmental period (Dorn and Whitley 1983). Like any other chemical dating technique, the cation method requires calibration, since the reaction rate of the cation exchange depends on petrographic features and environmental factors. Common calibration approaches include K–Ar ages of the host rock, uranium series ages of calcareous crusts and ^{14}C ages of the desert varnish itself. The calibration curves are valid only for special host rocks and for regions of comparable climate.

The applicability of cation ratio dating is restricted to such regions where desert varnishes actually occur. These also include dry high mountainous areas, in addition to deserts. For the formation of desert varnish the rock surfaces must remain stable over extended time periods and must not undergo any erosion or denudation. The method has been successfully applied to pieces of rock, rock faces, stone tools, to moraine blocks and to petroglyphs engraved upon rock surfaces already covered by desert varnish, and giving rise to a new episode of desert varnish formation on the freshly engraved surface. The age range, which can be estimated by the cation method, comprises several ka to Ma. Owing to fundamental uncertainties, the reliability of rock varnish cation ratio dating has been questioned (Bierman and Gillespie 1994).

8.6.1 Methodological Basis

Prominent constituents of desert varnish are oxides/hydroxides of iron and manganese; further components are the clay minerals illite and montmorillonite together with organic matter. Bacteria, lichen, algae and fungi participate in the formation of desert varnish (Krumbein 1971). The rind thickness of the desert varnish commonly comprises less than 100 µm. Its color varies from dark-brown to black. Usually, it is the clay fraction of the desert varnish that works as cation carrier. The clay has been blown out from soils by the action of wind, forming a dust which became redeposited onto the rock surface. In the course of time the cations of this clay fraction are affected by selective leaching processes. The diverse cations participating in these exchange processes possess different mobility. K^+, Na^+, Ca^{2+} and Mg^{2+}, for instance, are more readily leachable than Ti^{4+}, i.e., the ratio of mobile to immobile elements such as $K+Ca/Ti$ decreases with increasing age of the varnished surface. The selective leaching mechanism of the cations depends on environmental factors such as micromorphology, microclimate, lichen growth and others. However, measurements by Bierman and Gillespie (1994) on varnished chert from rock outcrops, artifacts and cobbles in the Mojave Desert do not confirm the relationship between the cation ratios and the relative ages of the varnished surfaces, probably precluding the simple chronometric interpretation of cation data.

The mobile cations K^+ and Ca^{2+} are analyzed for the age determination, since these cations can be readily measured by proton-induced X-ray emission (PIXE). Advantageous is the application of scanning electron microscopy (SEM) with energy dispersive wavelength X-ray fluorescence analysis (EDAX) as a technique enabling the non-destructive investigation of the desert varnish. Furthermore, SEM–EDAX-analysis may yield depth profiles of the cation concentrations through the desert varnish by increasing the acceleration voltage of the electron beam in steps of 5 keV commencing with 10 keV (Harrington and Whitney 1987). The minimum value of the K+Ca/Ti ratio measured along the profile is considered to be the basal layer of the desert varnish and used for the age calculation. The fact that the age determination requires only the cation ratio and not the absolute cation contents enables a high measurement precision with relatively low technical expenditure.

The cation method is most suitably calibrated by AMS-^{14}C data of the same desert-varnish sample (Dorn et al. 1986). For the dating range beyond 20 ka K–Ar ages of young volcanites (Dorn and Whitley 1983) and uranium series data of calcareous crusts are used for establishing the calibration curve (Harrington and Whitney 1987). Concerning volcanites, the assumption must be made that a stable rock surface carried the weathering crust ever since the solidification of the lava took place. The calibration diagrams are linear plots of the cation ratio vs. the logarithmic time axis, which result in straight lines. An example of a calibration curve is given in Fig. 138. It was established for the Mojave Basin, East California, on the basis of AMS-^{14}C and K–Ar ages. Such calibration curves are valid solely for restricted areas where constant ion exchange rates can be assumed as a result of rather uniform climatic conditions. But strong variations of the reaction rate may also occur within climatically closed systems, if any microclimatic deviations (niche, shady side, micro fauna, moisture, and others) are prevalent.

8.6.2 Practical Aspects

The choice of samples requires special care owing to the fact that the reliability of the cation ratio ages is limited first of all by the composition of the sample, but only secondarily from the analytical precision. Localities showing enhanced growth of microflora have to be omitted because of the risk of leading to anomalously young ages. The original crust of desert varnish may have been lost because of the effect of organic acids or wind abrasion. Mineral inclusions of the host rock can distort the real cation contents of the desert varnish. For this reason it is desirable to conduct a

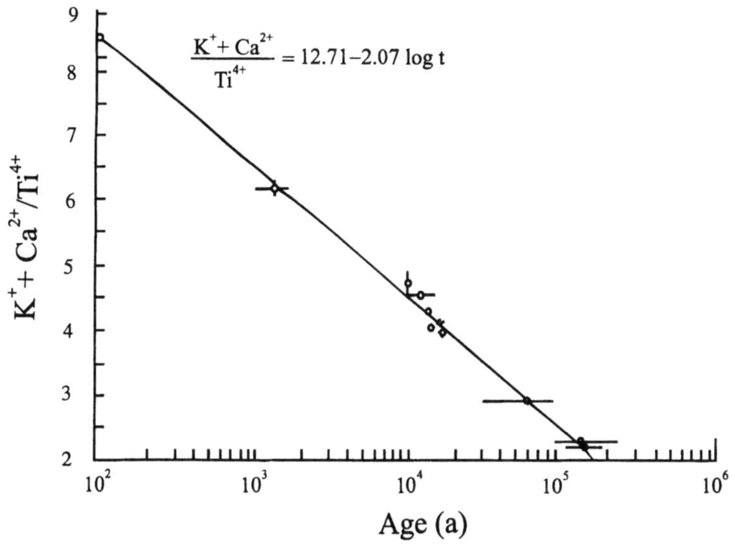

Fig. 138. Age calibration curve for cation ratios in desert varnish from the Mojave Basin. The $(K^+ + Ca^{2+})/Ti^{4+}$ ratios were measured on samples of known K–Ar or AMS-^{14}C ages. In order to determine the initial ratio (data point at *upper end*), recent aeolian dust deposits were analyzed. In the semi-logarithmic diagram the data points fall onto a straight line. Such calibration curves are only of regional validity. (Reprinted with permission from Science; Dorn et al. 1986; copyright American Association for the Advancement of Science)

thorough SEM–EDAX investigation on the same sample of desert varnish (Dorn et al. 1987). The cleaning of the rock surface is done by scrubbing in distilled water. For PIXE analysis the desert varnish is mechanically scratched off from the rock surface under the binocular using a tungsten carbide needle. The sample preparation for SEM–EDAX analysis of the cations is considerably less time-consuming. This approach only requires sampling a flat drill core of 2 cm in diameter from the rock surface and the host rock. The surface can be directly analyzed after it has been carbon-coated. Because of the semi-logarithmic presentation of the calibration diagrams, the error limits do not bracket the cation ratio age symmetrically, but show smaller values downwards and larger ones upwards.

8.6.3 Application

Rock Surface. Only rock surfaces which have not experienced denudation over longer periods possess desert varnish rinds. Accordingly, the cation ratio method determines the minimal duration over which the surface has been morphogenetically stable. This opens possibilities in arid and semi-arid areas of obtaining information on the stabilization of land surfaces and especially on the minimal ages of pediment surfaces, alluvial and colluvial fans, gravel terraces and aeolian deposits. This potential has not yet been evaluated in detail. First attempts in this direction have been undertaken in the Espanola Basin, New Mexico, and in the Yucca Mountains, Nevada (Harrington and Whitney 1987).

Moraines. Weathering rinds resembling desert varnish can also be formed in semi-arid high mountainous regions. This phenomenon enables the age determination of glacial episodes in the high mountains during the Pleistocene glaciation. During glacial transport, abrasive and grinding forces produce freshly polished rock surfaces. After their deposition in the moraine larger pieces of rocks are exposed to the atmosphere for long periods, resulting in the formation of desert varnish and leaching features. This however requires a fixed, stable position of the boulders, e.g., on the moraine crest. In a case study on the Tioga and Tahoe moraines in the eastern Sierra Nevada, California, Dorn et al. (1987) sampled aplitic and granodioritic blocks of sizes > 1 m, stemming from such favorable localities. Areas on the rock surface showing lichen growth and deviating microecology were omitted. In addition to the cation ratio, AMS-^{14}C dating was also applied to the weathering rinds, a considerably more elaborate technique in that it requires larger samples taken mechanically from the block surface (several 10^3 instead of ~20 cm^2 for the cation-ratio technique). The cation data were calibrated by a curve established from ^{14}C and K–Ar ages from the same region. For the Tioga glaciation the cation analyses yielded ages of 13.2 (+3.6/−2.9) (innermost moraine) and 18.5 (+3.5/−2.9) ka (outermost moraine). For the Tahoe glaciation two phases around 187 (+42/−34) and 143 (+17/−16) ka could be proved.

Petroglyphs. The engraving of petroglyphs in rock surfaces hitherto encrusted with desert varnish exposes fresh surfaces on which a new crust of desert varnish forms. The cation ratio method enables the dating of this secondarily formed desert varnish and hence the age evaluation of the petroglyphs. This technique was applied to petroglyphs of the Coso Range, southern California (Dorn and Whitley 1983). Five petroglyphs were chosen according to typological criteria and attention was paid to uniform microclimatic conditions at the individual sites. The K+Ca/Ti ratios determined by PIXE were compared to a calibration curve which was established from the K–Ar dates of volcanites from the same region. The cation ratio ages of the individual petroglyphs are presented in Fig. 139. The reported age ranges result from the errors inherent in the K + Ca/Ti deter-

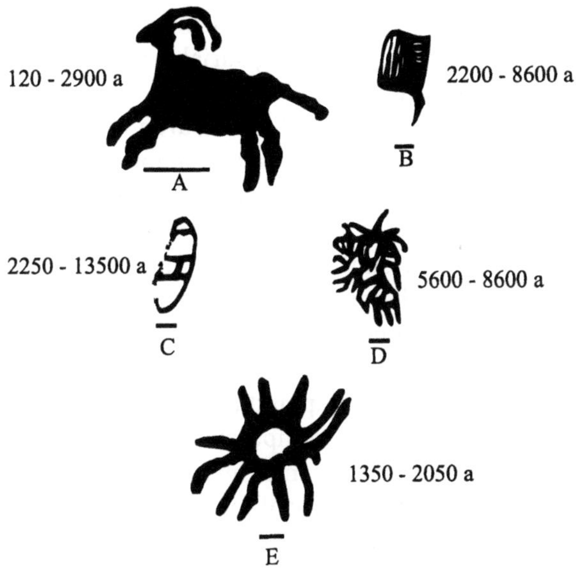

Fig. 139. Petroglyphs from the Coso Range, California, with their cation ages of desert varnish crusts which have newly formed on engraved positions. (Reprinted with permission from Nature; Dorn and Whitley 1983; copyright Macmillan Magazines Ltd.)

minations. The dating results attribute the petroglyphs to a significantly earlier period than had previously been assumed. On the other hand, the chronometric dating supported the previously hypothesized relative stylistic chronology.

Stone Artifacts. In the eastern Californian Mojave Basin a total of 167 stone flakes – consisting to a major part of chert – of six surface sites were dated by the cation ratio method (Dorn et al. 1986). Several subsamples were taken from each artifact and analyzed by the PIXE technique. The age calibration of the cation ratios was done by means of a curve representative of this region (Fig. 138). The comparison of chipped faces and natural surfaces of the same flake revealed substantial differences in the particular cation ratio ages. The oldest sequences of artifacts have cation ratio ages around 21, 16 and 14 ka. The average age precision for individual artifacts with sufficient varnish for multiple PIXE analyses is ~10%. These late Pleistocene cation ratio ages are among the oldest dates for surface artifacts from North America and may imply a pre-Clovis human occupation in southeastern California.

8.7 Fluorine-Uranium-Nitrogen Test

The collective term fluorine-uranium-nitrogen (FUN) covers three different techniques applicable to the relative dating of bones, antlers, ivory and teeth. They are all based on enrichment in fluorine as well as uranium and depletion in nitrogen during burial of mammal skeletons. Occasionally only a single element of the FUN-group is used for age estimation, accordingly the method is called the fluorine, uranium or nitrogen test, respectively. In the majority of cases, however, a combination of all three tests is applied.

Both the rate and extent to which the fluorine, uranium and nitrogen concentrations change in the buried fossils are dependent on manifold factors, namely on the type and structure of the fossil material, the burial medium, the chemical composition of the groundwater and the ambient temperature. Hence, at best *relative* age determinations can be obtained, i. e., statements can be made, whether a particular sample is younger, of the same age or older compared to related associated samples. Only the same types of samples are comparable, i. e., the same material type which has been exposed to the same environmental conditions. This requirement restricts the applicability of the method to finds essentially from the same spot. The age range covered by the FUN method falls between several 100 a and Ma, i. e., samples of the Holocene and Pleistocene periods.

Since the beginning of the 19th century fossil bones have been known to take up and become enriched in fluorine with increasing geological time (Fourcroy and Vauquelin 1806). The few studies on this subject carried out during the last century had been largely forgotten, when in the middle of this century a revival of fluorine, uranium and nitrogen analyses for the purpose of age evaluation and provenance of fossil skeletal finds commenced (Oakley 1948). Uranium and nitrogen analyses were not employed for dating purposes before the middle of this century.

8.7.1 Methodological Basis

Fluorine Method. Recent bones and other skeletal remains are extremely low in fluorine, as a rule $< 100\ \mu g\ g^{-1}$. Once buried in sediments and exposed to groundwater and moisture, the matrix of skeletal remains takes up water-dissolved fluorine. The major mineral component of bone is carbonate hydroxyapatite $Ca_5(PO_4, CO_3)_3OH$. Through ion exchange F^- replaces the OH^--group within the apatite structure, so that the more stable fluorapatite $Ca_5(PO_4, CO_3)_3F$ is formed. This ion exchange takes place gradually until finally the hydroxyapatite is completely transformed into fluorapatite – a reaction irreversible under normal groundwater pH conditions. It can be reversible only under strongly acid conditions. The stoichiometric fluorine content of fluorapatite comprises 3.8 %. Occasionally, bones with fluorine contents up to 5 % are found. Such high concentrations exceeding the theoretical maximum value are presumably caused by incorporation of

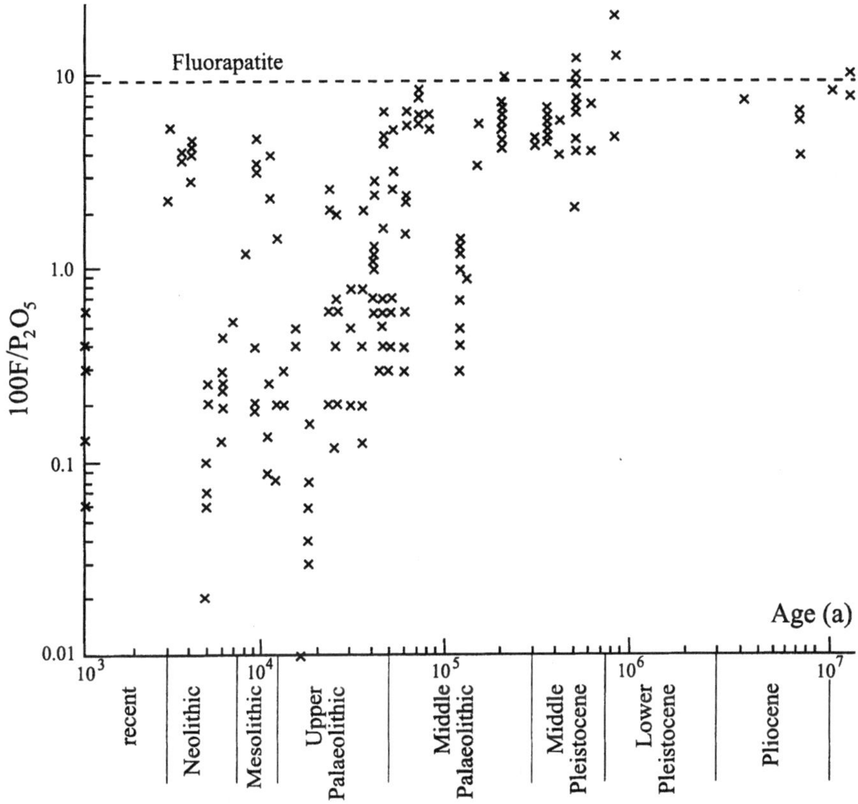

Fig. 140. The fluorine contents (normalized to phosphate) in fossil mammal skeletal remains of various European sites (excluding the British Isles) in dependence of the stratigraphic age (from Oakley 1980). Despite the broad scattering of the values, which is attributable to different environmental conditions, an ascending trend of the fluorine content with increasing age is recognizable

calcium fluoride. Since the fluorine is associated with the bone mineral apatite it is rather homogeneously distributed within the bone matrix (Oakley 1980). The fluorine concentrations are noted either in weight percent or as $F/P_2O_5 \times 100$, if normalized to the phosphate content (apatite). The rate of the fluorine incorporation depends both on the nature of the matrix and environmental parameters. Therefore a correlation between the burial period and fluorine content is only observable for the same sample materials of the same burial site. Skeletal remains of the same age found in different sites may contain strongly variable fluorine contents. Figure 140 includes a large number of skeletal samples of different European Quaternary sites demonstrating large scattering of the fluorine contents, but exhibiting also the age trend. This representation clearly shows that only relative ages can be determined by this method, which only works under favorable conditions.

Uranium Method. Modern bones and teeth contain low amounts of uranium ($<1\ \mu g\ g^{-1}$). After their burial in permeable sediments, such as sand and gravel, they are exposed to groundwater infiltration and thus take up uranium. Uranium occurs in two oxidation states with very different chemical behavior. Under oxidizing conditions the six-valent uranium is water soluble and mobile. In a reducing milieu, however, the four-valent uranium forms insoluble compounds. Uranium is mainly incorporated into the structure of the bone apatite. Apart from the length of burial period, the uranium content of a bone depends also on the redox conditions reigning in the sediment and the concentration of water-dissolved uranium. The uranium uptake of fossil bones and teeth is promoted by a reducing micro-milieu generated by degrading organic material. The uranium content of fossil bones can increase by three orders of magnitude to ~0.1%. The major uranium enrichment presumably takes place in an early stage of the fossilization. Through changes in the redox conditions, uranium already incorporated in bones can be remobilized, which finds its expression in a zoned uranium distribution (Williams and Marlow 1987).

Nitrogen Method. In vivo bones and dentine contain around 5% nitrogen in their organic matrix, the protein *collagen*. Post-mortem, when buried in the soil, the collagen decomposes into its amino acids. The amino acids are leached away from the bone and the nitrogen content gradually decreases, possibly obeying an exponential decomposition function (Ortner et al. 1972). The decomposition rate is influenced by the moisture and pH value of the soil or sediment. Under oxidizing conditions nitrogen is quickly lost. In addition, elevated temperatures and microorganisms favor the nitrogen depletion. In the course of time the nitrogen content decreases to non-detectable traces.

8.7.2 Practical Aspects

The FUN dating of bones, teeth, ivory and antlers serves the purpose of determining large age differences between individual find objects of the same site or similar adjacent neighboring ones. As a rule, the principle applies that only the same type of material stemming from the same sedimentological, hydrological and hydrochemical environment may be used for mutual comparison (e.g., dentine with dentine, or porous bone with other porous bones having the same burial history). Attention has to be paid to features signaling any spatial and temporal changes in the environmental redox potential. The skeletal remains should have been buried in permeable sediments, such as sand or gravel. Embedding materials such as clay and lime, are, however, less favorable owing to their reduced water circulation. Lack of soil moisture makes arid regions unsuitable for the fluorine and uranium tests. The fluorine test is, according to Oakley (1980), disturbed in volcanic areas, where groundwater contains high fluorine concentrations, especially when tropical temperatures strongly accelerate the ion-exchange rate. Sample sizes of 0.1–1 g are usually sufficient. In the case of teeth, dentine represents the preferable material.

The fluorine analysis can be conducted in different ways, namely wet-chemically employing titration, by X-ray diffraction (Oakley 1980) or by inelastic proton scattering (Haddy and Hanson 1982). X-ray diffraction analysis has the advantage of being fast, but is less accurate. Advantages of the proton scattering method are the high detection sensitivity ($<1\ \mathrm{mg\ g^{-1}}$) and the record of fluorine depth profiles. α- and γ-counting provides the easiest way to measure the uranium content and several grams of sample material are usually sufficient. These analyses can be also conducted non-destructively on the complete object. Uranium analysis by means of fission tracks is advantageous, since it not only yields the content of uranium, but also its microscopic distribution across polished sections (Chap. 6). The nitrogen can be chemically analyzed by the Kjeldahl method. It is essential to make sure that the sample material has not been impregnated with conservation materials or come in contact with any other solvents and glues. If possible, all three elements should be analyzed because they supplement each other for the age estimation.

8.7.3 Application

Bones and Teeth. Age determination of mammal skeletal constituents is mostly conducted by ^{14}C and uranium series. In comparison to these methods, the FUN test nowadays plays only a very modest role. Nevertheless, a considerable data set of thousands of F, U and N analyses on fossil and modern skeletal samples are available and through it interesting results on the age sequences have been obtained. Richter (1958), for instance, attempted to establish the chronology of the Pleistocene on the basis of the increasing fluorine content in Quaternary bones from northwestern Germany.

The most spectacular result seems to be the revealing of the Piltdown forgery. In 1911 an ape mandible associated with a modern human skull was allegedly found in lower to middle Pleistocene gravels near Piltdown in Sussex, England. The gravels also contained a Pleistocene mammal fauna. During the first half of the 20th century, the problematic of the Piltdown find had not only raised heated paleoanthropologic debates, but also doubts of its authenticity. The low fluorine content (< 0.0 3% mandible, 0.1% skull) and moderate nitrogen content (3.9 and 1.4%, respectively), in contrast to the Pleistocene mammal teeth (F: 1.9 – 2.65%, U: 9 – 82 $\mu g\, g^{-1}$, N: < detection limit), revealed that the Piltdown "hominid" fossils could not be older than upper Pleistocene and were thus in contradiction to the remaining find material (Weiner et al. 1953). The young age of the mandible and the skull was later confirmed by ^{14}C ages (500 ± 100 and 620 ± 100 a BP, respectively; Oakley 1980). On the other hand, the Swanscombe skull, which was discovered in 1935 in Holsteinian interglacial layers in Kent, England, and classified as *Homo sapiens steinheimensis*, yielded FUN values (F: 1.7 – 1.9%, U: 9 – 34 $\mu g\, g^{-1}$, N: 0.09 – 0.18%) in agreement with the analytical results obtained from the bones of the associated fauna (F: 1.7 – 2.3%, U: 8 – 40 $\mu g\, g^{-1}$, N: ≤ 0.2%; Oakley 1980).

The mandible of *Homo heidelbergensis*, which was found in 1907 in the middle Pleistocene Neckar sands of Mauer, southwestern Germany, and its associated mammalian fauna were subjected to FUN analyses by Oakley

Table 8. Fluorine, uranium and nitrogen analyses of bones and teeth from Mauer and Steinheim/Murr, Germany. (Oakley 1980)

Site	Material	F (%)	$F/P_2O_5 \times 100$	N (%)	U ($\mu g\, g^{-1}$)
Mauer	*Homo erectus heidelbergensis*, mandible	1.13	10	0.08	7
	Homo erectus heidelbergensis, premolar dentine	4.2	13	–	–
	Palaeoloxodon, skull	2.11	6.6	0.12	9
	Palaeoloxodon, tusk	1.67 – 2.36	4.8 – 7.1	≤ 0.07	13
	Dicerorhinus etruscus, molar	3.06	9.0	0	–
	Dicerorhinus etruscus, jaw	2.21	7.4	< 0.01	18
Steinheim	*Homo sapiens steinheimensis*, skull	1.2	10.2	0.37	2.5
	Bison, horn	1.88	5.9	–	–
	Homotherium, maxilla	1.40	4.3	0	–
	Palaeoloxodon, bones	1.98	6.4	0.23	11
	Mammuthus, tusk	1.55	4.7	–	–
	Elephas, tusk	1.10 – 2.20	5.5 – 7.0	–	–

(1980). The skull of the *Homo sapiens steinheimensis*, discovered in 1933, and the associated fauna of Steinheim, southwestern Germany, were sampled in the same way. It has to be emphasized that the data (Table 8) are not suitable for any useful age estimations; nevertheless, they testify to the consistency between both the hominid finds and the Middle Pleistocene faunas with respect to the FUN tests.

8.8 Racemization

Amino acids form the basic constituents of the proteins. They occur in two molecular configurations which are compositionally identical but which possess different, mirror-inverted symmetric structures. In analogy to the left and right hand, the two configurations are called L- and D-types. The amino acids of the proteins of all living organisms consist almost exclusively of molecules belonging to the L-configuration. A state in which only one of the both types is realized is thermodynamically unstable so that in course of time an equilibrium state is attained containing both configurations in equal amounts. This reaction is known as racemization. At a given reaction rate the degree of racemization furnishes a clue to the age.

The term racemization means the reaction during which an optically active mixture is converted into an optically inactive one ("racemic mixture" after the Latin *racemus* for the grape, in which the racemic acid was initially analyzed). In optically active mixtures one of the configurations is predominant, whereas inactive mixtures contain the D- and L-molecules in equal portions. Michael and Wing (1884) observed this phenomenon for the first time, when they transformed optically active into inactive aspartic acid by heating. Abelson (1954), who found amino acids in fossils, suggested to exploit the time dependence of the racemization process for dating purpose. Since the 1970s, this idea has increasingly found its application to the dating of bones and other materials (Bada and Schroeder 1975). The initial expectations, however, were not fulfilled, since the racemization process is only of limited applicability to numerical dating. As in any other chemical reactions, the reaction rate of the racemization, too, is controlled by physical and chemical environmental conditions, of which temperature dominates. Successful application of this dating method requires not only knowledge of the relevant reaction rates but also the temperature history. In the majority of cases the thermal history during the burial period of the fossil is not sufficiently known to allow an explicit age determination. On the other hand, samples with identical temperature history, e.g., stemming from the same site, can be relatively dated on the basis of their degree of racemization.

Amino acids are constituents of fossil skeletons. Particularly suitable for dating are bones, teeth, mollusk shells and avian eggshells. The age range depends on the reaction rates of the racemization, which are rather variable for individual amino acids. Consequently, depending on the type of amino acid analyzed, a range from a few years to several Ma is datable. The archaeological potential of amino acid racemization dating was summarized by Johnson and Miller (1997).

8.8.1 Methodological Basis

Proteins play an active role in the biomineralization of hard tissues. In living organisms ~20 amino acids participate in the formation of the proteins (polypeptides), bones, teeth, hair and hard shells. In the course of the diagenetic decomposition large proteins break down into smaller polypeptide chains, non-amino acid molecules and free amino acids. Most amino acids (α-amino acids) possess an asymmetric central carbon atom to which four atoms or side groups R are linked (Fig. 141). The most abundant asymmetric α-amino acids, characterized by the following R, are:

glycine	$-H$	leucine	$-CH_2-CH-(CH_3)_2$
alanine	$-CH_3$	isoleucine	$-CH-CH_3-C_2H_5$
valine	$-CH-(CH_3)_2$	glutamic acid	$-CH_2-CH_2-COOH$
aspartic acid	$-CH_2-COOH$	phenylalanine	$-CH_2-C_6H_5$

They form two optical isomers, called L- and D-enantiomers. The chemical characteristics of the two enantiomers are identical; however, in solution they rotate the polarization plane of the light either to the left or right hand according to their structure, a feature called optical activity. This optical phenomenon gave rise to the term L (for *levo*) or D (for *dextro*), respectively.

The gradual conversion (*racemization*) of optically active mixtures into optically inactive ones can be strongly accelerated by heating. It is a reversible first-order reaction:

L-amino acid $\leftrightarrow$ D-amino acid

with the specific rate coefficients a_1 and a_2, respectively, for interconversion of the amino acid enantiomers. In the course of time t a dynamic equilibrium between the two enantiomers is gradually attained. The kinetic equation for this reaction is

$$\ln[(1 + \mathrm{D}/\mathrm{L}) \div (1 - K \times \mathrm{D}/\mathrm{L})] - \ln[(1 + \mathrm{D}/\mathrm{L}) \div (1 - K \times \mathrm{D}/\mathrm{L})]_0 = (1 + K)\, a_1 \times t \quad (67)$$

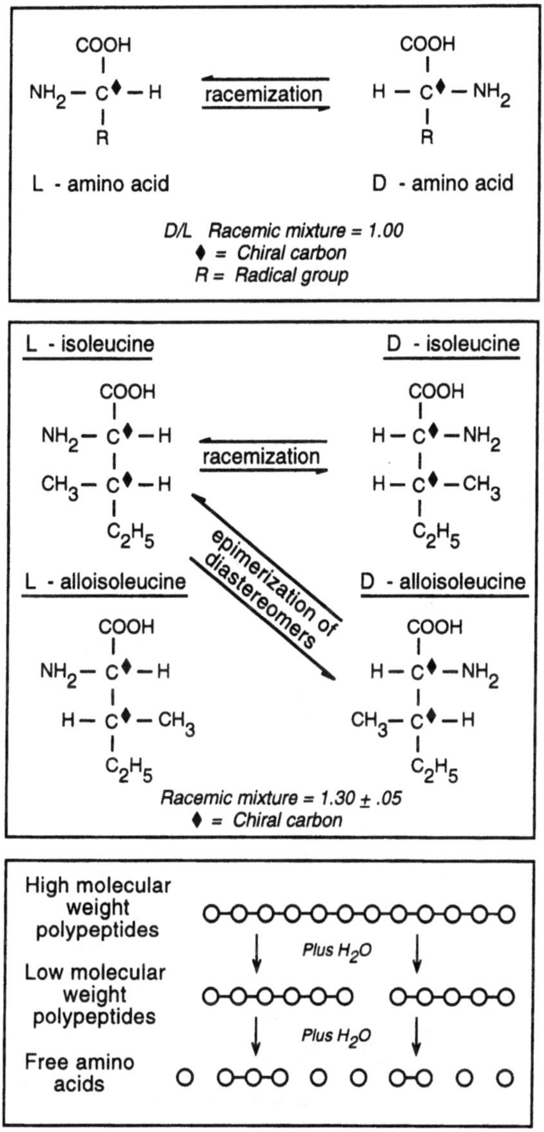

a b c

Fig. 141 a–c. Amino acids are organic acids with an amino-group (–NH_2). The so-called α-amino acids (e.g., aspartic acid) possess a carboxyl group (–COOH) at the asymmetric central C atom, a hydrogen atom and a radical R, so that two mirror-inverted configurations occur, which can be reversibly interconverted by racemization (**a**). Concerning isoleucine with two asymmetric C atoms, this reaction effects only the α-C atom. L-isoleucine is converted into D-alloisoleucine. This reversible reaction is called epimerization (**b**). The free amino acids of fossil skeletal remains are formed by hydrolytic decomposition of the proteins (**c**). (Reprinted from Miller and Brigham-Grette 1989; copyright 1989, with kind permission from Elsevier Science Ltd., The Boulevard, Langford Lane, Kidlington OX5 IGB, UK) where D/L is the amino enantiomeric ratio at a particular time t, and $K = a_2/a_1$ represents the equilibrium constant. The term $\ln[(1 + D/L) \div (1 - K \times D/L)]_0$ accounts for the possibility that the initial D/L-ratio may not have been zero at time $t = 0$, or that some degree of racemization may have been induced during the analysis. For the majority of amino acids (with the exception of the diastereomeric amino acids containing more than one asymmetric carbon atom, such as isoleucine and hydroxyproline, whose racemization is more properly referred to as epimerization) both reaction coefficients are equal, so that Eq. (67) can be simplified to

$$\ln[(1 + D/L) \div (1 - D/L)] - \ln[(1 + D/L) \div (1 - D/L)]_0 = 2a \times t. \quad (68)$$

The coefficient $2a$ is called racemization rate k_{rac} and – for defined environmental conditions (T, pH value) – can be expressed as half-life ($t_{1/2} = 0.693/k_{rac}$).

The time dependence of the racemization reaction forms the basis of the dating method. In living organisms the amino acids consist almost exclusively of L-enantiomers. After their death the racemization process starts. The L-enantiomers convert into D-enantiomers and vice versa until they are equally abundant. The consumption of the L-enantiomers and the production of the D-enantiomers follows the exponential functions characteristic of first-order kinetics, consequently, the degree of racemization D/L grows exponentially as well (Fig. 142). At $t_{1/2}$ the D/L-ratio is 0.33, after $5 \times t_{1/2}$ already

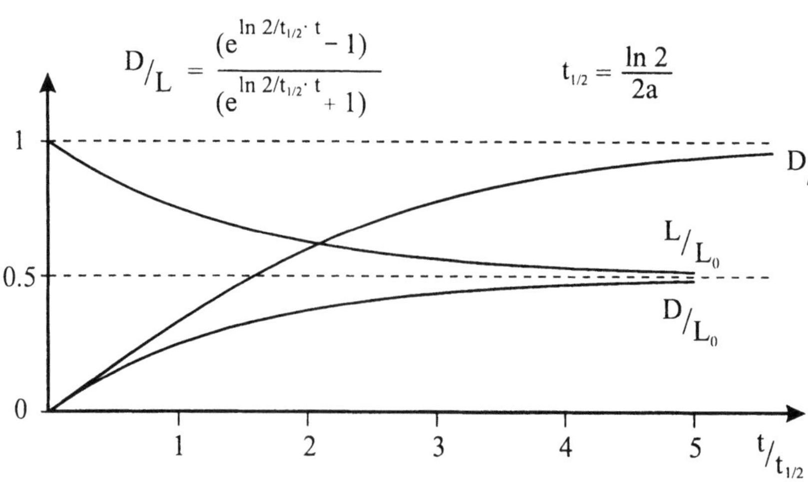

Fig. 142. Amino acids of living organisms consist only of the L-enantiomer. After the organism's death the amount of the L-enantiomer decreases exponentially with time (represented in units of the racemization half-life $t_{1/2}$) to half of its initial amount in favor of the D-enantiomer. The D/L-ratio of D- and L-enantiomers increases from zero to D/L = 1 and is a measure of the expired time

0.94. The elapsed time t since the organism's death can be deduced from the racemization D/L according to Eq. (68). The individual amino acids differ in their racemization rates (Table 9). Since ages up to $\sim 5 \times t_{1/2}$ can be determined, the various amino acids yield clues to different age ranges between 10 a and 1 Ma. Table 9 presents the half-lives of some commonly used amino acids under various environmental conditions. The extent of racemization of aspartic acid, alanine and leucine also provides criteria for assessing whether fossil bone material still preserves endogenous DNA which may be employed for paleogenetic studies (Poinar et al. 1996).

Instead of the linear plot of the racemization D/L vs. the time t a semi-logarithmic diagram is often used, in which ln [(1 + D/L)/(1 – D/L)] is plotted vs. the linear time scale. According to equation 68, the data points are situated on a straight line whose slope is equal to the racemization rate k_{rac} (Fig. 143). Apart from first-order kinetics, more complex reaction kinetics are observable, as for instance in the case of foraminifers. Although they possess a first-order kinetics above 130 °C in experiments, under natural environmental conditions the racemization is more complex (Wehmiller 1986). The linearity of the ln (D/L)-time relation is not realized if a reaction obeys any kinetic order other than the first-order kinetics. So far, only a few fundamental investigations have dealt with this kind of complication.

The major difficulty inherent in this dating approach is the strong temperature dependence of the racemization. According to the Arrhenius equation [Eq. (61)], the rate coefficient of the reaction increases with increasing temperature. The strong influence of the temperature on the racemization can be experimentally studied by means of samples which have been heated at different temperatures over variable time spans. Figure 121 shows the result of such an experiment on aspartic acid from bone tissue. The activation energy of 38 ± 3 kcal mol^{-1}, obtainable from k_{rac} according to Eq. (61), implies that a temperature difference of 1 °C causes a change of 25 % in the racemization rate! The marked temperature dependence results in the requirement that the temperature history of the whole burial period needs to be known with an accuracy of <1 °C, a requirement which is not fulfilled in the majority of cases. Because of the non-linearity of this relation, temperature peaks influence the racemization more than proportionally. Consequently, in the case of seasonal fluctuations, it is not the annual average temperature T_a of the air which controls the racemization rate, but an effective ambient temperature T_{eff} (Sect. 8.2.1) exceeding T_a. According to Lee (1969), T_{eff} is deducible from T_a and the temperature fluctuation R_T (difference between the highest and lowest monthly average temperature within 1 year) according to

$$T_{eff} = (T_a + 1.2316 + 0.1607 \times R_T)/1.0645 . \qquad (69)$$

Table 9. Racemization half-lives of some amino acids

Material	Environment	Aspartic acid	Alanine	Isoleucine	References
Free amino acid	100 °C, pH 7–8	30 d	120 d	300 d	Bada (1984)
Recent tooth	In vivo	630 a, 350 a			Gillard et al. (1991); Bada (1984)
Fossil bones	Egypt, Sudan Hungary	3.5 ka 13.5 ka	10 ka 32 ka	– 110 ka	Bada and Shou (1980); Csapo et al. 1991
	East Africa	5–200 ka	–	100–300 ka	Hare et al. (1978); Bada (1981)
	Californian coast	30 ka	100 ka	–	Ike et al. (1979); Bada et al. (1979 a)
Shells	South Florida		–	60 ka	Mitterer (1975)
Shells	Arctic	–	–	300 ka	Miller and Hare (1980)
Bristlecone pine	California	30 ka	–	–	Zumberge et al. (1980)
Carbonate	Deep-sea	100 ka	200 ka	200 ka	Kvenvolden et al. (1973); Bada and Man (1980)

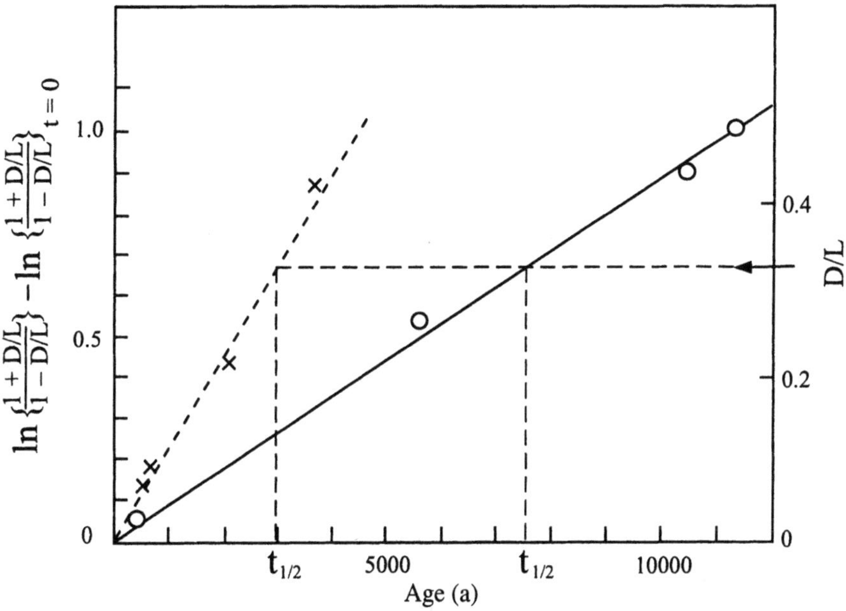

Fig. 143. Aspartic acid racemization in fossil bones depending on age (after Bada 1985). Shown are bones from two regions with different soil temperatures: Sudan and the Philippines with 26–27 °C (*crosses*), Arizona with 18 °C (*circles*). The racemization rate (slope of the *straight lines*) rises with the soil temperature. D/L = 1/3 (*arrowhead*) equals the respective $t_{1/2}$

In practice, the racemization method of numeric age determination may be applied in two ways. The first, independent approach requires experimental determination of k_{rac} (= $2a$) as a function of T for the analyzed amino acid. By inserting the value of k_{rac} that corresponds to T_{eff} of the sampling site into Eq. (68), the age t is obtained from the degree of racemization D/L. This approach requires knowledge of the reaction kinetics and the temperature history. Considerable age uncertainties may be associated with this direct approach, due to the strong influence of the largely unknown temperature history (50–100 %). In the second approach, the racemization rate of a given site is calibrated by means of samples of known age from the same site. This approach avoids the need of assessing the temperature history, which may bear great uncertainties due to daily, seasonal as well as long-term climatic fluctuations, in particular those caused by the Glacial and Interglacial periods. Once the racemization rate has been calibrated the racemization ages of samples of the same site can be calculated. If such independent age standards are not available, the D/L-ratio of a specific amino acid under identical environmental conditions can be used for the establishment of a relative age sequence. This methodological approach is called *amino-stratigraphy.*

Another disturbing environmental factor worth mentioning is the pH value, although its influence is less marked than that of the temperature. In acid soils, increased leaching of amino acids has to be expected. During burial, exchange of amino acids between the sample material and the environment takes place, a process in which groundwater plays an important role. Amino acids can be both leached out of the bone or brought into it from the environment. Infestation with bacteria forms another source through which the primary amino acid pattern may be altered. Such secondary amino acids are commonly enriched in L-enantiomers and, hence, represent a serious disturbance to the dating system. Another difficulty arises from the hydrolytic decomposition of the amino acids, which in turn influences the racemization rate. Studying Paleolithic skeletons, Bada (1985) observed that the racemization rate of aspartic acid increases with the degree of decomposition. Consequently, both, the samples of known and unknown ages should not only have been exposed to identical environmental conditions (concerning T and pH), but also should show similar degradation of their amino acid pattern. Furthermore, the racemization rate is influenced by the sample type. For instance two different mollusk genera of the same age and find layer may show different racemization rates (Wehmiller 1986).

8.8.2 Practical Aspects

Surface finds have to be rejected due to the strong temperature fluctuations. Hence, samples must be collected from deeper levels, at least 1 m below the surface. Well suited with regard to their temperature history are samples from caves which have experienced steady annual temperatures. In any case, samples displaying signs of heating, e.g., hearth-derived ones, must be strictly avoided. It is self explanatory that any later heating event has to be excluded. A further complication is contamination with recent amino acids. Samples bearing any indication of bacterial attack are unsuitable. Three to five samples should be collected from each stratigraphic level. The samples should be handled with caution; this applies in particular to the touching of the samples. Samples which have been treated with preservatives are not suitable. Owing to the influences of diagenetic degradation and sample type, several samples belonging to the same context should be collected, e.g., the sampling strategy should consider the possible occurrence of several mollusk genera at the site (Miller and Brigham-Grette 1989). Individual finds are less favorable than complete profiles consisting of different horizons and furnishing samples of different ages. A more comprehensive find context permits a check on the age – at least to a certain degree.

Sample sizes of several grams are sufficient. Preferable in any case are compact pieces of bones and mollusk shells, since they are less liable to exchange processes. It is just as important to consider the degree of degradation, which should be as low as possible. For marine bivalve *Chione* a variation of up to 30% in the isoleucine epimerization rate was observed among and within different layers comprising the shell, necessitating careful choice of the sampling position (Goodfriend et al. 1997). Subsequent to cleaning, 0.1 – 1 g of the sample is dissolved in hydrochloric acid for complete hydrolytic disintegration into free amino acids. The separation of the different amino acids is conducted chromatographically. By means of an admixture of optically active or inactive compounds, derivatives are formed from the enantiomers, which according to their volatility and polarity possess slightly different diffusion rates and thus can be separated chromatographically (Fig. 144). Various gas and liquid chromatography techniques are in use (Miller and Brigham-Grette 1989). D/L is determined from

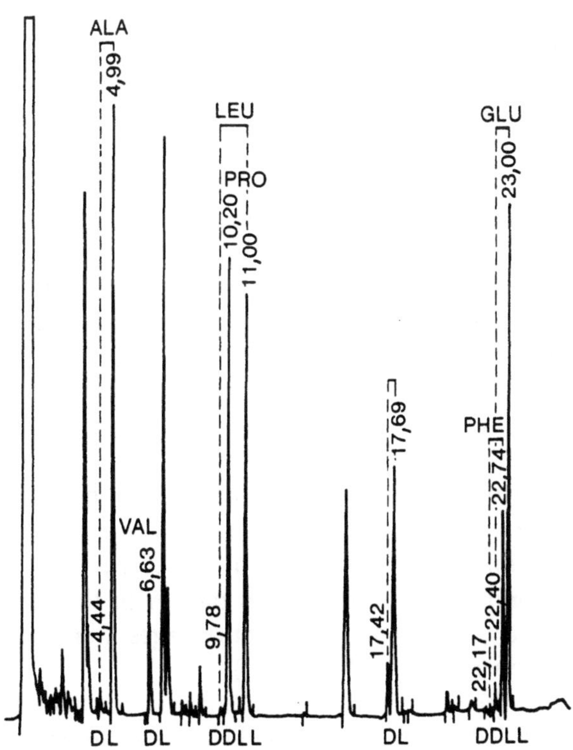

Fig. 144. Gas chromatographic separation of isopropylesters from amino acids occurring in wood (from Rutter et al. 1985). *Ala* Alanine; *Val* valine; *Leu* leucine; *Asp* aspartic acid; *Phe* phenylalanine; *Glu* glutamic acid; *Pro* proline the numbers are the diffusion times in minutes. The D/L-ratio of the individual amino acids is computed from the peaks

the signal intensities, usually with an accuracy of 2–10%. The analysis of several amino acids from the same sample is preferable. On the basis of their different racemization rates, the D/L values should decrease in the sequence aspartic acid → alanine ≅ glutamic acid → leucine ≅ alloisoleucine/isoleucine (Bada 1985). This sequence serves as a sample-inherent criterion for the absence of any disturbances caused by temperature effects or contaminants.

8.8.3 Application

Bones. Bones of living mammals contain about 25–30% organic material, whose most prominent constituent is protein. Upon the death of an organism the bone proteins start to undergo hydrolytic decomposition into free amino acids. Since bones do not represent closed systems, these amino acids are in exchange with the surrounding environment. Even under cool environmental conditions, collagen and its hydrolysis products are lost from the bone after ~1 Ma at the latest. The collagen-specific amino acid 4-hydroxyproline can be used as an indicator of the degree of collagen decomposition. Of the various amino acids, aspartic acid and isoleucine have been most widely used in the dating of bones. In this connection it has to be emphasized that due to their different half-lives, aspartic acid covers the age range up to 100 ka, whereas the effective dating range of isoleucine comprises up to ~1 Ma.

In the 1970s, the ages of several Paleo-Indian skeletons from California were deduced from the extent of aspartic acid racemization. The resulting data caused a controversy on the reliability of the racemization method. Human skeletons of the Del Mar and Sunnyvale sites were dated by this method to 40–60 ka (Bada et al. 1974). The age calculation was based on a racemization rate value, which was calibrated by means of a ^{14}C-dated skull from Laguna. The resulting extremely high racemization ages for the human bones on the North American continent significantly predated the opinion accepted so far on the earliest colonization ~13 ka ago. Comparative age determinations by uranium series and AMS-^{14}C dating on bones of the two find sites yielded, in contrast to the aspartic acid dates, consistent Holocene ages (Bischoff and Rosenbauer 1981; Taylor 1983). Meanwhile the racemization ages have been revised by using AMS-^{14}C-dated bones for calibrating the racemization rate. With the revised value of $k_{rac} = 6.0 \times 10^{-5}\ a^{-1}$, Holocene racemization ages were obtained for Del Mar and Sunnyvale as well (Bada 1985).

At the Olduvai Gorge in Tanzania, the aspartic acid k_{rac} values, which were deduced from ^{14}C-dated Holocene bones, were $\sim 7 \times 10^{-5}\ a^{-1}$; however, the Upper Pleistocene k_{rac} values of $\sim 1.7 \times 10^{-5}\ a^{-1}$ were significantly lower, which must be attributed to the lower annual average temperatures during the last glaciation. Usage of these racemization rates led to aspartic acid ages in good agreement with the ^{14}C data. Belluomini (1981) subjected bones of hominids and mammals from several Upper Paleolithic and Mesolithic sites in southern Italy to aspartic acid racemization dating. For each site a racemization rate was determined by means of the ^{14}C age of a bone, which was considered to properly reflect the specific temperature conditions. The k_{rac} values varied between 0.46×10^{-5} and $1.06 \times 10^{-5}\ a^{-1}$. The measured ages range around 5–25 ka with a good accuracy of around 8 %. Outliers with high ages could be attributed to heating of the bones.

In their study on fossil bones from the Upper Paleolithic Abri Pataud, France, El Mansouri et al. (1996) observed a poor correlation between the total aspartic acid racemization ratios and ^{14}C ages. The racemization rates turned out to be controlled by the type of sedimentary surroundings (with $k_{rac} = 0.44 \times 10^{-5}\ a^{-1}$ for fine and $k_{rac} = 0.9 \times 10^{-5}\ a^{-1}$ for coarse deposits). Racemization ratios of aspartic acid in a high molecular weight protein fraction showed much better correlation with the ^{14}C ages ($k_{rac} = 2.1 \times 10^{-5}\ a^{-1}$). It also appeared that the aspartic acid racemization occurs with a delay of several thousand years, presumably caused by a coupling between the degradation of proteins and the racemization of amino acids.

Csapo et al. (1991) subjected a larger number of ^{14}C-dated fossil bones from Hungarian sites to amino acid analyses. The D/L-measurements of

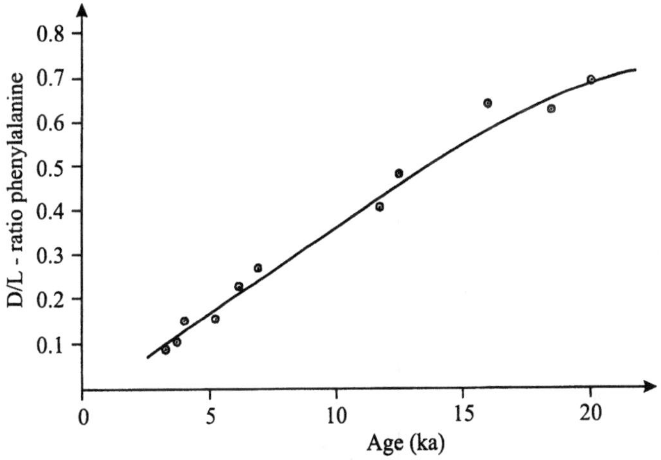

Fig. 145. Racemization degree D/L of phenylalanine in bone collagen vs. the independently determined age (after Csapo et al. 1991). The curve was established on bones from Hungarian sites and can be used as a regional standard curve for racemization dating. Analogous calibration curves exist for other amino acids

the different amino acids enabled the establishment of calibration curves valid for Hungarian soil conditions (Fig. 145). The individual amino acids are optimally suited for certain age ranges; to be precise, histidine ($t_{1/2}$ = 5.5 ka) for 2–12 ka, phenylalanine ($t_{1/2}$ = 8.5 ka) for 3–20 ka, aspartic acid ($t_{1/2}$ = 13.5 ka) for 5–35 ka, alanine ($t_{1/2}$ = 32 ka) for 10–80 ka, isoleucine ($t_{1/2}$ = 110 ka) for >30 ka and valine ($t_{1/2}$ = 180 ka) for >55 ka. For the dating of bone, Csapo et al. (1991) suggested analyzing two to three amino acids and calculating a mean racemization age from the results.

Teeth. The aspartic acid racemization rates of related teeth and bones are the same (Bada et al. 1979b). For Middle and Lower Pleistocene teeth the alloiso-/isoleucine epimerization ratio is used, which increases with stratigraphic ages up to 3 Ma (Bada 1985). The D-aspartic acid of modern teeth is suitable for determining the lifetime of individuals. The high body temperature of 36.5 °C results in a fast in vivo racemization with $t_{1/2}$ of around 500 a for aspartic acid. For extracted human premolars, Gillard et al. (1991) succeeded in proving a relationship, mathematically describable as a second-order polynomial, between the lifetime of the donor and the content of D-aspartic acid (Fig. 146). This curve allowed the determination of the age at death with an accuracy of 4 a. The method was tested on three individuals whose lifetime was unknown to the analyzers. In this way ages of 61.7, 33.5 and 43.5 a were determined from D-aspartic acid contents of 3.30, 2.27 and 2.67%, respectively; these results were in good agreement with the actual ages of 60.5, 31.0 and 42.0 a. Such results are of forensic interest. Upon the death of an individual, the racemization of the teeth is controlled by the cooler soil temperature resulting in a reduced progress by two orders of magnitude. Consequently, although less accurately, the method is applic-

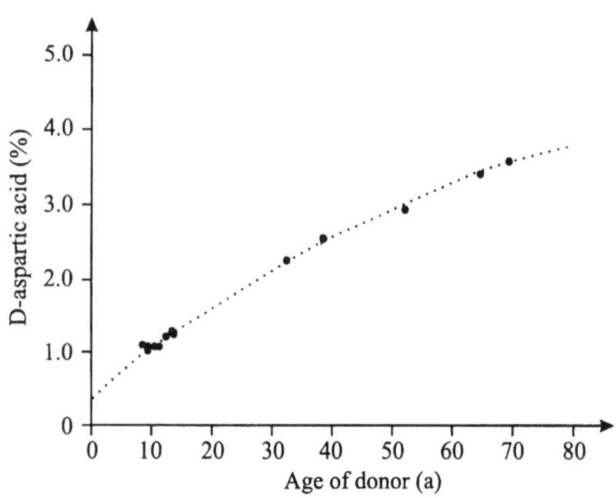

Fig. 146. Content of D-aspartic acid in collagen of human premolars depending on the donor's lifetime (from Gillard et al. 1991). The growth curve can be described as a 2nd-order polynomial. It enables the determination of the individual age at death

able to teeth of long deceased individuals, as has been demonstrated by investigations on documented 18th/19th centuries burials in the London Spitalfields (Gillard et al. 1991). This application might be of interest to anthropology and archaeology.

Mollusks. Shells of mussels and snails contain ~0.02% protein manifested in thin membranes between the calcareous matrix. Within the membranes the composition of the amino acids is the same, but may vary from one lamella to the next. Various genera are differently well suited for amino acid racemization dating (Miller and Brigham-Grette 1989). The amino acid isoleucine, with its D-alloisoleucine/L-isoleucine epimerization, is especially well suited for geochronologic studies of mollusk shells. It epimerizes relatively slowly, thereby providing a larger age range than other amino acids, is stable and is not created by the decomposition of more complex amino acids.

Fossil mussel shells of the littoral habitat are indicators of previous coastlines, which had been formed during interglacial sea level maxima and thus are of especial interest to the Quaternary chronology. Dating of such shells by racemization has been tried. Hearty et al. (1986) intended to reach a correlation of the Glacial seashores over large distances within the western Mediterranean region. For this purpose they collected fossil shells of *Glycimeris* and *Arca* on 46 coastal sites and subjected them to racemization analysis. To exclude the influence of daily temperature fluctuations, the samples were collected from depths >1 m below the soil surface. The D-alloisoleucine/L-isoleucine epimerization ratio was used as the age criterion for the mollusk shells. Up to 30% intrashell variation in the D/L-ratio was observed, so that the samples needed to be taken from the same spot within the shells. For specimens of equal age and from the same locality, no differences in the D/L-ratios in shells of the *Glycimeris* species *violescens, glycimeris* and *bimaculata* were observed, but a significant difference was found between the genera (the D/L-ratio of *Glycimeris* was 31% higher compared to shells of the genus *Arca*). Hearty et al. (1986) introduced the term *amino group* for the interpretation of the analytical results. An amino group represents a collection of equally aged shell deposits that, because of dissimilar thermal histories, yield D/L-ratios that follow the long-term regional temperature gradient. Proof of equal age was carried out by $^{230}Th/^{234}U$ dating on associated corals. Figure 147 shows the regional distribution of the 125-ka-old amino group of mollusk shells belonging to the Eemian sea level maximum. The degree of racemization decreases northward together with the mean annual temperature. Older amino groups belonging to earlier interglacials show parallel trends, but at higher D/L-ratios (Fig. 148). This regularity permits the amino-stratigraphic cor-

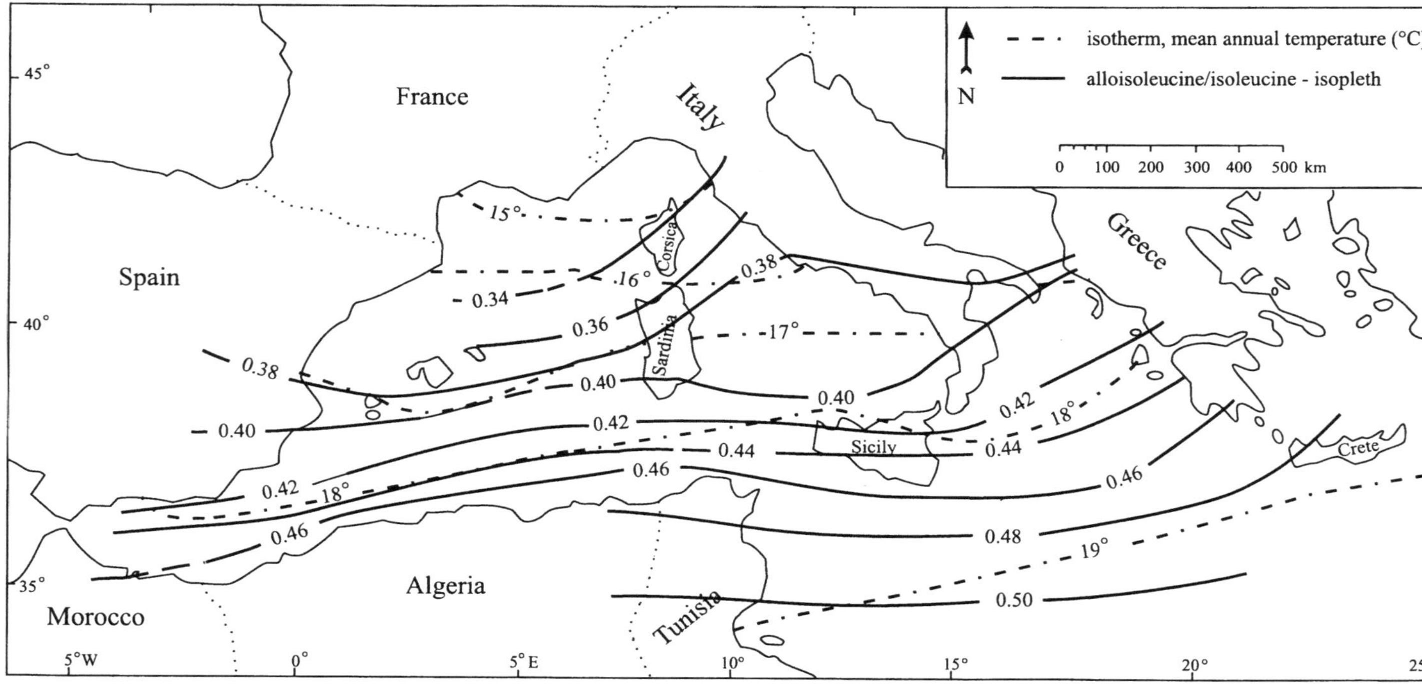

Fig. 147. Lines of identical epimerization in the fossil coastal mollusk *Glycimeris* from the sea level maximum during the last interglacial in the western Mediterranean. The epimerization increases with the annual mean temperature. (After Hearty et al. 1986)

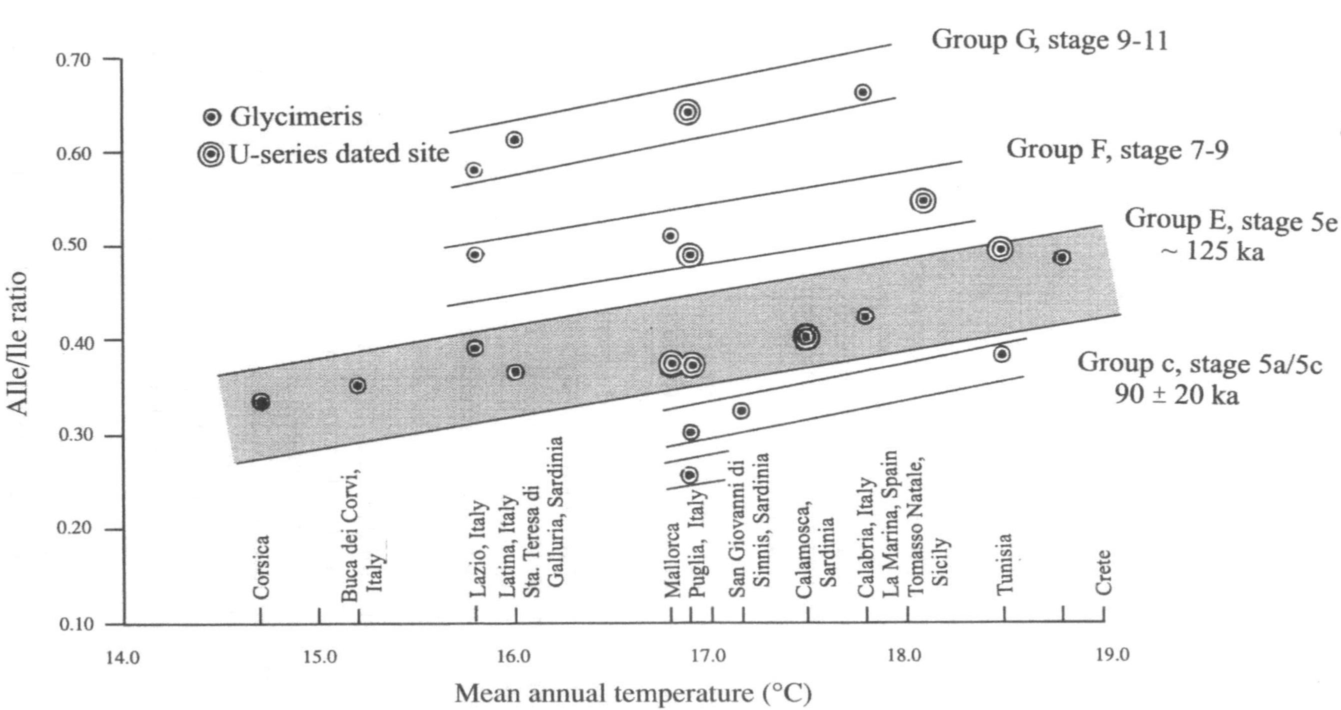

Fig. 148. Amino-stratigraphy of Quaternary coastlines of the western Mediterranean (after Hearty et al. 1986). The AIle/Ile (alloisoleucine/isoleucine)-values of mollusks of the same age, i.e., belonging to the same interglacial, from regions with different temperatures are classified into one amino group. The data presented in Fig. 147 correspond to the amino group E (Eemian interglacial). The amino groups with different epimerization degrees show a parallel trend and represent earlier or later interglacial sea level maxima

relation of the Pleistocene coastal lines in the Mediterranean and – if independently dated – their linkage to the marine $\delta^{18}O$ chronology (Sect. 10.2.2).

Oches and McCoy (1995) collected gastropod shells for amino-stratigraphic purposes from loess localities in the central European Danube Basin. The sampled sites include the stratigraphic and archaeological key sections of Stillfried, Lower Austria, and Dolni Vestonice, Moravia. The extent of isoleucine epimerization was determined in nearly 200 samples of the genera *Succinea*, *Pupilla*, *Trichia* and *Arianta*. The alloisoleucine/isoleucine ratios in the shells from the loess of the penultimate Glacial were significantly higher than those from the last glaciation. This difference in the epimerization ratio between the samples of the two loess groups reflects the difference in age as well as the thermal effect on the older group of the relatively warm last Interglacial. The gastropod alloisoleucine/isoleucine ratios also allow discrimination between the loesses of the third- and fourth-last glaciations, as long as the shells are preserved in these old loesses. The amino acid data support previous correlations based on pedostratigraphic, paleomagnetic and thermoluminescence dating studies of the loess in the Danube Basin.

Eggshell. The shells of eggs laid by ostrich are well suited for amino acid dating. They occur commonly at Paleolithic sites in Africa and Asia. The eggs served as food sources and their shells as water containers. Brooks et al. (1990) studied the dating potential of L-isoleucine to D-alloisoleucine epimerization in eggshells from various archaeological sites in Africa. The age estimates are consistent with independent evidence. Reliable dating reaching back to 200 ka seems reasonable, provided the local temperature history is sufficiently known. Also, Australian emu eggshells were successfully dated (Miller et al. 1997).

Foraminifers. The applicability of the racemization method to foraminifers is favored by steady temperatures around 2–4 °C on the ocean floor. Different genera covering the age range from the present to 5 Ma have been studied (Bada and Schroeder 1975; Muller 1984). The dating application is complicated through the non-linearity of the reaction kinetics which causes a reduction of the rate of the alloisoleucine/isoleucine epimerization by the factor 10 with increasing age. In addition, the racemization varies from genus to genus.

Corals. Coral reefs grow under steadily warm temperatures >20 °C closely beneath the sea surface. Upon their death they are subjected to larger temperature changes due to the variations in the sea level. Wehmiller et al.

(1976) investigated the racemization of several amino acids, especially the alloisoleucine/isoleucine epimerization, in fossil corals from Barbados, New Guinea, and the Ryukyu islands, with independent age control. The extent of epimerization of the amino acids did not show any unequivocal correlation with the age of the corals. On the other hand – using the coefficients determined on foraminifers under consideration of the different environmental temperatures – 16 of the 38 analyzed corals yielded alloisoleucine/isoleucine ages in agreement with the known ages between 80 and 350 ka. In the majority of cases the epimerization ages, however, were too young, a fact which is attributed to the influences of the skeletal structure on the epimerization rate. Disturbing factors are the leaching of the amino acids and the contamination with modern ones. Obviously, a better understanding of the kinetic and geochemical behavior of the amino acids in corals is needed.

Wood. The organic wood matrix consists predominantly of cellulose and to a lesser extent of a whole series of complex organic compounds including amino acids. Lee et al. (1976) studied recent and fossil wood for their amino acids and their degrees of racemization. Altogether 12 different amino acids were determined. With increasing age the total amount of amino acids decreases and the degree of racemization increases. The racemization rates are slower compared to those of bones. A calibration conducted on a ^{14}C-dated wood sample, permitted an age estimate of >110 ka from the D/L-ratio of the proline and from the allohydroxyproline/hydroxyproline epimerization of an Upper Acheulian wood artifact from Africa. On fossil pieces of wood from Yukon and Alaska, which were buried under permafrost conditions over the greater part of their geological history, a clear correlation between the degree of racemization of the aspartic acid and the geological age was found (Rutter and Crawford 1984). In this way the sediments of the last Glacial in the subarctic region can be amino-stratigraphically correlated.

9 Paleomagnetism

The Earth's magnetic field experiences temporal changes affecting both its direction and strength. The basis of archaeo- and paleomagnetism is the property of minerals, and thus of rocks too, to record and retain the Earth's magnetic field – as it was during their formation or subsequent heating – in the form of a remanent magnetization. If the history of the variation in the Earth's magnetic field is known from investigations of dated samples, a magnetization event becomes datable by deducing its age from the remanent magnetization. Therefore the method of magnetic dating is not independent but transfers already existing chronologies. Terminologically, a distinction is commonly made between archaeomagnetism and paleomagnetism. Archaeomagnetism refers to the remanent magnetization of artifacts, i. e., it covers the age range of about the last 10^5 a, whereas paleomagnetism usually deals in an analogous way with rocks, in particular from of older epochs. However, this terminological difference is insignificant to the aspects which will be discussed in this chapter. Monographs on paleomagnetism were published by Eighmy and Sternberg (1990) and Butler (1992).

Edmund Halley, in 1692, was the first to recognize the directional change of the Earth's magnetic field, known as secular variation, on the basis of data which had been recorded since the end of the 16th century. Measurements of the field intensity date back to Carl Friedrich Gauss around 1830. Archaeomagnetic research commenced in the 19th century (Gheradi 1862). Since the middle of the 20th century, paleomagnetic research has experienced great progress, a development which not only led to a gain in a basic knowledge of geomagnetism, but also revolutionized global-tectonic conceptions. The remanent magnetization of artifacts and rocks revealed that secular variation is a perpetual phenomenon of the Earth's magnetic field, not at all restricted to the historically recorded period (Johnson et al. 1948). Furthermore, in the early 1960s, it became known that the geomagnetic field experiences polarity reversals within irregularly long intervals. Both phenomena are applied with great success to chronometric questions.

The term thermoremanence was coined for thermally aquired magnetization. This type of magnetization is observed in ceramics, heated rocks and volcanites. Detrital remanence is aquired during clastic deposition and enables the dating of marine and continental sedimentary layers. The determination of archaeomagnetic age values requires the availability of standard curves for the particular time span. Such standard curves need to be a priori established using samples of known age. Magnetic dating furnishes very accurate, although not independent data. Paleomagnetism enables also the magnetostratigraphic correlation of rocks with high temporal resolution power. In contrast to the phenomenon of secular variation, which is only of regional character, polarity reversals are phenomena of worldwide contemporaneous extent.

9.1 Methodological Basis

A freely hanging compass needle aligns itself parallel to the lines of force of the geomagnetic field. In central Europe, for instance, the compass needle directs its north pole towards the north and is dipped downwards with an angle of ca. 65°. The directional deviation of the Earth's magnetic field vector of the *intensity F* from geographic north is called *declination D* and the angle of dip *inclination I* (Fig. 149). The three parameters *D*, *I* and *F* define the geomagnetic field vector at each locality. The conduction of

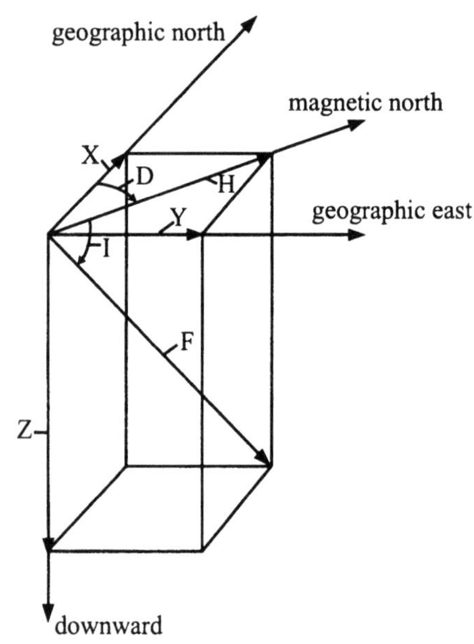

Fig. 149. The magnetic field is describable at any place by the field vector of intensity *F*, whose directional deviation from geographic north is called declination *D* and its angle of dip inclination *I*

such determinations at many places on the Earth's surface yields the pattern of the geomagnetic field which can be approximately described by a *geocentric dipole* with an axis roughly parallel to the one of the Earth's rotation. Consequently, on the Earth's surface, the geomagnetic field possesses in good approximation a geometry as if it were generated by a fictive bar magnet ("dipole"), situated in the center of the Earth, with a magnetic moment of ca. 8×10^{22} A m^2 (Fig. 150). Roughly 80 % of the overall magnetic field can be described by this dipole component and the remaining portion is called the non-dipole field. The points at which the dipole axis intersects the Earth's surface are called geomagnetic poles and the geomagnetic equatorial plane is geocentrically perpendicular to the dipole axis. At present, the dipole axis is inclined to the rotation axis at about 11° towards the SW, if observed from the North Pole. The intensity F decreases half as much from the geomagnetic poles to the equator and at present comprises ca. 48 μT in central Europe.

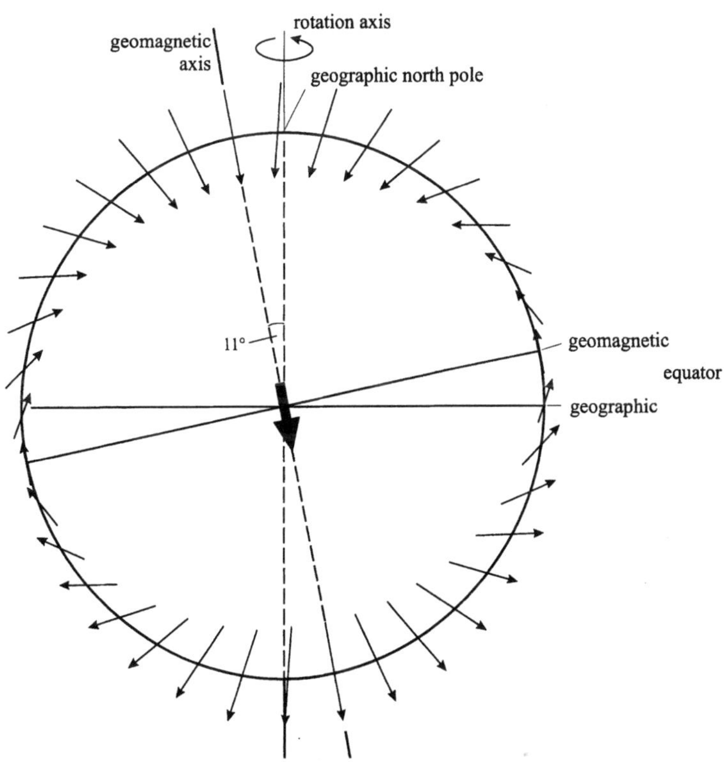

Fig. 150. The Earth's magnetic field possesses in good approximation a geometry as if it were produced by a fictive bar magnet situated in the Earth's center (geocentric dipole) with an axis inclined 11° to the rotation axis

The continuous record of the magnetic field reveals its temporal variation in respect of its direction and intensity – a phenomenon known as *secular variation*. The secular variation is expressed in a western drift of the Earth's magnetic field. An observer who studies the magnetic field from a fixed point recognizes apparently, but not strictly periodic variations of D, I and F. The amplitude of the directional variations falls within the range of ca. 40° and the irregular period length comprises several hundred years, an observation that gave rise to the expression"secular" variation (Fig. 151). The secular variation is of regional character, i.e., uniform D and I values are only met in restricted regions. It is mainly a feature of the non-dipole field.

Each material when exposed to a magnetic field experiences complex interaction, with the result that the intensity of the magnetic field becomes either intensified or decreased. This also applies to rocks penetrated by the Earth's magnetic field. The term *rock magnetism* covers the magnetic properties of minerals and rocks. The phenomenon of magnetism ultimately arises from the motion of the electrons. The electron obtains a magnetic moment through its spinning about its axis and its orbital motion about its

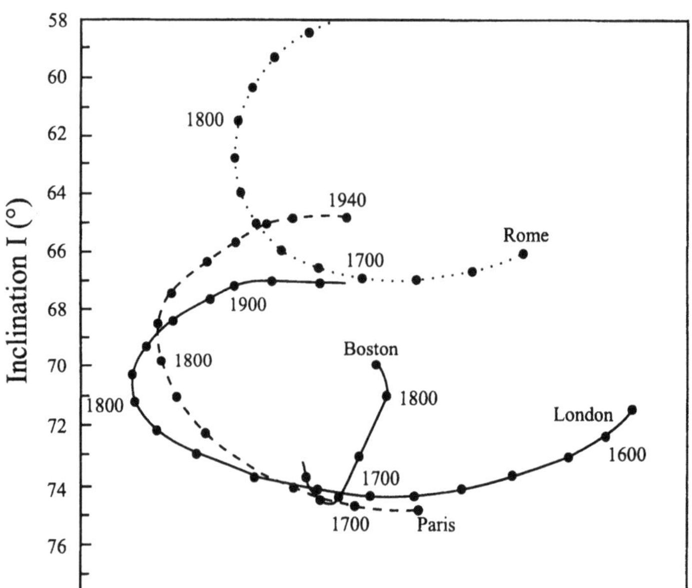

Fig. 151. The historical record of declination and inclination values at different places reveals the quasi-periodic and regional character of the secular variation. (After Aitken 1990, reprinted with permission of Addison Wesley Longman Ltd.)

atomic nucleus. The resulting magnetic moments of the atoms largely experience mutual compensation by anti-parallel or statistically disordered orientations. Incomplete compensation finds its material expression in the *magnetization J*. The magnetization of a material can be induced by an external magnetic field H (*induced magnetization* J_i) or be retained as *remanent magnetization* J_r after the magnetic field has been switched off. J_i grows proportionally at small field intensities H. The *susceptibility k* ($k = J_i/H$) is a measure of the extent to which a material is magnetized. It is dimension-less but possesses, if normalized to the density (*specific susceptibility*), the dimension $kg^{-1}\,m^3$. At high field intensities, when all magnetic moments of the material are aligned, saturation of the magnetization is reached. According to their type of susceptibility, one distinguishes diamagnetic (k is negative and very small), paramagnetic (k is positive and very small) and in a broader sense ferromagnetic (k is positive and large) minerals. The ferromagnetic minerals intensify an externally applied magnetic field. Even minor amounts of admixed ferromagnetic minerals dominate the rock susceptibility, since the susceptibility of ferromagnetic minerals exceeds that of paramagnetic or diamagnetic rock constituents by four orders of magnitude. Ferromagnetism disappears above the *Curie temperature* T_c. Important ferromagnetic minerals are the iron and titanium oxides magnetite (Fe_3O_4), ulvospinel (Fe_2TiO_4), maghemite (γ-Fe_2O_3), hematite (Fe_2O_3), and ilmenite ($FeTiO_3$); the iron hydroxide goethite ($FeO(OH)$); and the iron sulfide pyrrhotite ($Fe_{1-x}S$). They are the carriers of natural remanence in rocks and artifacts.

Natural Remanent Magnetization (NRM). Many rocks possess a natural remanent magnetization NRM. It is found in igneous, metamorphic and sedimentary rocks, but also in ceramic sherds and bricks. The NRM is produced by minor admixtures of ferromagnetic minerals. The processes leading to the generation of remanent magnetization are complex and can be merely touched upon within this treatise (compare e.g., Soffel 1991; Butler 1992). The various types of NRM differ from each other in the imprint mechanism and depend on the type and grain size of the ferromagnetic mineral constituents of the rock and the duration, direction and intensity of the magnetic field. Two important NRM types are displayed in Fig. 152.

Thermoremanent magnetization (*TRM*) occurs as soon as a rock cools below T_c in the Earth's magnetic field (Thellier 1941). As a result of this the magnetic moments align themselves with the direction of the field. The thermoremanent magnetization of the rock reflects the direction and intensity of the magnetic field which was prevalent during the cooling event. The magnetic moments, which at first are mobile, become blocked in the acquired magnetic direction at the blocking temperature T_b, which lies just

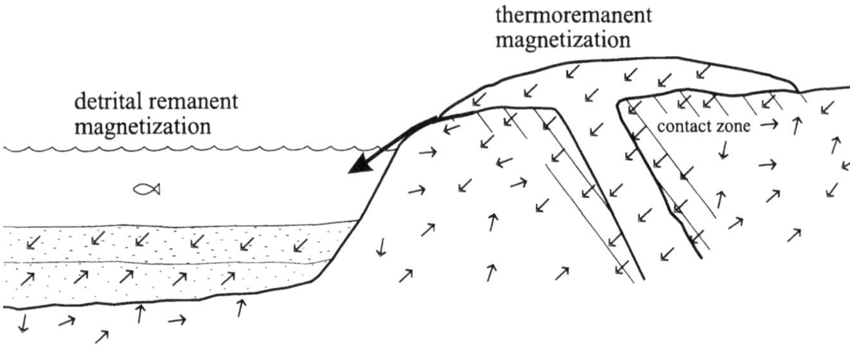

Fig. 152. Schematic representation of thermoremanent magnetization in a volcanic rock and its baked contact zone, as well as detrital remanent magnetization in sediments of different ages

below T_c. The values of T_c and T_b for the different ferromagnetic minerals cover a wide range < 675 °C, whereby stable TRM requires blocking temperatures T_b > 200 – 300 °C. In this "frozen" state the remanent magnetization shows resistance to the re-orientating magnetic fields, the *coercivity*. A crucial prerequisite for dating applications is the TRM stability. In this context the grain size plays an essential role. Since magnetite is a major TRM carrier, high heating temperatures (> 580 °C) are favorable for the formation of a strong and stable natural remanence. A partial thermoremanent magnetization (PTRM) will occur, if the heating temperature exceeds some but not all the T_c values of the ferromagnetic mineral components. TRM is essentially found in basalt and other volcanic rocks, burned soils, kilns and ceramic objects. These materials gained the TRM during defined heating events at their formation or a later stage, which are of interest for manifold dating problems.

If they settle through quiet water, ferromagnetic mineral grains can more or less align with the direction of the magnetic field and retain this orientation in the sediment. Subsequent compaction can stabilize this grain orientation. In this way the sediment gains a natural remanent magnetization, the so-called *detrital remanent magnetization* (*DRM*) which records the field direction prevalent during the deposition event. Flowing water, however, disturbs the aligning process of the magnetic mineral grains so that at flow velocities > 10^{-2} m s^{-1} no significant remanence is created. The possibility of acquiring DRM in clastic sediments decreases with increasing grain size. Well suited are clays, whereas this rarely can be said of sand. The grain shape also has an influence. The orientation of platy or elongated grains follows the bedding plane of the sediment. Consequently these grain shapes do allow the manifestation of the declination;

however, they suppress a recording of the inclination. Prior to compaction a fine-grained sediment can acquire the so-called *post-depositional remanent magnetization* (*PDRM*) by rotation of the magnetic grains within the water-filled pore spaces aligning the magnetic moments into the field direction. This orientation can be preserved through subsequent diagenetic compaction of the sediment (Verosub 1977). Unlike DRM, in the case of PDRM, the grains can more easily achieve alignment parallel to the inclination, thus reducing the risk of an inclination error. PDRM, however, yields a more or less delayed record of the magnetic field with respect to the sedimentation process. Hence, both remanence types can yield systematically different information if the magnetic field has changed during the diagenetic stage. Due to bioturbation, which affects a several-cm-thick upper sediment layer, the paleomagnetic fine structure becomes blurred. Thus bioturbation limits the temporal resolution of DRM depending on the sedimentation rate (Hambach 1992). Generally, DRM enables the dating of fluvial, limnic and pelagic sediments deposited under calm conditions. Also in the aeolian loess, DRM – more likely PDRM – occurs. The DRM of sediments is weaker than the TRM of volcanic rocks.

If ferromagnetic minerals – frequently hematite and goethite – are freshly formed within the pore space of a rock, the magnetic moments of the growing crystals become aligned parallel to the external magnetic field (Kobayashi 1959). This stable component, called *chemical remanent magnetization* (*CRM*), preserves the field direction prevailing at the time of a secondary event and hence may disturb the dating based on TRM or DRM. It is frequently met in weathered rocks. It may also have been imprinted during a metamorphic event. In "red beds" the hematite skins were formed during deposition or soon after under oxidizing conditions, resulting in the record of the synsedimentary magnetic field. Furthermore, in all rocks *viscous remanent magnetization* (*VRM*), which has been gradually formed over time by weak magnetic fields, is observable and has to be attributed to the long-term effect of the recent magnetic field. This remanence component possesses only a weak stability and can be magnetically or thermally removed, otherwise it would disturb the dating. It might, however, possess a certain dating potential for burned clay (Atkinson and Shaw 1991). Further remanence types exist, in addition to those discussed.

The various co-existing NRM types of a rock always concern only a part of its constituting ferromagnetic minerals. However, only a single NRM type is relevant to particular questions, e.g., TRM or DRM to dating problems. Therefore attempts are made in the laboratory to separate the *characteristic remanent magnetization* (*ChRM*) – being of relevance to a specific problem – from other NRM types. This is achieved by magnetic and thermal demagnetizing techniques, which make use of the different stabilities of the individual NRM components, in combination with mineralogical examination.

On condition that the age of a sample is independently known, the direction and intensity of the ancient geomagnetic field for a specific site and time can be deduced from the paleomagnetic data. Hence, on the basis of many paleomagnetic data, the history of the Earth's magnetic field can be reconstructed far beyond the historic records of the last four centuries. Once the exact temporal course of the field variation has been uncovered, this knowledge can be used in turn for dating of remanently magnetized samples of unknown age. Secular variations and polarity changes represent phenomena essential to magnetic dating.

Paleo-Secular Variation. Because of the relatively fast rate of secular variation its paleomagnetic detection requires NRM carriers possessing a high temporal resolution power and high data density (ca. five data points per century). In addition, the ages of the ChRM-imprinting events need to be known independently, and that with an absolute error lying significantly below the variation period of several hundred years. Ceramic objects frequently fulfill this requirement, but rocks rarely do. Since the events of firing and TRM acquisition coincide for ceramic objects, and since their production dates are frequently fairly well known from historical contexts, it is possible to reconstruct the course of the secular variation over the last 2500 a for Europe. For the preceding prehistoric millennia the ages of ceramics and kilns must be deduced from the find contexts with calibrated ^{14}C dates. The temporal resolution power of TL ages is insufficient for this purpose. Archaeomagnetic curves, reaching differently far back in time, exist for the Britain (Tarling 1991), central Europe (Thellier 1981; Thouveny and Williamson 1991; Reinders et al. 1997), southeast Europe (Kovacheva and Zagniy 1985; Marton 1991), the Ukraine (Rusakov and Zagniy 1973a), Anatolia (Becker 1979; Korfmann 1987), China (Wei et al. 1981), North America (Wolfman 1984, 1990a; Sternberg and McGuire 1990; Labelle and Eighmy 1997) and Mesoamerica (Wolfman 1990b). At a first approximation, the amplitude of the directional variation swings periodically. Such standard curves of the secular variation serve the archaeomagnetic dating. The inclination and declination values of the TRM are measured for the samples in question and linked with the standard curve. However, for a given value of D or I, several age solutions are possible. In order to ease this ambiguity problem, the availability of at least two or all three magnetic field parameters is desirable. The secular variation possesses regional character, i.e., similar D and I curves are only met in limited regions with extensions of about 1000 km, a fact to be taken into consideration, if standard curves are used for the magnetic dating (Tarling 1991).

For periods prior to the occurrence of ceramics (around 5500 BC in central Europe), sediments with sufficiently high deposition rates (> 0.1 mm a^{-1}) can be used for the reconstruction of the secular variation on the basis of their DRM and PDRM. Especially favorable in this context are the annually formed varves in freshwater lakes. Even if their temporal placement has not been evaluated, the annual resolution enables the recording of the secular variation for more remote epochs. The independent dating of the sediments is – apart from varve counting (Sect. 10.1.1) – furthermore done by ^{14}C dating of organic components (Sect. 5.4.3), pollen analysis (Sect. 10.2.4) and biostratigraphy. The curves of secular variation obtained for the Holocene and the last Glacial period show similar behavior as the ones deduced for the archaeological and historical periods (Haverkamp and Beuker 1993). Loess, too, is suitable for recording the paleo-secular variation (Fig. 153). Instead of using curves of *D* and *I*, the secular variation can be presented as a *virtual geomagnetic pole VGP*,

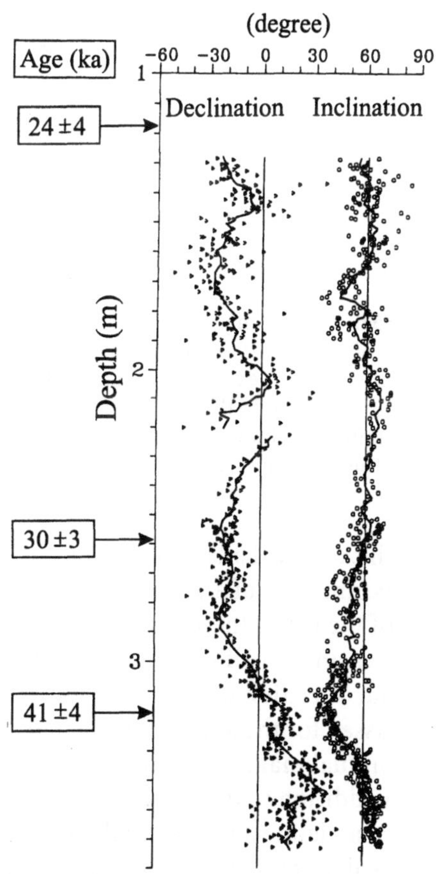

Fig. 153. Declination and inclination of the characteristic remanent magnetization of the Würmian loess profile at Böckingen, southern Germany (from Limbrock 1992). The rapid loess deposition of 0.2 mm a^{-1} and the dense sampling enables a high temporal resolution of the paleo-secular variation. The ages were derived by loess TL dating (Zöller and Wagner 1989)

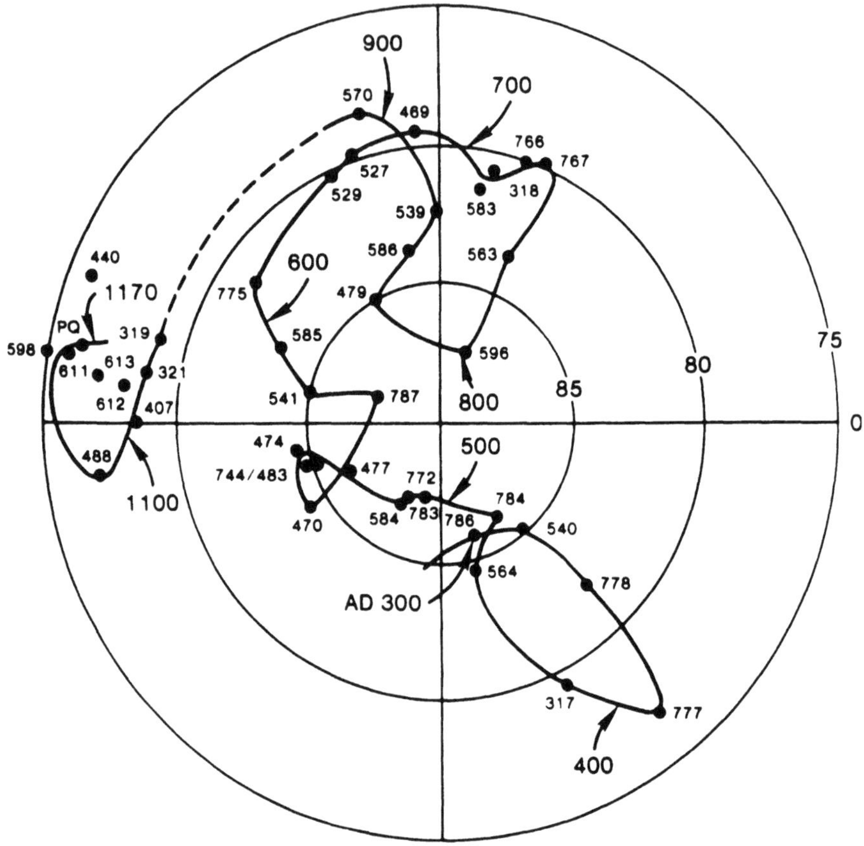

Fig. 154. Archaeomagnetic field direction (standard curve) for Mesoamerica between 300 and 1170 AD as virtual geomagnetic pole VGP, based on numerous data points with sample number (from Wolfman 1990b). In the VGP diagram, the geographic north pole is in the *center*, surrounded by the 85°, 80° and 70° latitude circles; the 0° longitude is indicated at the *right*. This representation of the secular variation eases the interregional comparison

which can be calculated from the NRM direction parameters *D* and *I* under the assumption of a geocentric dipole (Fig. 154). With increasing geological age of the layers these natural records become fragmentary due to missing profiles and insufficient dating. Within limited regions the pattern of the secular variation can be used for dating sediments – as has been already discussed for the archaeological period. The secular variation also enables the mutual magnetostratigraphic correlation of different sequences.

In addition to the directional variation, a variation of the field intensity was also delineated, essentially from TRM measurements on fired archaeological objects, in central Europe by Thellier (1977) and Bucha (1971), in

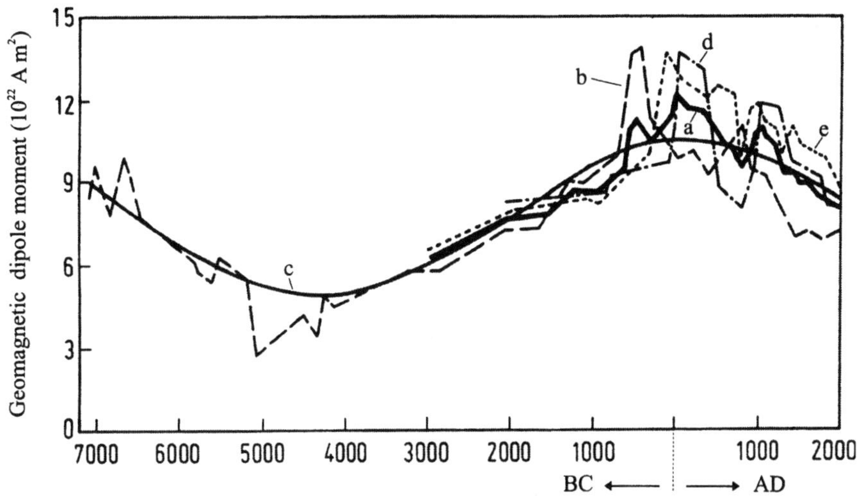

Fig. 155. Average geomagnetic dipole moment (*a*) since 7000 BC from archaeomagnetic intensity measurements on ceramics, kilns and burned sediments in Europe (*b*), southwestern North America (*d*) and Japan (*e*). The curve (*c*) is smoothed. The similarity of the curves from different continents supports the global character of the dipole change. (After Bucha 1971)

eastern Europe by Rusakov and Zagnyi (1973b), in southeast Europe by Kovacheva and Veljovich (1977), in the eastern Mediterranean by Hassan (1983) and Aitken et al. (1989) and in Peru by Gunn and Murray (1980). The paleo-intensity data of the last 9 ka show in good approximation that the geomagnetic moment was subject to a long-period sine-shaped trend with variations of (5–10) × 10^{22} A m^2 (Fig. 155). Paleo-intensity measurements were extended to older fire places, volcanites (Barbetti and Flude 1979; Brassart et al. 1997) and sediments (Guyodo and Valet 1996; Schneider and Mellow 1996). The data also reveal the long-term variation of the dipole field. Meynadier et al. (1992) observed strong quasi-cyclic intensity variations in sediment cores from the Indian Ocean covering the last 140 ka (Fig. 156). These oscillations are presumably of global character, as suggested by data obtained from the Mediterranean, and exhibit periods of 100, 43, 24 and 19 ka. Surprisingly enough, these values concur with those of the Earth's orbital parameters which are basic to the Milankovitch theory (Sect. 10.2).

Geomagnetic Polarity Time Scale. Over time periods of several 10^4 a the directional changes of the secular variation compensate each other, so that on average the geomagnetic poles coincide with the Earth's rotation pole. Apparently, the geomagnetic dipole axis is oriented along the Earth's rota-

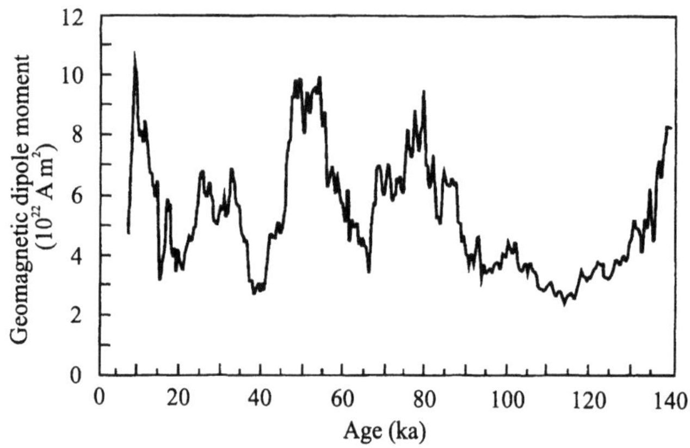

Fig. 156. Fluctuation of the geomagnetic dipole moment during the last 140 ka. The curve is based on NRM data of deep-sea cores from the Indian Ocean and the Mediterranean Sea. (Reprinted from Meynadier et al. 1992; copyright 1992, with kind permission from Elsevier Science Ltd., The Boulevard, Langford Lane, Kidlington OX5 IGB, UK)

tion axis. This coupling, however, can be parallel or antiparallel. *Polarity changes* occur at irregular intervals, giving rise to a reversal of the geomagnetic field. Consequently, the axial geocentric dipole can flip into the opposite direction. Brunhes (1906) observed on Tertiary volcanites and heated soils from the Massif Central, France, magnetizations which were parallel (normal) as well as anti-parallel (reversed) in relation to the present geomagnetic field. Matuyama (1929) observed normal magnetization for the younger and a reverse one for the older successions of Pleistocene basalt flows of known ages. Apart from volcanic rocks, alternating polarities have been found in marine and continental sediments as well. The polarity reversal is attributed to magneto-hydrodynamic processes in the Earth's fluid outer core. The global duration of transitional field direction takes place within several thousand years, as has been shown, for instance, for the Matuyama–Brunhes boundary (M/B boundary) by Singer and Pringle (1996). The acceleration of the Earth's rotation, caused by the lowering of the sea level during glaciations, has been suggested as a possible mechanism for field reversals, since reversals were apparently coupled to cold stages during the last 2.6 Ma (Worm 1997).

The repeated and worldwide synchronous reversals of the axial geomagnetic dipole field furnish the basis of the geomagnetic polarity time scale, which was first established by Cox et al. (1963) for the past 3 Ma using K–Ar-dated basalts. Improved polarity time scales are based on precise K–Ar dating, especially by applying the $^{39}Ar-^{40}Ar$ laser technique to volcanites, and on magnetostratigraphic correlation with deep-sea sediments.

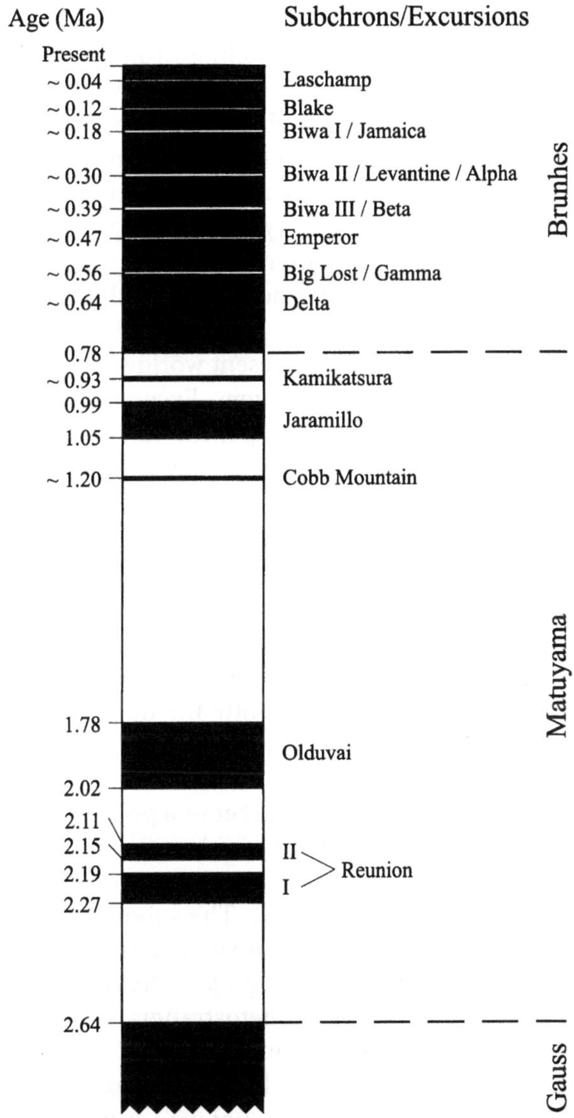

Fig. 157. Geomagnetic polarity time scale of the last 2.8 Ma. The scale is based on NRM polarity measurements of volcanites and deep-sea sediments. The temporal placement is derived from K–Ar-dated continental volcanites and their magnetostratigraphic correlation with the deep-sea record as well as from astronomic dating of deep-sea sediments. (After Champion et al. 1988; Hilgen 1991; McDougall et al. 1992; Tauxe et al. 1992, 1996; Baksi 1993; Nowaczyk et al. 1994)

These sediments have the advantage over the episodically formed basalt flows in that they provide a continuous record of the paleomagnetic field. Dating of deep-sea sediments is conducted by linear interpolation between known time markers, assuming constant sedimentation rates, and by astronomical data (Sect. 10.2.1).

The polarity time scale (Fig. 157) is divided into differently long (~1 Ma and longer) intervals of uniform polarity, the *chrons*. Counting of chrons begins with the recent chron C1. The last four chrons are named after the prominent researchers of geomagnetism *Brunhes*, *Matuyama*, *Gauss* and *Gilbert*. The presently prevailing Brunhes chron (Ç1) of normal polarity commenced 778.7 ± 1.9 ka ago (Singer and Pringle 1996). It is preceded by the reversed Matuyama chron (C2). The chrons are interrupted by short (several 10^3–10^5 a) global episodes of reversed polarity, the *subchrons* or *events*. The subchrons are named after their *locus typicus*, e.g., the Olduvai subchron (normal polarity between 1.97–1.78 Ma within the reversed Matuyama chron) called after the Olduvai Gorge in East Africa. It is not at all trivial to verify global synchronism with an accuracy of 10^4 a by independent dating. For this and other reasons, certain subchrons have not yet been proved to really represent worldwide polarity reversals or rather so-called geomagnetic *excursions*. Excursions are intermediary VGP deviations from the longer-lasting mean value, exceeding the extent of a secular variation but not themselves representing polarity reversals. Excursions do not possess a global character, but nevertheless have a strong regional effect on direction and intensity of the magnetic field and may mimic field reversals. The subchrons and excursions with inferred ages of less than 500 ka were compiled by Nowaczyk et al. (1994). Paleo-intensity determinations on deep-sea sediments, covering the last 4 Ma, show that the polarity reversals coincide with and may be triggered by field-intensity minima and that subsequently the intensity regenerates within a few ka (Valet and Meynadier 1993).

The polarity time scale can, in favorable cases, be used for numerical dating, namely if the time marker of a polarity change, e.g., the Olduvai subchron or the M/B boundary, is identified in a rock. Usually, however, a relative dating is conducted by comparison of a measured polarization pattern with the polarity time scale. The subdivision and correlation of rock sequences on the basis of their successive normal and reversed polarity pattern is called *magnetostratigraphy*. Provided the polarity pattern has been independently dated, magnetostratigraphy may enable the temporal placement of volcanic and sedimentary rock sequences. In deep-sea sediments the polarity scale can be linked also to biostratigraphic and lithostratigraphic characteristics. The unique advantage which magnetostratigraphy has over bio- and lithostratigraphy is that the feature on which it is based has a global synchronism. For this reason the beginning of the Middle Pleistocene is magnetostratigraphically placed at the M/B boundary.

9.2 Practical Aspects

The first step in the sampling procedure is the assessment of whether the material is suitable for magnetic dating with respect to its mineralogical composition and NRM type. For the period of the last 2 Ma suitable materials are above all volcanic rocks and burned soils retaining TRM, as well as clastic sediments possessing DRM or PDRM. The sampling strategy must account for the relationship of the paleomagnetic age data to the geological or archaeological event of interest. Weathered samples are inappropriate because of their CRM. Fine-grained clastic rocks, such as clay and silt, are preferable to sands of increasing grain size. If the age is to be evaluated from the pattern of the secular variation, as for instance in the case of limnic sediments, a continuous sample series is required that sufficiently covers the course of the magnetic field variations. Drilled cores, for example, serve this purpose. Exposed sites on hill tops, where NRM contributions by lightning strokes have to be expected, should not be sampled. For paleomagnetic dating the samples must be taken in an oriented way. Only outcropping rocks are suitable. Tectonic movements which took place after the NRM acquisition must be known and accounted for. Kilns and burned soils usually occur in their primary position which they had during heating. Very rarely ceramic objects are found to be still in their original position – like vessels which remained in situ in the kiln. Whether burned clays had been sufficiently heated can be judged on the basis of their lacking plasticity when wetted. A summary of all aspects of sampling for paleomagnetic dating was presented by Eighmy (1990).

Usually the sample orientation is to be determined with a magnetic compass. The directions of dip and strike can be directly marked on a flat surface of the sample. On uneven samples plain surfaces can be artificially created with gypsum plaster on which the north direction is inscribed. The use of a sun compass is appropriate if magnetic disturbances occur in rocks adjacent to the sample site. Instead of collecting hand samples, oriented cores (of 2.20 cm length and 2.54 cm diameter) can be drilled from outcrops by portable coring drills. Such cores have the advantage of immediately fitting into the magnetometer without further preparation in the laboratory. Exposed deposits of non-consolidated sediments are most suitably sampled by means of a cube-shaped plastic container which is inserted into the sediment starting from an oriented plane. Non-consolidated sediments from the subsoil (including underwater sediments) are sampled by pushing a plastic coring tube into the ground. Drilled samples may present difficulties in the determination of the azimuth direction, rather than the vertical one. If the azimuth is not known, apart from the paleo-intensity only the paleo-inclination can be measured, which, however, is sufficient for polarity evaluation (exceptional cases are those sites of low geomagnetic latitudes). Small cylinders or cubes (2.1 cm edge length) are usually prepared in an oriented way for the magnetometer measurements. The orientation error inherent in the sampling procedure should be kept within 1–2°. The errors of measurement commonly comprise 1–2° for the direction and 2–3% for the intensity (Tarling 1991). The evaluation of the paleomagnetic parameters requires a total of at least seven samples per interesting level – this applies to both cores and fully oriented hand samples.

A variety of instruments, which differ in respect to the principle, sensitivity, precision and expense of measurement, are employed for the measurement of remanent magnetization (Collinson 1983). For strongly magnetized volcanites the spinner magnetometer – working on the induction principle – is sufficient. Inside this instrument the samples are subjected to fast rotation. On the other hand, weakly magnetized sediments require the more sensitive cryogen magnetometer. Measuring the sample in different orientations allows the determination of the direction and intensity of the NRM and thus the paleomagnetic field parameters.

Two magnetic properties of rocks are used for separating the NRM into its different types. First of all, the different NRM types are always linked with particular ferromagnetic mineral components of the sample, with the grain size playing a fundamental role. Secondly, the grain size influences both the coercivity and the blocking temperature, so that the various grain sizes differentially withstand the magnetic and thermal NRM demagnetization. Demagnetization of the sample is achieved either in an alternating magnetic field of a certain intensity or during heating. By stepwise increase of the field intensity or of the temperature, the NRM components of lower

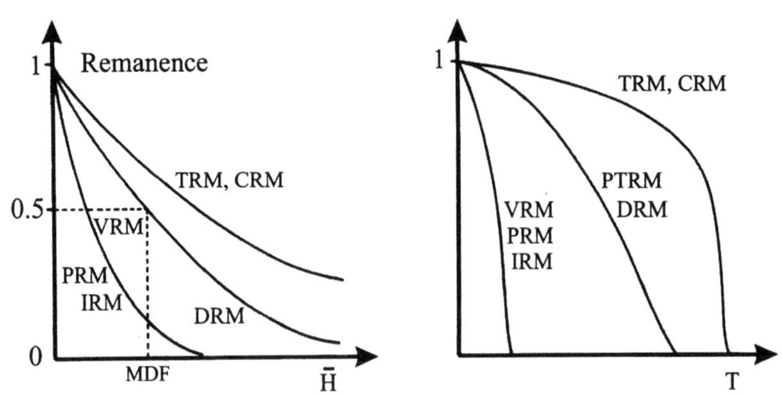

Fig. 158. Schematic of the magnetic (on the *left*) and thermal (on the *right*) demagnetization of the natural remanent magnetization NRM (from Soffel 1991). Owing to their different stability, individual NRM components can be separated (*TRM* thermoremanent magnetization; *CRM* chemical remanent magnetization; *DRM* detrital remanent magnetization; *PTRM* partial thermoremanent magnetization; *VRM* viscose remanent magnetization)

coercivity forces and blocking temperatures, respectively, are removed at first, until finally the more stable residual types of remanence fully disappear (Fig. 158). The different NRM types greatly differ in their demagnetization behavior, with the TRM and CRM being the most stable. In case these demagnetization procedures do not work, further techniques are available (Soffel 1991). The identification of ferromagnetic minerals of the sample is done by reflected light microscopy, X-ray diffraction, scanning and transmission electron microscopy as well as by Mössbauer spectroscopy. In addition, the magnetic characteristics of rocks are exploited for this purpose.

9.3 Application

Volcanites. Basalts and other volcanic rocks contain rather high admixtures of ferromagnetic minerals and are thus appropriate candidates for archaeomagnetic measurements. Their susceptibility decreases with increasingly acid rock character. The rocks solidify at temperatures of 700–1000 °C thus exceeding the Curie temperatures of the ferromagnetic minerals. These minerals are thermoremanently magnetized during cooling of the hot, but for the most part already solidified rock. Tuffaceous rocks are less well suited for paleomagnetic measurements, firstly because the volcanic fragments frequently have already cooled below the Curie temperature prior to their deposition, and secondly because their deposition is too fast to allow the acquisition of sedimentation remanence. Exceptions are the hot-deposited ignimbrites (Kent et al. 1981). Redeposited tuffs can show DRM and CRM (Hillhouse et al. 1986).

The historically precisely known eruptions of Mount Etna, Italy, enable the establishment of a standard curve of the secular variation for Sicily from 1169 AD onwards (Rolph et al. 1987). Some lava samples from Etna of doubtful historical ages could be dated by their paleomagnetic *D* and *I* parameters. Another example is the decline of the Late Minoan culture on Crete around 1500 BC. In this context it is not simply the chronology but the contemporaneity which is in particular interesting. The suddenness of the Minoan culture's cessation gave rise to the hypothesis that the tremendous natural catastrophe of the Late Minoan eruption of the Thera (Santorini) volcano was the responsible factor (Sect. 5.4.3). The Late Minoan culture on Crete was assumed to have been extinguished by the destructive effects of ash layers, earthquakes and tidal waves originating from the 120 km distant Thera island. On Thera mighty ash layers of the upper pumice sequence record two eruptive episodes, with the pumice (basal Plinian pumice) representing the first phase and the overlying cross-bedded pumice tuffs originating from the second phase (Pichler 1973). Both ash complexes were sampled for paleomagnetic investigation (Downey and Tarling 1984). On Crete the ash layers are only a few centimeters thick; consequently, they are not suitable for paleomagnetic examination. On Crete the most suitable sample material for archaeomagnetic studies are fired mud bricks of the Late Minoan palaces, since the destruction was accompanied by a tremendous conflagration. Figure 159 displays the results of the paleomagnetic study. The systematic differences in the paleomagnetic direction (5.2°) and intensity imply a significant hiatus between the first and second phases. Supposing a directional change of 0.07 $°a^{-1}$, the two eruptions are separated by an interval of 20 ± 10 a. Each one of the burned layers of the Cretan palaces displays an archaeomagnetic direction that is identical with the direction of one of the two ash layers on Santorini. Identity with the pumice

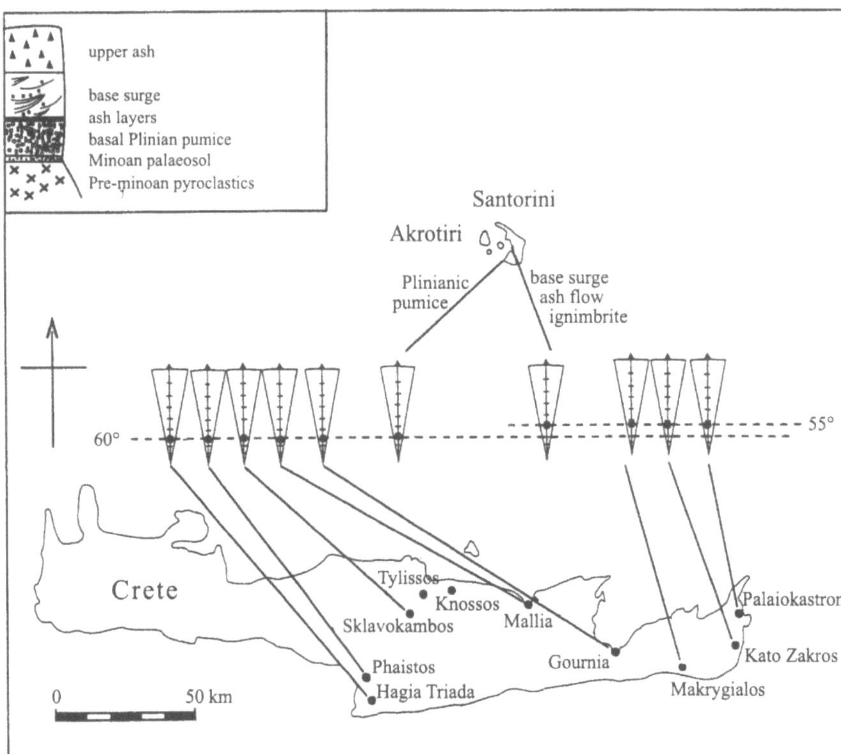

Fig. 159. Archaeomagnetic directions of Late Minoan volcanic ashes on Thera (Santorini) and burned destruction horizons of the Late Minoan palaces on Crete. The mean direction is represented by the dot in the stereographic projection, the central line indicates north, the inclinations are steeper on the left (60°) than those on the right (55°). (After Downey and Tarling 1984)

of the first eruptive phase was found for central Crete and with the ashes of the second eruption phase for eastern Crete. The thus deducible contemporaneity supports the idea of a causality between the Thera eruptions and the decline of the Minoan culture.

The geomagnetic polarity time scale plays an essential role in the magnetostratigraphic dating of early hominid finds in East Africa. At Lake Turkana (former Lake Rudolf) in northern Kenya, numerous fossil skeletal remains and stone tools attributed to early hominids – among others *Homo erectus* – were found in the 450 m mighty Pliocene/Pleistocene Koobi Fora formation. Intercalated in the sedimentary sequence are several tuffaceous layers forming the tephrochronological backbone of the stratigraphic correlation of the find horizons in the Turkana basin. Of particular concern within this sequence is the KBS Tuff. The paleomagnetic investigation of tuffs by Hillhouse et al. (1986) proved that the KBS Tuff falls into the Olduvai subchron (Fig. 160). This magnetostratigraphic dat-

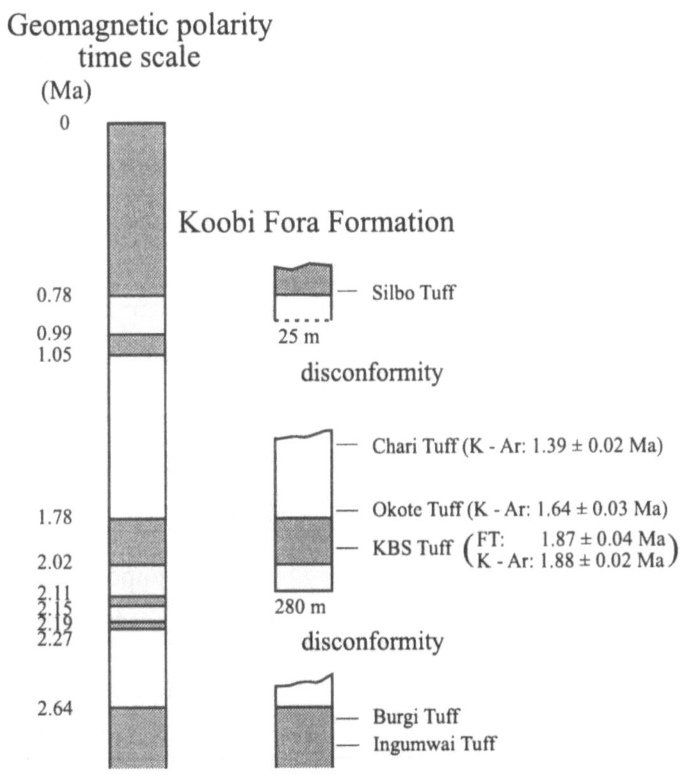

Fig. 160. Magnetostratigraphy of the Koobi Fora Formation at Lake Turkana, Kenya (after Hillhouse et al. 1986). The paleomagnetic dating is in agreement with fission track (Gleadow 1980) and K–Ar ages (McDougall et al. 1980, 1985; McDougall 1985)

ing is in agreement with the K–Ar and fission track ages of this tuff. This result helped to turn the controversy on the temporal placement of the KBS Tuff in favor of the younger chronological scheme (Sect. 6.1.3). The acquisition of the characteristic remanence of these mostly redeposited tuffs is attributable to DRM and CRM.

Kilns and Burned Soil. Fireplaces frequently show in situ preservation, i.e., the burned ground is still found in the same positional context as it was during the last firing. Such hearths provide inclination and declination as well as intensity values, and are thus datable. However, it has to be ascertained whether the samples being used for evaluating the TRM acquisition have been exposed to sufficiently high firing temperatures (> 400 °C). The strong temperature gradient prevalent in kiln walls demands sampling just from the face of heat contact (Eighmy 1990). On the basis of the paleomagnetic *D*, *I* and *F* parameters Kovacheva (1991) dated fired clay from several Bulgarian Neolithic sites with an accuracy of 300 a (Fig. 161). Two Early Medieval kilns for quicklime production from Herrsching, Bavaria, were archaeomagnetically dated with a high temporal resolution to 670 ± 30 AD

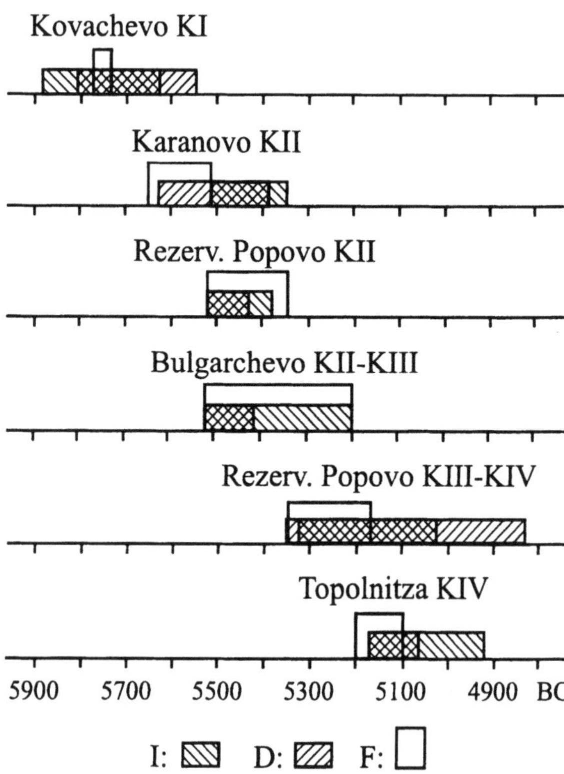

Fig. 161. Archaeomagnetic data from Neolithic sites in Bulgaria. The declination *D* and inclination *I* as well as the intensity *F* of the NRM were determined on in situ kiln structures and burned sediments and compared with standard curves. (After Kovacheva 1991)

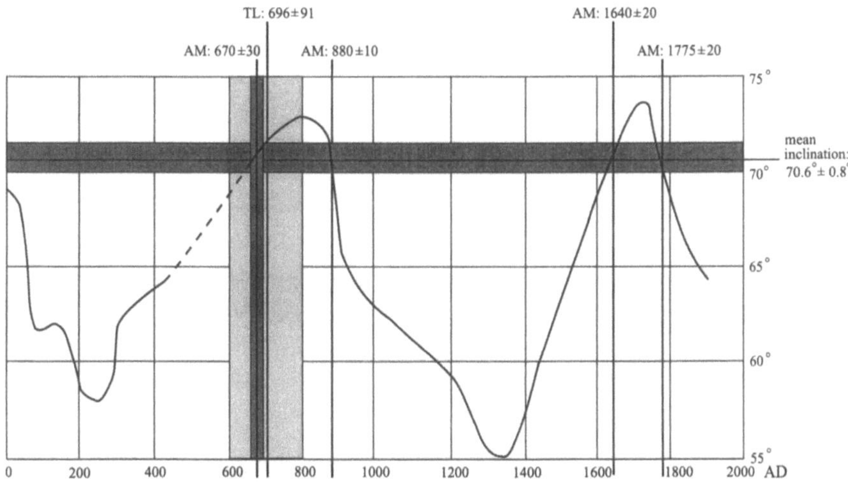

Fig. 162. Archaeomagnetic dating of a kiln for quicklime production in Herrsching, southern Germany (after Becker et al. 1994). Matching of the measured paleo-inclination to the standard curve (cf. Fig. 164) furnishes four possible archaeomagnetic ages *AM*. The ambiguity can be resolved in favor of AM = 670 ± 30 AD by TL dating. This combined methodological approach increases the dating accuracy

by Becker et al. (1994), who employed inclination measurements in combination with a TL age (Fig. 162). A large number of archaeomagnetic data on kilns and burned soils exist for Mesoamerica. Wolfman (1990b) sampled pre-Columbian sites of the last 7 ka mainly in southern Mexico. The regional standard curve (compare Fig. 154) covers the period 1–1200 AD. For this time range, Wolfman (1990b) conducted a whole series of archaeomagnetic age determinations, which in part are given with an accuracy of 20 a.

Ceramics and Bricks. Ceramic clays contain small amounts of iron-bearing minerals. During the firing process the water is eliminated from the hydroxide and, depending on the redox conditions, two- or three-valent iron oxides are formed. In this way the susceptibility of the material is further enhanced. Dark colours indicate reducing and red ones oxidizing firing conditions. Sufficient heating of the clay can readily be recognized from its non-plasticity in the presence of moisture. Usually firing temperatures of 700–800 °C were attained, and these are in excess of the T_c values of the ferromagnetic minerals. Lower firing temperatures (> 400 °C) only result in gaining PTRM by those mineral components with lower T_c. The prevailing field is thermoremanently imprinted during cooling. The fact that TRM acquisition is contemporaneous with the production of the object al-

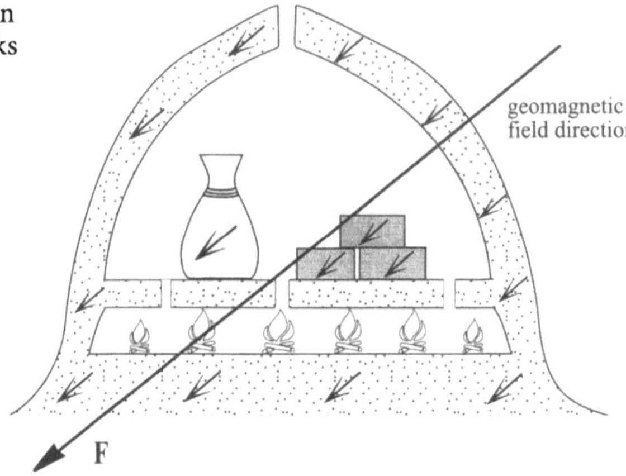

Fig. 163. Typical orientation of ceramic vessels and bricks during firing in the kiln

lows a straightforward interpretation of the archaeomagnetic age. Both the archaeomagnetic direction parameters and the field intensity are used. Sherds and bricks are usually found in any other orientation but the original one they had during firing, and hence the problem is the determination of their orientation during heating. Since pots are usually fired in an upright position and bricks lie on one of their sides during firing (Fig. 163), only the archaeomagnetic inclination and intensity can be used for dating. Their azimuth is no longer reproducible, which prevents the use of the declination for dating. Kiln remains, however, are frequently found in situ, but attention has to be paid to the potential occurrence of disturbances as a result of decomposition and other related features subsequent to the heating event. Another problem which sometimes arises is the refraction of the field lines affecting the inclination and declination. The extent of refraction depends on the position of the specimen relative to the ambient field as well as the size and the shape of the heated object (Evans and Hoye 1991; Reinders et al. 1997).

Dating requires the existence of standard curves of the archaeomagnetic field. Reinders et al. (1997) compiled such a calibration curve for central Europe on the basis of numerous D and I data (Fig. 164). Obviously, the standard curve does not always permit an unequivocal solution. Ambiguities can frequently be settled on the basis of archaeological and typological criteria as well as independent age data (Fig. 162). Once established for a certain region, the standard curves permit a dating accuracy of 25–50 a (Tarling 1991). The paleo-intensity is of minor use for age determination of fired clay objects, due to its slower variation and lower measurement precision, and to the present lack of appropriate standard curves. Paleo-intensity, however, may serve as a decision aid for the ambiguities mentioned above.

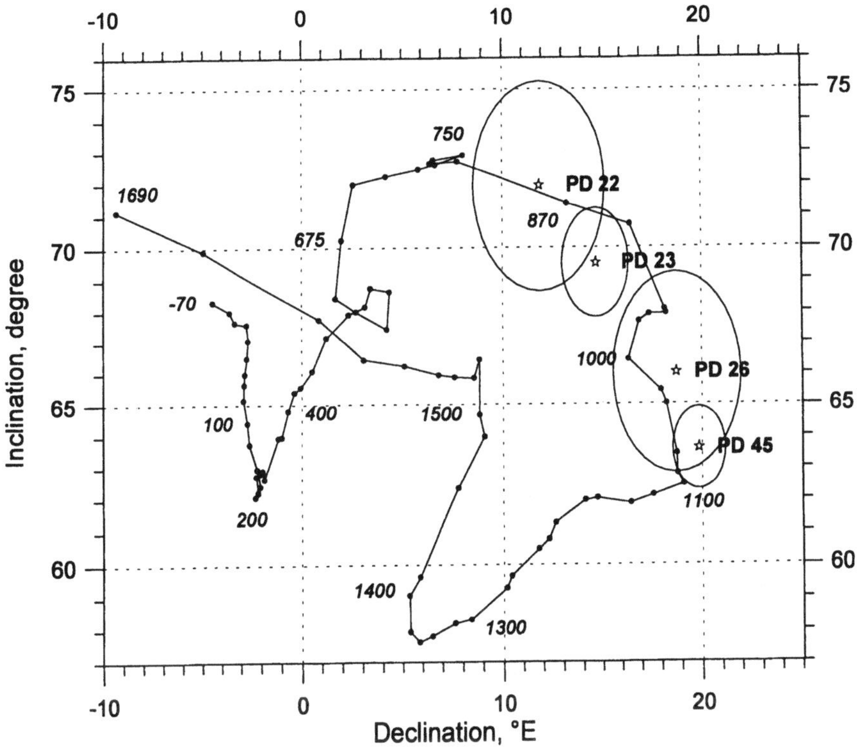

Fig. 164. Standard curve of the secular variation of declination and inclination during the last two millennia for western central Europe (from Reinders et al. 1997). The curve is based on direct measurements for the last 400 a and on archaeomagnetic measurements prior to this. The numbers along the curve denote years AD or BC (minus). The *stars* are archaeomagnetic data of several medieval kilns with the *circles* representing the 95% confidence limits

Deep-Sea Sediments. Deep-sea sediments form an almost ideal archive for the history of the magnetic field. Their continuous growth ensures a complete record of the temporal variation of the magnetic field. Their wide occurrence enables investigations over vast areas. The polarity pattern and the intensity variation of the NRM were observed in deep-sea sediment cores from all major oceans (Opdyke 1972). The polarity pattern enables the correlation of deep-sea sediments of very different facies (Fig. 165). Paleo-intensity variations (cf. Fig. 156) are also employed to an increasing extent for magnetostratigraphic correlation (Guyodo and Valet 1996). In combination with oxygen isotopy (Sect. 10.2.2) and micropaleontology, paleomagnetism – being based on the polarity reversals – of deep-sea sediments forms the framework of marine Quaternary stratigraphy. When linked to radiometric age data on continental volcanites, paleomagnetism

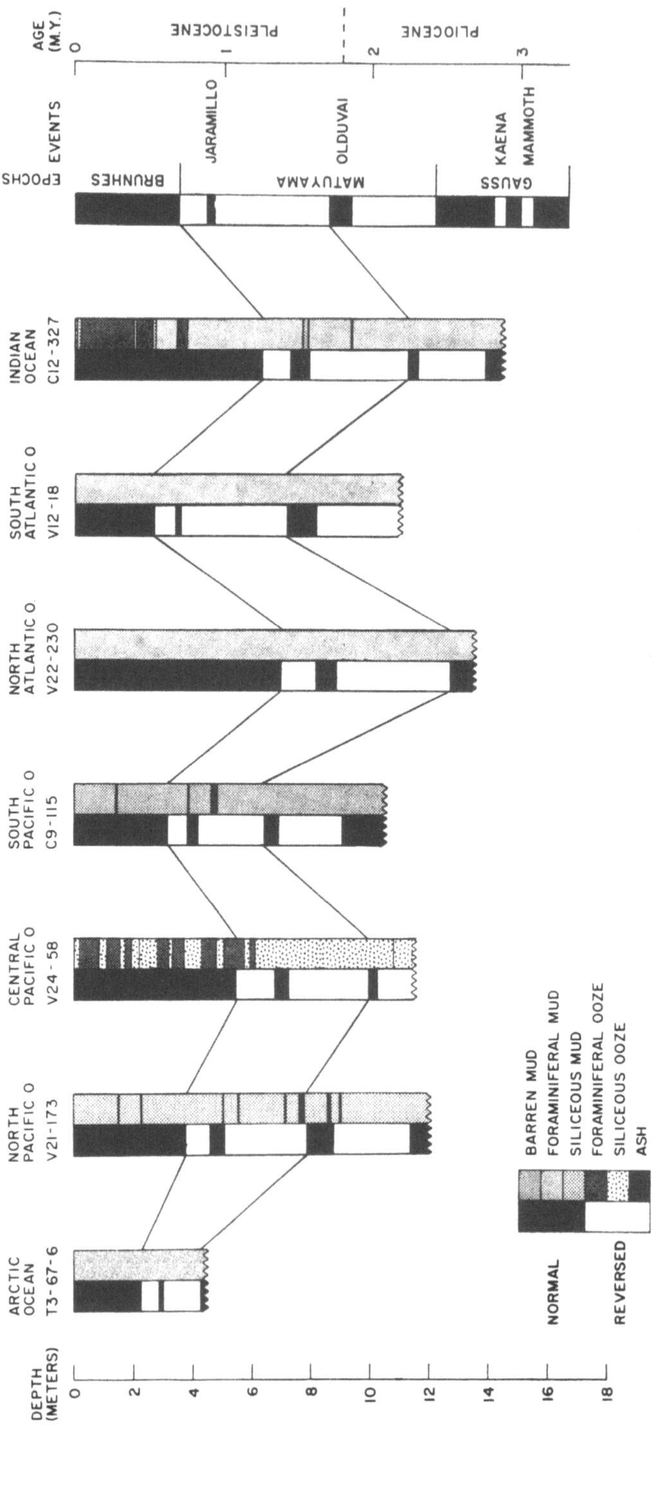

Fig. 165. Paleomagnetic correlation of deep-sea sediment cores of major oceans (after Opdyke 1972). The polarity pattern seems to be independent of the very different lithological composition of the deposits

furnishes chronometric time markers for the marine stratigraphic system. And vice versa, the continental successions can be coupled with this system by magnetostratigraphic connection.

The stable remanent magnetization of the deep-sea sediments is caused predominantly by small admixtures of titanomagnetite. The microscopically small (< 10 μm) ore grains presumably become blown by wind action over long distances onto the ocean surface, from where they slowly sink to the sea floor. Calm depositional conditions permit the ore grains to attain a magnetic orientation thus giving rise to DRM acquisition. As long as the sediment remains unconsolidated, the ferromagnetic minerals can adapt to changing field directions by re-orientation. In this way the sediment does not acquire a PDRM before reaching a certain burial depth, probably within a few cm on average below the sediment–water interface (Tauxe et al. 1996). Consequently, depending on the sedimentation rate (typically in the range of mm ka^{-1} - cm ka^{-1} for deep-sea sediments), a significantly long interval (10^3–10^4 a) may exist between the deposition event and the remanence imprint, resulting in the delayed recording of the magnetic field and a reduced temporal resolution. A further disturbance arises from organisms inhabiting the ocean floor and bringing about redepositional effects. This so-called bioturbation may affect a layer several cm thick beneath the sediment surface corresponding up to several 10^4 a. Bioturbation complicates the remanence acquisition within this time span and results in a temporal blurring of the NRM signal.

The dating of the Southeast Asian–Australian microtektites demonstrates exemplarily the high resolution power of magnetostratigraphy for deep-sea sediments. The occurrence of the microtektites in the deep-sea sediments is restricted to a layer a few cm thick adjacent to the M/B boundary. Since the formation of the microtektites is thought to be associated with a giant meteorite impact, speculations were made that the reversals of the magnetic field might have been induced by such a tremendous impact. Accordingly, it is thinkable that the M/B polarity change was the result of the impact that produced the Asian–Australian tektites (Muller and Morris 1986). The impact-induced dust which was ejected into the atmosphere would have led to a global cooling which in turn would have triggered increased growth of the polar ice caps. The resulting reduction of the Earth's momentum of inertia would have given rise to a faster rotation of the Earth and exert a strong influence on the hydromagnetic processes in the outer core, leading finally to a polarity reversal within less than 1 ka after the impact. With this possibility in mind, Schneider et al. (1992) analyzed azimuth-oriented deep-sea cores from the Indochinese Archipelago for their remanent intensity, declination and inclination. The M/B boundary is unequivocally recognizable by a very low field intensity and abrupt

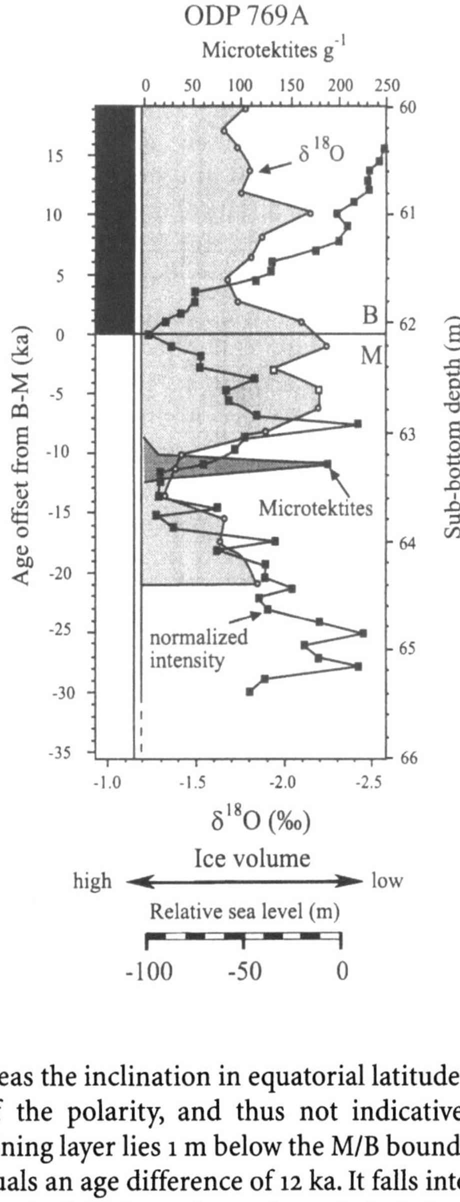

Fig. 166. Comparison between the NRM intensity, the $\delta^{18}O$ values and the microtektite abundance in deep-sea sediments from the Indochinese Archipelago. The microtektites are ~12 ka older than the M/B boundary, which is characterized by a very low magnetic field intensity. The microtektite event falls into a period of increasing field strength and commencing warming (Reprinted from Schneider et al. 1992; copyright 1992, with kind permission from Elsevier Science Ltd., The Boulevard, Langford Lane, Kidlington OX5 IGB, UK)

declination change of 180° (whereas the inclination in equatorial latitudes is always flat, independently of the polarity, and thus not indicative; Fig. 166). The microtektite-containing layer lies 1 m below the M/B boundary, which in these sediments equals an age difference of 12 ka. It falls into a period of increasing geomagnetic field intensity and a relatively low intensity is already observable 6 ka before the impact event. In addition, the oxygen isotopes of foraminifera shells indicate that this event fell into a period of commencing warming (at the end of the cold stage 20). This thoroughly reconstructed sequence of events is incompatible with the suggested chain of cause and effect *impact → climatic deterioration → polarity reversal.*

Limnic and Fluvial Sediments. The sediments of freshwater lakes belong to the most important climatic proxies of the continental Quaternary; this especially applies to the upper Pleistocene and Holocene. Similar to deep-sea sediments, the sedimentation in freshwater lakes is a continuous process. The period, however, is limited by silting-up but, nevertheless, may extend over several 100 ka and even several Ma, such as in Lake Baikal. Climatic changes express themselves in both the lithology and the fossil – primarily pollen – content, allowing the correlation of the lake sediments with other deposits. In addition to the *D*, *I* and *F* parameters, the susceptibility can also be used for correlation (e.g., Haverkamp and Beuke 1993), because this property depends on the proportion of ferromagnetic mineral components, which, in turn, is environmentally controlled. Calm depositional conditions favor DRM acquisition. Because of their high water content the non-consolidated sediments tend to temporally delay PDRM development. In contrast to deep-sea sediments, limnic deposition is fast with typical sedimentation rates of 0.1–1 mm a^{-1}. Therefore they have the capacity to record short-term secular variation.

On drill cores, taken from the bottom of the Lac du Bouchet, a maar lake in the French Massif Central, the secular variation was traced back until the end of the last interglacial 120 ka ago (Thouveny et al. 1990). Such curves of the *D*, *I* and *F* variation can be used for magnetostratigraphic correlation and – if fixed in time – they may serve as regional standard curves for paleomagnetic dating. In the case of the Lac du Bouchet, the deepest layers even record the polarity reversal of the Blake subchron. It is noteworthy that the Laschamp excursion, which was first observed in volcanic rocks of the Massif Central, is not recognizable in these sediments. Analogous investigations were conducted on sediments of the maar lakes of the Eifel, Germany (Haverkamp and Beuker 1993). The sedimentary deposits of Lake Baikal, Siberia, bear the longest paleomagnetic record for freshwater sediments, extending at least over the last 5 Ma and presumably back to the Miocene. The paleo-intensity record of the last 84 ka correlates with that of the Mediterranean Sea and the Indian Ocean, suggesting a strong global component and, consequently, its use for the correlation of continental and marine sequences with a potential resolution of 7 ka (Peck et al. 1996).

Fine-grained fluvial sediments which were deposited within stack water zones of rivers are also suitable for paleomagnetic examination, as for instance the clays in and under the find layer of the *Homo heidelbergensis* of the former Neckar meander at Mauer, Germany. The normal polarity places the find layer into the Brunhes chron, where the M/B boundary with 0.778 Ma represents a *terminus post quem* for this hominid (Hambach et al. 1992). Due to the periglacial pollen assemblages of the underlying clays

(Urban 1992), at least one glacial period must be intercalated between the M/B boundary and the find layer. In the Gona sequence, Ethiopia, Oldowan stone tools were discovered in sediments of a floodplain environment. Paleomagnetic investigation of fine-grained fluvial deposits revealed that these artifacts are the oldest known ones from anywhere in the world (Semaw et al. 1997). The archaeological find horizon is stratigraphically bracketed by the underlying Gauss/Matuyama boundary (2.6 Ma ago) and an overlying tuff dated by $^{39}Ar-^{40}Ar$ (2.52 ± 0.08 Ma, Sect. 3.1.3).

Loess. In the mighty loess blankets of the glacial climate periods a natural remanence is frequently observable. The formation of the natural remanence in loess is not yet fully understood, since it is hard to imagine that a sufficient magnetic alignment of the ferromagnetic dust fraction takes place during wind deposition with high sedimentation rates. As has been shown by thermal and alternating-field cleaning, magnetite carries a primary PDRM and hematite a secondary CRM of early diagenetic origin (Reinders and Hambach 1995). The fast deposition rate (>0.1 mm a^{-1}) of loess is a good prerequisite for reconstructing the secular variation with a high temporal resolution. Limbrock (1992) conducted a detailed paleomagnetic survey along a 4-m-high profile of Würmian loess at Böckingen, southern Germany, that covers the age range of 41–24 ka, according to TL dates (Zöller and Wagner 1989). Low values of susceptibility were observed in humic loess layers. The paleomagnetic D and I variations (Fig. 153) show many similarities to comparable profiles, e.g., the Lac du Bouchet, indicating the significance of the paleomagnetic loess data. A polarity reversal of possible correspondence to the Laschamp event was not established in Böckingen.

Studies on Chinese loess profiles yielded a strong correlation of the susceptibility k with both the pedostratigraphy, whereby k is higher in soil than in loess, and the $\delta^{18}O$ curve of the deep-sea sediments, whereby k negatively correlates with $\delta^{18}O$ (Kukla et al. 1988). Obviously, the susceptibility of Chinese loess varies depending on the local and global climate. Possible origins of the climatically controlled susceptibility changes could be the iron oxidation stage (magnetite possesses 10^2-fold susceptibility compared to hematite) or dilution effects by non-magnetic sediment components during glacial periods (Heller and Liu 1986; Beer et al. 1993). The susceptibility curves in combination with the polarity of the ChRM are used for the stratigraphic correlation of loess sequences (Fig. 167). However, a detailed analysis of the position of the M/B boundary within Asian loess sections implies that their paleomagnetic interpretation may be more complicated than hitherto supposed (Tauxe et al. 1996). In most of these loess sections the M/B boundary appears within a loess interval, presum-

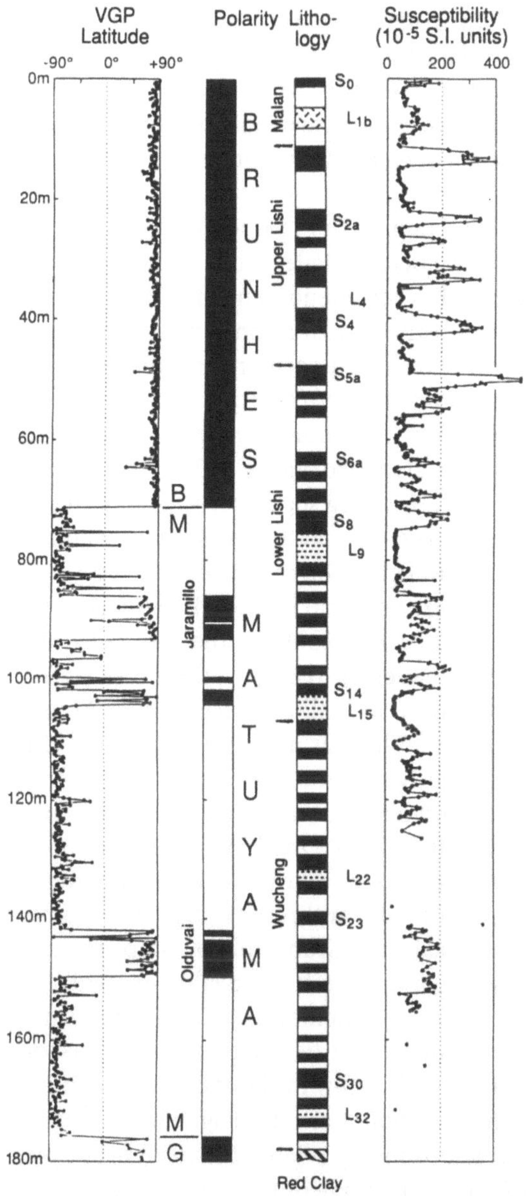

Fig. 167. The polarity (presented as VGP) of the remanent magnetization and the susceptibility in comparison to the pedostratigraphy (in *white*: loess, in *black*: paleosols) of the loess profile at Xifeng in China. High susceptibility is clearly associated with soils. This behavior enables the linkage between pedostratigraphy and magnetostratigraphy and their connection to the marine $\delta^{18}O$ curve via the polarity of the loess. In this way a numerical dating is achieved. (Reprinted from Heller et al. 1991; copyright 1991, with kind permission from Elsevier Science Ltd., The Boulevard, Langford Lane, Kidlington OX5 1GB, UK)

ably associated with a glacial period, whereas in marine records this boundary is firmly placed within the interglacial $\delta^{18}O$-stage 19. It is not yet clear if this inconsistency is due to some form of delayed acquisition of the remanence in these loesses compared to deep-sea sediments, or to a climatic phase lag between oceans and continents.

Cave Sediments. A challenging but frequently difficult task is the dating of cave sediments and associated Paleolithic finds. The remanence of cave sediments is attributed to the phenomena of PDRM or CRM, with bacteria being also involved. For the case of a continuous sedimentation, the deposits can be dated by means of the secular variation. First experiments were undertaken by Creer and Kopper (1974) who examined Upper Paleolithic layers of the Tito Bustillo cave, northern Spain. The remanent *D*, *I* and *F* data were linked to a standard curve which had been established on English lake sediments. By doing so, a layer – correlated with Magdalenian murals – was dated to 11.6–11.2 ka. Likewise, archaeomagnetic investigations were carried out in Petralona cave, northern Greece, which had furnished the skull of an early hominid in 1960 (Papamarinopoulos et al. 1987). The 4.5-m-thick profile, which contains the find layer at 2 m depth, was sampled with only considering the soft and moist layers. The measurements included *D*, *I* and *F* as well as susceptibility. Standard profiles for comparison are not available, thus inhibiting magnetic dating. The normal inclination, however, indicates that the whole sediment complex of this profile is younger than the M/B boundary. This result is in agreement with uranium series and ESR dating (~200 ka) on the Petralona skull (Sects. 4.3, 7.3.3). Some non-consistent *D* and *I* values are attributed to the activities of prehistoric Man and animals (anthropoturbation and bioturbation).

At the archaeological site Gran Dolina, Atapuerca, northern Spain, paleomagnetic investigation of clays and silts revealed the M/B boundary in the lower part of this sedimentary section (Pares and Perez-Gonzalez 1995). The hominid- and artifact-bearing Aurora layer is located within the Matuyama chron not much below the boundary, implying an age of 780 ka, near the time of the M/B boundary. This age makes these the oldest European hominids so far, probably related to later Europeans, such as *Homo heidelbergensis*. Analogously, paleomagnetic investigation of clay samples from the Longgupo Cave, Sichuan, China, placed the layer with fossil finds of an early *Homo* and associated artifacts within the Olduvai subchron, implying the very early arrival of the genus *Homo* in Asia (Wanpo et al. 1995).

Calcareous Cave Deposits. Calcareous cave sinters are only weakly, though measurably magnetized. Their low natural remanence is of chemical origin (CRM) and caused by the presence of traces of magnetite and hematite. Investigations of early Holocene stalagmites of Kingsdale, England, showed the D and I data of the stable NRM component to be in agreement with the expected values. Because of this, Latham et al. (1979) considered the material to be suitable for magnetic dating. On a 72-cm-high stalagmite of the Sotano del Arroyo cave, Mexico, the secular variation was reconstructed in detail over a period of 1200 a – the overall period of the stalagmite formation as being deduced by $^{230}Th/^{234}U$ and ^{14}C ages (Latham et al. 1986). The data are consistent with the archaeomagnetic standard curve of the declination and inclination valid for southwestern North America. This observation demonstrates that calcareous sinters from caves can be accurately dated by paleomagnetism with the aid of regional standard curves.

Sun-Dried Bricks. Sun-dried bricks possess a weak natural remanence whose cause has not been fully clarified. Presumably, it is induced during the shaping of the bricks. Sun-dried bricks are produced by filling clay into a mold and subsequent sun-drying. This NRM type, the so-called shear or shock remanent magnetization (SRM), is used for the determination of the paleo-intensity (Games 1977).

10 Earth's Orbit, Climate and Age

Over 99.9% of the energy balance on the Earth's surface is supplied by the sun which controls both weather and climate. Temporal variations in the insolation result in periodic temperature fluctuations. While the Earth's rotation around its axis causes diurnal temperature changes, the Earth's orbital movement around the sun determines the annual oscillation. Long-term variations (20–100 ka) in the Earth's orbit cause climatic fluctuations, the Milankovitch cycles, which are held responsible for the Quaternary glacial/interglacial cycles. The annual and long-term climatic fluctuations leave significant traces in the geological and biological records which in turn can be used for dating. The methods based on climatic cycles are only briefly treated in this context. Due to their high accuracy, some of them play a fundamental role in Holocene and Pleistocene chronometry.

10.1 Annual Cycles

Periodic differences in insolation between both hemispheres cause seasonal temperature variations. These in turn trigger off geological and biological processes that find their expression in the annual layering of sediments, annual ice layers and annual growth rings in trees, corals (Knutson et al. 1972; Beck et al. 1997) and stalactites (Baker et al. 1993). Various dating methods are based on counting the annual formation of new material. Since climatic conditions do not exactly resemble one another from year to year, the annual growth layers express specific properties, e.g., the thickness of a tree ring reflects one particular summer's weather. Gradually, the signatures of such features form a specific pattern thus enabling the correlation of sequences. Individual dating methods, which can only be touched on here, are – due to their very high temporal resolution – of particular importance to the late Glacial and Holocene period with its drastic climatic changes.

10.1.1 Varve Chronology

Melting of glacial ice follows the seasonal temperature fluctuation. The meltwaters produced in summer transport relatively high loads of coarse-grained clastic rock material. With decreasing temperature their transportation power decreases. Hence, in the periglacial lakes rhythmic annual deposits, *varves*, form, showing a graded bedding from sand or silt to clay fractions (De Geer 1912). Each subsequent varve abruptly sets in with coarse sedimentation. Depending on the summer temperature, layers of various thickness (mm–cm) are produced resulting in a characteristic varve succession that reflects the climatic history. The varves are dated by counting and correlating on the basis of their thickness signature. A periglacial position close to the glacier represents only a temporary state during the late Glacial ice retreat. To establish a varve chronology for as long as possible, the overlapping profiles need to follow the ice retreat. For example, in Sweden the oldest varves are found in the south and the younger ones towards the north along the Gulf of Bothnia (Fig. 168). The Swedish varve chronology comprises more than 13,000 varve years (BP, i.e., before the reference year 1950) with potential errors of around +10, –180 a (Björck et al. 1992; Andren et al. 1996). For the Younger Dryas/Preboreal boundary (identical with the Pleistocene/Holocene boundary), Strömberg (1985) states a figure of 10,700 + 50, – 150 varve years BP which is about 800 a less – probably due to missing varves – than the corresponding dendro-age (Björk et al. 1996).

Annually laminated sediments are met beyond the fluvioglacial depositional environment, too, where their presence in many cases can only be detected on a microscopic level. Such non-glacial varves were, for instance, observed in the late Glacial and Holocene sediments of Soppensee, Switzerland, where light calcite-bearing summer layers alternate with dark autumn and winter layers (Lotter et al. 1992). The sediments, formed during the past 15 ka, permit the linking of varve counting with pollen analysis and AMS ^{14}C dating. This is not only of chronological but also of paleo-ecological interest. Annually layered limnic sediments are also described for the Meerfelder Maar and the Holzmaar in the Eifel, Germany (Zolitschka et al. 1992). The microscopic annual layers of these maar sediments consist of a light spring and summer layer containing planktonic diatoms and a dark autumn and winter layer of clastic and organogenic detritus. Starting from the top of the sedimentary column in the Holzmaar, a total of 12,794 Holocene and late Glacial varves were counted. The varve age for the intercalated tuff of the Laacher See volcano, which erupted ca. 12.9 ka ago, is 11,323 ± 224 BP and for the Pleistocene/Holocene boundary 10,630 ± 183 BP.

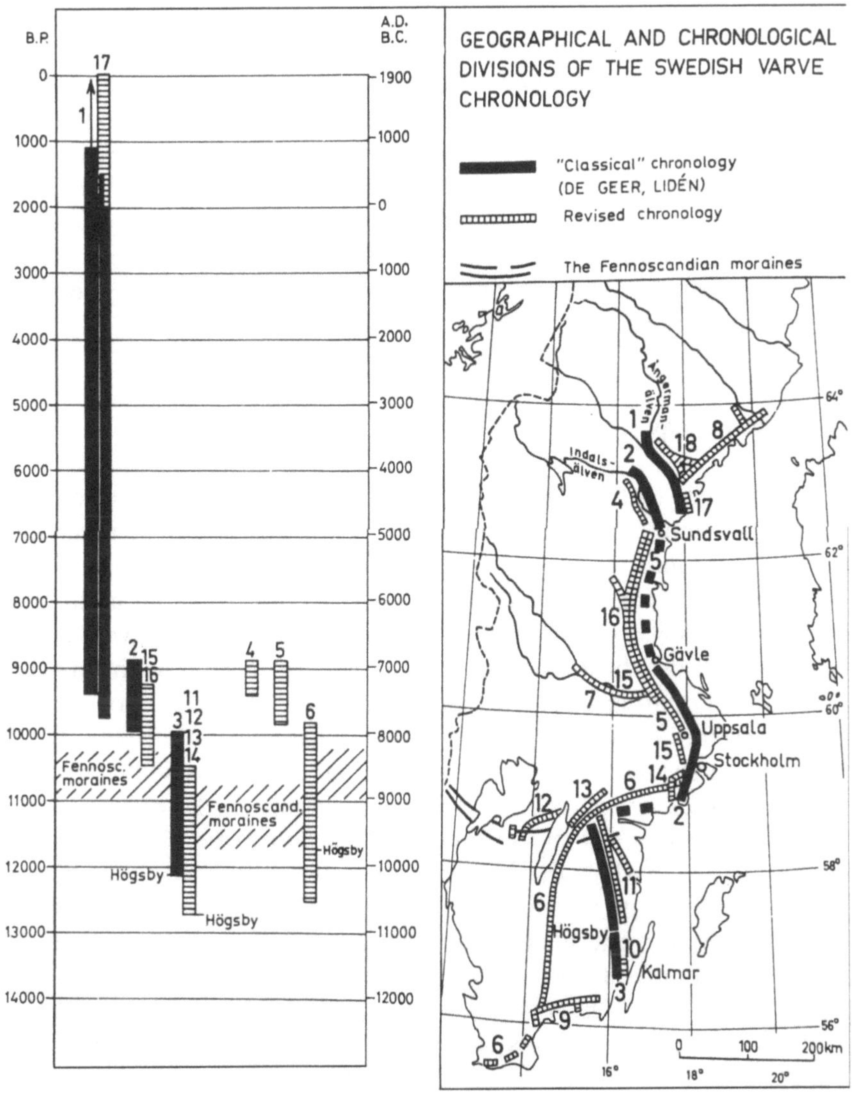

Fig. 168. Swedish varve chronology covering the last 12,800 a BC (i.e., varve years before 1950; from Björck et al. 1992). The overlapping of the individual chronologies (*1–18*) permits the establishment of a continuous chronology for the late and post-Glacial melting of the Scandinavian inland ice. Due to the regression of the ice the younger varve profiles are found further to the north

These ages are definitely too low compared to calibrated ^{14}C, dendro and ice core data (Björk et al. 1996). AMS ^{14}C dating of macrofossils from the Holzmaar sediments revealed that 878 varve years are missing in the younger section (Hajdas et al. 1995). This illustrates that when applying varve chronometry one must always bear in mind the possibility of a hiatus which may not be observable from the sedimentary record; in other words, that varve ages are minimum ages. Except for that difficulty, laminated lake sediments may enable the extension of the ^{14}C calibration curve into the late Glacial age range beyond the period covered by dendrochronology (Stuiver and Kra 1986).

10.1.2 Dendrochronology

At intermediate latitudes with a pronounced seasonal climate, the annual tree-ring structure is largely controlled by the periodicity of vegetative growth. Cambium tissue situated beneath the tree bark forms during spring and summer. After autumn and winter dormancy, an additional woody ring starts to grow in spring, enveloping the older wood. The thickness, but also the density, of each annual ring depends on the prevailing weather conditions, primarily on the amount of precipitation and on the temperature, during the growing season. Consequently, each individual tree ring bears an annual record of the weather conditions, so that wide or narrow rings follow each other in a climatically controlled succession, leaving a characteristic thickness signature. This information is recorded as *tree-ring curves* in which the thickness for each ring is successively plotted against the year of growth. Such curves enable the correlation of tree-ring sequences among one another (*cross-dating*). Overlapping sections of fossil wood of different ages are useful in assembling continuous dendrochronological curves, thus providing a record of the tree-ring signature, which is extended to periods far beyond the oldest living tree material (Fig. 169).

Due to site-specific microclimatic influences, individual tree-ring curves need to be merged into a regional *master curve*. Tree-ring sequences of unknown age can be matched to such averaged regional master curves and, hence, become dendrochronologically datable. The annual ring sequence of a wood sample, however, is never in full accord with the standard curve due to differing site conditions. Its adaptation to the master curve requires statistical analysis. In addition to its chronological application, dendrochronology furnishes important information for paleo-climatology, which can be further enhanced by accompanying isotope analyses of the wood (Becker et al. 1991). Ecological influences, including recent

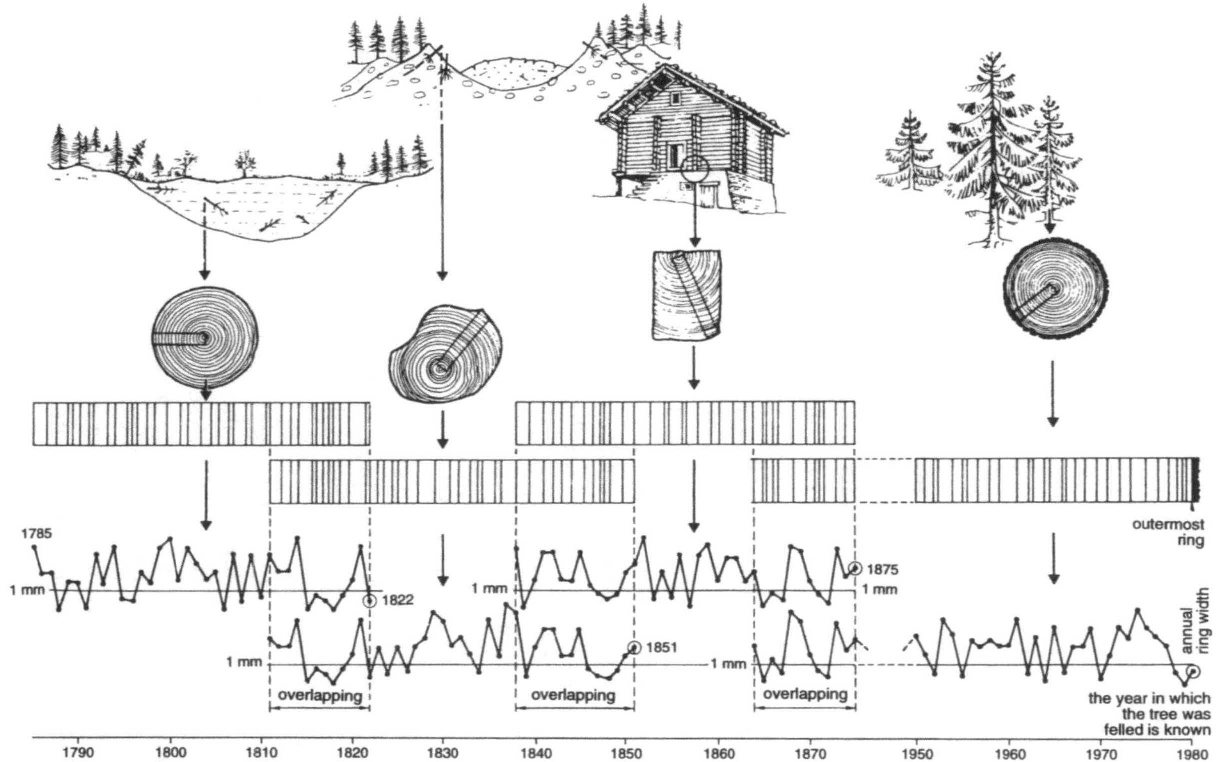

Fig. 169. Schematic of the dendrochronological overlapping of wood samples of different ages on the basis of characteristic signatures of successive tree rings (from Schweingruber 1988). In this way the continuous dendrochronological calendar is reconstructed back to the end of the last Glacial 11.5 ka ago

forest damage, are also recognizable in the tree-rings (Klein and Eckstein 1988). The beginnings of dendrochronology date back to Douglass (1921) in North America and Huber (1941) in central Europe. The method is comprehensively treated by Schweingruber (1988).

In North America a Holocene tree-ring chronology was established using bristlecone pine, which inhabits high mountains and reaches ages of up to 4000 a. In central and western Europe, oak and pine are the most important trees for dendrochronology. More than 6000 individual tree-ring curves form the framework of the German oak chronology. Historic and prehistoric wood as well as fossil trunks recovered from the terraces of the Rhine, Main and Danube rivers serve this purpose (Becker et al. 1991). This oak chronology ("Hohenheim chronology") comprised – without any interruptions – a total of 9928 annual rings and dates back to 7938 BC or 9887 dendro years BP. It was discovered that in this chronology 41 rings are missing at 5200 BC (Björk et al. 1996). An additional backward shift of 54 rings at 7800 BC was necessary (Spurk et al. 1998). The revised Hohenheim chronology extends back to 10,429 dendro years BP (8480 BC). It is the longest dendrochronological calendar commencing with the occurrence of the mixed oak forest which replaced the Preboreal and Boreal pine forest. For a further dendrochronological extension – including the Preboreal and late Glacial – pine has to be used. A pine curve comprising 1941 annual rings is now available for this age range. This has been tentatively dendrochronologically connected with the oak curve resulting in a combined oak/pine dendrochronology that extends back to ~11,900 dendro years BP (Becker 1993; Björk et al. 1996; Spurk et al. 1998). Ring widths in German pines increase markedly at 11,510±20 dendro years BP indicating the late Glacial/Holocene boundary.

The felling date of a tree can be determined with an accuracy of 1 year or, in some cases, even a few summer months, provided that the bark with the most recent growth ring is still preserved. In the absence of the bark, the number of missing rings cannot be exactly determined. However, if the outer sapwood is present, it is possible to estimate the number of missing sapwood rings. The dating of wood samples requires ca. 100 or more successive tree-rings. Datable wooden materials include fossil trunks found in sediments, timbers of historic buildings, panels of paintings (Hillam and Tyers 1995), and medieval and modern stringed instruments (Klein and Eckstein 1988). On the basis of Irish and German oak chronologies, oak samples from western and central Europe falling into the last 10 ka are precisely datable. Dendro and ^{14}C ages, for instance, demonstrate that the beginning of the Early Bronze Age in southwestern Germany has to be placed into the 23rd century BC (Becker et al. 1989). In addition to oak and pine, samples of beech, spruce and fir wood are sometimes used. An attempt was made to date the Late Minoan volcanic eruption of Thera (Sect. 5.4.3) dendrochronologically by associating anomalous tree-rings ("frost rings") from North America and Ireland (LaMarche and Hirschboek 1984; Baillie and Munro 1988) with an eruption-induced world-wide climatic deterioration. This approach yielded an eruption date of 1628/1627 BC. For the Aegean a standard curve is under development, which makes use of cedar, juniper and fir, in addition to oak and pine (Kuniholm and Striker 1987). A floating tree-ring sequence of 1503 years of this chronology was tied within narrow limits by ^{14}C wiggle-matching to the ^{14}C calibration curve beginning ~2230 BC and running to ~730 BC. An even more accurate temporal placement of this sequence was proposed by correlating a remarkable growth anomaly within the 17th century BC with the above mentioned major growth anomalies at 1628/1627 BC in the dendrochronologies of Europe and North America (Kuniholm et al. 1996). The entire 1503-year chronology thus allows accurate dendrochronological cross-dating of wood and charcoal samples from the eastern Mediterranean.

Dendrochronology represents the most precise calendric calibration tool for the ^{14}C time scale, and thus has played a substantial role in the development of the ^{14}C dating method (Sect. 5.4.1). Therefore, the age calibration of the Holocene is based almost entirely upon the tree-ring calendar. The most important dendrochronological master curves were established on the North American bristlecone pines and Irish as well as German oaks (Stuiver et al. 1993). The correspondence of these calibration curves is a remarkable scientific result, considering that they were established on different types of trees, stemming from rather different climatic and geographic environments.

10.1.3 Ice Layer Counting

Snow deposited on the ice caps of the Earth shows seasonal variation in composition. In the case of continuous snowfall – with an ice accumulation rate of >25 cm a^{-1} – annual layers are formed within the ice deposits. These prerequisites are better fulfilled in Greenland than in the Antarctic (Dansgaard 1981). An annual rhythm is observable in chemical and isotopic variations, mainly in the degree of acidity, the dust content and the isotopic composition $\delta^{18}O$ (winter snow is isotopically lighter). The causes of these variations are complex (Herron 1982). With increasing ice load the ice diagenetic processes involved cause the layers to become gradually thinned until the layered structure finally gets blurred. The age is derived by counting the annual layers downward from the surface. In the Summit core (GRIP) in central Greenland the counting of annual layers was conducted on the basis of several parameters, namely $\delta^{18}O$ as well as the contents of Ca^{2+}, NH_4^+, nitrate and microparticles (Johnsen et al. 1992). This approach was applicable at least to the last 14.5 ka of the core. In this way the onset of the Bölling interstadial was determined as 14,410 ± 200 a BP, the onset of the Younger Dryas as 12,660 ± 100 a BP and the Dryas/Preboreal boundary as 11,510 ± 70 a BP, after being corrected to the conventional reference year 1950 AD.

High peaks in the acidity occasionally occur in the ice cores and are attributed to large volcanic eruptions, which liberate considerable amounts of sulfuric acid into the atmosphere in the form of an aerosol. Acid ice layers may also be noticeable by their increased dust content. One acidity peak in the Dye-3 core which was dated as 1645 ± 20 BC (Fig. 170) on the basis of ice layer counting, is associated with the atmospheric pollution generated through the Late Minoan eruption of Thera (Sect. 5.4.3; Hammer et al. 1987). The Summit core yields the more precise age determination of 1645 ± 7 BC (Johnsen et al. 1992). In the GISP2 ice core, drilled at Summit, Greenland, volcanic ash layers were identified by SO_4^{2-} as well as by several cation concentrations (Zielinski et al. 1994). The annual layers were continuously scanned from the present back to 7000 BC, whereby 69 events were identified. Many of the events were matched to documented volcanic eruptions. The chemical analyses and the dust content in this ice core also allowed the retrieval of information on the dynamics of the Holocene climate (O'Brien et al. 1995).

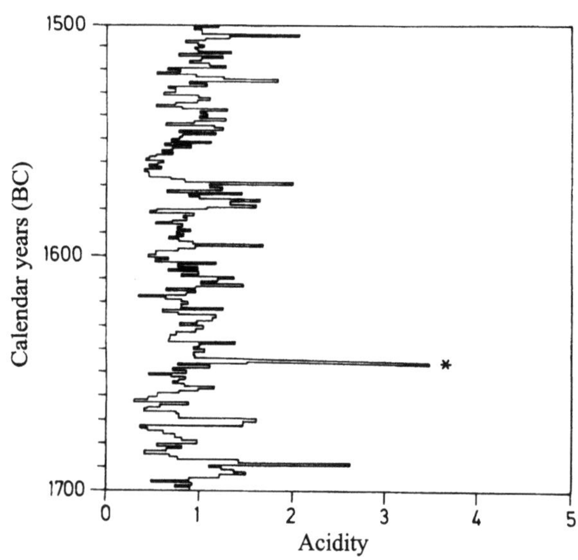

Fig. 170. Annual variation of the acidity degree in ice core Dye-3 from South Greenland. The acidity peak (*) at 1645 ± 20 ice-layer years BC is attributed to the liberated sulfuric acid during the Late Minoan eruption on the Aegean island of Thera. (Hammer et al. 1987; reprinted with permission from Nature, Macmillan Magazines Ltd.)

10.2 Milankovitch Cycles

Since insolation exerts a dominating influence on the climate, it seemed reasonable to assume that fluctuations in the solar energy flux are responsible for the long-term climatic oscillations of the Ice Age. Approaches in this direction date back to the last century (Adhemar 1842; Croll 1875) and were further pursued by Pilgrim (1904). It is largely the achievement of Milankovitch (1941) to have developed a solid theory capable also of explaining the geologically observable, long-term climatic variation of the Ice Age, under the precondition of constant solar radiation. The Milankovitch theory states that minor periodic variations of the Earth's orbit in the course of time – caused by gravitational interactions between Earth, moon, sun and planets – significantly alter the seasonal insolation at high geographic latitudes, thus forcing the ice-age cyclicity.

The periodicity inherent in the orbital parameters of the Earth is accurately computable and, hence, definite age constraints exist for paleoclimate modelling based on insolation variations. Hence, an indisputable correlation of these variations with the recorded climatic fluctuations during the Ice Age is a test for the validity of the Milankovitch theory. However, as long as Quaternary chronometry lacked sufficient precision, this correlation could not be proved. This is the reason why the Milankovitch theory experienced such a delayed acceptance. It was not until the end of the 1960s that this attitude changed in favor of the Milankovitch theory; this change was particularly promoted by investigations on fossil coral tracts and deep-sea sediments, combined with chronometric data.

The climatic fluctuations during the Quaternary are attributed to the periodic variations of three orbital parameters of the Earth (Fig. 171):

1. *Obliquity of the ecliptic.* This term describes the tilt of the Earth's rotation axis and is defined as angle between the plane of the Earth's equator and that of the Earth's orbit. At present, the obliquity of the ecliptic comprises ~23.5° and varies within a 40-ka cycle between 22.1° and 24.4°. The obliquity of the ecliptic causes variably strong insolation on the northern and southern hemispheres in a semi-annual change, thus giving rise to the seasons. With growing obliquity of the ecliptic the seasonal effects become more emphasized.
2. *Eccentricity of the Earth's orbit.* The Earth moves in an elliptical orbit around the sun, which is one of the ellipse centers. The eccentricity describes the deviation from a circular orbit. Presently, it comprises 1.7% and varies with a 100-ka cycle between 0.5 and 6%; superimposed is an additional cycle of 400 ka. The eccentricity variation causes differ-

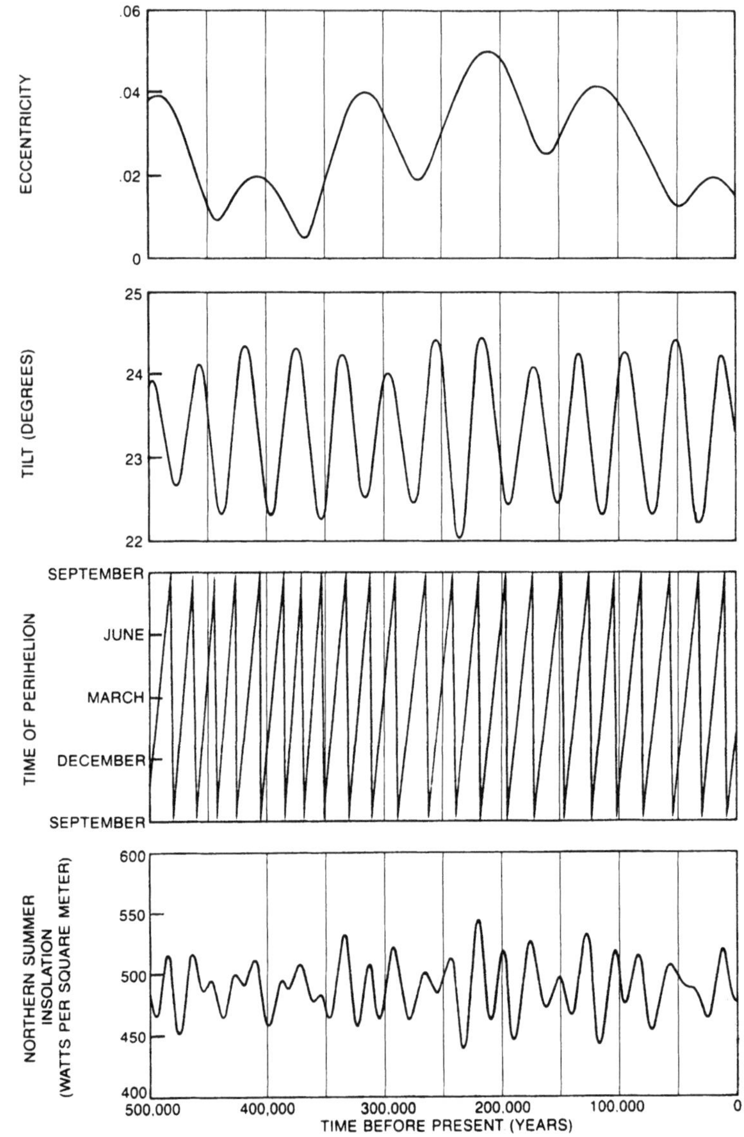

Fig. 171. According to the Milankovitch theory, the climate fluctuation during the Quaternary is the effect of the cyclicity of three orbital elements of the Earth, the obliquity of the ecliptic, the eccentricity and the precession (from Covey 1984). The obliquity of the ecliptic varies with a periodicity of 40 ka; the eccentricity varies with a periodicity of 100 ka; the precession varies with a periodicity of 20 ka. The periodic variation of the orbital elements results in periodic variations of the insolated energy flow for a given geographic latitude and season, here (*bottom*) for the month of July at 60–70° northern latitude. In the northern summer a maximum insolation occurs for the concurrence of high obliquity of the ecliptic, high eccentricity and July perihelion

ent insolation on the Earth between the perihelial (point in the orbit of the Earth where it is closest to the sun) and aphelial positions (point where it is most distant). At present the perihelion is reached on January 3rd. This causes the northern winters to be milder and summers to be cooler than they would be in the absence of eccentricity.

3. *Precession of the equinoxes.* The Earth's rotational axis describes a revolving precession movement relative to the firmament, under maintenance of the eccentricity. The precession has a 26-ka cycle, and the rotational axis points at present to the North Star. The precession causes wandering of the perihelial and aphelial dates through the seasons within a 20-ka cycle.

The periodicity of the orbital parameters does not influence the average global energy flow reaching the Earth from the sun, but it does influence the hemispheric and seasonal distribution of the insolation. It results in periodical insolation fluctuations for given geographic latitudes and seasons. Figure 171 shows the insolation variation for the month of July at 60–70° northern latitude; this variation may at most comprise 20 % in the long run. At first, it is not obvious how such regional insolation oscillations can exert climatic influences on a global scale.

According to the Milankovitch mechanism, the high northern latitudes possess a key function for global climatic development. The extensive continental masses of North America and Eurasia are located at the intermediate and higher northern latitudes. During the northern winter, a closed snow blanket is deposited on them, and this is largely independent of the insolation, i. e., in severe winters as well as in mild ones. To what degree the snow survives during summer depends on the summer insolation. In hot summers it completely melts away, whereas in cool summers remnants of the snow blanket are preserved. A succession of many cool summers during an insolation minimum gives rise to gradual growth of an ice sheet, which even increases the tendency towards lower temperatures by its albedo effect, and in this way the glacial scenario takes its course. On the other hand, the ice will melt again in the course many years of strong insolation during the northern summer. Hence, it is the summer insolation in the northern hemisphere that possesses the triggering effect on the global climatic development during the Quaternary.

Despite the presumably higher complexity of the causality between orbital parameters and climate, it is of essential consequence for the principal validity of the Milankovitch theory that certain properties, which are manifested in the Quaternary sedimentary record, show periodic variation in correspondence with the computed insolation curve of the northern summer. The $^{18}O/^{16}O$ ratio of foraminifera shells in deep-sea sediments

(Sect. 10.2.2) reflects the continental ice volume, and its variation shows the same periods of 100, 40 and 20 ka as the Milankovitch insolation curve. In arctic deep-sea sediments the variation of the ^{10}Be content is synchronous with the glacial/interglacial rhythm (Sect. 5.3.3). Correlation with the insolation curve is also observable in the sea level fluctuation as recognized through uranium series dating on fossil coral reef tracts (Sect. 4.3), the ^{10}Be content (Fig. 41) and the susceptibility (Fig. 167) in loess, and the isotopic signature in ice cores. In all these phenomena the 100-ka-cycle dominates, caused by eccentricity variation. These observations are suggestive of the glacial/interglacial cycles being driven by the Earth's orbital variations. Recently, this was doubted through oxygen isotopic investigation – combined with mass spectrometric ^{230}Th/^{234}U dating – on calcite from the Devils Hole vein, Nevada (Ludwig et al. 1992; Winograd et al. 1992). In this calcite the timing for the shift from glacial to interglacial conditions obtained from δ^{18}O data (~140 ka ago) precedes that of the insolation by ~13 ka. The correctness of the Devils Hole chronology was confirmed by ^{231}Pa/^{235}U dating (Edwards et al. 1997). However, a transfer of a continental curve, reflecting the precipitation history of Nevada, to a marine δ^{18}O curve, controlled by the global ice volume, is not legitimate, according to the view of Imbrie et al. (1993).

10.2.1 Astronomical Dating

If one accepts that the well known orbital cyclicity drives the Quaternary climatic fluctuation, the obvious consequence is its application for chronometry, so that lithologically or isotopically recorded oscillations, particularly the δ^{18}O variation in deep-sea sediments, can be *astronomically* tuned by their matching to the Milankovitch curve. One has to realize the procedural sequence involved. In order to prove the astronomic forcing of the climate, the first need is to confirm by chronometric dating that the cyclicity of the geological record essentially correlates with that of the Earth's orbit. In the next step, the precisely known cyclicity of orbital parameters is in turn used to achieve a dating of the geological record that is more accurate than that possible by most chronometric approaches.

A complication arises from the systematic delay with which the climatic-induced records lag behind the driving variation of insolation (e.g., build-up or melting of ice caps). This delay amounts to 4–8 ka for the marine δ^{18}O record and thus the accuracy of the astronomic dating is estimated at ± 5 ka. For the last 300 ka a high resolution astronomic time scale of the marine δ^{18}O stages was established by Martinson et al. (1987). The

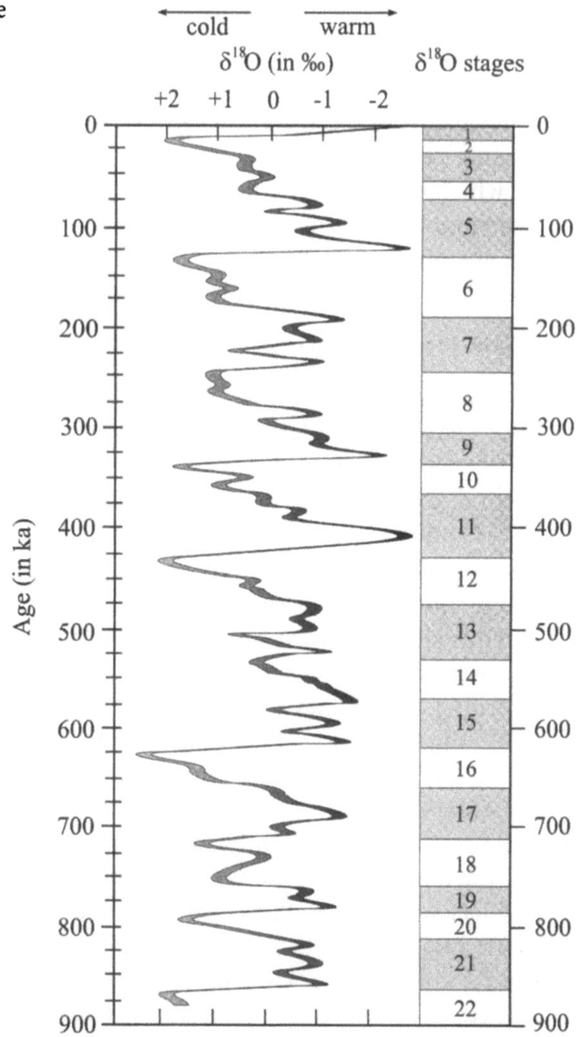

Fig. 172. Marine $\delta^{18}O$ curve with the astronomically dated $\delta^{18}O$ stages. (After Bassinot et al. 1994)

astronomic chronologies of Imbrie et al. (1984) and Bassinot et al. (1994) cover the range back to 800 ka (Fig. 172). Generally, excellent agreement is met for the time marks whose validity can be checked by chronometric techniques. The boundary between the $\delta^{18}O$ stages 6 and 5 (termination II) possesses an astronomic age of 127 ka and uranium series ages of 127–128 ka (Edwards et al. 1986/87; Zhu et al. 1993). The astronomic age of 777.9 ± 1.8 ka for the Matuyama/Brunhes boundary compares well with the K–Ar age of 778.2 ± 3.5 ka (Tauxe et al. 1996). Principally, astronomic dating is extendable over the whole Quaternary.

10.2.2 Oxygen Isotopes

Isotope analyses on calcareous shells of fossil foraminifera from deep-sea sediments reveal long-term variations of the $^{18}O/^{16}O$ ratio of the ocean water which correlate with the climatic history of the Quaternary. Observable are high values during cold and low values during warm periods. The ratios are given as $\delta^{18}O$ values, which express the deviation [‰] of a sample from a standard and are calculated according to

$$\delta^{18}O = 1000\,[(^{18}O/^{16}O)_{sample} - (^{18}O/^{16}O)_{standard})]/(^{18}O/^{16}O)_{standard} \quad (70)$$

V-SMOW, which stands for "Vienna Standard Mean Ocean Water", is usually taken as standard. The observable $\delta^{18}O$ variability in deep-sea cores is a few ‰ (Fig. 172).

Natural oxygen consists of three stable isotopes, ^{16}O (99.76%), ^{17}O (0.04%) and ^{18}O (0.20%), of which the relative abundances are variable to some extent. In the course of physical (e.g., evaporation of water), chemical (e.g., precipitation of calcareous matter) and biological (e.g., secretion of calcareous shells) processes fractionation occurs, as a result of the relatively large mass differences between the isotopes. Since the fractionation is – according to the laws of reaction kinetics – temperature-dependent, it was initially thought that the observed $\delta^{18}O$ variation in calcareous foraminifera shells has to be attributed to the temperature fluctuations of the ocean water during the ice-age cycles (Emiliani 1955). The observation that variations also occur in benthic foraminifera, i.e., living on the ocean floor, where the temperature maintains 4 °C independently of the temperature of the surface water, led to the assumption that the $\delta^{18}O$ variation of the foraminifers mainly reflects that of the ocean water, which in turn reflects the global ice volume (Shackleton 1967; Shackleton and Opdyke 1973).

The vapour phase above the ocean surface is isotopically lighter than the ocean water (Fig. 173). The moisture-rich air prevalent above the ocean surface is blown by the wind across the continents, where the air moisture precipitates. In interglacial periods most of the precipitation returns via the hydrological cycle back to the ocean. In glacial periods, however, a significant fraction of the precipitation is fixed in the growing ice caps. The continental glaciation is associated with a lowering of the sea level of up to 120 m. Since the ice is isotopically lighter than the ocean water from which it stems, the ocean water experiences a depletion of ^{16}O with increasing ice volume, i.e., the $\delta^{18}O$ value of the remaining ocean water increases. The foraminifers incorporate oxygen from the ocean water into their calcitic shells. This process gives rise to a further, though smaller, but also temperature-dependent oxygen isotopic fractionation. In this way the foramini-

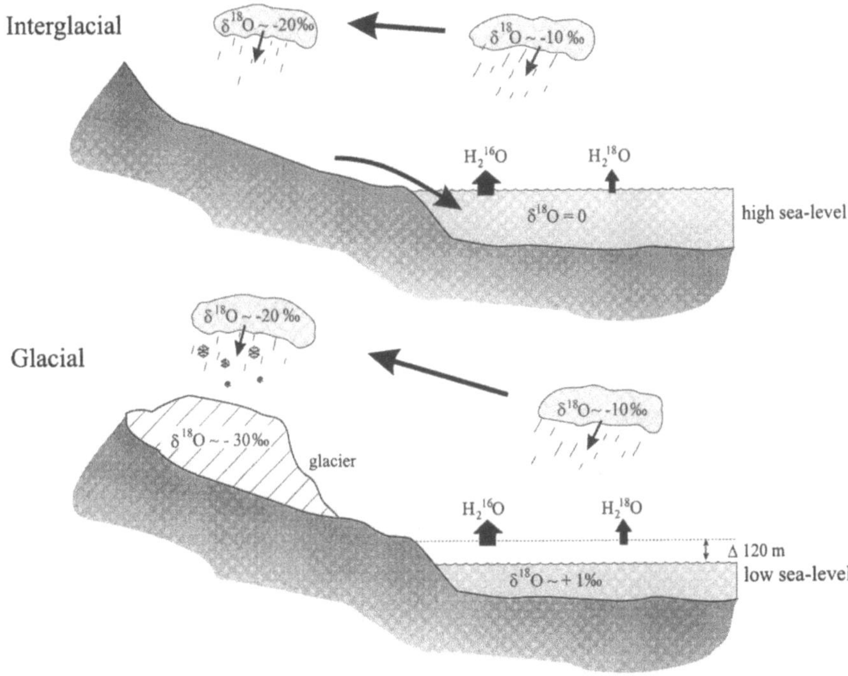

Fig. 173. Quaternary continental glaciation as cause of the $\delta^{18}O$ variation in ocean water and deep-sea sediments, as well as of the sea level variation

fera shells conserve the oxygen isotope ratio of the ocean water for the time being. The fossil foraminifers, found in deep-sea sediments, record the climatic fluctuation and thus enable the construction of a $\delta^{18}O$ curve (Fig. 172).

The marine $\delta^{18}O$ curve exhibits a distinct signature which allows its division into stages. According to Emiliani (1955) the $\delta^{18}O$ stages are counted from the present backward, commencing with uneven numbers for the warm and even ones for the cold periods. Accordingly, the warm periods are stage 1 designating the Holocene, stage 5 for the last Interglacial, stage 7 for the second last Interglacial etc. Stages 2–4 represent the last Glacial period, where for historic reasons its interstadial is stage 3. The second last Glacial period is stage 6 etc. Substages are designated with letters, e.g., 5e (= 5.5 in the decimal system) for the Eemian. The expression *termination* specifies the rapid $\delta^{18}O$ decline at the Glacial/Interglacial transition, e.g., termination II is the transition from stage 6 to stage 5 as opposed to the gradual increase from the Interglacial to the Glacial ("saw-tooth" pattern of the $\delta^{18}O$ curve, Broecker and van Donk 1970).

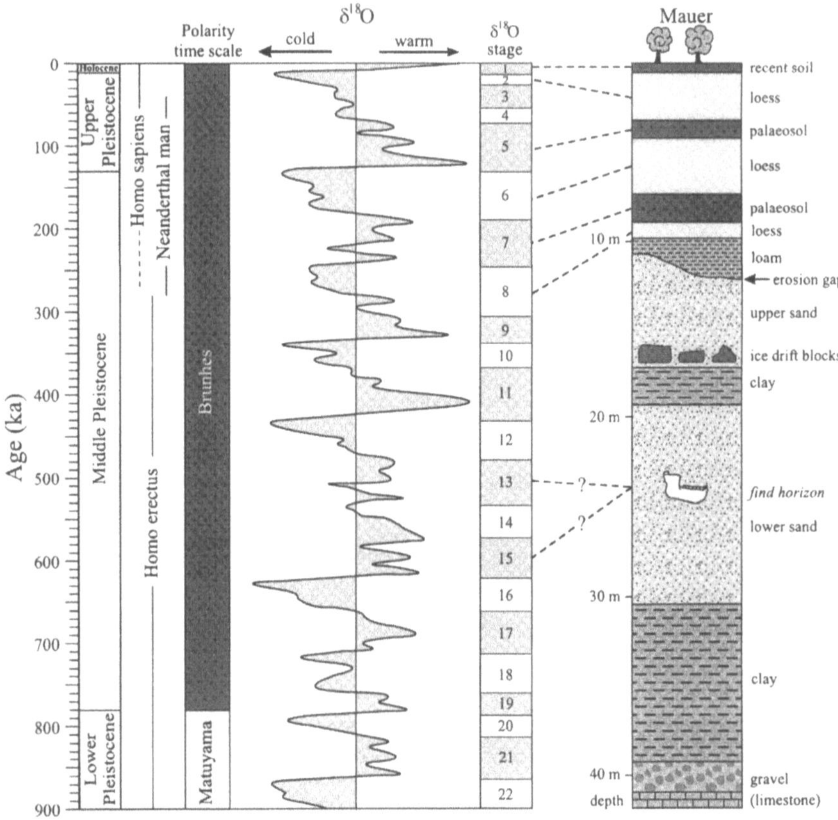

Fig. 174. Middle and upper Pleistocene with the geomagnetic polarity time scale and marine $\delta^{18}O$ curve. Through the paralleling of Interglacial and Glacial layers in the Quaternary profile at Mauer, southwestern Germany, with the $\delta^{18}O$ curve, the find horizon of *Homo heidelbergensis* of Mauer can be correlated with the $\delta^{18}O$ stages 13 or 15. (Wagner et al. 1997)

The time scale of the marine $\delta^{18}O$ curve (Fig. 174) can be fixed by paleomagnetic data (Hays et al. 1969; Opdyke 1972), where the K–Ar ages are magnetostratigraphically transferred from continental volcanites, especially the age of the M/B boundary. Between such time marks the scale is interpolated under the assumption of a constant sedimentation rate. Uranium series dating provides ages of deep-sea sediments and corals (Broecker and van Donk 1970; Edwards et al. 1986/87; Edwards et al. 1997). The youngest section of the scale is covered by ^{14}C ages (Andree et al. 1984a; Bond et al. 1992; Duplessy et al. 1992). The striking resemblance between curves of the marine $\delta^{18}O$ and the variation in insolation allows the precise astronomic dating of the $\delta^{18}O$ stages (Sect. 10.2.1).

The marine $\delta^{18}O$ chronology forms the backbone of Quaternary chronology, mainly for two reasons. Firstly, the $\delta^{18}O$ variation in the ocean water takes place practically at the same time and worldwide, thus allowing $\delta^{18}O$ stratigraphic correlation of marine sediments (Pisias et al. 1984). Secondly, the marine $\delta^{18}O$ curve is derived from continuously deposited sediments as opposed to the more or less interrupted continental sequences (Fig. 6). In spite of this shortcoming, the continental climatic development seems to correlate well with the marine $\delta^{18}O$ curve. The proof of this synchronism is extraordinarily difficult. It was attempted in continuous loess sections (Kukla et al. 1988; Heller et al. 1991; Shen et al. 1992) and limnic sediments (Woillard and Mook 1982; Tzedakis et al. 1997). Also pollen occurrences in deep-sea sediments near the coast prove the temporal comparability of the marine and continental paleoclimatic records (Mangerud et al. 1979). The $\delta^{18}O$ chronology can be directly used for the dating of marine sediments and intercalated volcanic ash layers, as for instance the Toba eruption on Sumatra at the transition of the $\delta^{18}O$ stages 5a/4 73.5 ka ago (Rampino and Self 1992).

Variation in $\delta^{18}O$ has also been observed in continental deposits. In the calcareous sinter from the Devils Hole vein, Nevada, whose continuous formation over the period of 550–60 ka was dated mass spectrometrically by $^{234}U/^{238}U$, $^{230}Th/^{234}U$ and $^{231}Pa/^{235}U$ (Ludwig et al. 1992; Edwards et al. 1997), $\delta^{18}O$ variations occur (Fig. 175), showing a pattern that resembles the marine $\delta^{18}O$ curve. Therefore, a paleoclimatic-driven $\delta^{18}O$ variation has to be assumed in this case also (Sect. 10.2). But in contrast to the marine curve, the $\delta^{18}O$ value of calcareous sinters increases with rising temperature. It probably reflects the temperature prevailing during the precipitation period in the catchment area of the groundwater stream from which the calcareous sinter was precipitated.

The correlation of continental deposits with the marine $\delta^{18}O$ chronology allows the dating of Quaternary sediments principally by paralleling the continental warm and cold periods with the $\delta^{18}O$ stages. This is unequivocally achieved until the Eemian (= stage 5e). There is much debate about the correlation for older glacial-interglacial cycles, e.g., whether the Holsteinian Interglacial (between the Elsterian and Saalian glaciations) belongs to stages 7, 9, 11 or 13 is still under discussion (Sarnthein et al. 1986). An example of dating a Quaternary section by linking it to the $\delta^{18}O$ chronology is that of Mauer, southwestern Germany, where *Homo heidelbergensis* was found in 1907. The mandible was recovered – associated with a rich Interglacial mammal fauna – in sands of a previous meander of Neckar river. These fluvial sediments are overlain by a loess sequence with intercalated paleosols (Fig. 174). Combined application of paleontology, pollen analysis, pedostratigraphy, paleomagnetism as well as TL, ESR and

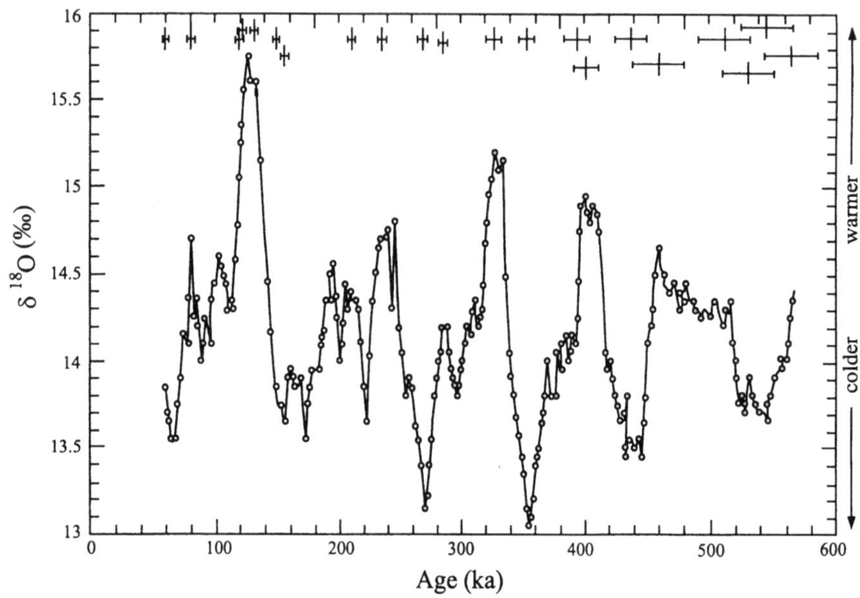

Fig. 175. $\delta^{18}O$ variation along a 36-cm long core drilled through a calcitic vein at Devils Hole, Nevada. The continuous calcareous precipitation from groundwater over a period of 500 ka is verified by mass spectrometric uranium series dates (*bars at the top*). (Winograd et al. 1992; reprinted with permission from Science, American Association for the Advancement of Science)

uranium series dating suggested to correlate the find horizon with the $\delta^{18}O$ stages 15 or 13 (Wagner et al. 1997).

10.2.3 Ice Core Stratigraphy

Long-term climatic trends left their imprint in the ice caps of Greenland and Antarctica. Deep ice cores, drilled from the mighty ice deposits, form a continuous paleoclimatic archive of the last 100–200 ka. The isotopic variations of oxygen ($\delta^{18}O$), hydrogen ($\delta^2H = \delta D$) and beryllium (^{10}Be) are especially useful climatic indicators, as are the dust content and the CO_2 incorporated in the ice. The reconstruction of the depositional history of a vertical ice column is complicated by thinning and lateral flowing as a result of increasing ice load. Because the age estimations of the ice cores are based on snow accumulation rates, glaciologic models are required. The climatically controlled isotopic signature can be used for the correlation not only among ice cores themselves, but also with the marine $\delta^{18}O$ curve. Thus deep-sea $\delta^{18}O$ chronology is transferable to the ice cores for the purpose of dating.

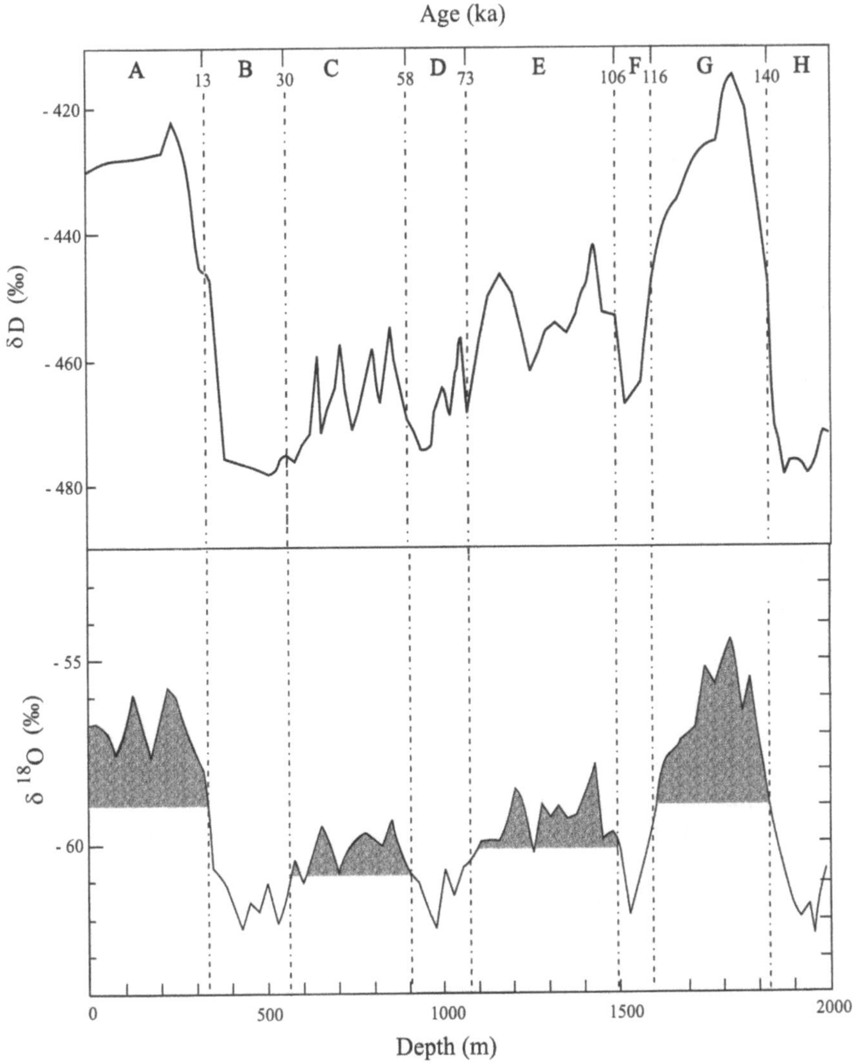

Fig. 176. Variation of the deuterium (δD) and the oxygen-18 ($\delta^{18}O$) in the east Antarctic Vostok ice core (after Lorius et al. 1985; Jouzel et al. 1987). They reflect the long-term fluctuation of the mean annual air temperature during the last 160 ka

The fractionation of the oxygen isotopes during evaporation at the ocean surface results in a $\delta^{18}O$ decrease of ~30‰ in the ice compared to the ocean water (Fig. 173). However, the $\delta^{18}O$ value varies within the ice as well. For instance, in the 2083-m-deep Vostok ice core from Antarctica, it varies by up to 7‰ (Fig. 176). The $\delta^{18}O$ variation is essentially attributed to variations in the air temperature during snow formation (Lorius et al.

1985), where the $\delta^{18}O$ value rises with temperature. The variation is interpreted as maximum temperature difference of ~10 °C during the last 160 ka at this east Antarctic site at 3500 m altitude. The climatic variations from the second last Glacial until the Holocene clearly appear in the $\delta^{18}O$ curve of the Vostok ice core. Despite the fact that there are many differences in detail, the $\delta^{18}O$ variation in ice cores can be correlated with the marine $\delta^{18}O$ curve, although in the reversed sense. In the meantime the Vostok record covers the last 400 ka back to $\delta^{18}O$ stage 11 (Petit et al. 1997). In four Greenland ice cores, including the Summit core, the $\delta^{18}O$ depth profiles over the last 40 ka reflect several episodes of mild climate (interstadial) of short duration between 55–2000 a, with abrupt warm-up and gradual cooling in each case (Johnsen et al. 1992). In the 3029-m-deep Summit core the climatic history covering the last 250 ka appears to be characterized by a general instability, with the exception of the Holocene (Dansgaard et al. 1993).

The deuterium content, too, varies with depth in the ice cores, in the Vostok core within somewhat more than 60‰ (Jouzel et al. 1987). The variation of δD is similar to that of $\delta^{18}O$ in the Vostok core, showing high δD values during the Interglacial and low ones during the Glacial periods. In recent snow precipitates of east Antarctica a linear correlation between the surface temperature (between –20 °C and –50 °C) and the δD value is observable with 6‰ δD per °C. Thus the δD variation found in the Vostok core can be interpreted as a long-term variation in the mean annual air temperature of up to 11 °C during the last 160 ka. For kinetic reasons the δD value is considered to be a better temperature indicator than the $\delta^{18}O$ value. The δD variation possesses a strong similarity to the marine $\delta^{18}O$ curve, which together with the $\delta^{18}O$ variation proves the global paleoclimatic significance of the Antarctic ice cores. The spectral analysis of the δD variation allows the recognition of the periods of the precession and obliquity of the ecliptic – a further indicator of the climatic control over the isotopic composition.

During cold and dry periods, the cosmogenic ^{10}Be input to the ice becomes less strongly diluted by snow. Consequently, the ^{10}Be content in the ice core shows climate-induced variation, with high values during cold and low ones during warm periods. The ^{10}Be content of the Vostok core shows negative correlation with $\delta^{18}O$ (Raisbeck et al. 1987). The ^{10}Be curves of ice can be globally correlated (Beer et al. 1987, 1992). Not only the isotopic variation, but also the CO_2 and dust content in the Vostok core are climatically controlled (Barnola et al. 1987; Petit et al. 1990).

10.2.4
Pollen Analysis

Vegetation reacts sensitively to climatic conditions and was unable to adapt to the fast climatic changes of the Quaternary. Whole plant assemblages disappeared and were replaced by others. Hence, the investigation of fossil plants is a key for the reconstruction of the paleoclimate of the continental Quaternary. The climate-induced temporal succession of plant assemblages can be used for biostratigraphic dating.

Pollen and spores essentially constitute the preserved plant remains in sediments. The study of their qualitative and quantitative composition is called pollen analysis. By considering the various pollen production and distribution mechanisms for individual plant species, study of the pollen content in a sediment allows the deduction of the plant assemblage of the nearby environment. The pollen composition in successive layers is presented in so-called pollen diagrams. They reflect the climatically controlled temporal changes of the vegetation of the environment. High proportions of tree pollen are indicative of a warm climate, and even higher time resolution is achievable by investigating the participating species. The disappearance of certain plants occurs synchronously with climatic deterioration. Subsequent climatic improvement lets only a part of these plants reappear with some delay. They return – if they can – only gradually from their niches, in which they survived. Therefore, the European flora north of the Alps experienced an increasing impoverishment of species during the Quaternary. Hence, it is possible to achieve relative age information on the basis of the last occurrence of a plant species.

The fossil pollen assemblages of central and northern Europe show marked changes since the end of the Pliocene. The vegetation history of the Quaternary indicates repeated changes from warm to cool periods enabling the division into Interglacial/Glacial complexes. For the lower Pleistocene the Tegelen, Eburon, Waal and Bavel complexes can be differentiated. The middle Pleistocene starts in Dutch profiles with the Cromer complex, consisting of at least four Interglacial/Glacial cycles. Following Zagwijn et al. (1971), the stratigraphically important time mark of the paleomagnetic M/B boundary is situated between the Interglacials Cromer I and Cromer II. In this case the Cromer I, however, has to be placed into the lower Pleistocene instead of the middle Pleistocene. The Cromer complex of northern Europe is followed by the Elsterian Glacial, the Holsteinian Interglacial, the Saalian complex and finally in the upper Pleistocene the Eemian Interglacial and Weichselian Glacial. The Holocene also can be furthermore subdivided by pollen analysis (Table 3).

Pollen analysis enables the biostratigraphic correlation of Quaternary continental deposits. Defined pollen zones play an important role in continental Quaternary chronology, but complications for biostratigraphic division may arise from the time-transgressive character of the pollen zones. The linking of the continental chronologies with the marine $\delta^{18}O$ chronology is one of the great challenges of Quaternary research. For marine sediments along the Norwegian coast, Mangerud et al. (1979) succeeded in correlating the Eemian with the $\delta^{18}O$ stage 5e by pollen analysis. The pollen

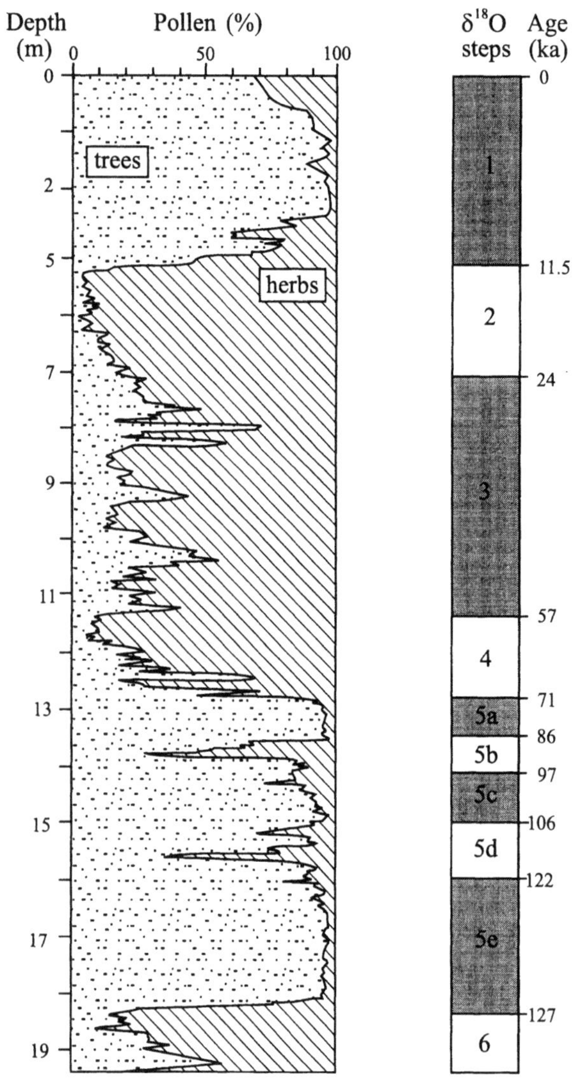

Fig. 177. Correlation of the pollen diagrams of Grande Pile in the Vosges Mountains, eastern France, with the marine $\delta^{18}O$ chronology (after Woillard and Mook 1982). The ages for the $\delta^{18}O$ stages are given after Bassinot et al. (1994)

diagram of Grande Pile in the Vosges Mountains shows a good correlation with marine chronology (Woillard and Mook 1982). This sedimentary profile yields a continuous pollen record for the last 140 ka (Fig. 177). The sediment ages are established for the last 70 ka by ^{14}C dating, whereby the high ages were obtained by the isotope enrichment technique (Sect. 5.4.2). The minima in the tree pollen record coincide with temperature minima of the marine curve ($\delta^{18}O$ stages 2, 4, 5b, 5d and 6) demonstrating the correlation between continental and marine chronology. A 125-ka-long pollen record from the eastern Cascade Range, North America, correlates well with the marine $\delta^{18}O$ curve, providing evidence that the vegetational changes on millennial time scales are primarily controlled by global climate variations (Whitlock and Bartlein 1997). Comparison of the four longest European pollen records in Lac du Bouchet (France), Valle di Castiglione (Italy), Ioannina and Tenagi Philippon (both in Greece), which cover the last 500 ka, shows good agreement with the marine $\delta^{18}O$ curve (Tzedakis et al. 1997). Such concordance between terrestrial and marine records is fundamental not only for establishing a general and reliable Quaternary chronology, but also for the better understanding of the climatic oscillations during the past 2 million years.

References

Abelson PH (1954) Amino acids in fossils. Science 119:576

Adhemar (1842) Les revolutions de la mer, deluges periodiques. Paris

Aitken MJ (1978) Radiation loss evaluation by alpha counting. PACT 2:104–117

Aitken MJ (1985) Thermoluminescence dating. Academic Press, London

Aitken MJ (1990) Science based dating in archaeology. Longman, London

Aitken MJ (1992) Optical dating. Quat Sci Rev 11:127–131

Aitken MJ, Smith BW (1988) Optical dating: recuperation after bleaching. Quat Sci Rev 7:387–393

Aitken MJ, Xie J (1992) Optical dating using infrared diodes: young samples. Quat Sci Rev 11:147–152

Aitken MJ, Allsop AL, Bussell GC, Winter MB (1989) Geomagnetic intensity variations during the last 4000 years. Phys Earth Planet Inter 56:49–58

Alburger DE, Harbottle G, Norton EF (1986) Half-life of ^{32}Si. Earth Planet Sci Lett 78:168–176

Aldrich LT, Nier AO (1948) Argon 40 in potassium minerals. Phys Rev 74:876–877

Ambrose W (1976) Intrinsic hydration rate dating of obsidian. In: Taylor RE (ed) Advances in obsidian glass studies. Noyes, Park Ridge, pp 81–105

Ambrose W (1980) Monitoring long-term temperature and humidity. Inst Conserv Cult Mater Bull 1:36–42

Amin BS, Kharkar DP, Lal D (1966) Cosmogenic ^{10}Be and ^{26}Al in marine sediments. Deep-Sea Res 13:805–824

Amin BS, Lal D, Somayajulu BLK (1975) Chronology of marine sediments using the ^{10}Be method: intercomparison with other methods. Geochim Cosmochim Acta 39:1187–1192

Anderson EC, Libby WF, Weinhouse S, Reid AF, Kirshenbaum AD, Grosse AV (1947) Natural radiocarbon from cosmic radiation. Phys Rev 72:931–936

Andree M, Beer J, Oeschger H, Broecker W, Mix A, Ragano M, O'Hara P, Bonani G, Hofmann HJ, Morenzoni E, Nessi M, Suter M, Wölfli W (1984a) ^{14}C measurements on foraminifera of deep sea core V28–238 and their preliminary interpretation. Nucl Instrum Methods Phys Res B5:340–345

Andree M, Moor E, Beer J, Oeschger H, Staufer B, Bonani G, Hofmann HJ, Morenzoni E, Nessi M, Suter M, Wölfli W (1984b) ^{14}C dating of polar ice. Nucl Instrum Methods Phys Res B5:385–388

Andren T, Brunnberg L, Ringberg B (1996) The reactivated Geochronological Institute in Stockholm – the first clay-varve database. PACT 50:213–218

Andres G, Geyh M (1972) Paläohydrologische Studien mit Hilfe von ^{14}C über den pleistozänen Grundwasserhaushalt in Mitteleuropa (Südliche Frankenalb). Naturwissenschaften 59:418

Andrews JN, Kay RL (1982) Natural production of tritium in permeable rocks. Nature 298:361–363
Andrews JN, Davis SN, Fabryka-Martin J, Fontes JC, Lehmann BE, Loosli HH, Michelot JL, Moser H, Smith B, Wolf M (1989) The in situ production of Radioisotopes in rock matrices with particular reference to the Stripa granite. Geochim Cosmochim Acta 53:1803–1815
Anthony EY, Poths J (1992) ^{3}He surface exposure dating and its implications for magma evolution in the Potrillo volcanic field, Rio Grande Rift, New Mexico, USA. Geochim Cosmochim Acta 56:4105–4108
Appleby PG, Oldfield F (1992) Application of lead-210 to sedimentation studies. In: Ivanovich M, Harmon RS (eds) Uranium-series disequilibrium: applications to earth, marine, and environmental sciences. Clarendon, Oxford, pp 731–778
Aquirre E, Pasini G (1985) The Pliocene-Pleistocene boundary. Episodes 8:116–120
Arias C, Bigazzi G, Bonadonna FP (1981) Size corrections and plateau age in glass shards. Nucl Tracks 5:129–136
Arnold JR (1956) Beryllium-10 produced by cosmic rays. Science 124:584–585
Arnold JR, Libby WF (1949) Age determination by radiocarbon content: checks with samples of known age. Science 110:678–680
Asfaw B, Beyene Y, Suwa G, Walter RC, White TD, WoldeGabriel G, Yemane T (1992) The earliest Acheulean from Konso-Gardula. Nature 360:732–735
Atkinson D, Shaw J (1991) Magnetic viscosity dating. In: Pernicka E, Wagner GA (eds) Archaeometry '90. Birkhäuser, Basel, pp 533–540
Atwater BF, Stuiver M, Yamaguchi K (1991) Radiocarbon test of earthquake magnitude at the Cascadia subduction zone. Nature 353:156–158
Aumento F (1969) The Mid-Atlantic ridge near 45°N: fission track and ferro-manganese chronology. Can J Earth Sci 6:1431–1440
Azzaroli A, De Guili C, Ficcarelli G, Torre D (1988) Late Pliocene to early Mid-Pleistocene mammals in Eurasia: faunal succession and dispersion events. Palaeogeogr Palaeoclimatol Palaeoecol 66:77–100
Bada JL (1981) Racemization of amino acids in fossil bones and teeth from the Olduvai Gorge region, Tanzania, East Africa. Earth Planet Sci Lett 55:292–298
Bada JL (1984) Racemization of amino acids. In: Barrett GC (ed) Chemistry and biochemistry of amino acids. Chapman and Hall, London, pp 399–414
Bada JL (1985) Amino acid racemization dating of fossil bones. Annu Rev Earth Planet Sci 13:241–268
Bada JL, Man EH (1980) Amino acid diagenesis in deep sea drilling project cores: kinetics and mechanisms of some reactions and their implications in geochronology and in paleotemperature and heat flow determinations. Earth Sci Rev 16:21–25
Bada JL, Schroeder RA (1975) Amino acid racemization reactions and their geochemical implications. Naturwissenschaften 62:71–79
Bada JL, Shou MY (1980) Kinetics and mechanism of amino acid racemization in aqueous solutions and in bones. In: Behrensmeyer AK, Hill AP (eds) Fossils in the making: vertebrate taphonomy and paleoecology. University of Chicago Press, Chicago, pp 235–255
Bada JL, Hoopes E, Darling D, Dungworth G, Kessels HJ (1979a) Amino acid racemization dating of fossil bones – interlaboratory comparisons of racemization measurements. Earth Planet Sci Lett 43:265–268
Bada JL, Masters PM, Hoopes E, Darling D (1979b) The dating of fossil bones using amino acid racemization. In: Berger R, Suess HE (eds) Radiocarbon dating. University of California Press, Los Angeles, pp 740–756

Bada JL, Schroeder RA, Carter G (1974) New evidence for the antiquity of man in North America deduced from aspartic acid racemization. Science 194:791–793

Bailiff IK (1994) The pre-dose technique. Radiat Meas 23:471–479

Baillie MGL, Munro MAR (1988) Irish tree rings, Santorini and volcanic dust veils. Nature 332:344–346

Baker A, Smart PL, Ford DC (1993a) North-west European palaeoclimate as indicated by growth frequency variations of secondary calcite deposits. Palaeogeogr Palaeoclimatol Palaeoecol 100:291–301

Baker A, Smart PL, Edwards RL, Richards DA (1993b) Annual growth banding in a cave stalagmite. Nature 364:518–520

Baksi AJ (1993) A geomagnetic polarity time scale for the period 0–17 Ma, based on $^{40}Ar/^{39}Ar$ plateau age for selected field reversals. Geophys Res Lett 20:1607–1610

Baksi AK, Hsu V, McWilliams MO, Farrar E (1992) $^{40}Ar/^{39}Ar$ dating of the Brunhes-Matuyama geomagnetic boundary field reversal. Science 256:356–357

Balescu S, Packman SC, Wintle AG, Grün R (1992) Thermoluminescence dating of the Middle Pleistocene raised beach of Sangatte (northern France). Quat Res 37:390–396

Banerjee D, Singhvi AK, Bagati TN, Mohindra R (1997) Luminescence chronology of seismites at Sumdo (Spiti valley) near Kaurik-Chango fault, northwestern Himalaya. Curr Sci 73:276–281

Barabas M (1989) ESR-Datierung von Karbonaten: Grundlagen, Systematik, Anwendungen. Dissertation, Universität Heidelberg

Barabas M, Mangini A, Sarnthein M, Stremme H (1988) The age of the Holstein interglaciation: a reply. Quat Res 29:80–84

Baray M, Zöller L (1993) Aspekte der Thermolumineszenzdatierung an spätglazial-holozänen Dünen im Oberrheingraben und in Brandenburg. Berl Geogr Arbeit 78:1–33

Barbetti M, Flude K (1979) Geomagnetic variation during the late Pleistocene period and changes in the radiocarbon time scale. Nature 279:202–205

Bard E, Hamelin B, Fairbanks R (1990a) U-Th ages obtained by mass spectrometry in corals from Barbados: sea level during the past 130,000 years. Nature 346:455–458

Bard E, Hamelin B, Fairbanks R, Zindler A (1990b) Calibration of the ^{14}C timescale over the past 30,000 years using mass spectrometric U-Th ages from Barbados corals. Nature 345:405–410

Bard E, Fairbanks R, Hamelin B, Zindler A, Hoang CT (1991) Uranium-234 anomalies in corals older than 150,000 years. Geochim Cosmochim Acta 55:2385–2390

Bard E, Fairbanks RG, Arnold M, Hamelin B (1992) $^{230}Th/^{234}U$ and ^{14}C ages obtained by mass spectrometry on corals from Barbados (West Indies), Isabela (Galapagos) and Mururoa (French Polynesia). In: Bard E, Broecker WS (eds) The last deglaciation: absolute and radiocarbon chronologies. Springer, Berlin Heidelberg New York, pp 103–110

Barg E, Lal D, Pavich MJ, Caffee MW, Southon JR (1997) Beryllium geochemistry in soils: evaluation of the $^{10}Be/^{9}Be$ ratios in authigenic minerals as a basis for age models. Chem Geol 140:237–258

Barnes JW, Lang EJ, Portratz KA (1956) Ratio of ionium to thorium in coral limestone. Science 124:175–176

Barnola JM, Raynaud D, Korotkevich YS, Lorius C (1987) Vostok ice core provides 160,000-year record of atmospheric CO_2. Nature 329:408–414

Bar-Yosef O, Vandermeersch B (1993) Koexistenz von Neandertaler und modernem *Homo sapiens*. Spektr Wiss 1993/6:32–39
Bassinot FC, Labeyrie LD, Vincent E, Quidelleur X, Shackelton NJ, Lancelot Y (1994) The astronomical theory of climate and the age of the Brunhes-Matuyama magnetic reversal. Earth Planet Sci Lett 126:91–108
Beck JW, Recy J, Taylor F, Edwards RL, Cabioch G (1997) Abrupt changes in early Holocene tropical sea surface temperature derived form coral records. Nature 385:705–707
Becker B (1992) The history of dendrochronology and radiocarbon calibration. In: Taylor RE, Long A, Kra RS (eds) Radiocarbon after four decades – an interdisciplinary perspective. Springer, Berlin Heidelberg New York, pp 34–49
Becker B (1993) An 11,000-year German oak and pine dendrochronology for radiocarbon calibration. Radiocarbon 35:201–231
Becker B, Kromer B (1990) A 9927-years Holocene tree-ring chronology: dendrodating and ^{14}C-calibration. In: Pernicka E, Wagner GA (eds) Abstr Int Symp Archaeometry, MPI Kernphysik, Heidelberg, p 192
Becker B, Krause R, Kromer B (1989) Zur absoluten Chronologie der Frühen Bronzezeit. Germania 67:421–442
Becker B, Kromer B, Trimborn P (1991) A stable-isotope tree-ring timescale of the Late Glacial/Holocene boundary. Nature 353:647–649
Becker H (1979) Archaeomagnetic investigations in Anatolia from prehistoric and Hittite sites (first preliminary results). Archaeo-Physika 10:382–387
Becker H, Göksu HY, Regulla DF (1994) Combination of archaeomagnetism and thermoluminescence for precision dating. Quat Sci Rev 13:563–567
Beer J, Oeschger H, Finkel RC, Castagnoli G, Bonino G (1985) Accelerator measurements of ^{10}Be: the 11 year solar cycle from 1180–1800 A.D. Nucl Instrum Methods Phys Res B10/11:415–418
Beer J, Bonani G, Hofmann HJ, Suter M, Synal A, Wölfli W, Oeschger H, Siegenthaler U, Finkel RC (1987) ^{10}Be measurements on polar ice: comparison of arctic and antarctic records. Nucl Instrum Methods Phys Res B29:203–206
Beer J, Johnsen SJ, Bonani G, Finkel RC, Langway CC, Oeschger H, Stauffer B, Suter M, Wölfli W (1992) ^{10}Be peaks as time markers in polar ice. In: Bard E, Broecker WS (eds) The last deglaciation: absolute and radiocarbon chronologies. Springer, Berlin Heidelberg New York, pp 141–153
Beer J, Shen C, Heller F, Liu T, Bonani G, Dittrich B, Suter M, Kubik PW (1993) ^{10}Be and magnetic susceptibility in Chinese loess. Geophys Res Lett 20:57–60
Belluomini G (1981) Direct aspartic acid racemization dating of human bones from archaeological sites of central southern Italy. Archaeometry 23:125–137
Bender ML (1973) Helium-uranium dating of corals. Geochim Cosmochim Acta 37:1229–1247
Benett CL, Beukers RP, Clover MR, Gove HH, Liebert RB, Litherland AE, Purser KK, Sondheim NE (1977) Radiocarbon dating using electrostatic accelerators: negative ions provided the key. Science 198:508–509
Bentley HW, Phillips FM, Davis SN, Gifford S, Elmore D, Tubbs LE, Grove HE (1982) Thermonuclear ^{36}Cl pulse in natural waters. Nature 300:737–740
Bentley HW, Phillips FM, Davis SN (1986) Chlorine-36 in the terrestrial environment. In: Fritz P, Fontes JC (eds) Handbook of environmental isotope geochemistry, vol 2. The terrestrial environment. Elsevier, Amsterdam, pp 427–480
Berger GW (1984) Thermoluminescence dating studies of rapidly deposited silts from South-Central British Columbia. Can J Earth Sci 22:704–710

Berger GW (1991) The use of glass for dating volcanic ash by thermoluminescence. J Geophys Res 96:19705–19720
Berger GW, Easterbrook DJ (1993) Thermoluminescence dating tests for lacustrine, glaciomarine, and floodplain sediments from western Washington and British Columbia. Can J Earth Sci 30:1815–1828
Berger GW, Huntley DJ (1994) Tests for optically stimulated luminescence from tephra glass. Quat Sci Rev 13:509–511
Berger GW, Pillans BJ, Palmer AS (1992) Dating loess up to 800 ka by thermoluminescence. Geology 20:403–406
Berger R (1990) Early medieval Irish buildings: radiocarbon dating of mortar. PACT 29:415–422
Beukens RP (1992) Radiocarbon accelerator mass spectrometry: background, precision and accuracy. In: Taylor RE, Long A, Kra RS (eds) Radiocarbon after four decades – an interdisciplinary perspective. Springer, Berlin Heidelberg New York, pp 230–239
Bezborodov MA (1975) Chemie und Technologie der antiken und mittelalterlichen Gläser. Philipp von Zabern, Mainz
Bierman P, Turner J (1995) ^{10}Be an ^{26}Al evidence for exceptionally low rates of Australian bedrock erosion and the likely existence of pre-Pleistocene landscapes. Quat Res 44:378–382
Bierman PR (1994) Using in situ produced cosmogenic isotopes to estimate rates of landscape evolution: a review from the geomorphic perspective. J Geophys Res 99:13885–13896
Bierman PR, Gillespie AR (1994) Evidence suggesting that methods of rock varnish cation-ratio dating are neither comparable nor consistently reliable. Quat Res 41:82–90
Bierman PR, Gillespie AR, Caffee MW (1995) Cosmogenic ages for earthquake recurrence intervals and debris flow fan deposition, Owens Valley, California. Science 270:447–450
Bigazzi G (1996) Archaeological application of fission-track (FT) dating: an overview. Proc 13th Int Congr Prehist Protohist Sci, Forli, Coll III, pp 39–54
Bigazzi G, Bonadonna FP, Belluomini G, Malperi L (1971) Studi sulle ossidiane italiane IV: datazione con il metodo delle trace di fissione. Boll Soc Geol Ital 90:469–480
Bill J, Keller WA, Erne RB, Bonani G, Wölfli W (1984) ^{14}C dating of small archaeological samples: neolithic to iron age in the central Alpine region. Nucl Instrum Methods Phys Res B5:317–320
Bischoff JL, Rosenbauer RJ (1981) Uranium series dating of human skeletal remains from the Del Mar and Sunnyvale sites, California. Science 213:1003–1005
Björck S, Cato I, Brunnberg L, Strömberg B (1992) The clay-varve based Swedish time scale and its relation to the late Weichselian radiocarbon chronology. In: Bard E, Broecker WS (eds) The last deglaciation: absolute and radiocarbon chronologies. Springer, Berlin Heidelberg New York, pp 25–44
Björck S, Kromer B, Johnsen S, Bennike O, Hammarlund D, Lemdahl G, Possnert G, Rasmussen TL, Wohlfarth B, Hammer CU, Spurk M (1996) Synchronized terrestrial-atmospheric deglacial records around the North Atlantic. Science 274: 1155–1160
Blackwell B, Schwarcz HP (1986) U-series analysis of the lower travertine at Ehringsdorf, DDR. Quat Res 25:215–222
Bollong CA, Vogel JC, Jacobson L, van der Westhuizen WA, Sampson CG (1993) Direct dating and identity of fibre temper in pre-contact Bushman (Basarwa) pottery. J Archaeol Sci 20:41–55

Bond G, Heinrich H, Broecker W, Labeyrie L, McManus J, Andrews J, Huon S, Jantschik R, Clasen S, Simet C, Tedesco K, Klas M, Bonani G, Ivy S (1992) Evidence for massive discharges of icebergs into the North Atlantic ocean during the last glacial period. Nature 360:245–213

Bonhommet N, Zähringer J (1969) Paleomagnetism and potassium argon age determinations of the Laschamp geomagnetic polarity event. Earth Planet Sci Lett 6:43–46

Bötter-Jensen L, Mejdahl V (1988) Assessment of beta dose-rate using a GM multicounter system. Nucl Tracks Radiat Meas 14:187–191

Bötter-Jensen L, Duller GAT (1992) A new system for measuring OSL from quartz samples. Nucl Tracks Radiat Meas 20:349–550

Bowen DQ (1978) Quaternary geology. Pergamon, Oxford

Bowman SGE (1979) Phototransferred thermoluminescence in quartz and its potential use for dating. PACT 3:381–400

Bowman SGE (1990) Radiocarbon dating. British Museum, London

Brassart J, Tric E, Valet JP, Herrero-Bervera E (1997) Absolute paleointensity between 60 and 400 ka from the Kohala Mountain (Hawaii). Earth Planet Sci Lett 148:141–156

Brewster D (1863) On the structure and optical phenomena of ancient decomposed glass. Trans R Soc Edinb 23:193–204

Brill RH (1965) Applications of fission-track dating to historic and prehistoric glasses. Archaeometry 7:51–57

Brill RH, Hood HP (1961) A new method for dating ancient glass. Nature 189:12–14

Broecker WS, van Donk J (1970) Insolation changes, ice volumes, and the O^{18} record in deep sea cores. Rev Geophys Space Phys 8:169–198

Brook EJ, Kurz MD, Ackert RP, Raisbeck G, Yiou F (1995) Cosmogenic nuclide exposure ages and glacial history of late Quaternary Ross Sea drift in McMurdo Sound, Antarctica. Earth Planet Sci Lett 131:41–56

Brookes IA (1985) Weathering. In: Rutter NW (ed) Dating methods of pleistocene deposits and their problems. Geoscience Canada Reprint Ser 2, Edmonton, pp 61–71

Brooks AS, Hare PE, Kokis JE, Miller GH, Ernst RD, Wendorf F (1990) Dating Pleistocene archaeological sites by protein diagenesis in ostrich eggshell. Science 248: 60–64

Brooks AS, Helgren DM, Cramer JS, Franklin A, Hornyak W, Keating JM, Klein RG, Rink WJ, Schwarcz H, Smith JNL, Stewart K, Todd NE, Verniers J, Yellen JE (1995) Dating and context of three Middle Stone Age sites with bone points in the Upper Semliki Valley, Zaire. Science 268:548–553

Brown ET, Stallard RF, Larsen MC, Raisbeck GM, Yiou F (1995) Denudation rates determined from the accumulation of in situ-produced ^{10}Be in the Luquillo Experimental Forest, Puerto Rico. Earth Planet Sci Lett 129:193–202

Brown L (1987) ^{10}Be as a tracer of erosion and sediment transport. Chem Geol 65:189–196

Brown RM (1970) Distribution of hydrogene isotopes in Canadian waters. Isotope hydrology 1970. IAEA, Vienna, pp 3–21

Brown TA, Nelson DE, Mathewes RW, Vogel JS, Southon JS (1989) Radiocarbon dating of pollen by accelerator mass spectrometry. Quat Res 32:205–212

Brückner H, Halfar RA (1994) Evolution and age of shorelines along Woodfiord, northern Spitsbergen. Z Geomorph NF Suppl-Bd 97:75–91

Brunhes B (1906) Recherches par la direction d'aimantation des roches volcaniques. J Phys Ser 4(5):705–724

Brunnacker K, Jäger KD, Hennig GJ, Preuss J, Grün R (1983) Radiometrische Untersuchungen zur Datierung mitteleuropäischer Travertinvorkommen. Ethnogr-Archäol Z 24:217–266
Bruns M, Levin I, Münnich KO, Hubberten HW, Fillipakis S (1980) Regional sources of vulcanic carbon dioxide and their influence on ^{14}C content of present-day plant material. Radiocarbon 22:532–536
Bucha V (1971) Archaeomagnetic dating. In: Michael HN, Ralph EK (eds) Dating techniques for the archaeologist. MIT, Cambridge, Massachusetts, pp 57–117
Buhay W, Schwarcz HP, Grün R (1988) ESR dating of fault gouge: the effect of grain size. Quat Sci Rev 7:515–522
Bull WB, Brandon MT (1998) Lichen dating of earthquake-generated regional rockfall events. Southern Alps, New Zealand. Bull Geol Soc Am 110:60–84
Burleigh R, Baynes-Cope AD (1983) Possibilities in the dating of writing materials and textiles. Radiocarbon 25:669–674
Burnett WC, Veeh HH (1992) Uranium-series studies of marine phosphates and carbonates. In: Ivanovich M, Harmon RS (eds) Uranium-series disequilibrium: applications to earth, marine, and environmental sciences. Clarendon, Oxford, pp 486–512
Butler RF (1992) Paleomagnetism: magnetic domains to geologic terranes. Blackwell Scientific, Boston
Cann JR, Dixon JE, Renfrew C (1969) Obsidian analysis and the obsidian trade. In: Brothwell D, Higgins E (eds) Science in archaeology. Thames and Hudson, London, pp 578–591
Cassignol C, Gillot PY (1982) Range and effectiveness of unspiked potassium-argon dating: experimental groundwork and applications. In: Odin GS (ed) Numerical dating in stratigraphy. Wiley, New York, pp 159–173
Cerling TE, Craig H (1994) Geomorphology and in-situ cosmogenic isotopes. Annu Rev Earth Planet Sci 22:273–317
Cernohouz J, Solc I (1966) Use of sandstone wanes and weathered basaltic crust in absolute chronology. Nature 212:806–807
Champion DE, Lanphere MA, Kuntz MA (1988) Evidence for a new geomagnetic reversal from lava flows in Idaho: discussion of short polarity reversals in the Brunhes and late Matuyama polarity chrones. J Geophys Res 93:11667–11680
Chawla S, Singhvi AK (1989) Thermoluminescence dating of archaeological sediments. Naturwissenschaften 76:416–418
Chawla S, Dhir RP, Singhvi AK (1992) Thermoluminescence chronology of sand profiles in the Thar desert and their implications. Quat Sci Rev 11:25–32
Chen CH, Kramer DS, Allman SL, Hurst GS (1984) Selective counting of krypton atoms using resonance ionization spectroscopy. Appl Phys Lett 44:640–642
Chen JH, Edwards RL, Wasserburg GJ (1986) ^{238}U, ^{234}U and Th in seawater. Earth Planet Sci Lett 80:241–251
Chen JH, Edwards RL, Wasserburg GJ (1992) Mass spectrometry and applications to uranium-series disequilibrium. In: Ivanovich M, Harmon RS (eds) Uranium-series disequilibrium: applications to earth, marine, and environmental sciences. Clarendon, Oxford, pp 174–206
Chen T, Quan Y, Wu E (1994) Antiquity of *Homo sapiens* in China. Nature 368:55–56
Chen Y, Smith EP, Evensen NM, York D, Lajoie KR (1996) The edge of time: dating young volcanic ash layers with the ^{40}Ar-^{39}Ar laser probe. Science 274: 1176–1178
Clausen HB (1973) Dating of polar ice by ^{32}Si. J Glaciol 12:411–416

Clausen HB, Buchmann B, Ambach B (1967) Si^{32} dating of an alpine glacier. Proc IUGG, Snow and Ice Commission, Bern, pp 135–140

Clottes J, Chauvet JM, Brunel-Deschamps E, Hillaire C, Daugas JP, Arnold M, Cachier H, Evin J, Fortin P, Oberlin C, Tisnerat, N, Valladas H (1995) The Palaeolithic paintings of the Chauvet-Pont-D'Arc cave, at Vallon-Pont D'Arc (Ardeche, France): direct and indirect Radiocarbon dating. CR Acad Sci Paris 320:1133–1140

Collinson DW (1983) Methods in rock magnetism and palaeomagnetism: techniques and instrumentation. Chapman and Hall, London

Colman SM, Pierce KL (1981) Weathering rinds on andesitic and basaltic stones as a Quaternary age indicator, western United States. US Geol Surv Profess Pap 1210

Condomines M, Allegre CJ (1980) Ages and magmatic evolution of Stromboli volcano from ^{230}Th-^{238}U disequilibrium data. Nature 288:354–357

Covey C (1984) The earth's orbit and the ice ages. Sci Am 250/2:42–50

Cox A, Doell RR, Dalrymple GB (1963) Geomagnetic polarity epochs and Pleistocene geochronology. Nature 198:1049–1051

Craig H, Lupton JE (1976) Primordial neon, helium and hydrogen in oceanic basalts. Earth Planet Sci Lett 31:369–385

Crawford RW, Trole J, Baxter MS, Thomson J (1985) A comparison of the particle track and alpha-spectrometric techniques in excess thorium-230 dating of eastern Atlantic pelagic sediments. J Environ Radioact 2:135–144

Creer KM, Kopper JS (1974) Paleomagnetic dating of cave paintings in Tito Bustillo cave, Asturias, Spain. Science 186:348–350

Croll J (1875) Climate and time in their geological relations. Appleton, New York

Csapo J, Csapo-Kiss Z, Költö L, Papp I (1991) Age determination of fossil bone samples based on the ratio of amino acid racemization. In: Pernicka E, Wagner GA (eds) Archaeometry '90. Birkhäuser, Basel, pp 627–635

Curtis GH, Drake T, Cerling TE, Cerling BL, Hampel JH (1975) Age of KBS tuff in Koobi Fora formation, East Rudolf, Kenya. Nature 258:395–398

Dalrymple GB, Lanphere MA (1969) Potassium-argon dating. Freeman, San Francisco

Damon PE, Donahue DJ, Gore BH, Hatheway AL, Jull AJT, Linick TW, Sercel PJ, Toolin LJ, Bronk CR, Hall ET, Hedges REM, Housley R, Law A, Perry C, Bonani G, Trumbore S, Woelfli W, Ambers JC, Bowman SGE, Leese MN, Tite MS (1989) Radiocarbon dating of the Shroud of Turin. Nature 337:611–615

Daniels F, Boyd CA, Saunders DF (1953) Thermoluminescence as a research tool. Science 117:343–349

Dansgaard W (1981) Ice core studies: dating the past to find the future. Nature 290:360–361

Dansgaard W, Johnsen SJ, Clausen HB, Dahl-Jensen D, Gundestrup NS, Hammer CU, Hvidberg CS, Steffensen JP, Sveinbjörnsdottir AE, Jouzel J, Bond G (1993) Evidence for general instability of past climate from a 250-kyr ice-core record. Nature 364:218–220

Davis R, Schaeffer OA (1955) Chlorine-36 in nature. Ann NY Acad Sci 62:105–122

De Corte F, Van den Haute P, De Wispelaere A, Jonckheere R (1991) Calibration of the fission-track dating method: is Cu useful for an absolute thermal neutron fluence monitor? Chem Geol (Isotop Geosci Sect) 86:187–194

De Geer G (1912) A geochronology of the last 12000 years. Compte Rendue 1, 11 Congr Geol Int Stockholm 1910, pp 241–253

De Master DJ, Cochran JK (1982) Particle mixing rates in deep sea sediments determined from excess ^{210}Pb and ^{32}Si profiles. Earth Planet Sci Lett 61:257–271

de Vries H (1958) Variation of the concentration of radiocarbon with time and location on earth. Kon Ned Akad Wetensch Proc Ser B 61:267–281
Debenham NC, Aitken MJ (1984) Thermoluminescence dating of stalagmitic calcite. Archaeometry 26:155–170
Deevey Jr ES, Gross MS, Hutchinson GE, Kraybill HL (1954) The natural ^{14}C contents of materials from hard-water lakes. Proc Natl Acad Sci Wash 40:285–288
Delmas RJ (1989) Silicon-32 and argon-39. In: Roth E, Poty B (eds) Nuclear methods of dating. Kluwer, Dordrecht, pp 455–472
Denell RW, Rendell H, Hailwood E (1988) Early tool-making in Asia: two-million-year-old artefacts in Pakistan. Antiquity 62:98–106
Ditlefsen C (1992) Bleaching of K-feldspars in turbid water suspensions: a comparison of photo- und thermoluminescence signals. Quat Sci Rev 11:33–38
Dorale JA, Gonzalez LA, Reagan MK, Pickett DA, Murrell MT, Baker RG (1992) A high-resolution record of Holocene climate change in speleothem calcite from Cold Water Cave, northeast Iowa. Science 258:1626–1630
Dorn RI, Whitley DS (1983) Cation-ratio dating of petroglyphs from the Western Great Basin, North America. Nature 302:816–818
Dorn RI, Bamforth DB, Cahill TA, Dohrenwend JC, Turrin BD, Donahue DJ, Jull AJT, Long A, Macko ME, Weil EB, Whitley DS, Zabel TH (1986) Cation-ratio and accelerator radiocarbon dating of rock varnish on Mojave artifacts and landforms. Science 231:830–833
Dorn RI, Turrin BD, Jull AJT, Linick TW, Donahue DJ (1987) Radiocarbon dating and cation-ratio ages for rock varnish on Tioga and Tahoe morainal boulders of Pine Creek, eastern Sierra Nevada, California, and their paleoclimatic implications. Quat Res 28:38–49
Dorn RI, Jull AJT, Donahue DJ, Linick TW, Toolin LJ (1989) Accelerator mass spectrometry radiocarbon dating of rock varnish. Geol Soc Am Bull 101:1363–1372
Douglass AE (1921) Dating our prehistoric ruins. Nat Hist 21:27–30
Downey B, Tarling DH (1984) Archaeomagnetic dating of Santorini volcanic eruptions and fired destruction levels of late Minoan civilization. Nature 209: 519–523
Duller GA, Wintle AG (1991) On infrared stimulated luminescence at elevated temperatures. Nucl Tracks Radiat Meas 18:379–384
Duller GAT (1991) Equivalent dose determination using single aliquots. Nucl Tracks Radiat Meas 18:371–378
Duller GAT (1992) Comparison of equivalent doses determined by thermoluminescence and infrared stimulated luminescence for dune sands in New Zealand. Quat Sci Rev 11:39–43
Duller GAT (1994) Luminescence dating of poorly bleached sediments from Scotland. Quat Sci Rev 13:521–524
Duller GAT (1995) Luminescence dating using single aliquots: methods and applications. Radiat Meas 24:217–226
Duller GAT, Bötter-Jensen L (1992) Luminescence from potassium feldspars stimulated by infrared and green light. Radiat Prot Dosim 47:683–688
Duplessy JC, Labeyrie L, Arnold M, Paterne M Duprat C, van Weering TCE (1992) Changes in surface salinity of the North Atlantic ocean during the last deglaciation. Nature 358:485–487
Edwards RL, Chen JH, Wasserburg GJ (1986/87) ^{238}U-^{234}U-^{230}Th-^{232}Th systematics and the precise measurement of time over the past 500,000 years. Earth Planet Sci Lett 81:175–192

Edwards RL, Beck JW, Burr GS, Donahue DJ, Chappell JMA, Bloom AL, Druffel ERM, Taylor FW (1993) A large drop in the $^{14}C/^{12}C$ and reduced melting in the Younger Dryas, documented with ^{230}Th ages of corals. Science 260:962–968
Edwards RL, Cheng H, Murrell MT, Goldstein SJ (1997) Protactinium-231 dating of carbonates by thermal ionization mass spectrometry: implications for Quaternary climate change. Science 276:782–786
Edwards SR (1993) Luminescence dating of sand from the Kelso dunes, California. In: Pye K (ed) The dynamics and environmental context of aeolian sedimentary systems. Geol Soc Lond Spec Publ 72:59–68
Eighmy JL (1990) Archaeomagnetic dating: practical problems for the archaeologist. In: Eighmy JL, Sternberg RS (eds) Archaeomagnetic dating. University of Arizona Press, Tucson, pp 33–64
Eighmy JL, Sternberg RS (eds) (1990) Archaeomagnetic dating. University of Arizona Press, Tucson
Eisenhauer A, Gögen K, Pernicka E, Mangini A (1992) Climatic influences on the growth rates of Mn crusts during the late Quaternary. Earth Planet Sci Lett 109:25–36
Eisenhauer A, Wasserburg GJ, Chen JH, Bonani G, Collins LB, Zhu ZR, Wyrwoll KH (1993) Holocene sea-level determinations relative to the Australian continent: U/Th (TIMS) and ^{14}C (AMS) dating of coral cores from the Abrolhos Islands. Earth Planet Sci Lett 114:529–547
Eisenhauer A, Spielhagen RF, Frank M, Hentzschel G, Mangini A, Kubik PW, Dittrich-Hannen B, Billen T (1994) ^{10}Be records of sediment cores from high northern latitudes: implications for environmental and climatic changes. Earth Planet Sci Lett 124:171–184
El Mansouri M, El Fouikar A, Saint-Martin B (1996) Correlation between ^{14}C ages and aspartic acid racemization at the Upper Palaeolithic site of the Abri Pataud (Dordogne, France). J Archaeol Sci 23:803–809
Elitzsch C, Pernicka E, Wagner GA (1983) Thermoluminescence dating of archaeometallurgical slags. PACT 9:271–286
Elmore D, Conard NJ, Kubik PW, Gove HE, Beer J, Suter M (1987) ^{36}Cl and ^{10}Be profiles in Greenland ice – dating and production rate variations. Nucl Instrum Methods Phys Res B 29:207–217
Emiliani C (1955) Pleistocene temperatures. J Geol 63:538–578
Ericson JE, Mackenzie JD, Berger R (1976) Physics and chemistry of the hydration process in obsidians I: theoretical implications. In: Taylor RE (ed) Advances in obsidian glass studies. Noyes, Park Ridge, pp 25–45
Evans M E, Hoye GS (1991) Magnetic refraction and archaeomagnetic fidelity. In: Pernicka E, Wagner GA (eds) Archaeometry '90. Birkhäuser, Basel, pp 551–558
Evin J, Marechal J, Pachiaudi C, Puissegur JJ (1980) Conditions involved in dating terrestrial shells. Radiocarbon 22:545–555
Fabryka-Martin J, Davis SN, Elmore D (1987) Applications of ^{129}I and ^{36}Cl in hydrology. Nucl Instrum Methods Phys Res B29:361–371
Fairbanks RG (1989) A 17,000-year glacio-eustatic sea level record: influence of glacial melting rates on the Younger Dryas event and deep-ocean circulation. Nature 342:637–642
Fanale FP, Schaeffer OA (1965) Helium-uranium ratios for Pleistocene and Tertiary fossil aragonites. Science 149:312–316
Ferreira MP, Macedo R, Costa V, Reynolds JH, Riley JE Jr, Rowe MW (1975) Rare gas dating II: attempted uranium-helium dating of young volcanic rocks from the Madeira Archipelago. Earth Planet Sci Lett 25:142–150

Fink D, Klein J, Middleton R (1990) ^{41}Ca: past, present and future. Phys Res B52:572–582
Fireman EL, Norris TL (1982) Preliminary study on dating polar ice by carbon-14 and radon-222. In: Currie LA (ed) Nuclear and chemical dating techniques. ACS Symp Ser 176:319–329
Fitch FJ, Miller JA (1970) New hominid remains and early artefacts from northern Kenya. Nature 226:223–228
Fleischer RL, Price PB (1964) Fission track evidence for the simultaneous origin of tektites and other natural glasses. Geochim Cosmochim Acta 28:755–766
Fleischer RL, Price PB, Walker RM (1965) Effects of temperature, pressure, and ionization on the formation and stability of fission tracks in minerals and glasses. J Geophys Res 70:1497–1502
Fleischer RL, Viertl JRM, Price PB, Aumento F (1968) Mid-Atlantic ridge:age and spreading rates. Science 161:1339–1342
Fleischer RL, Price PB, Walker RM (1975) Nuclear tracks in solids. University of California Press, Berkeley
Flisch M (1986) K-Ar dating of Quaternary samples. In: Hurford AJ, Jäger E, Ten Cate JAM (eds) Dating young sediments. CCOP Technical Secretariat, Bangkok, pp 299–323
Fontes JC (1992) Chemical and isotopic constraints on ^{14}C dating of groundwater. In: Taylor RE, Long A, Kra RS (eds) Radiocarbon after four decades. Springer, Berlin Heidelberg New York, pp 242–261
Forman SL (1988) The solar resetting of thermoluminescence of sediments in a glacier-dominated fiord environment in Spitsbergen: geochronologic implications. Arct Alp Res 20:243–253
Forman SL (1989) Applications and limitations of thermoluminescence to date Quaternary sediments. Quat Int 1:47–59
Forster M, Ramm K, Maier P (1992) Argon-39 dating of groundwater and its limiting conditions. Proc Int Symp Isotope techniques in water resources development 1991. IAEA-SM-319/13, Wien, pp 203–214
Fourcroy AF, Vauquelin LN (1806) Experiments made upon new ivory, fossil ivory and the enamel of teeth, in order to ascertain if these substances contained fluoric acid. Phil Mag 25:265–268
Fowler J (1880) On the process of decay in glass and, incidentally, on the composition and texture of glass at different periods, and the history of its manufacture. Archaeologia 46:65–162
Frank N (1996) Anwendung der Thermionen-Massenspektrometrie zur Uranreihen-Datierung pleistozäner, mitteleuropäischer Travertinvorkommen. Dissertation, Universität Heidelberg, 136 S
Franke T, Fröhlich K, Gellermann R, Hebert D (1986) ^{32}Si in precipitation of Freiberg. J Radiat Nucl Chem Lett 103:11–18
Frechen J, Lippolt HJ (1965) Kalium-Argon-Daten zum Alter des Laacher Vulkanismus, der Rheinterrassen und der Eiszeiten. Eiszeitalter u Gegenwart 16:5–30
Friedman I (1977) Hydration dating of volcanism at Newberry Crater, Oregon. J Res US Geol Soc 5:337–342
Friedman I, Long W (1976) Hydration rate of obsidian. Science 191:347–352
Friedman I, Obradovich J (1981) Obsidian hydration dating of volcanic events. Quat Res 16:37–47
Friedman I, Smith RL (1960) A new dating method using obsidian, part I: The development of the method. Am Antiquity 25:476–493

Friedman I, Trembour FW (1978) Obsidian: the dating stone. Am Sci 66:44–51
Friedman I, Pierce KL, Obradovich J, Long WD (1973) Obsidian hydration dates glacial loading? Science 180:733–734
Friedman I, Trembour FW, Smith FL, Smith GI (1994) Is obsidian dating affected by relative humidity? Quat Res 41:185–190
Fuhrmann U, Lippolt HJ (1987) Excess argon and dating of Quaternary Eifel volcanism: III. Alkali basaltic rocks of the central West Eifel/FR Germany. Neues Jahrb Geol Palaeontol Monatsh 1987/4:213–236
Fukuchi T (1989) Increase of radiation sensitivity of ESR centers by faulting and criteria of fault dates. Earth Planet Sci Lett 94:109–122
Fukuchi T (1992) A reply to comments by W.M. Buhay on "Increase of radiation sensitivity of ESR centers by faulting and criteria of fault dates". Earth Planet Sci Lett 114:211–213
Fullagar RLK, Price DM, Head LM (1996) Early human occupation of northern Australia: archaeology and thermoluminescence dating of Jinmium rock-shelter, Northern Territory. Antiquity 70:751–773
Gallup CD, Edwards RL, Johnson RG (1994) The timing of high sea levels over the past 200 000 years. Science 263:796–799
Games K (1977) The magnitude of the palaeomagnetic field: a new non-thermal, non-detrital method using sun-dried bricks. Geophys J R Astron Soc 48:315–319
Garrison EG, McGimsey III CR, Zinke OH (1978) Alpha-recoil tracks in archaeological ceramic dating. Archaeometry 20:39–46
Gehrke E (1930) Die Patina auf Quarzen als Zeitmesser. Forsch Fortschr 6:406–407
Geilmann W (1956) Beiträge zur Kenntnis alter Gläser IV: Die Zersetzung der Gläser im Boden. Glastech Ber 29:145–168
Geiss J, Oeschger H, Schwarz U (1962) The history of cosmic radiation as revealed by isotopic changes in the meteorites and on the earth. Space Sci Rev 1:197–223
Geiss J, Buhler F, Cerutti H, Eberhardt P, Filleux CH (1972) Solar wind composition experiment. Apollo 16 Preliminary Science Report. NASA Spec Publ 315: 14.1–14.10
Gemell AMD (1988) Thermoluminescence dating of glacially transported sediments: some considerations. Quat Sci Rev 7:277–285
Gentner W, Lippolt HJ, Schaeffer OA (1961) Das Kalium-Argon-Alter einer Glasprobe vom Nördlinger Ries. Z Naturforsch 16a:184
Gentner W, Lippolt HJ, Schaeffer OA (1963) Das Kalium-Argon Alter der Gläser des Nördlinger Rieses und der böhmisch-mährischen Tektite. Geochim Cosmochim Acta 27:191–200
Gentner W, Lippolt HJ, Müller O (1964) Das Kalium-Argon-Alter des Bosumtwi-Kraters in Ghana und die chemische Beschaffenheit seiner Gläser. Z Naturforsch 19a:150–153
Gentner W, Storzer D, Wagner GA (1969) Das Alter von Tektiten und verwandten Gläsern. Naturwissenschaften 56:255–261
Gentner W, Glass BP, Storzer D, Wagner GA (1970) Fission track ages and ages of deposition of deep-sea microtektites. Science 168:359–361
Gentner W, Kirsten T, Storzer D, Wagner GA (1973) K-Ar and fission track dating of Darwin Crater glass. Earth Planet Sci Lett 20:204–210
Getty RS, DePaolo DJ (1995) Quaternary geochronology using U-Th-Pb method. Geochim Cosmochim Acta 59:3267–3272
Geyh MA (1970) Zeitliche Abgrenzung von Klimaänderungen mit ^{14}C-Daten von Kalksinter und organischen Substanzen. Beih Geol Jahrb 98:15–22

Geyh MA (1983) Physikalische und chemische Datierungsmethoden in der Quartärforschung. Clausthaler Tektonische Hefte 19. Ellen Pilger, Clausthal-Zellerfeld

Geyh MA (1992) Numerical modeling with groundwater ages. In: Taylor RE, Long A, Kra RS (eds) Radiocarbon after four decades. Springer, Berlin Heidelberg New York, pp 276–287

Geyh MA, Franke HW (1970) Zur Wachstumsgeschwindigkeit von Stalagmiten. Atompraxis 16:1–3

Geyh MA, Hennig GJ (1986) Multiple dating of a long flowstone profile. Radiocarbon 28(2 A):503–509

Geyh MA, Schleicher H (1990) Absolute age determination: physical and chemical dating methods and their application. Springer, Berlin Heidelberg New York

Geyh MA, Röthlisberger F, Gellalty A (1985) Reliablity tests and interpretation of ^{14}C dates from palaeosols in glacier environments. Z Gletscherkunde Gazialgeol 21:275–281

Gheradi S (1862) Sul magnetismo polare de palazzi et altri edifizi in Torino. Il Nuovo Cimento 16:384–404

Gill JB, Pyle DM, Williams RW (1992) Igneous rocks. In: Ivanovich M, Harmon RS (eds) Uranium-series disequilibrium: applications to earth, marine, and environmental sciences. Clarendon, Oxford, pp 207–258

Gillard RD, Hardman SM, Pollard AM, Sutton PA, Whittaker DK (1991) Determinations of age at death in archaeological populations using the D/L ratio of aspartic acid in dental collagen. In: Pernicka E, Wagner GA (eds) Archaeometry '90. Birkhäuser, Basel, pp 637–644

Gläser R (1991) Bemerkungen zur absoluten Datierung des Beginns der westlichen Linearbandkeramik. Banatica 11:53–64

Gläser R (1996) Zur absoluten Chronologie der Vinca-Kultur anhand von ^{14}C-Daten. Proc Int Symp Vinca Culture, Museum of Banat, Timisoara, Romania, pp 175–212

Gleadow AJW (1980) Fission track age of the KBS tuff and associated hominid remains in northern Kenya. Nature 284:225–230

Gleadow AJW (1981) Fission track dating methods:what are the real alternatives? Nucl Tracks 5:3–14

Godfrey-Smith DI (1994) Thermal effects in the optically stimulated luminescence of quartz and mixed feldspars from sediments. J Phys D Appl Phys 27:1737–1746

Godfrey-Smith DI, Huntley DJ, Chen WH (1988) Optical dating studies of quartz and feldspar sediment extracts. Quat Sci Rev 7:373–380

Goetz C, Hillaire-Marcel C (1992) U-series disequilibria in early diagenetic minerals from Lake Magadi sediments, Kenya: dating potential. Geochim Cosmochim Acta 56:1331–1341

Göksu HY, Wieser A, Regulla DF, Schramel P, Vogenauer A, Wagner U (1992) Datierung mit Thermolumineszenz und Bestimmung der Erhitzungstemperatur. In: Behm-Blancke MR (Hrsg) Hassek Höyük. Istanbuler Forsch 38:140–147

Goldberg ED (1963) Geochronology with lead-210. In: Radioactive dating. IAEA, Vienna, pp 121–131

Goldstein SJ, Murrell MT, Williams RW (1993) ^{231}Pa and-^{230}Th chronology of mid-ocean ridge basalts. Earth Planet Sci Lett 115:151–159

Goodfriend GA, Flessa KW, Hare PE (1997) Variation in amino acid epimerization rates and amino acid composition among shell layers in the bivalve *Chione* from the Gulf of California. Geochim Cosmochim Acta 61:1487–1493

Goslar T, Arnold M, Bard E, Kuc T, Pazdur MF, Ralska-Jasiewiczowa M, Rozanski K, Tisnerat M, Walanus A, Wicik B, Wieckowski K (1995) High concentration of atmospheric ^{14}C during the Younger Dryas cold episode. Nature 377:414-417
Graf T, Kohl CP, Marti K, Nishiizumi K (1991) Cosmic-ray produced neon in Antarctic rocks. Geophys Res Lett 18:203-206
Griffiths DR, Bergman CA, Clayton CJ, Robins GV, Seeley NJ (1986) Experimental investigation of the heat treatment of flint. In: Sieveking G de G, Hart MB (eds) The human uses of flint and chert. Cambridge University Press, Cambridge, pp 43-52
Grögler N, Houtermans FG, Stauffer H (1958) Radiation damage as a research tool for geology and prehistory. Convegno sulle dotazioni con metodi nucleari. 5ª Rass. Internazion Elettr Nucl Sezione Nucleare Roma, pp 5-15
Grootes PM (1978) Carbon-14 time scale extended: comparison of chronologies. Science 200:11-15
Grosse AV (1934) An unknown radioactivity. J Am Chem Soc 56:1922-1923
Grün R (1986) ESR-dating of a flowstone core from Cova de Sa Bassa Blanca (Mallorca, Spain). ENDINS 12:19-23
Grün R (1989) Die ESR-Altersbestimmungsmethode. Springer, Berlin Heidelberg New York
Grün R (1991) Potentials and problems of ESR dating. Nucl Tracks Radiat Meas 18:155-161
Grün R, Stringer CB (1991) Electron spin resonance dating and the evolution of modern humans. Archaeometry 33:153-199
Grün R, Schwarcz HP, Ford DC, Hentzsch B (1988) ESR dating of spring deposited travertines. Quat Sci Rev 7:429-432
Grün R, Radtke U, Omura A (1992) ESR and U-Th analyses on corals from Huon Peninsula, New Guinea. Quat Sci Rev 11:197-202
Gu ZY, Lal D, Liu TS, Southon J, Caffee MW, Guo ZT, Chen MY (1996) Five million year ^{10}Be record in Chinese loess and red-clay:climate and weathering relationships. Earth Planet Sci Lett 144:273-287
Guerin G, Valladas G (1980) Thermoluminescence dating of volcanic plagioclases. Nature 286:697-699
Guichard F, Reyss JL, Yokoyama Y (1978) Growth rate of manganese nodule measured with ^{10}Be and ^{26}Al. Nature 272:155-156
Guillou, H, Turpin L, Garnier F, Charbit S, Thomas DM (1997) Unspiked K-Ar dating of Pleistocene tholeiitic basalts from the deep core SOH-4, Kilauea, Hawaii. Chem Geol 140:81-88
Gunn NM, Murray AS (1980) Geomagnetic field magnitude variations in Peru derived from archaeological ceramics dated by thermoluminescence. Geophys J R Astron Soc 62:345-366
Guo SL, Liu SS, Sun FS, Zhang F, Zhou SH, Hao XH, Hu RY, Meng W, Zhang PF, Liu JF (1991) Age and duration of Peking man site by fission track method. Nucl Tracks Radiat Meas 19:719-724
Gustavsson K, Erämetsä P, Sonninen E, Jungner H (1990) Radiocarbon dating of mortar and thermoluminescence dating of bricks from Aland. PACT 29:203-212
Guyodo V, Valet JP (1996) Relative variations in geomagnetic intensity from sedimentary records: the past 200,000 years. Earth Planet Sci Lett 143:23-36
Haddy A, Hanson A (1982) Nitrogene and fluorine dating of Moundville skeletal samples. Archaeometry 24:37-44
Hahn J (1991) Erkennen und Bestimmen von Stein- und Knochenartefakten. Archaeologica Venatora, Tübingen

Hajdas I, Zolitschka B, Ivy-Ochs SD, Beer J, Bonani G, Leroy SAG, Negendank JW, Ramrath M, Suter M (1995) AMS Radiocarbon dating of annually laminated sediments from Lake Holzmaar, Germany. Quat Sci Rev 14:137–143

Hall CM, Walter RC, Westgate JA, York D (1984) Geochronology, stratigraphy and geochemistry of Cindery tuff in Pliocene hominid-bearing sediments in the Middle Awash, Ethiopia. Nature 308:26–31

Hambach U (1992) Magnetostratigraphie in der borealen Kreide. Uni Press Hochschulschriften 38. Lit, Münster

Hamelin B, Bard E, Zindler A, Fairbanks RG (1991) $^{234}U/^{238}U$ mass spectrometry of corals: how accurate is the U-Th age of the last interglacial period? Earth Planet Sci Lett 106:169–180

Hammer CU, Clausen HB, Friedrich WL, Tauber H (1987) The Minoan eruption of Santorini in Greece dated to 1645 B.C.? Nature 328:517–519

Hampel W, Takagi J, Sakamoto K, Tanaka S (1975) Measurement of muon-induced ^{26}Al in terrestrial silicate rocks. J Geophys Res 80:3757–3760

Harbottle G, Sayre EV, Stoenner RW (1979) Carbon-14 dating of small samples by proportional counting. Science 206:683–685

Hare PE, Turnbull HF, Taylor RE (1978) Amino acid dating of Pleistocene fossil materials: Olduvai Gorge, Tanzania. In: Freeman GL (ed) Views of the past: essays in old world prehistory and paleoanthropology. Mouton, The Hague, pp 7–12

Harland WB, Armstrong RL, Cox AV, Craig LE, Smith AG, Smith DG (1990) A geologic time scale 1989. Cambridge University Press, Cambridge

Harmon RS, Glazek J, Nowak K (1980) $^{230}Th/^{234}U$ dating of travertine from the Bilzingsleben archaeological site. Nature 284:132–135

Harrington CD, Whitney JW (1987) Scanning electron microscope method for rock-varnish dating. Geology 15:967–970

Hashemi-Nezhad SR, Durrani SA (1981) Registration of alpha-recoil tracks in mica: the prospect for alpha-recoil dating method. Nucl Tracks 5:189–205

Hassan AG (1983) Archaeomagnetic investigations in Egypt: inclination and field intensity determinations. J Geophys Res 53:131–140

Haverkamp B, Beuker Th (1993) A palaeomagnetic study of maar-lake sediments from the Westeifel. In: Negendank JFW, Zolitschka B (eds) Paleolimnology of European maar lakes. Springer, Berlin Heidelberg New York, pp 349–365

Hays JD, Saito T, Opdyke ND, Burckle LH (1969) Pliocene-Pleistocene sediments of the Equatorial Pacific: their palaeomagnetic, biostratigraphic and climatic record. Bull Geol Soc Am 80:1481–1514

Head MJ, Zhou W, Zhou M (1988) Evaluation of ^{14}C ages of organic fractions of paleosols from loess-paleosol sequences near Xian, China. Radiocarbon 31: 680–696

Hearty PJ, Miller GH, Stearns CE, Szabo BJ (1986) Aminostratigraphy of Quaternary shorelines in the Mediterranean basin. Bull Geol Soc Am 97:850–858

Hebert D (1990) Tritium in der Atmosphäre – Quellen, Verteilung, Perspektive. Freiberger Forschungshefte C 443

Hedges REM, Gowlett JAJ (1986) Radiocarbon dating by accelerator mass spectrometry. Sci Am 254,1:82–89

Hedges REM, Van Klinken GJ (1992) A review of current approaches in the pretreatment of bone for radiocarbon dating by AMS. Radiocarbon 34:279–291

Hedges REM, Housley RA, Ramsey CB, Van Klinken GJ (1994) Radiocarbon dates from the Oxford AMS system: archaeometry datelist 18. Archaeometry 36: 337–374

Heijnis H, van der Plicht J (1992) Uranium/thorium dating of Late Pleistocene peat deposit in NW Europe, uranium/thorium isotope systematics and open-system behaviour of peat layers. Chem Geol (Isot Geosc Sect) 94:161–171
Heinermeier J, Hornshöj P, Nielsen HL, Rud N, Thomsen MS (1987) Accelerator mass spectrometry applied to $^{22,24}Na$, $^{31,32}Si$, and ^{14}C. Nucl Instrum Methods Phys Res B29:110–123
Heller F, Liu T (1986) Palaeoclimatic and sedimentary history from magnetic susceptibility of loess in China. Geophys Res Lett 13:1169–1172
Heller F, Xiuming L, Tungsheng L, Tongchun X (1991) Magnetic susceptibility of loess in China. Earth Planet Sci Lett 103:301–310
Henderson GM, O'Nions RK (1995) $^{234}U/^{238}U$ ratios in Quaternary planktonic foraminifera. Geochim Cosmochim Acta 59:4685–4694
Henderson GM, Cohen AS, O'Nions RK (1993) $^{234}U/^{238}U$ ratios and ^{230}Th ages for Hateruma Atoll corals: implications for coral diagenesis and seawater $^{234}U/^{238}U$ ratios. Earth Planet Sci Lett 115:65–73
Hennig GJ, Grün R (1983) ESR dating in Quaternary geology. Quat Sci Rev 2:157–238
Hennig GJ, Herr W, Weber E, Xirotiris NI (1981) Petralona cave dating controversy. Nature 299:281–282
Hennig GJ, Grün R, Brunnacker K (1983) Speleothems, travertines and paleoclimates. Quat Res 20:1–29
Henning W, Bell WA, Billquist BJ, Glagola BG, Kutschera W (1987) Calcium-41 concentration in terrestrial materials: prospects for dating Pleistocene samples. Science 236:725–727
Henry DO (1992) The impact of radiocarbon dating on Near Eastern prehistory. In: Taylor RE, Long A, Kra RS (eds) Radiocarbon after four decades – an interdisciplinary perspective. Springer, Berlin Heidelberg New York, pp 324–334
Herron MM (1982) Glaciochemical dating techniques. In: Currie LA (ed) Nuclear and chemical dating techniques. Am Chem Soc, Washington, pp 303–318
Hilgen FJ (1991) Astronomical calibration of Gauss to Matuyama sapropels in the Mediterranean and implication for the geomagnetic polarity time scale. Earth Planet Sci Lett 104:226–244
Hillam J, Tyers I (1995) Reliability and repeatability in dendrochronological analysis: tests using the Fletcher archive of panel-dating data. Archaeometry 37:395–405
Hille P (1979) An open system model for uranium series dating. Earth Planet Sci Lett 42:138–142
Hillhouse JW, Cerling TE, Brown FH (1986) Magnetostratigraphy of the Koobi Fora Formation, Lake Turkana, Kenya. J Geophys Res 91, B11:11581–11595
Holmes A (1937) The age of the earth. Nelson, London
Holzhauser H (1984) Rekonstruktion von Gletscherschwankungen mit Hilfe fossiler Hölzer. Geograph Helvet 9:3–15
Hopf M (1969) Plant remains and early farming in Jericho. In: Ucko PJ, Dimbledy FW (eds) The domestication and exploitation of plants and animals. Duckworth, London, pp 355–357
Horn P, Müller-Sohnius D (1988) A differential etching and magnetic separation approach to whole-rock potassium-argon dating of basaltic rocks. Geochem J 22:115–128
Horn P, Müller-Sohnius D, Storzer D, Zöller L (1993) K-Ar-, fission-track-, and thermoluminescence ages of Quaternary volcanic tuffs and their bearing on Acheulian artifacts from Bori, Kukdi valley, Pune district, India. Z Dt Geol Ges 144: 326–329

Hu Q, Smith PE, Evensen NM, York D (1994) Lasing the Holocene: extending the ^{40}Ar-^{39}Ar laser probe method into the ^{14}C age range. Earth Planet Sci Lett 123: 331–336
Huang PH, Jin SH, Peng ZC, Liang RY, Quan JC, Wang ZR (1988) ESR dating and trapped electrons lifetime in quartz grains in loess of China. Quat Sci Rev 7:533–536
Huang WH, Walker RM (1967) Fossil alpha-particle recoil tracks: a new method of age determination. Science 155:1103–1106
Huber B (1941) Aufbau einer mitteleuropäischen Jahrring-Chronologie. Mitt Akad Dtsch Forstwiss 1:110–125
Huntley DJ, Godfrey-Smith DI, Thewalt MLW (1985) Optical dating of sediments. Nature 313:105–107
Huntley DJ, Godfrey-Smith DI, Haskell EH (1991) Light-induced emission spectra from some quartz and feldspars. Nucl Tracks Radiat Meas 18:127–131
Huntley DJ, Hutton JT, Prescott JR (1993a) The stranded beach-dune sequence of south-east South Australia: a test of thermoluminescence dating, 0–800 ka. Quat Sci Rev 12:1–20
Huntley DJ, Hutton JT, Prescott JR (1993b) Optical dating using inclusions within quartz. Geology 21:1087–1090
Hurford AJ (1990) Standardization of fission track dating calibration: recommendation by the Fission Track Working Group of the I. U. G. S. Subcommission on Geochronology. Chem Geol (Isotop Geosci Sect) 80:171–178
Hurford AJ, Green PF (1983) The zeta calibration of fission track dating. Isotope Geosci 1:285–317
Hurford AJ, Gleadow AJW, Naeser CW (1976) Fission-track dating of pumice from the KBS Tuff, East Rudolf, Kenya. Nature 263:738–740
Hütt G, Jungner H (1992) Optical and TL dating of glaciofluvial sediments. Quat Sci Rev 11:161–163
Hütt G, Jaek I, Tchonka J (1988) Optical dating: K-feldspars optical response stimulation spectra. Quat Sci Rev 7:381–385
Hütt G, Jungner H, Kujansuu R, Saarnisto M (1993) OSL and TL dating of buried podsols and overlying sands in Ostrobothnia, western Finland. J Quat Sci 8:125–132
Huxtable J, Aitken MJ (1985) Thermoluminescence dating results for the palaeolithic site Maastricht-Belvedere. An Praehist Leidensia 18:41–44
Huxtable J, Aitken MJ, Hedges JW, Renfrew AC (1976) Dating a settlement pattern by thermoluminescence: the burnt mounds of Orkney. Archaeometry 18:5–17
Ike D, Bada JL, Masters PM, Kennedy G, Vogel JD (1979) Aspartic acid racemization and radiocarbon dating of an early milling stone horizon burial in California. Am Antiq 44:524–530
Ikea M (1981) Paramagnetic alanine molecular radicals in fossil shells and bones. Naturwissenschaften 67:474
Ikea M, Miki T, Tanaka K (1982) Dating of fault by electron spin resonance on intrafault materials. Science 215:1392–1393
Imai N, Shimokawa K, Hirota M (1985) ESR dating of volcanic ash. Nature 314:81–83
Imbrie J, Hays JD, Martinson DG, McIntyre A, Mix AC, Morley JJ, Pisias NG, Prell WL, Shackleton NJ (1984) The orbital theory of Pleistocene climate: support from a revised chronology of the marine $\delta^{18}O$ record. In: Berger AL, Imbrie J, Hays J, Kukla G, Saltzman B (eds) Milankovitch and climate. Reidel, Dordrecht, pp 269–305
Imbrie J, Mix AC, Martinson DG (1993) Milankovitch theory viewed from Devils Hole. Nature 363:531–533

Ivanovich M, Harmon S (1992) Uranium-series disequilibrium: applications to earth, marine, and environmental sciences. Clarendon, Oxford

Ivanovich M, Murray A (1992) Spectroscopic methods. In: Ivanovich M, Harmon RS (eds) Uranium-series disequilibrium: applications to earth, marine, and environmental sciences. Clarendon, Oxford, pp 126–173

Jäger E, Chen WJ, Hurford AJ, Liu RX, Hunziker JC, Li DM (1985) BB-6: A Quaternary age standard for K-Ar-dating, Chem Geol (Isotope Geosci Sect) 52:275–279

Jenkins WJ, Beg MA, Clarke WB, Wangersky PJ, Craig H (1972) Excess ^{3}He in the Atlantic Ocean. Earth Planet Sci Lett 16:122–126

Johnsen SJ, Clausen HB, Dansgaard W, Fuhrer K, Gundestrup N, Hammer CU, Iversen P, Jouzel J, Stauffer B, Steffensen JP (1992) Irregular glacial interstadials in a new Greenland ice core. Nature 359:311–313

Johnson BJ, Miller GH (1997) Archaeological applications of amino acid racemization. Archaeometry 39:265–287

Johnson EA, Murphy T, Torreson OW (1948) Prehistory of the earth's magnetic field. Terrestr Magnet Atmos Electr 53:349–372

Johnson GD, Zeitler P, Naeser CW, Johnson NM, Summers DM, Frost CD, Opdyke ND, Tahirkheli RAK (1982) The occurrence and fission-track ages of late Neogene and Quaternary volcanic sediments, Siwalik Group, northern Pakistan. Palaeogeogr Palaeoclimatol Palaeoecol 37:63–69

Joly J (1908) The radioactivity of sea-water. Philos Mag J Sci Lond 6:385–393

Jouzel J (1989) The tritium content and tritium-helium-3 methods. In: Roth E, Poty B (eds) Nuclear methods of dating. Kluwer, Dordrecht, pp 493–510

Jouzel J, Merlivat L, Pourchet M, Lorius C (1979) A continuous record of artificial tritium fallout at the South Pole (1954–1978). Earth Planet Sci Lett 45:188–200

Jouzel J, Lorius C, Petit JR, Genthon C, Barkov NI, Kotlyakov VM, Petrov VM (1987) Vostok ice core: a continuous isotope temperature record over last climatic cycle (160,000 years). Nature 329:403–408

Jull AJT, Donahue DJ, Damon PE (1996) Factors affecting the apparent radiocarbon age of textiles: a comment on "Effects of fire and biofractionation of carbon isotopes on results of radiocarbon dating of old textiles: the Shroud of Turin", by D. A. Kouznetsov et al. J Archaeol Sci 23:157–160

Kaneoka I (1969) The use of obsidian for K-Ar dating. Mass Spectrosc 17:514–521

Kasuya M, Ikeda S (1991) Radioactive disequilibrium dating of corals by nuclear track detection. Earth Planet Sci Lett 102:455–459

Katsui Y, Kondo Y (1976) Variation in obsidian hydration rates for Hokkaido, northern Japan. In: Taylor RE (ed) Advances in obsidian glass studies. Noyes, Park Ridge, pp 120–140

Kaufman A (1993) An evaluation of several methods for determining $^{230}Th/U$ ages in impure carbonates. Geochim Cosmochim Acta 57:2303–2317

Kaufman A, Ku TL, Luo S (1995) Uranium-series dating of carnotites: concordance between ^{230}Th and ^{231}Pa ages. Chem Geol 120:175–181

Kaufman S, Libby WF (1954) The natural distribution of tritium. Phys Rev 93:1337–1344

Keisch B (1968) Dating works of art through their natural radioactivity: improvements and applications. Science 160:413–415

Keisch B, Feller RL, Levine AS, Edwards RR (1967) Dating and authenticating works of art by measurement of natural alpha emitters. Science 155:1238–1242

Kent DV, Ninkovich D, Pescatore P, Sparks SRJ (1981) Palaeomagnetic determination of emplacement temperature of Vesuvius AD 79 pyroclastic deposits. Nature 290:393–396

Kharkar DP, Turekian KK, Scott MR (1969) Comparison of sedimentation rates obtained by ^{32}Si and uranium decay series determinations in some siliceous Antarctic cores. Earth Planet Sci Lett 6:61–68

Kim KH, McMutry G (1991) Radial growth rates and ^{210}Pb ages of hydrothermal massive sulfides from the Juan de Fuca Ridge. Earth Planet Sci Lett 104:299–314

Klein P, Eckstein D (1988) Die Dendrochronologie und ihre Anwendung. Spektrosc Wiss 1988/1:56–68

Knauss K (1981) Uranium-series dating of pedogenic carbonates from the Livermore valley, California. Lawrence Livermore Labor, Rep UCRL-53095

Knutson DW, Buddemeier RW, Smith SV (1972) Coral chronometers: seasonal growth bands in reef corals. Science 177:270–272

Kobayashi K (1959) Chemical remanent magnetization of ferromagnetic minerals and its application to rock magnetism. J Geomagn Geoelectr 10:99–117

Koeberl C, Storzer D, Reimold WU (1994) The age of the Saltpan impact crater, South Africa. Meteoritics 29:374–426

Koeberl C, Bottomley R, Glass BP, Storzer D (1997) Geochemistry and age of Ivory Coast tektites. Geochim Cosmochim Acta 61:1745–1772

Korfmann M (1987) Vorwort des Herausgebers. In: Korfmann M (ed) Demircihüyük Bd. II: Naturwissenschaftliche Untersuchungen, Philipp v Zabern, Mainz, S 7–19

Korfmann M, Kromer B (1993) Demircihüyük, Besik-Tepe, Troia – Eine Zwischenbilanz zur Chronologie dreier Orte in Westanatolien. Studia Troica 3: 135–172

Kouznetsov DA, Ivanov AA, Veletsky PR (1996) Effects of fire and biofractionation of carbon isotopes on results of radiocarbon dating of old textiles: the Shroud of Turin. J Archaeol Sci 23:109–121

Kovacheva M (1991) Prehistoric sites from Bulgaria studied archaeomagnetically. In: Pernicka E, Wagner GA (eds) Archaeometry '90. Birkhäuser, Basel, pp 559–567

Kovacheva M, Veljovich D (1977) Geomagnetic field variations in southeastern Europe between 6500 and 100 years B.C. Earth Planet Sci Lett 37:131–138

Kovacheva M, Zagniy G (1985) Archaeomagnetic results from some prehistoric sites in Bulgaria. Archaeometry 27:179–184

Krbetschek MR, Rieser U, Stolz W (1996) Optical dating: some luminescence properties of natural feldspars. Radiat Prot Dosim 66:407–412

Krishnaswami S, Lal D, Prahbu NR, MacDougall D (1974) Characteristics of fission tracks in zircon: applications to geochronology and cosmology. Earth Planet Sci Lett 22:51–59

Kromer B, Münnich KO (1992) CO_2 gas proportional counting in radiocarbon dating – review and perspective. In: Taylor RE, Long A, Kra RS (eds) Radiocarbon after four decades – an interdisciplinary perspective. Springer, Berlin Heidelberg New York, pp 185–197

Kromer B, Rhein M, Bruns M, Schoch-Fischer H, Münnich KO, Stuiver M, Becker B (1986) Radiocarbon calibration data for the 6th to the 8th millenia BC. Radiocarbon 28 (2B):954–960

Kronfeld J, Vogel JC, Talma AS (1994) A new explanation for extreme $^{234}U/^{238}U$ disequilibria in a dolomitic aquifer. Earth Planet Sci Lett 123:81–93

Krumbein WE (1971) Biologische Entstehung von Wüstenlack. Umschau 1971: 240–241

Ku TL (1976) The uranium-series methods of age determination. Annu Rev Earth Planet Sci 4:347–379

Ku TL, Broecker WS (1967) Uranium, thorium and protactinium in manganese nodule. Earth Planet Sci Lett 2:317–321

Ku TL, Buhl WB, Freeman ST, Knauss KG (1979) Th^{230}-U^{234} dating of pedogenic carbonates in gravelly desert soils of Vidal Valley, southeastern California. Bull Geol Soc Am 90:1063–1073

Kukla G, Heller F, Liu XM, Xu TC, Liu TS, An ZS (1988) Pleistocene climates in China dated by magnetic susceptibility. Geology 16:811–814

Kuniholm PI, Striker CI (1987) Dendrochronological investigations in the Aegean and neighbouring regions. J Field Archaeol 14:385–398

Kuniholm PI, Kromer B, Manning SW, Newton M, Latini CE, Bruce MJ (1996) Anatolian tree rings and the absolute chronology of the eastern Mediterranean, 2220–718 BC. Nature 381:780 783

Kunz ML, Reanier RE (1994) Paleoindians in Beringia: evidence from Arctic Alaska. Science 263:660–662

Kurz MD (1986a) Cosmogenic helium in a terrestrial rock. Nature 320:435–439

Kurz MD (1986b) In situ production of terrestrial cosmogenic helium and some applications to geochronology. Geochim Cosmochim Acta 50:2855–2862

Kurz MD, Colodner D, Trull TW, Moore RB, O'Brien K (1990) Cosmic ray exposure dating with in situ produced cosmogenic ^{3}He: results from young Hawaiian lava flows. Earth Planet Sci Lett 97:177–189

Kvenvolden KA, Peterson E, Wehmiller J, Hare PE (1973) Racemization of amino acids in marine sediments determined by gas chromatography. Geochim Cosmochim Acta 37:2215–2225

Labelle JM, Eighmy JL (1997) Additional archaeomagnetic data on the south-west USA master geomagnetic pole curve. Archaeometry 39:431–439

Lal D (1986) On the study of continental erosion rates and cycles using cosmogenic ^{10}Be and ^{26}Al and other isotopes. In: Hurford AJ, Jäger E, Ten Cate JAM (eds) Dating young sediments. CCOP Technical Secretariat, Bangkok, pp 285–298

Lal D (1988) In situ-produced cosmogenic isotopes in terrestrial rocks. Annu Rev Earth Sci 16:355–388

Lal D (1991) Cosmic ray labeling of erosion surfaces: in situ production rates and erosion models. Earth Planet Sci Lett 104:424–439

Lal D, Peters B (1967) Cosmic ray produced radioactivity on the earth. In: Flügge S (ed) Handbuch der Physik. Springer, Berlin Heidelberg New York, pp 551–612

Lal D, Goldberg ED, Koide M (1960) Cosmic ray-produced silicon-32 in nature. Science 131:332–337

Lal D, Nijampurkar VN, Rama S (1970) Silicon-32 hydrology. Isotopes in hydrology. IAEA, Vienna, pp 847–861

Lal D, Nijampurkar VN, Somayajulu BLK, Goldberg ED (1976) Silicon-32 specific activities on coastal waters of the world oceans. Limnol Oceanogr 21:285

Lally AE (1992) Chemical procedures. In: Ivanovich M, Harmon RS (eds) Uranium-series disequilibrium: applications to earth, marine, and environmental sciences. Clarendon, Oxford, pp 95–126

LaMarche VC Jr, Hirschboek KK (1984) Frost rings in trees as records of major volcanic eruptions. Nature 307:121–126

Lanford WA (1977) Glass hydration: a method for dating glass objects. Science 196:975–977

Lanford WA, Trautvetter HP, Ziegler J, Keller J (1976) New precision technique for measuring the concentration versus depth of hydrogen in solids. Appl Phys Lett 28:566–568

Lang A (1996) Die Infrarot-Stimulierte-Lumineszenz als Datierungsmethode für holozäne Lössderivate. Heidelb Geogr Arb 103

Lang A, Wagner GA (1996) Infrared stimulated luminescence dating of archaeosediments. Archaeometry 38:129–141

Lang A, Wagner GA (1997) Infrared stimulated luminescence dating of Holocene colluvial sediments using the 410 nm emission. Quat Sci Rev 16:393–396

Lang A, Zöller L, Schukraft G (1992) Thermolumineszenz-Untersuchungen an Auensedimenten. Flensburger Regionale Studien, Sonderheft 2:99–130

Lao Y, Anderson RF, Broecker WS, Trumbore SE, Hofmann HJ, Wölfli W (1992a) Increased production of cosmogenic ^{10}Be during the Last Glacial Maximum. Nature 357:576–578

Lao Y, Anderson RF, Broecker WS, Trumbore SE, Hofmann HJ, Wölfli W (1992b) Transport and burial rates of ^{10}Be and ^{231}Pa in the Pacific Ocean during the Holocene period. Earth Planet Sci Lett 113:173–189

Latham AG (1997) Uranium-series dating of bone by gamma-ray spectrometry: comment. Archaeometry 31:217–219

Latham AG, Schwarcz HP (1992) The Petralona hominid site: uranium-series RE-analysis of "layer 10" calcite and associated paleomagnetic analyses. Archaeometry 34:135–140

Latham AG, Schwarcz HP, Ford DC, Pearce GW (1979) Paleomagnetism of stalagmite deposits. Nature 280:383–385

Latham AG, Schwarcz HP, Ford DC (1986) The paleomagnetism and U-Th dating of Mexican stalagmite. Earth Planet Sci Lett 79:195–207

Lee C, Bada JL, Peterson E (1976) Amino acids in modern and fossil woods. Nature 259:183–186

Lee HK, Schwarcz HP (1994) ESR plateau dating of fault gouge. Quat Sci Rev 13:629–634

Lee R (1969) Chemical temperature integration. J Appl Meteorol 8:423–430

Lee RR, Leich DA, Tombrello TA, Ericson JE, Friedman I (1974) Obsidian hydration profile measurements using nuclear reaction technique. Nature 250:44–47

Lehmann BE, Oeschger H, Loosli H, Hurst GS, Allman SL, Chen CH, Kramer SD, Payne MG, Phillips RC, Willis RD, Thonnard N (1985) Counting ^{81}Kr atoms for analysis of groundwater. J Geophys Res 90, B13:11547–11551

Leitner-Wild E, Steffan T (1993) Uranium-series dating of fossil bones from Alpine caves. Archaeometry 35:137–146

Lerman JC, Mook WG, Vogel JC (1970) ^{14}C in tree rings from different localities. In: Olsson IU (ed) Radiocarbon variations and absolute chronology. Almquist, Wicksell, Uppsala, pp 275–301

Levin I, Bösinger R, Bonani G, Francey RJ, Kromer B, Münnich KO, Suter M, Trivett NBA, Wölfli W (1992) Radiocarbon in atmospheric dioxide and methane: global distribution and trends. In: Taylor RE, Long A, Kra RS (eds) Radiocarbon after four decades. Springer, Berlin Heidelberg New York, pp 503–518

Li SH, Wintle AG (1992) A global view of the stability of luminescence signals from loess. Quat Sci Rev 11:133–137

Libby WF (1946) Atmospheric helium-three and radiocarbon from cosmic radiation. Phys Rev 69:671–672

Libby WF (1952) Radiocarbon dating. University of Chicago Press, Chicago

Limbrock K (1992) Hochauflösende gesteins- und paläomagnetische Untersuchungen am Würmlößprofil Böckingen (Heilbronn). Diplomarbeit, Universität Münster

Lin JC, Broecker WS, Anderson RF, Hemming S, Rubenstone JL, Bonani G (1996) New $^{230}Th/U$ and ^{14}C ages from Lake Lahontan carbonates, Nevada, USA, and a discussion of the origin of initial thorium. Geochim Cosmochim Acta 60:2817–2832

Linke G, Katzenberger O, Grün R (1985) Description and ESR dating of the Holstein interglaciation. Quat Sci Rev 4:319–331

Lippolt HJ, Weigel E (1988) ^{4}He diffusion in ^{40}Ar-retentive minerals. Geochim Cosmochim Acta 52:1449–1458

Lister G, Kelts K, Schmid R, Bonani G, Hofmann H, Morenzoni E, Nessi M, Suter M, Wölfli W (1990) Correlation of the paleoclimatic record in lacustrine sediment sequences: ^{14}C dating by AMS. Nucl Instrum Methods Phys Res B5:389–393

Long A, Hendershott RB, Martin PS (1983) Radiocarbon dating of fossil eggshell. Radiocarbon 25:533–539

Loosli HH (1983) A dating method with ^{39}Ar. Earth Planet Sci Lett 63:51–62

Loosli HH, Oeschger H (1968) Detection of ^{39}Ar in atmospheric argon. Earth Planet Sci Lett 5:191–198

Loosli HH, Lehmann BE, Balderer W (1989) Argon-39, argon-37 and krypton-85 isotopes in Stripa groundwaters. Geochim Cosmochim Acta 53:1825–1829

Lorenz IB (1988) Thermolumineszenz-Datierung an alten Kupferschlacken. Dissertation, Universität Heidelberg

Lorius C, Jouzel J, Ritz C, Merlivat L, Barkov NI, Korotkevich YS, Kotlyakov VM (1985) A 150,000 year climatic record from Antarctic ice. Nature 316:591–596

Lotter AF, Ammann B, Beer J, Hajdas I, Sturm M (1992) A step towards an absolute time-scale for the late-glacial: annually laminated sediments from Soppensee (Switzerland). In: Bard E, Broecker WS (eds) The last deglaciation: absolute and radiocarbon chronologies. Springer, Berlin Heidelberg New York, pp 45–68

Loy TH (1991) Prehistoric organic residues: recent advances in identification, dating and their antiquity. In: Pernicka E, Wagner GA (eds) Archaeometry '90. Birkhäuser, Basel, pp 645–656

Ludwig KR, Simmons KR, Szabo BJ, Winograd IJ, Landwehr JM, Riggs AC, Hoffman RJ (1992) Mass-spectrometric ^{230}Th-^{234}U-^{238}U dating of the Devils Hole calcite vein. Science 258:284–297

Macdougall D (1976) Fission track annealing and correction procedures for oceanic basalt glasses. Earth Planet Sci Lett 30:19–26

Mamyrin BA, Tolstikhin IN (1984) Helium isotopes in nature. Elsevier, Amsterdam, 273 pp

Mangerud J, Sönstegaard E, Sejrup HP (1979) Correlation of the Eemian (interglacial) stage and the deep-sea oxygen-isotope stratigraphy. Nature 277: 189–192

Mangini A (1986) application of ^{230}Th, ^{231}Pa and ^{10}Be radioisotopes in sedimentary geology. In: Hurford AJ, Jäger E, Ten Cate JAM (eds) Dating young sediments. CCOP Technical Secretariat, Bangkok, pp 52–71

Mangini A, Pernicka E, Wagner GA (1983) Dose-rate determination by radiochemical analysis. PACT 9:49–56

Mankinen EA, Dalrymple BG (1979) Revised geomagnetic polarity time scale for the interval 0–5 my BP. J Geophys Res 84:615–626

Manning SW (1990) The Thera eruption: the third congress and the problem of the date. Archaeometry 32:91–100

Manning SW, Weninger B (1992) A light in the dark: archaeological wiggle matching and the absolute chronology of the close of the Aegean Late Bronze Age. Antiquity 66:636–663

Marti K, Craig H (1987) Cosmic-ray-produced neon and helium in the summit lavas of Maui. Nature 325:335–337
Martinson DG, Pisias NG, Hays JD, Imbrie J, Moore TC Jr, Shackleton NJ (1987) Age dating and the orbital theory of the ice ages: development of a high-resolution 0 to 300,000-year chronostratigraphy. Quat Res 27:1–29
Marton P (1991) Archaeomagnetic directional data from Hungary. some new results. In: Pernicka E, Wagner GA (eds) Archaeometry '90. Birkhäuser, Basel, pp 569–576
Masarik J, Reedy RC (1995) Terrestrial cosmogenic-nuclide production systematics calculated from numerical simulations. Earth Planet Sci Lett 136:381–395
Mathewes RW, Westgate JA (1980) Bridge River tephra: revised distribution and significance for detecting old carbon errors in radiocarbon dates of limnic sediments in southern British Columbia. Can J Earth Sci 17:1454–1461
Matuyama M (1929) On the direction of magnetization of basalt in Japan, Tyosen and Manchuria. J P N Acad Proc 5:203–205
Mauz B, Buccheri G, Zöller L, Greco A (1997) Middle to Upper Pleistocene morphostructural evolution of the NW-coast of Sicily: thermoluminescence dating and palaeontological-stratigraphical evaluations of littoral sediments. Palaeogeogr Palaeoclimatol Palaeoecol 128:269–285
May RJ (1977) Thermoluminescence dating of Hawaiian alkalic basalts. J Geophys Res 82:3023–3029
May RJ (1979) Thermoluminescence dating of Hawaiian basalt. US Geol Surv Profess Paper 1095, 47 S
McDermott F, Grün R, Stringer CB, Hawkesworth CJ (1993) Mass-spectrometric U-series dates for Israeli Neanderthal/early modern hominid sites. Nature 363:252–255
McDougall I (1985) K-Ar and $^{40}Ar/^{39}Ar$ dating of the hominid-bearing Pliocene-Pleistocene sequence at Koobi Fora, Lake Turkana, northern Kenya. Bull Geol Soc Am 96:159–175
McDougall I, Polach HA, Stipp JJ (1969) Excess radiogenic argon in young subaerial basalts from Auckland volcanic field. Geochim Cosmochim Acta 33:1485–1520
McDougall I, Maier R, Sutherland-Hawkes P, Gleadow AJW (1980) K-Ar age estimates for the KBS Tuff, East Turkana, Kenya. Nature 283:230–234
McDougall I, Davies T, Maier R, Rudowski R (1985) Age of the Okote Tuff Complex at Koobi Fora, Kenya. Nature 316:792–794
McDougall I, Brown FH, Cerling TE, Hillhouse JW (1992) A reappraisal of the geomagnetic polarity time scale to 4 Ma using data from the Turkana Basin, East Africa. Geophys Res Lett 19:2349–2352
McKeever SWS (1991) Mechanisms of thermoluminescence production: some problems and a few answers? Nucl Tracks Radiat Meas 18:5–12
Meighan CW (1976) Empirical determination of obsidian hydration rates from archaeological evidence. In: Taylor RE (ed) Advances in obsidian glass studies. Noyes, Park Ridge, pp 106–119
Meighan CW, Scalise JL (1988) Obsidian dates IV: a compendium of the obsidian hydration determinations made at the UCLA obsidian hydration laboratory. Monograph 29, Inst Archaeol Univ Calif, Los Angeles
Mejdahl V (1988) Long-term stability of the TL signal in alkali feldspars. Quat Sci Rev 7:357–360
Mejdahl V, Bötter-Jensen L (1997) Experience with the SARA OSL method. Radiat Meas 27:291–294

Mercier N, Valladas H, Bar-Yosef O, Vandermeersch B, Stringer C, Joron JL (1993) Thermoluminescence date for the Mousterian burial site of Es-Skhul, Mt. Carmel. J Archaeol Sci 20:169–174

Mercier N, Valladas H, Valladas G, Reyss JL, Jelinek A, Meignen L, Joron JL (1995) TL dates of burnt flints from Jelinek's excavations at Tabun and their implications. J Archaeol Sci 22:495–X509

Merrihue C, Turner G (1976) Potassium-argon dating by activation with fast neutrons. J Geophys Res 71:2852–2857

Mesollela KJ, Matthews RK, Broecker WS, Thurber DL (1969) The astronomical theory of climatic change: Barbados data. J Geol 77:205–274

Meyer K (1968) Physikalisch-chemische Kristallographie. Grundstoffindustrie, Leipzig

Meynadier L, Valet JP, Weeks R, Shackleton NJ, Hagee VL (1992) Relative geomagnetic intensity of the field during the last 140 ka. Earth Planet Sci Lett 114: 39–57

Miallier D, Fain J, Sanzelle S, Pilleyre T, Montret M, Soumana S, Falgueres C (1994) Attempts at dating pumice deposits around 580 ka by use of red TL and ESR of xenolithic quartz inclusions. Radiat Meas 23:399–404

Michael A, Wing JF (1884) Ueber die optisch-inaktive Asparaginsäure. Ber Dtsch Chem Ges 17:2984

Michels JW, Tsong IST (1980) Obsidian hydration dating – a coming of age. Adv Archaeol Methods Theor 3:405–444

Michels JW, Tsong IST, Nelson CM (1983a) Obsidian dating and East African archaeology. Science 219:361–366

Michels JW, Tsong IST, Smith GA (1983b) Experimentally derived hydration rates in obsidian dating. Archaeometry 25:107–117

Middleton R, Klein J (1987) ^{26}Al: measurements and applications. Philos Trans R Soc Lond A323:121–143

Milankovitch MM (1941) Kanon der Erdbestrahlung und seine Anwendung auf das Eiszeitenproblem. Königl Serb Akad Spez Publ 133, Belg Radiat

Millard AR, Hedges REM (1995) The role of the environment in uranium uptake by buried bone. J Archaeol Sci 22:239–250

Miller DS, Wagner GA (1981) Fission-track ages applied to obsidian artifacts from South America using the plateau-annealing and track-size age-correction techniques. Nucl Tracks 5:147–155

Miller GH, Brigham-Grette J (1989) Amino acid geochronology: resolution and precision in carbonate fossils. Quat Int 1:111–128

Miller GH, Hare PE (1980) Amino acid geochronology: integrity of the carbonate matrix and potential for molluscan fossils. In: Behrensmeyer AK, Hill AP (eds) Fossils in the making: vertebrate taphonomy and paleoecology. University of Chicago Press, Chicago, pp 415–443

Miller GH, Magee JW, Jull AJT (1997) Low latitude, low elevation southern hemisphere ice-age cooling deduced from amino acid racemization in emu eggshell. Nature 358:241–244

Mitterer RM (1975) Ages and diagenetic temperatures of Pleistocene deposits of Florida based on isoleucine epimerization in Mercenaria. Earth Planet Sci Lett 28:275–282

Monaghan MC, Krishnaswami S, Thomas JH (1983) ^{10}Be concentrations and the long-term fate of particle-reactive nuclides in five soil profiles from California. Earth Planet Sci Lett 65:51–60

Monaghan MC, Krishnaswami S, Turekian KK (1985/86) The global-average production rate of ^{10}Be. Earth Palnet Sci Lett 76:279–287
Mook WG, Streurman HJ (1983) Physical and chemical aspects of radiocarbon dating. PACT 8:31–55
Mook WG, Waterbolk HT (1985) Radiocarbon dating. Handbooks for archaeologists 3. ESF, Strassbourg, 65 pp
Morgenstein M, Rosendahl P (1976) Basaltic glass hydration dating in the Hawaiian Islands. In: Taylor RE (ed) Advances in obsidian glass studies. Noyes, Park Ridge, pp 141–164
Morgenstern U, Gellermann R, Hebert D, Börner I, Stolz W, Vaikmäe R, Rajamäe R, Putnik H (1995) ^{32}Si in limestone aquifers. Chem Geol 120:127–134
Morgenstern U, Taylor CB, Parrat Y, Gäggeler HW, Eichler B (1996) ^{32}Si in precipitation: evaluation of temporal and spatial variation and as dating tool for glacial ice. Earth Planet Sci Lett 144:289–296
Morris JD (1991) Applications of cosmogenic ^{10}Be to problems in the earth sciences. Annu Rev Earth Planet Sci 19:313–350
Moser H, Rauert W (1980) Isotopenmethoden in der Hydrologie. Borntraeger, Stuttgart, 400 S
Mudelsee M, Barabas M, Mangini A (1992) ESR dating of the Quaternary deep-sea sediment core RC17-177. Quat Sci Rev 11:181–189
Muhs DR, Rosholt JN, Bush CA (1989) The uranium-trend dating method: principles and application for southern California marine terrace deposits. Quat Int 1: 19–34
Mulholland SC, Prior C (1993) AMS radiocarbon dating of phytoliths. MASCA Res Pap Sci Archaeol 10, 3:21–23
Müller P, Schvoerer M (1993) Factors affecting the viability of thermoluminescence dating of glass. Archaeometry 35:299–304
Muller PJ (1984) Isoleucine epimerization in Quaternary planktonic foraminifera: effects of diagenetic hydrolysis and leaching, and Atlantic-Pacific intercore correlation. Meteor Forschungsergeb C38:25–47
Muller RA (1977) Radioisotope dating with a cyclotron. Science 196:489–494
Muller RA, Morris DE (1986) Geomagnetic reversals from impacts on the earth. Geophys Res Lett 13:1177–1180
Münnich KO (1968) Isotopen-Datierung von Grundwasser. Naturwissenschaften 55:158–163
Murray AS (1996) Developments in optically stimulated luminescence and phototransferred luminescence dating of young sediments: application to a 2000-year sequence of flood deposits. Geochim Cosmochim Acta 60:565–576
Murray AS, Aitken MJ (1988) Analysis of low-level natural radioactivity in small mineral samples for use in thermoluminescence dating, using high-resolution gamma spectrometry. Appl Radiat Isot 39:145–158
Murray AS, Roberts AG, Wintle AG (1997) Equivalent dose measurement using a single aliquot of quartz. Radiat Meas 27:171–184
Naeser CW, Faul H (1969) Fission track annealing in apatite and sphene. J Geophys Res 74:705–710
Naeser CW, Gleadow AJW, Wagner GA (1979) Standardization of fission-track data reports. Nucl Tracks 3:133–136
Nagpaul KK, Mehta PP, Gupta ML (1974) Annealing studies of radiation damages in biotite, apatite and sphene and corrections to fission track ages. Pure Appl Geophys 112:131–139

Nambi KSV, Aitken MJ (1986) Annual dose conversion factors for TL and ESR dating. Archaeometry 28:202–205

Nelson DE, Loy TH, Vogel JS, Southon JR (1986) Radiocarbon dating blood residues. Radiocarbon 28:170–174

Nelson DE, Ckaloupka G, Chippindale C, Alderson MS, Southon JR (1995) Radiocarbon dates for beeswax figures in the prehistoric art of northern Australia. Archaeometry 37:151–156

Netterberg F(1978) Dating and correlation of calcretes and other pedocretes. Trans Geol Soc S Afr 81:379–391

Newton RG (1966) Some problems in the dating of ancient glass by counting the layers in the weathering crust. Glass Technol 7:22–25

Newton RG (1971) The enigma of the layered crusts on some weathered glasses, a chronological account of the investigations. Archaeometry 13:1–9

Niedermann S, Graf Th, Kim JS, Kohl CP, Marti K, Nishiizumi K (1994) Cosmic-ray produced ^{21}Ne in terrestrial quartz: the neon inventory of Sierra Nevada quartz separates. Earth Planet Sci Lett 125:341–355

Ninagawa K, Adachi K, Uchimura, N, Yamamoto I, Wada T, Yamashita Y, Takashima I, Sekimoto K, Hasegawa H (1992) Thermoluminescence dating of calcite shells in the pectinidae familiy. Quat Sci Rev 11:121–126

Ninagawa K, Matsukuma Y, Fukuda T, Sato A, Hoshinoo N, Nakagawa M, Yamamoto I, Wada T, Yamashita Y, Takashima I, Sekimoto K, Komura K (1994) Thermoluminescence dating of calcite shell, *Crassostrea gigas* (Thunberg) in the Ostreidae familiy. Quat Sci Rev 13:589–593

Nishiizumi K, Lal D, Klein J, Middleton R, Arnold JR (1986) Production of ^{10}Be and ^{26}Al by cosmic rays in terrestrial quartz in situ and implications for erosion rates. Nature 319:134–136

Nishiizumi K, Winterer EL, Kohl CP, Klein J, Middleton R (1989) Cosmic ray produced rates of ^{10}Be and ^{26}Al in quartz from glacially polished rocks. J Geophys Res 97:17907–17915

Nishiizumi K, Kohl CP, Shoemaker EM, Arnold JR, Klein J, Fink D, Middleton R (1991) In situ ^{10}Be-^{26}Al exposure ages at Meteor Crater, Arizona. Geochim Cosmochim Acta 55:2699–2703

Nowaczyk NR, Frederichs TW, Eisenhauer A, Gard G (1994) Magnetostratigraphic data from late Quaternary sediments from the Yermak Plateau, Arctic Ocean: evidence for four geomagnetic polarity events within the last 170 ka of the Brunhes Chron. Geophys J Int 117:453–471

Oakley KB (1980) Relative dating of the fossil hominids of Europe. Bull Brit Mus Nat Hist (Geol) 34:1–63

Oakley KP (1948) Fluorine and the relative dating of bones. Adv Sci 4:336–337

O'Brien GWO, Veeh HH, Cullen DJ, Milnes AR (1986) Uranium series isotopic studies of marine phosphorites and associated sediments from the East Australian continental margin. Earth Planet Sci Lett 80:19–35

O'Brien SR, Mayewski PA, Meeker LD, Meese DA, Twickler MS, Whitlow S (1995) Complexity of Holocene climate as reconstructed from a Greenland ice core. Science 270:1962–1964

Oches EA, McCoy WD (1995) Amino acid geochronology applied to the correlation and dating of central European loess deposits. Quat Sci Rev 14: 767–782

Oeschger H (1987) Accelerator mass spectrometry and ice core research. Nucl Instrum Methods Phys Res B29:196–202

Oeschger H, Schotterer U (1986) Dating of water and ice. In: Hurford AJ, Jäger E, Ten Cate JAM (eds) Dating young sediments, CCOP Technical Secretariat, Bangkok, pp 111–125
Oeschger H, Alder B, Loosli H, Langway CC Jr, Renaud A (1966) radiocarbon dating of ice. Earth Planet Sci Lett 1:49–54
Olsson IU (1979) The radiocarbon contents of various reservoirs. In: Berger R, Suess HE (eds) Radiocarbon dating. University of California Press, Los Angeles, pp 613–618
Opdyke NP (1972) Paleomagnetism of deep-sea cores. Rev Geophys Space Phys 10:213–349
Ortner DJ, von Endt DW, Robinson MS (1972) The effect of temperature on protein decay in bone: its significance in nitrogen dating of archaeological samples. Am Antiq 37:514–520
Otlet RL, Polach HA (1990) Improvements in the precision of radiocarbon dating through recent developments in liquid scintillation counters. PACT 29:225–238
Packman SC, Grün R (1992) TL analysis of loess samples from Achenheim. Quat Sci Rev 11:103–107
Papamarinopoulos S, Readman PW, Maniatis Y, Simopoulos A (1987) Paleomagnetic and mineral magnetic studies of sediments from Petralona Cave, Greece. Archaeometry 29:50–59
Pavich MJ, Brown L, Harden J, Klein J, Middleton R (1986) ^{10}Be distribution in soils from Merced terraces, California. Geochim Cosmochim Acta 50:1727–1735
Pares JM, Perez-Gonzalez A (1995) Paleomagnetic age for hominid fossils at Atapuerca archaeological site, Spain. Science 269:830–832
Partridge TC (1997) The Plio-Pleistocene boundary. Quat Int 40:100 pp
Paytan A, Moore WS, Kastner M (1996) Sedimentation rate as determined by ^{226}Ra activity in marine barite. Geochim Cosmochim Acta 60:4313–4319
Pearson GW (1986) Precise calendrical dating of known-growth period samples using a 'curve fitting' technique. Radiocarbon 28 (2 A) 229–299
Pearson GW (1987) How to cope with calibration. Antiquity 61:98–103
Pearson GW, Stuiver M, (1993) High-precision bidecadal calibration of the radiocarbon time scale, 500–2500 BC. Radiocarbon 35:23–33
Peck JA, King JW, Colman SM, Kravchinsky VA (1996) An 84-kyr paleomagnetic record from the sediments of Lake Baikal, Siberia. J Geophys Res 101, B5:11365–11385
Perkin NK, Rhodes EJ (1994) Optical dating of fluvial sediments from Tattershall, U.K. Quat Sci Rev 13:517–520
Pernicka E, Wagner GA (1983/84) Datierung neolithischer Erdwerke mittels Thermolumineszenz: Bestätigung der langen Chronologie. Mitt Österr Arbeitsgem Ur-Frühgesch 33/34:247–267
Petit JR, Mounier L, Jouzel J, Korotkevich YS, Lorius C (1990) Palaeoclimatological and chronological implications of the Vostok core dust record. Nature 343: 56–58
Petit JR, Baisle I, Leruyuet A, Raynaud D, Lorius C, Jouzel J, Stievenard M, Lipenkov VY, Barkov NI, Kudryashov BB, Davis M, Saltzman E Kotlyakov V (1997) Four climate cycles in Vostok ice core. Nature 387:359–360
Petrasch J, Kromer B (1989) Aussagemöglichkeiten von ^{14}C-Daten zur Verfüllungsgeschichte prähistorischer Gräben am Beispiel der mittelneolithischen Kreisgrabenanlage von Künzing-Unternberg, Ldkr. Deggendorf. Archäol Korrbl 19: 231–238

Pettersson H (1937) Das Verhältnis Thorium zu Uran in dem Gestein und im Meer. Anz Akad Wiss Wien Math-Naturw Kl: 127

Phillips FM, Smith SS, Bentley HW, Elmore D, Grove HE (1983) Chlorine-36 dating of saline sediments: preliminary results from Searles Lake, California. Science 222:925–927

Phillips FM, Bentley HW, Davis SN, Elmore D, Swanick GB (1986a) Chlorine-36 dating of very old groundwater 2. Milk River aquifer, Alberta, Canada. Water Resour Res 22:2003–2016

Phillips FM, Leavy BD, Jannik NO, Elmore D, Kubik PW (1986b) The accumulation of cosmogenic chlorine-36 in rocks: a method for surface exposure dating. Science 231:41–43

Phillips FM, Zreda MG, Smith SS, Elmore D, Kubik PW, Sharma P (1990) Cosmogenic chlorine-36 chronology for glacial deposits at Bloody Canyon, eastern Sierra Nevada. Science 248:1529–1532

Phillips FM, Zreda MG, Smith SS, Elmore D, Kubik PW, Dorn RI, Roddy DJ (1991) Age and geomorphic history of Meteor Crater, Arizona from cosmogenic ^{36}Cl and ^{14}C in rock varnish. Geochim Cosmochim Acta 55:2695–2698

Phillips FM, Zreda MG, Ku TL, Luo S, Huang Q, Elmore D, Kubik PW, Sharma P (1993) $^{230}Th/^{234}U$ and ^{36}Cl dating of evaporite deposits from western Qaidam basin, China: implications for glacial dust export from central Asia. Bull Geol Soc Am 105:1606–1616

Phillips FM, Zreda MG, Benson LV, Plummer MA, Elmore D, Sharma P (1996) Chronology for the fluctuations in late Pleistocene Sierra Nevada glaciers and lakes. Science 274:749–751

Pichler H (1973) "Base surge"-Ablagerungen auf Santorin. Naturwissenschaften 60:198

Picket DA, Murrell MT (1997) Observation of $^{231}Pa/^{235}U$ disequilibrium in volcanic rocks. Earth Planet Sci Lett 148:273–285

Piggot CS, Urry WMD (1942) Time relations in ocean sediments. Bull Geol Soc Am 53:1187–1210

Pilgrim L (1904) Versuch einer rechnerischen Behandlung des Eiszeitproblems. Jahreshefte Verein vaterländ Naturkunde Württ 1904:24–117

Pilleyre T, Montret M, Fain J, Miallier D, Sanzelle S (1992) Attempts at dating ancient volcanoes using the red TL from quartz. Quat Sci Rev 11:13–17

Pisias NG, Martinson DG, Moore TC Jr, Shackleton NJ, Prell W, Hays J, Boden G (1984) High resolution stratigraphic correlation of benthic oxygen isotopic records spanning the last 300,000 years. Mar Geol 56:119–136

Plummer MA, Phillips FM, Fabryka-Martin J, Turin HJ, Wigand PE, Sharma P (1997) Chlorine-36 in fossil rat urine: an archive of cosmogenic nuclide deposition during the past 40,000 years. Nature 277:538–541

Poinar HN, Höss M, Bada JL, Pääbo S (1996) Amino acid racemization and the preservation of ancient DNA. Science 272:864–866

Polach HA (1992) Four decades of progress in ^{14}C dating by liquid scintillation counting and spectrometry. In: Taylor RE, Long A, Kra RS (eds) Radiocarbon after four decades - an interdisciplinary perspective. Springer, Berlin Heidelberg New York, pp 198–213

Poolton NRJ, Bailiff IK (1989) The use of LEDs as an exitation source for photoluminescence dating of sediments. Ancient TL 7:18–20

Porat N, Schwarcz HP (1991) Use of signal subtraction methods in ESR dating of burned flint. Nucl Tracks Radiat Meas 18:203–212

Porat N, Schwarcz HP, Valladas H, Bar-Yosef O, Vandermeersch B (1994) Electron spin resonance dating of burned flint from Kebara Cave, Israel. Geoarchaeology 9:393–407

Prescott JR, Hutton JT (1988) Cosmic ray and gamma ray dosimetry for TL and ESR. Nucl Tracks Radiat Meas 14:223–227

Prescott JR, Hutton JT (1994) Cosmic ray contributions to dose rates for luminescence and ESR dating: large depths and long-term time variations. Radiat Meas 23:497–500

Price PB, Walker RM (1962) A new detector for heavy particle studies. Phys Lett 3:113–115

Price PB, Walker RM (1963) Fossil tracks of charged particles in mica and the age of minerals. J Geophys Res 68:4747–4862

Prior CA, Piperno DR (1996) Radiocarbon dates on phytoliths from the Vegas site, Ecuador. Int Symp Archaeometry, Urbana-Champaign, Abstracts 86

Purdy BA, Clark DE (1987) Weathering of inorganic materials: dating and other applications. Adv Archaeol Methods Theor 11:211–253

Questiaux DG (1991) Optical dating of loess: comparisons between different grain size fractions for infrared and green excitation wavelengths. Nucl Tracks Radiat Meas 18:133–139

Radicati di Brozolo F, Hunecke JC, Papanastassiou DA, Wasserburg GJ (1981) ^{39}Ar-^{40}Ar and Rb-Sr age determination on Quaternary volcanic rocks. Earth Planet Sci Lett 53:445–456

Radtke U, Grün R, Schwarcz HP (1988) ESR dating of Pleistocene coral reef tracts of Barbados (W.I.). Quat Res 29:197–215

Rae AM, Ivanovich M, Green HS, Head MJ (1987) A comparative dating study of bones from Little Hoyle cave, South Wales, UK. J Archaeol Sci 14:243–250

Raisbeck GM, Yiou F (1979) Possible use of ^{41}Ca for radioactive dating. Nature 277:42–44

Raisbeck GM, Yiou F (1980) Progress report on the possible use of ^{41}Ca for radioactive dating. Rev D'Archeometrie 4:121–125

Raisbeck GM, Yiou F, Klein J, Middleton R (1983) Accelerator mass spectrometry measurement of cosmogenic ^{26}Al in terrestrial and extraterrestrial matter. Nature 301:690–692

Raisbeck GM, Yiou F, Bourles D, Lorius C, Jouzel J, Barkov NI (1987) Evidence for two intervals of enhanced ^{10}Be deposition in Antarctic ice during the last glacial period. Nature 326:273–277

Ralph EK (1971) Carbon-14 dating. In: Michael HN, Ralph EK (eds) Dating techniques for the archaeologist. MIT, Cambridge, Massachusetts, pp 1–48

Rampino MR, Self S (1992) Volcanic winter and accelerated glaciation following the Toba super-eruption. Nature 359:50–52

Ravichandran M, Baskaran M, Santschi PH, Bianchi PS (1995) Geochronology of sediments in the Sabine-Neches estuary, Texas, USA. Chem Geol 125:291–306

Reagan MK, Volpe AM, Cashman KV (1992) ^{238}U- and ^{232}Th-series chronology of phonolite fractionation at Mount Erebus, Antarctica. Geochim Cosmochim Acta 56:1401–1407

Rees-Jones J, Tite MS (1997) Optical dating results from British archaeological sites. Archaeometry 39:177–187

Reid MR, Coath CD, Harrison TM, McKeegan KD (1997) Prolonged residence times for the youngest rhyolites associated with the Long Valley Caldera: ^{230}Th-^{238}U ion microprobe dating of young zircons. Earth Planet Sci Lett 150:27–39

Reinders J, Hambach U (1995) A geomagnetic event recorded in loess deposits of the Tönchesberg (Germany): identification of the Blake magnetic polarity episode. Geophys J Int 122:407–418

Reinders J, Hambach U, Krumsiek KAO, Strack N (1997) An archaeomagnetic study on pottery kilns from Brühl-Pingsdorf (Germany). Archaeometry (in press)

Reller A, Wilde PM, Wiedemann HG, Hauptmann H (1992) Comparative studies of ancient mortars from Gizah, Egypt, and Nevali Cori, Turkey. Mat Res Soc Symp Proc 267, pp 1007–1011

Rendell HM, Hailwood EA, Denell RW (1987) Magnetic polarity stratigraphy of upper Siwalik sub-group, Soan valley, Pakistan: implications for early human occupance of Asia. Earth Planet Sci Lett 85:488–496

Rendell H, Webster SE, Sheffer NL (1994) Underwater bleaching of signals from sediment grains: new experimental data. Quat Sci Rev 13:433–435

Renfrew C (1976) Before civilization – the radiocarbon revolution and prehistoric Europe. Penguin, Harmondsworth

Reyes AO, Moore WS, Stakes DS (1995) $^{228}Th/^{228}Ra$ ages of a barite-rich chimney from the Endeavour Segment of the Juan de Fuca Ridge. Earth Planet Sci Lett 131: 99–113

Reyss JL, Yokoyama Y, Tanaka S (1976) Aluminum-26 in deep-sea sediment. Science 193:1119–1120

Reyss JL, Yokoyama Y, Duplessy JC (1978) A rapid determination of oceanic sedimentation rates by non-destructive gamma-gamma coincidence spectrometry. Deep-Sea Res 25:491–498

Rhodes EJ (1988) Methodological considerations in the optical dating of quartz. Quat Sci Rev 7:395–400

Richter DG (1997) Thermolumineszenzdatierungen erhitzter Silices aus paläolithischen Fundstellen. Dissertation, Universität Tübingen

Richter K (1958) Fluortest quartärer Knochen in ihrer Bedeutung für die absolute Chronologie des Quartärs. Eiszeitalter Gegenwart 9:18–27

Rieser U, Krbetschek MR, Stolz W (1994) CCD-camera based high sensitivity TL/OSL-spectrometer. Radiat Meas 23:523–528

Rink WJ, Odom AL (1991) Natural alpha recoil particle radiation and ionizing radiation sensitivities in quartz detected with EPR: implications for geochronometry. Nucl Tracks Radiat Meas 18:163–173

Rink WJ, Schwarcz HP, Lee HK, Cabrera Valdes V, Bernaldo de Quiros F, Hoyos M (1996) ESR dating of tooth enamel: comparison with AMS ^{14}C at El Castillo Cave, Spain. J Archaeol Sci 23:945–951

Roberts RG, Jones R, Smith MA (1990) Thermoluminescence dating of a 50,000-year-old human occupation site in northern Australia. Nature 345:153–156

Roberts RG, Jones R, Spooner NA, Head MJ, Murray AS, Smith MA (1994) The human colonisation of Australia: Optical dates of 53,000 and 60,000 years bracket human arrival at Deaf Adder Gorge, Northern Territory. Quat Sci Rev 13:575–583

Roberts RG, Walsh G, Murray AS, Olley J, Jones R, Morwood M, Tuniz C, Lawson E, Macphail M, Bowdery D, Naumann I (1997) Luminescence dating of rock art and past environments using mud-wasp nests in northern Australia. Nature 387:696–699

Robins D (1991) Archäologie mit Elektronen – Die Vermessung der Vergangenheit mit physikalisch-chemischen Hilfsmitteln. Mannheimer Forum 90/91:195–245

Rolph TC, Shaw J, Guest JE (1987) Geomagnetic field variations as a dating tool: application to Sicilian lavas. J Archaeol Sci 14:215–225

Roosevelt AC, da Costa ML, Machado CL, Michab M, Mercier N, Valladas H, Feathers J, Barnett W, da Silveira MI, Henderson A, Sliva J, Chernoff B, Reese DS, Holman JA, Toth N, Schick K (1996) Paleoindian cave dwellers in the Amazon: the peopling of the Americas. Science 272:373–384

Rosholt JN (1957) Quantitative radiochemical methods for the determination of the source of natural radioactivity. Ann Chem 29:1398–1408

Rosholt JN, Emiliani C, Geiss J, Kozy FF, Wangersky JP (1961) Absolute dating of deep-sea core by the $^{231}Pa/^{230}Th$ method. J Geol 69:162–185

Rosholt JN, Doe BR, Tatsumoto M (1966) Isotopic composition of uranium and thorium in soil profiles. Bull Geol Soc Am 77:987–1004

Rosholt JN, Bush CA, Shroba RR, Pierce KL, Richmond GM (1985) Uranium-trend dating and calibrations for Quaternary sediments. US Geol Surv Open-File Rep 85–299, 85 S

Rottländer RCA (1977) Schwierigkeiten bei der Datierung von Silices. In: Hennicke HW (Hrsg) Mineralische Rohstoffe als kulturhistorische Informationsquelle. Verein Deutsche Emailfachleute, Hagen, pp 191–199

Rottländer RCA (1989) Verwitterungserscheinungen an Silices und Knochen. Archaeologica Venatoria, Tübingen

Rowlett RM, Pearsall DM (1993) Archaeologigal age determinations derived from opal phytoliths by thermoluminescence. MASCA Res Pap Sci Archaeol 10(3): 25–29

Rusakov OM, Zagniy GF (1973a) Archaeomagnetic secular variation study of the Ukraine and Moldavia. Archaeometry 15:153–157

Rusakov OM, Zagniy GF (1973b) Intensity of geomagnetic field in the Ukraine and Moldavia during the past 6000 years. Archaeometry 15:275–285

Rutter NW, Crawford RJ (1984) Utilizing wood in amino acid dating. In: Mahaney WC (ed) Quaternary dating methods. Elsevier, Amsterdam, pp 195–209

Rutter NW, Crawford RJ, Hamilton RD (1985) Amino acid racemization dating. In: Rutter NW (ed) Dating methods of pleistocene deposits and their problems. Geoscience Canada, Rep 2, Geological Association of Canada, Toronto, pp 23–30

Sackett WM (1960) The protactinium-231 content of ocean water and sediments. Science 132:1761–1762

Sales KD, Robins GV, Oduwole D (1989) Electron spin resonance study of bones from paleolithic site at Zhoukoudian, China. In: Allen RO (ed) Archaeological chemistry. American Chemical Society, Washington, pp 353–368

Saliege JF, Person A, Paris F (1995) Preservation of $^{13}C/^{12}C$ original ratio and ^{14}C dating of the mineral fraction of human bones from Saharan tombs, Niger. J Archaeol Sci 22:302–312

Salvamoser J (1982) ^{85}Kr im Grundwasser – Messmethodik, Modellüberlegungen und Anwendungen auf natürliche Grundwassersysteme. Dissertation, Universität München

Sanderson DCW, Placido F, Tate JO (1988) Scottish vitrified forts: TL results from six study sites. Nucl Tracks Radiat Meas 14:307–316

Sarda P, Staudacher T, Allegre CJ, Lecomte A (1993) Cosmogenic neon and helium at reunion: measurement of erosion rate. Earth Planet Sci Lett 119:405–417

Sarnthein M, Stremme H, Mangini A (1986) The Holstein interglaciation: time-stratigraphic position and correlation to stable-isotope stratigraphy of deep-sea sediments. Quat Res 26:283–298

Schaeffer OA (1966) Tektites. In: Schaeffer OA, Zähringer J (eds) Potassium argon dating. Springer, Berlin Heidelberg New York, pp 162–173

Scharpenseel HW, Schiffmann H (1977) Radiocarbon dating of soils – a review. Z Pflanzenernähr Bodenk 140:159–174

Schneider DA, Mello GA (1996) A high-resolution marine sedimentary record of the geomagnetic intensity during the Brunhes Chron. Earth Planet Sci Lett 144: 297–314

Schneider DA, Kent DV, Mello GA (1992) A detailed chronology of the Australasian impact event, the Brunhes-Matuyama geomagnetic polarity reversal, and global climate change. Earth Planet Sci Lett 111:395–405

Schönhofer F (1989) Determination of ^{14}C in alcoholic beverages. Radiocarbon 31: 777–784

Schwarcz HP (1989) Uranium series dating of Quaternary deposits. Quat Int 1:7–17

Schwarcz HP, Grün R (1988) Comment on: Sarnthein M, Stremme H, Mangini A (eds) The Holstein interglaciation: time-stratigraphic position and correlation to stable-isotope stratigraphy of deep-sea sediments. Quat Res 29:75–79

Schwarcz HP, Skoflek I (1982) New dates for the Tata, Hungary archaeological site. Nature 295:590–591

Schwarcz HP, Grün R, Latham AG, Mania D, Brunnacker K (1988) The Bilzingsleben archaeological site: new dating evidence. Archaeometry 30:5–17

Schwarcz HP, Buhay WM, Grün R, Valladas H, Tchernov E, Bar-Yosef O, Vandermeersch B (1989) ESR dating of the Neanderthal site, Kebara Cave, Israel. J Archaeol Sci 16:653–659

Schweingruber FH (1988) Tree rings: basics and applications of dendrochronology. Kluwer Academic, Dordrecht, 276 S

Segl M, Mangini A, Bonani G, Hofmann HJ, Morenzoni E, Nessi M, Suter M, Wölfli W (1984) ^{10}Be dating of the inner structure of Mn-encrustations applying the Zürich tandem accelerator. Nucl Instrum Methods Phys Res B5:359–364

Selo M, Storzer D (1981) Uranium distribution and age pattern of some deep-sea basalts from the Entrecasteau area, south-western Pacific: a fission-track analysis. Nucl Tracks 5:137–145

Semaw S, Renne P, Harris JWK, Feibel CS, Bernor RL, Fesseha N, Mowbray K (1997) 2.5-million-year-old stone tools from Gona, Ethiopia. Nature 385:333–336

Seward D, Kohn BP (1997) New zircon fission-track ages from New Zealand Quaternary tephra: an interlaboratory experiment and recommendations for the determination of young ages. Chem Geol 141:127–140

Shackleton NJ (1967) Oxygen isotope analyses and Pleistocene temperatures reassessed. Nature 215:15–17

Shackleton NJ, Opdyke ND (1973) Oxygene isotope and paleomagnetic stratigraphy of equatorial Pacific core V28–238: oxygen istotope temperatures and ice volumes on a 10^5 year and 10^6 year scale. Quat Res 3:39–55

Shen C, Beer J, Liu T, Oeschger H, Bonani G, Suter M, Wölfli W (1992) ^{10}Be in Chinese loess. Earth Planet Sci Lett 109:169–177

Shephard MK, Arvidson RE, Caffee M, Finkel R, Harris L (1995) Cosmogenic exposure ages of basalt flows: Lunar Crater volcanic field, Nevada. Geology 23: 21–24

Shimokawa K, Imai N (1987) Simultaneous determination of alteration and eruption ages of volcanic rocks by electron spin resonance. Geochim Cosmochim Acta 51:115–119

Siegenthaler U, Oeschger H, Tongiorgi E (1970) Tritium and Oxygen-18 in natural water samples from Switzerland. Isotope Hydrology 1970, IAEA, Vienna, pp 373–385

Sigmarsson O, Codomines M, Morris JD, Harmon RS (1990) Uranium and ^{10}Be enrichments by fluids in Andean arc magmas. Nature 346:163–165
Singer BS, Pringle MS (1996) Age und duration of the Matuyama-Brunhes geomagnetic polarity reversal from ^{40}Ar/^{39}Ar incremental heating analyses of lavas. Earth Planet Sci Lett 139:47–61
Singhvi AK, Wagner GA (1986) Thermoluminescence dating and its applications to young sedimentary deposits. In: Hurford AJ, Jäger E, Ten Cate JAM (eds) Dating young sediments. CCOP Technical Secretariat, Bangkok, pp 159–197
Singhvi A, Banerjee D, Pande K, Gogte V, Valdiya KS (1994) Luminescence studies on neotectonic events in south-central Kumaun Himalaya – a feasibilty study. Quat Sci Rev 13:595–600
Singhvi A, Banerjee D, Ramesh R, Rajaguru SN, Gogte V (1996) A luminescence method for dating "dirty" pedogenic carbonates for paleoenvironmental reconstruction. Earth Planet Sci Lett 139:321–332
Slowey NC, Henderson GM, Curry WB (1996) Direct U-Th dating of marine sediments from the two most recent interglacial periods. Nature 383:242–244
Smethie WM Jr, Östlund HG, Loosli HH (1986) Ventilation of the deep Greenland and Norwegian seas: evidence from krypton-85, tritium, carbon-14 and argon-39. Deep-Sea Res 33:675–703
Smith BW, Rhodes EJ, Stokes S, Spooner NA (1990a) The optical dating of sediments using quartz. Radiat Prot Dosim 34:75–78
Smith BW, Rhodes EJ, Stokes S, Spooner NA, Aitken MJ (1990b) Optical dating of sediments: initial quartz results from Oxford. Archaeometry 32:19–31
Smith CA, Bonsall C (1990) AMS radiocarbon dating of British Late Upper Palaeolithic and Mesolithic artefacts: preliminary results. PACT 29:259–268
Smits F, Gentner W (1950) Argonbestimmungen an Kaliummineralen, I: Bestimmungen an tertiären Kalisalzen. Geochim Cosmochim Acta 1:22–27
Soffel HC (1991) Paläomagnetismus und Archäomagnetismus. Springer, Berlin Heidelberg New York
Sonntag C (1980) Isotopenhydrologie. Sitzungsber Heidelb Akad Wiss Math Naturwiss Kl 1979/80, 4. Abh:243–249
Spooner NA, Questiaux DG (1989) Optical dating – Achenheim beyond the Eemian using green and red stimulation. In: Long and short range limits in luminescence dating. Res Labor Archaeol Hist Art Occas Publ 9, Oxford
Spooner NA, Aitken MJ, Smith BW, Franks M, McElroy C (1990) Archaeological dating by infrared-stimulated luminescence using a diode array. Radiat Prot Dosim 34:83–86
Spurk M, Hofmann J, Friedrich M, Remmele S, Leuschner HH, Kromer B (1998) Revision and extension of the Hohenheim oak and pine chronologies – new evidence about timing of the Younger Dryas/Preboreal transition. Calibration issue. Radiocarbon (in press)
Srdoc D, Chafetz H, Utech N (1989) Radiocarbon dating of travertine deposits, Arbuckle Mountains, Oklahoma. Radiocarbon 31:619–626
Stafford TW Jr, Jull AJT, Brendel K, Duhamel RC, Donahue D (1987) Study of bone radiocarbon dating accuracy at the University of Arizona NSF accelerator facility for radioisotope analysis. Radiocarbon 29:24–44
Staudacher T, Allegre CJ (1991) Cosmogenic neon in ultramafic nodules from Asia and in quarzite from Antarctica. Earth Planet Sci Lett 106:87–102
Staudacher T, Allegre CJ (1993) Ages of the second caldera of Piton de la Fournaise volcano (Reunion) determined by cosmic ray produced ^{3}He and ^{21}Ne. Earth Planet Sci Lett 119:395–404

Sternberg RS, McGuire RH (1990) Archaeomagnetic secular variation in the American southwest, AD. 700–1450. In: Eighmy JL, Sternberg RS (eds) Archaeomagnetic dating. University of Arizona Press, Tucson, pp 199–225
Stevenson CM, Carpenter J, Scheetz BE (1989) Obsidian dating: recent advances in the experimental determination and application of hydration rates. Archaeometry 31:193–206
Stokes S, Thomas DSG, Washington R (1997) Multiple episodes of aridity in southern Africa since the last interglacial period. Nature:154–158
Stone JO, Allan GL, Fifield LK, Cresswell RG (1996) Cosmogenic chlorine-36 from calcium spallation. Geochim Cosmochim Acta 60:679–692
Storzer D, Poupeau G (1973) Geochronologie: ages plateaux de mineraux et verres par la methode des traces de fission. CR Acad Sci Paris D276:137–139
Storzer D, Selo M (1976) Uranium contents and fission track ages of some basalts from the FAMOUS area. Bull Soc Geol Fr 18:807–810
Storzer D, Selo M (1978) Fission track age of magnetic anomaly M-Zero and some aspects of sea-water weathering. In: Donelly T (ed) Initial reports of the deep sea drilling project. 51/53, Washington, pp 1129–1133
Storzer D, Wagner GA (1969) Correction of thermally lowered fission track ages of tektites. Earth Planet Sci Lett 5:463–468
Storzer D, Wagner GA (1977) Fission track dating of meteorite impacts. Meteoritics 12:368–369
Storzer D, Horn P, Kleinmann B (1971) The age and the origin of Köfels structure, Austria. Earth Planet Sci Lett 12:238–244
Storzer D, Jessberger EK, Klay N, Wagner GA (1984) $^{40}Ar/^{39}Ar$ evidence for two discrete tektite-forming events in the Australian-Southeast Asian area. Meteoritics 19:317 (Abstr)
Street M, Baales M, Weninger B (1994) Absolute Chronologie des späten Paläolithikums und des Frühmesolithikums im nördlichen Rheinland. Archäol Korrespondenzbl 24:1–28
Stringer CB, Grün R, Schwarcz HP, Goldberg P (1989) ESR dates for the hominid burial site of Es Skhul in Israel. Nature 338:756–758
Strömberg B (1985) Revision of the Late-Glacial Swedish varve chronology. Boreas 14:101–105
Strutt RJ (1905) On the radioactive minerals. Proc R Soc Lond A 76:88–101
Stuiver M (1978) Radiocarbon timescale tested against magnetic and other dating methods. Nature 273:271–274
Stuiver M (1989) Dating proxy data. In: Berger A, Schneider S, Duplessey J (eds) Climate and geosciences. Cl NATO ASI Math Phys Sci 285:39–45
Stuiver M (1990) Timescales and telltale corals. Nature 345:387–388
Stuiver M (1993) Single year calibration of the radiocarbon time scale, AD 1510–1954. Radiocarbon 35:67–72
Stuiver M, Becker B (1993) High-precision decadal calibration of the radiocarbon time scale, AD 1950–6000 BC. Radiocarbon 35:35–65
Stuiver M, Braziunas TF (1993) Modelling atmospheric ^{14}C influences and ^{14}C ages of marine samples to 10,000 BC. Radiocarbon 35:137–189
Stuiver M, Kra R (1986) Calibration issue. Radiocarbon 28 (B2), 805–1030
Stuiver M, Pearson GW (1993) High-precision bidecadal calibration of the Radiocarbon time scale, AD 1950–500 BC and 2500–6000 BC. Radiocarbon 35:1–23
Stuiver M, Reimer PJ (1993) Extended ^{14}C data base and revised CALIB 3.0 ^{14}C age calibration program. Radiocarbon 35:215–230

Stuiver M, Heusser CJ, Yang IC (1978) North American glacial history is extended to 75,000 years ago. Science 200:16–21
Stuiver M, Long A, Kra RS (1993) Calibration 1993. Radiocarbon 35
Suess HE (1965) Secular variations in the cosmic-ray produced carbon-14 in the atmosphere and their interpretation. J Geophys Res 70:5937–5952
Suess HE (1979) Ist die Sonnenaktivität für Klimaschwankungen verantwortlich? Umschau 79:312–317
Sutton SR (1984) Thermoluminescence age of Meteor Crater, Arizona. Meteoritics 19:317
Sutton SR (1985) Thermoluminescence measurements on shock-metamorphosed sandstone and dolomite from Meteor Crater, Arizona. 2. Thermoluminescence age of Meteor Crater. J Geophys Res 90:3690–3700
Suzuki M (1970) Fission track dating and uranium contents of obsidian. J Anthropol Soc Nippon 78:50–58
Suzuki M (1973) Chronology of prehistoric human activity in Kanton, Japan. Part I: framework for reconstructing prehistoric human activity in obsidian. J Fac Sci Tokyo Univ 4:241–318
Swisher CC, Curtis GH Jacob T Getty AG, Suprijo AW (1994) Age of the earliest known hominids in Java, Indonesia. Science 263:1118–1121
Swisher CC, Rink WJ, Anton SC, Schwarcz HP, Curtis GH, Suprijo AW (1996) Latest *Homo erectus* of Java: potential contemporaneity with *Homo sapiens* in Southeast Asia. Science 274:1870–1874
Szabo BJ (1979) Dating fossil bone from Cornelia, Orange Free State, South Africa. J Archaeol Sci 6:201–203
Szabo BJ (1980) ^{230}Th and ^{231}Pa dating of unrecrystallized fossil mollusks from marine terrace deposits in west-central California. Isochron/West 27:3–4
Szabo BJ, Butzer KW (1979) Uranium-series dating of lacustrine limestone from Pan deposits with final Acheulian assemblage at Rooidam, Kimberley District, South Africa. Quat Res 11:257–260
Tanaka S, Sakamoto K, Takagi J, Tschuchimoto M (1968) Search for the ^{26}Al induced by cosmic ray muons in terrestrial rocks. J Geophys Res 73:3303–3309
Tarling DH (1991) Archaeomagnetism and palaeomagnetism. In: Göksu HY, Oberhofer M, Regulla D (eds) Scientific dating methods. Kluwer Academic, Dordrecht, pp 217–250
Tauxe L, Deino AD, Behrensmeyer AK, Potts R (1992) Pinning down the Brunhes/Matuyama and upper Jaramillo boundaries: a reconciliation of orbital and isotopic time scales. Earth Planet Sci Lett 109:561–572
Tauxe L, Herbert T, Shackleton NJ, Kok YS (1996) Astronomical calibration of the Matuyama-Brunhes boundary: consequences for magnetic remanence acquisition in marine carbonates and Asian loess sequences. Earth Planet Sci Lett 140:133–146
Taylor RE (1975) Fluorine diffusion: a new dating method for chipped lithic materials. World Archaeol 7:125–135
Taylor RE (1983) Non-concordance of radiocarbon and amino acid racemization deduced age estimates on human bone. Radiocarbon 25:647–654
Taylor RE (1987a) Radiocarbon dating – an archaeological perspective. Academic Press, Orlando
Taylor RE (1987b) AMS ^{14}C dating of critical bone samples: proposed protocol and criteria for evaluation. Nucl Instrum Methods Phys Res B29:159–163
Taylor RE (1991) Radioisotope dating by accelerator mass spectrometry: archaeological and paleoanthropological perspectives. In: Göksu HY, Oberhofer M, Regulla D (eds) Scientific dating methods. Kluwer Academic, Dordrecht, pp 37–54

Taylor RE, Slota PJ, Henning W, Kutschera W, Paul M (1989) Radiocalcium dating: potential applications in archaeology and paleoanthropology. In: Allen RO (ed) Archaeological chemistry. American Chemical Society, Washington, pp 321–335

Templer RH (1993) Auto-regenerative thermoluminescence dating using zircon inclusions. Archaeometry 35:117–136

Thellier E (1941) Sur les propriétés l'aimantation thermorémanente des terres cuites. CR Acad Sci Paris 213:1019–1022

Thellier E (1977) Early research on the intensity of the ancient geomagnetic field. Phys Earth Planet Inter 13:241–244

Thellier E (1981) Sur la direction du champ magnetique terrestre, en France, durant les deux derniers millenaires. Phys Earth Planet Inter 24:89–132

Thouveny N, Williamson D (1991) Palaeomagnetic secular variation as a chronological tool for the Holocene. In: Frenzel B (ed) Evaluation of climate proxy data in relation to the European Holocene. Gustav Fischer, Stuttgart, pp 13–27

Thouveny N Creer KM, Blank I (1990) Extensions of Lac du Bouchet paleomagnetic record over the last 120,000 years. Earth Planet Sci Lett 97:140–161

Thurber DL (1962) Anomalous U^{234}/U^{238} in nature. J Geophys Res 67:4518–4520

Torgersen T, Top Z, Clarke WB, Jenkins WJ, Broecker WS (1977) A new method for physical limnology: tritium-helium-3 age results for lakes Erie, Huron and Ontario. Limnol Oceanogr 22:181–193

Törnqvist TE, de Jong AFM, van der Borg K (1990) Comparison of AMS ^{14}C ages of organic deposits and macrofossils: a progress report. Nucl Instrum Methods Phys Res B52:442–445

Toyoda S, Ikeya M (1991) ESR dating of quartz and plagioclase from volcanic ashes using E', Al and Ti centers. Nucl Tracks Radiat Meas 18:179–184

Trull TW, Brown ET, Marty B, Rausbeck GM, Yiou F (1995) Cosmogenic ^{10}Be and ^{3}He accumulation in Pleistocene beach terraces in Death Valley, California, U.S.A.: implications for cosmic-ray exposure dating of young surfaces in hot climate. Chem Geol 119:191–207

Tsong IST, Houser CA, Yusef NA, Messier RF, White WB, Michels JW (1978) Obsidian hydration profiles measured by sputter-induced optical emission. Science 201:339–341

Tuniz C (1996) AMS ^{14}C dating: Impact on prehistoric science. Proc 13th Int Congr Prehist Protohist Sci, Coll. III, Forli, pp 55–64

Turekian KK, Kharkar DP, Funkhouser J, Schaeffer OA (1970) An evaluation of the uranium-helium method of dating fossil bones. Earth Planet Sci Lett 7:420–424

Tzedakis PC, Andrieu V, de Beaulieu JL, Crowhurst S, Follieri M, Hooghiemstra H, Magri D, Reille M, Sadori L, Shackleton NJ, Wijmstra TA (1997) Comparison of terrestrial and marine records of changing climate of the last 500,000 years. Earth Planet Sci Lett 150:171–176

Uerpmann HP (1990) Radiocarbon dating of shell middens in the Sultanate of Oman. PACT 29:335–347

Unger H (1993) Alpha-Rückstoßspuren in Glimmern und ihre Anwendung zur Datierung. Diplomarbeit, Universität Heidelberg

Urbach F (1930) Zur Lumineszenz der Alkalihalogenide. Sitzungsber Akad Wiss Wien Naturwiss Math Kl 139:353–373

Valdes VC, Bischoff JL (1989) Accelerator ^{14}C dates for Early Upper Paleolithic (Basal Aurignacien) at El Castillo Cave (Spain). J Archaeol Sci 16:577–584

Valet J-P, Meynadier L (1993) Geomagnetic field intensity and reversals during the past four million years. Nature 366:234–238

Valladas G, Gillot PY (1978) Dating of the Olby lava flow using heated quartz pebbles: some problems. PACT 2:141–150

Valladas H (1992) Thermoluminescence dating of flint. Quat Sci Rev 11:1–5

Valladas H, Reyss JL, Joron JL, Valladas G, Bar-Yosef O, Vandermeersch B (1988) Thermoluminescence dating of Mousterian 'Proto-Cro-Magnon' remains from Israel and the origin of modern man. Nature 331:614–616

van den Bogaard P (1995) $^{40}Ar/^{39}Ar$ ages of sanidine phenocrysts from Laacher See Tephra (12,900 yr BP): chronostratigraphic and petrological significance. Earth Planet Sci Lett 133:163–174

van den Bogaard P, Schmincke HU (1988) Aschenlagen als quartäre Zeitmarken in Mitteleuropa. Geowissenschaften 6:75–84

van den Haute P, Jonckheere R, De Corte F (1988) Thermal neutron fluence determination for fission-track dating with metal activation monitors: a re-investigation. Chem Geol (Isotop Geosci Sect) 73:233–244

van der Plicht J, Mook WG, Hasper H (1990) Automatic calibration of radiocarbon ages. PACT 29:81–94

Veil S, Breest K, Höfle H-C, Meyer H-H, Plisson H, Urban-Küttel B, Wagner GA, Zöller L (1994) Ein mittelpaläolithischer Fundplatz aus der Weichsel-Kaltzeit bei Lichtenberg, Lkr. Lüchow-Dannenberg. Germania 72(1):1–66

Verosub K (1977) Depositional and postdepositional processes in the magnetization of sediments. Rev Geophys Space Phys 15:129–143

Vogel JC (1983) ^{14}C variations during the upper Pleistocene. Radiocarbon 25:213–218

Volpe AM, Goldstein SJ (1993) ^{226}Ra-^{230}Th disequilibrium in axial and off-axis mid-ocean ridge basalts. Geochim Cosmochim Acta 57:1233–1241

Voltaggio M, Andretta D, Taddeucci A (1994) ^{230}Th-^{238}U-data in conflict with $^{40}Ar/^{39}Ar$ leucite ages for Quaternary volcanic rocks of the Alban Hills, Italy. Eur J Miner 6:209–216

Voltaggio M, Branca M, Tucciomei P, Tecce F (1995) Leaching procedure used in dating potassic volcanic rocks by the $^{226}Ra/^{230}Th$ method. Earth Planet Sci Lett 136:123–131

von Koenigswald GHR, Gentner W, Lippolt HJ (1961) Age of the basalt flow at Olduvai, East Africa. Nature 192:720–721

von Weizsäcker CF (1937) Über die Möglichkeit eines dualen Beta-Zerfalls von Kalium. Physik Z 38:623–624

von Werlhof J, Casey H, Dorn RI, Jones GA (1995) AMS ^{14}C age constraints in the lower Colorado River region, Arizona and California. Geoarchaeology 10:257–273

Waddell C, Fountain JC (1984) Calcium diffusion: a new dating method for archeological materials. Geology 12:24–26

Wagner GA (1968) Fission track dating of apatites. Earth Planet Sci Lett 4:411–415

Wagner GA (1976) Strahlenschäden zur Datierung von Gesteinen und Artefakten. Endeavour 35:3–8

Wagner GA (1978) Archaeological applications of fission track dating. Nucl Tracks Detect 2:51–63

Wagner GA (1980a) Dose-rate evaluation for thermoluminescence dating by fission track counting. Proc 16th Int Symp Archaeometry and Archaeological Prospection, Edinburgh 1976. National Museum of Antiquities of Scotland, Edinburgh, pp 393–402

Wagner GA (1980b) Thermolumineszenz und Altersbestimmung. Naturwissenschaften 67:216–226

Wagner GA (1980c) Thermolumineszenz-Datierungen am Töpferofen Koberg 15 in Lübeck. Lübecker Schriften z Archäologie u Kulturgesch 3:83–87

Wagner GA, Lorenz IB (1997) Thermolumineszenz-Datierung an bandkeramischen Scherben von Lamersdorf (Aldenhovener Platte). In: Lüning J (Hrsg) Studien zur neolithischen Besiedlung der Aldenhovener Platte und ihrer Umgebung. Rheinische Ausgrabungen 43:747–754

Wagner GA, Storzer D (1970) Die Interpretation von Spaltspurenaltern (fission track ages) am Beispiel von natürlichen Gläsern, Apatiten und Zirkonen. Eclogae Geol Helv 63/1:335–344

Wagner GA, Van den haute P (1992) Fission-track-dating. Ferdinand Enke, Stuttgart; Kluwer Academic, Dordrecht

Wagner G, Wagner I (1994) Thermolumineszenz-Datierung an einem Verhüttungsplatz im Gewann Kurleshau, nordöstlich von Metzingen. Beiträge zur Eisenverhüttung auf der Schwäbischen Alb, Konrad Theiss, Stuttgart, pp 264–265

Wagner GA, Reimer GM Carpenter BS, Faul H, Van der Linden R, Gibbels R (1975) The spontaneous fission rate of U-238 and fission track dating. Geochim Cosmochim Acta 39:1279–1286

Wagner GA, Storzer D, Keller J (1976) Spaltspurendatierung quartärer Gesteinsgläser aus dem Mittelmeerraum. Neues Jahrb Miner Monatsh 1976:84–94

Wagner GA, Göksu HY, Regulla DF, Vogenauer A, Lorenz IB (1992) Thermolumineszenzdatierungen an Keramik. In: Behm-Blancke MR (Hrsg) Hassek Höyük. Istanbuler Forsch 38:148–158

Wagner GA, Fezer F, Hambach U, von Koenigswald W, Zöller L (1997) Das Alter des *Homo heidelbergensis* aus Mauer. In: Wagner GA, Beinhauer KW (Hrsg) *Homo heidelbergensis* aus Mauer: Das Auftreten des Menschen in Europa. C. Winter, Heidelberg, pp 124–143

Wagner I, Wagner GA, Kind CJ (1993) Thermolumineszenz-Datierungen an Bandkeramik der Grabung Ulm-Eggingen. Fundber Baden-Württemberg 18: 53–60

Wallner G, Wild E, Aref-Azar H, Hille P, Schmidt WFO (1990) Dating of Austrian loess deposits. Radiat Prot Dosim 34:69–72

Walter RC, Manega PC, Hay RL, Drake RE, Curtis GH (1991) Laser fusion $^{40}Ar/^{39}Ar$ dating of Bed I, Olduvai Gorge, Tanzania. Nature 354:145–149

Walther R (1996) Age estimates of the Neckar sediments of *Homo erectus heidelbergensis* with the ESR-method. Mannheimer Geschichtsblätter, Beiheft 1:51

Walther R, Zilles D (1994) ESR studies on bleached sedimentary quartz. Quat Sci Rev 13:611–614

Walther R, Barabas M, Mangini A (1992) Basic ESR studies on recent corals. Quat Sci Rev 11:191–196

Wang L, Ku T, Luo S, Suothon JR, Kusakabe M (1996) ^{26}Al-^{10}Be systematics in deep-sea sediments. Geochim Cosmochim Acta 60:109–119

Wanpo H, Ciochon R, Yumin G, Larick R, Qiren F, Schwarcz HP, Yonge C, de Vos J, Rink W (1995) Early Homo and associated artefacts from Asia. Nature 378:275–278

Ward GK, Wilson SR (1978) Procedures for comparing and combining radiocarbon age determinations: a critique. Archaeometry 20:19–31

Warner RB (1990) A proposed adjustment for the "old wood effect". PACT 29: 159–172

Waters MR, Forman S, Pierson JM (1997) Diring Yuriakh: a Lower Paleolithic site in central Siberia. Science 275:1281–1284

Watt S, Durrani SA (1985) Thermal stability of fission tracks in apatite and sphene: using confined track length measurements. Nucl Tracks 10:349–357

Wehmiller JF (1986) Amino acid racemization geochronology. In: Hurford AJ, Jäger E, Ten Cate JAM (eds) Dating young sediments, CCOP Technical Secretariat, Bangkok, pp 139–158

Wehmiller JF, Hare PE, Kujala GA (1976) Amino acids in fossil corals: racemization (epimerization) reactions and their implications for diagenetic models and geochronological studies. Geochim Cosmochim Acta 40:763–776

Wei QY, Li TC, Chao GY, Chang WS, Wang SP (1981) Secular variation of the direction of the ancient geomagnetic field for Loyang region, China. Phys Earth Planet Inter 25:107–112

Weiner JS, Oakley JP, Le Gros Clark WE (1953) The solution of the Piltdown problem. Bull Brit Mus Nat Hist (Geol) 2:139–146

Wendorf F (1987) The advantages of AMS to field archaeologists. Nucl Instrum Methods Phys Res B29:155–158

Weninger B (1986) High-precision calibration of archaeological radiocarbon dates. Acta Interdiscipl Archaeol 4:11–53

Westgate JA (1988) Isothermal plateau fission-track age of the Late Pleistocene Old Crow Tephra, Alaska. Geophys Res Lett 15:376–379

Whitlock C, Bartlein PJ (1997) Vegetation and climate change in northwest America during the past 125 kyr. Nature 388:57–61

Wiedemann E, Schmidt GC (1895) Ueber Lumineszenz. Ann Phys Chem 54: 604–625

Wiggenhorn H (1994) The effect of infrared light on TL and of temperature on IRSL of K-feldspars and its consequences for dating. Radiat Meas 23:387–391

Wiggenhorn H, Rieser U (1996) Analysis of natural IRSL and TL emission spectra of potassium-rich feldspars with regard to dating applications. Radiat Prot Dosim 66:403–406

Williams CT, Marlow CA (1987) Uranium and thorium distributions in fossil bones from Olduvai Gorge, Tanzania and Kanam, Kenya. J Archäeol Sci 14: 297–309

Willkomm H (1992) Radiokohlenstoffdatierungen. In: Behm-Blancke MR (Hrsg) Hassek Höyük. Istanbuler Forsch 38:135–139

Winog IJ, Coplen TB, Landwehr JM, Riggs AC, Ludwig KB, Szabo BJ, Kolesar PT, Revesz KM (1992) Continuous 500,000-year climate record from vein calcite in Devils Hole, Nevada. Science 258:255–260

Winsborough BM, Caran SC, Neely JA, Valstro S Jr (1996) Calcified microbial mats date prehistoric canals – radiocarbon assay of organic extracts from travertine. Geoarchaeology 11:37–50

Wintle AG (1973) Anomalous fading of thermoluminescence in mineral samples. Nature 245:143–144

Wintle AG (1993) Luminescence dating of aeolian sands: an overview. In: Pye K (ed) The dynamics and environmental context of aeolian sedimentary systems. Geol Soc Lond Spec Publ 72:49–58

Wintle AG, Catt JA (1985) Thermoluminescence dating of soils developed in late Devensian loess at Pegwell Bay, Kent. J Soil Sci 36:293–298

Woillard GM, Mook WG (1982) Carbon-14 dates at Grand Pile: correlation of land and sea chronologies. Science 215:159–161

WoldeGabriel G, White TD, Suwa G, Renne P, de Heinzelin J, Hart WK, Heiken G (1994) Ecological and temporal placement of early Pleistocene hominids at Aramis. Nature 371:330–333

Wölfli W (1987) Advances in accelerator mass spectrometry. Nucl Instrum Methods Phys Res B29:1–13

Wolfman D (1984) Geomagnetic dating methods in archaeology. Adv Archaeol Methods Theor 7:363–458

Wolfman D (1990a) Archaeomagnetic dating in Arkansas and the border areas of the adjacent states. In: Eighmy JL, Sternberg RS (eds) Archaeomagnetic dating. University of Arizona Press, Tucson, pp 237–260

Wolfman D (1990b) Mesoamerican chronology and archaeomagnetic dating, A.D. 1–1200. In: Eighmy JL, Sternberg RS (eds) Archaeomagnetic dating. University of Arizona Press, Tucson, pp 261–308

Wolfman D, Rolniak TM (1978) Alpha-recoil track dating: problems and prospects. Archaeo-Physika 10:512–521

Worm HU (1997) A link of geomagnetic reversals and events to glaciations. Earth Planet Sci Lett 147:55–67

Xirotiris NJ, Henke W, Hennig GJ (1982) Die phylogenetische Stellung des Petralona Schädels auf Grund computertomographischer Analysen und der absoluten Datierung mit der ESR-Methode. Humanbiologia Budapestinensis 9:89–94

Yechieli Y, Ronen D, Kaufman A (1996) The source and age of groundwater brines in the Dead Sea area, as deduced from ^{36}Cl and ^{14}C. Geochim Cosmochim Acta 60: 1909–1916

Yiou F, Raisbeck GM, Bourles D, Lorius C, Barkov NI (1985) ^{10}Be in ice at Vostok Antarctica during the last climatic cycle. Nature 316:616–617

Yokoyama Y, Quaegebeur JP, Bibron R, Leger C, Chappaz N, Michelot C, Chen GJ, Nguyen HV (1983) ESR dating of stalagmites of the Caune de l'Arago, the Grotte du Lazaret, the Grotte Valonet, and the Abri Pie Lombard: a comparison with the U-Th method. PACT 9:381–389

York D, Hall CM, Yanase Y, Hane JA, Kenyon WJ (1981) $^{40}Ar/^{39}Ar$ dating of terrestrial minerals with a continuous laser. Geophys Res Lett 8:1136–1138

Zagwijn WH, van Montfrans HM, Zandstra JG (1971) Subdivision of the "Cromerian" in the Netherlands; pollen analysis, paleomagnetism and sedimentary petrology. Geol Mijnbouw 50:41–58

Zeller EJ, Levy PW, Mattern PL (1967) Geologic dating by electron spin resonance. IAEA, Vienna, pp 531–540

Zhu ZR, Wyrwoll KH, Collins LB, Chen JH, Wasserburg GJ, Eisenhauer A (1993) High-precision U-series dating of Last Interglacial events by mass spectrometry: Houtman Abrolhos Islands, Western Australia. Earth Planet Sci Lett 118:281–293

Zielinski GA, Mayewski PA, Meeker LD, Whitlow S, Twickler MS, Morrison M, Meese DA, Gow AJ, Alley RB (1994) Record of volcanism since 7000 B.C. from the GISP2 Greenland ice core and implications for the volcano-climate system. Science 264:948–951

Zito R, Donahue DJ, Davis SN, Bentley HW, Fritz P (1980) Possible sun surface production of carbon-14. Geophys Res Lett 7:235–238

Zolitschka B, Haverkamp B, Negendank JFW (1992) Younger Dryas oscillation – varve dated microstratigraphic, palynological and palaeomagnetic records from lake Holzmaar, Germany. In: Bard E, Broecker WS (eds) The last deglaciation: absolute and radiocarbon chronologies. Springer, Berlin Heidelberg New York, pp 81–101

Zöller L (1989) Das Alter des Mosenberg-Vulkans in der Vulkaneifel. Die Eifel 84:415–418

Zöller L (1994) Würm- und Rißlöß-Stratigraphie und Thermolumineszenz-Datierung in Süddeutschland und angrenzenden Gebieten. Habilitationsschrift, Universität Heidelberg

Zöller L, Wagner GA (1989) Strong or partial thermal washing in TL-dating of sediments? In: Aitken M (ed) Long and short range limits in luminescence dating. RLA Occas Publ 9, Oxford University

Zöller L, Stremme H, Wagner GA (1988) Thermolumineszenz-Datierung an Löß-Paläoboden-Sequenzen von Nieder-, Mittel- und Oberrhein/Bundesrepublik Deutschland. Chem Geol 73:39–62

Zöller L, Conard NJ, Hahn J (1991) Thermoluminescence dating of Middle Palaeolithic open air sites in the Middle Rhine valley/Germany. Naturwissenschaften 78:408–410

Zreda MG, Phillips FM, Elmore D, Kubik PW Sharma P, Dorn RI (1991) Cosmogenic chlorine-36 production rates in terrestrial rocks. Earth Planet Sci Lett 105:94–109

Zreda MG, Phillips FM, Kubik PW Sharma P, Elmore D (1993) Cosmogenic ^{36}Cl dating of a young basaltic eruption complex, Lathorp Wells, Nevada. Geology 21:57–60

Zumberge JE, Engel MH, Nagy B (1980) Amino acids in bristlecone pine: an evaluation of factors affecting racemization rates and paleothermometry. In: Behrensmeyer AK, Hill AP (eds) Fossils in the making: vertebrate taphonomy and paleoecology. University of Chicago Press, Chicago, pp 503–525

Subject Index

MIX
Papier aus verantwortungsvollen Quellen
Paper from responsible sources
FSC® C105338

FSC
www.fsc.org

If you have any concerns about our products,
you can contact us on
ProductSafety@springernature.com

In case Publisher is established outside the EU,
the EU authorized representative is:
Springer Nature Customer Service Center GmbH
Europaplatz 3, 69115 Heidelberg, Germany

Printed by Libri Plureos GmbH
in Hamburg, Germany